WATER CONSERVANCY AND CIVIL CONSTRUCTION
VOLUME 2

Water Conservancy and Civil Construction gathers the most cutting-edge research on:

- Water Conservancy Projects
- Civil Engineering
- Construction Technology and Process

The book is aimed at academics and engineers in water and civil engineering.

PROCEEDINGS OF THE 4TH INTERNATIONAL CONFERENCE ON HYDRAULIC, CIVIL AND CONSTRUCTION ENGINEERING (HCCE 2022), HARBIN, CHINA, 16–18 DECEMBER 2022

Water Conservancy and Civil Construction Volume 2

Edited by

Saheed Adeyinka Oke
Central University of Technology Free State, South Africa

Fauziah Ahmad
Universiti Sains Malaysia

CRC Press is an imprint of the
Taylor & Francis Group, an **informa** business

A BALKEMA BOOK

First published 2024
by CRC Press/Balkema
4 Park Square, Milton Park, Abingdon, Oxon, OX14 4RN

and by CRC Press/Balkema
2385 NW Executive Center Drive, Suite 320, Boca Raton FL 33431

CRC Press/Balkema is an imprint of the Taylor & Francis Group, an informa business

© 2024 selection and editorial matter, Saheed Adeyinka Oke & Fauziah Ahmad; individual chapters, the contributors

The right of Saheed Adeyinka Oke & Fauziah Ahmad to be identified as the authors of the editorial material, and of the authors for their individual chapters, has been asserted in accordance with sections 77 and 78 of the Copyright, Designs and Patents Act 1988.

All rights reserved. No part of this book may be reprinted or reproduced or utilised in any form or by any electronic, mechanical, or other means, now known or hereafter invented, including photocopying and recording, or in any information storage or retrieval system, without permission in writing from the publishers.

Although all care is taken to ensure integrity and the quality of this publication and the information herein, no responsibility is assumed by the publishers nor the author for any damage to the property or persons as a result of operation or use of this publication and/or the information contained herein.

British Library Cataloguing-in-Publication Data
A catalogue record for this book is available from the British Library

Library of Congress Cataloging-in-Publication Data
A catalog record has been requested for this book

SET
ISBN: 978-1-032-51440-6 (hbk)
ISBN: 978-1-032-51464-2 (pbk)

Volume 1
ISBN: 978-1-032-58614-4 (hbk)
ISBN: 978-1-032-58615-1 (pbk)
ISBN: 978-1-003-45081-8 (ebk)
DOI: 10.1201/9781003450818

Volume 2
ISBN: 978-1-032-58618-2 (hbk)
ISBN: 978-1-032-58619-9 (pbk)
ISBN: 978-1-003-45083-2 (ebk)
DOI: 10.1201/9781003450832

Typeset in Times New Roman
by MPS Limited, Chennai, India

Water Conservancy and Civil Construction – Oke & Ahmad (Eds)
© 2024 The Editor(s), ISBN: 978-1-032-58618-2

Table of Contents

Preface	xi
Committee Members	xiii

VOLUME 2

Intelligent building and automatic control system

Nano-indentation technique-based evaluation method of rock mechanical parameters for tight oil
Junwei Wang, Mingguang Che, Xiao Song, Xian Shi & Jian Su — 3

Overall design and key technologies of extra-wide deck space special-shaped arch bridge
Jie Liu & Li-xin Wei — 11

Research on the algorithm of tooling position and attitude measurement based on a global camera system
Haifei Wang, Shaozhong Cao, Wei Zhao & Zhenqing Gao — 21

Rigid-flexible coupling analysis of worm and gear reducer based on ANSYS and ADAMS
Yanting Yu, Meili Song & Xuejie Chen — 27

Research on the optimization of auto parts inventory cost based on an improved demand forecasting method
Weining Zhu & Cunrong Li — 34

Design and identification management strategy of privately owned public space in NYC and its enlightenment on architecture public open space in China
Ying Shi, Wan Ying Wu & Ying Pan — 44

Summer thermal environment of building atrium in hot summer and warm winter areas
Jiating Chen & Yan Liu — 50

Surface settlement prediction of composite stratum shield construction based on e-GRA-LSTM
Lili Feng, Changming Hu & Liang Li — 57

Numerical analysis of wall shedding based on thermal coupling
Hongli Zhang & Changming Hu — 65

Study on properties of permeable brick with recycled aggregate from construction waste
Xiaoya Huang, Junxia Sun, Ting Hu, Yan Zhang & Limei Wang — 74

Analysis of ships vibration causes and application of active vibration control technology
Xiaohai Qi, Jincai Huang, Hao Wang, Xun Zhao & Wangxuan Zhang — 82

Hysteretic behavior analysis of new energy dissipation fuse system 92
Dingyu Chen, Xiaotong Peng & Chen Lin

SVR algorithm-based prediction of scraper wear for super-large-diameter
slurry shield 98
Bowen Tao, Lei Zhang, Jian Li & Zhipeng Shi

Study on dynamic stability of geogrid-reinforced cracked slope based on finite
element limit analysis 110
Cheng Li, Llidan Li, Xingqian Xu, Xin Qu, Haijun Wang & Junduo Yang

Research on the real-time scheduling model of pumping station group based on
differential evolution algorithm – take the Jin'an river basin as an example 124
*Yang Chen, Chao Wang, Shuanglin Jiang, Chen Ji, Dianchang Wang,
Ruozhu Shen & Haocheng Huang*

Study on the alignment characteristics of high bridge-tunnel ratio road sections
based on Catmull-Rom spline curve—take the section of Baima-Longtan
expressway as an example 132
Yudan Liu & Xianghong Li

Study on smart aggregate for monitoring asphalt pavement health system 139
*Shengchao Cui, Yanxiang Zhang, Guangzheng Wang, Qingna Ma &
Feng Sun*

Process analysis for the demolition work of mid-partition walls in
complex foundation pits in super high-rise buildings 150
Jianping Zhu, Yalong Liu & Jiao Zhang

Study on the technology of lime expanded pile foundation treatment for
airport runway in high-temperature permafrost regions 156
Danze Wu, Youshi Hao, Guoguang Liu, Li Feng & Yifan Wu

Study of tunnel excavation position on the stability of the overlying loess 171
Peihao Li & Jinhai Gao

Research and engineering application of roadbed settlement prediction
method based on measured data 177
Zhi Liu

The application of traditional elements in modern architectural design 185
Rui Wang, Jiang Jin, Yuetao Li & Mengyan Tan

Structural design of the power distribution building 192
Zhen Xie, Xilong Chen & Yan Yan

Calculation and analysis of foundation pit excavation near river embankment 199
Xu Tao, Gong Chunjuan & Wang Keliang

Analysis of the influence of different control water levels on the
upstream drainage of Hongze lake in flood season 204
Dawei Zhu & Xiaolin Zhong

Construction technology analysis of 80 meters deep underground diaphragm
wall under complex geology 208
Li Yefu, Wang Rong & Zhang Shoujia

Foundation trench excavation and basin dredging technology in port waterway construction 217
Xingquan Jiang

Study on the interaction and construction time sequence between interchange construction and subway construction 223
Lei Chen & Shihu Qi

Research on key technologies of underbalanced drilling 231
Jianqiu Feng, Xiangbing Pei, Tan Zhang, Ke Jiang & Xiaonan Wang

Study on optimal and fast drilling technology of extended reach wells 238
Hongwei Liu, Peng Wei & Mengyi Liu

Stability analysis of the foundation pit based on the three-dimensional limit equilibrium method 245
Haitao Wei, Ning Jia & Bin Jia

A mechanical constitutive damage model for non-penetrating crack rock-like specimen 252
Cheng Pu, Yuetao Li & Jiaxin Wen

Smart city and environmental planning and management

Contributing factors to urban transport carbon dioxide emissions and reduction measures 261
Yue Hao, Yugang Liu, Shuai Zheng & Linqiang Fan

Design and application of a new gas-bearing gas film in the closed coal yard 274
Xiaolong Sheng, Honghui Wang, Ming Shang, Weisheng Li & Sensen Wang

Design and construction of urban ventilation corridors based on wind circulation in the Liling ceramic valley area 280
Shan Guan, Ana Hao & Rujiao Cao

Study on the accumulation and spatial distribution characteristics of sediment in a reservoir 289
Hongbo Tao, Shuai Song, Yun Long & Yue Long

Effects of water and nitrogen coupling on potato yield, water and nitrogen utilization and soil environment 294
Chao Liang, Shouchao Yu & Hengjia Zhang

Research on the evaluation index system for Yangtze River waterway maintenance effectiveness under the new situation 299
Liu Lei, Duan Guo-sheng & Mu Bao-yin

Basic prototype and evolutionary logic of clan shrines in northern Guangdong 305
Ziru Ren, Zhaohui Tang & Yao Wang

Location planning method of flood control and drainage pumping station based on improved genetic algorithm 313
Wei Zhang, Yan Liang & Zhen Liu

The application of BIM+AR smart construction technology in municipal
engineering 319
Liguo Liu

Two-Dimensional numerical analysis of the Nanjiang gate water diversion to
improve rivernet water environment 325
Di Hao, Dongfeng Li, Zihao Li, Donghui Hu, Aijun Sun & Zhenghao Li

Study on the urban design guidelines of waterfront space in Pearl River
Delta—taking Foshan city as an example 332
Liang Liyun & Liu Hui

Effect analysis of water environment improvement in Sanhui District based on
mike21 ECO-LAB modular 338
He Huan, Yanfen Yu, Zihao Li, Donghui Hu, Aijun Sun & Zhenghao Li

Study on the optimal scheduling of the Jiangsu section of the south to north
water transfer project east line in ten days 345
Ke Xu, Chao Wang, Linan Xue, Chen Ji & Hao Wang

Evaluation and analysis of low carbon traffic in Jiaozuo city based on
comprehensive empowerment method 353
Zhezhe Zhang

Structural design and mechanical properties of assembled right-of-way in
ecologically sensitive areas 362
*Jing Guo, Hongzhang Wang, Wei Qiao, Xinglin Lin, Yuan Tian,
Shanjun Yang, Fei Li & Libin Zhang*

Research on the strategic choice of logistics development in Urumqi
based on AHP-SWOT 371
Xuegang Liang, Lan Li & Chengming Zhu

Analysis of impact indicators for dynamic management of on-street parking
spaces based on hierarchical analysis 378
Chunmei Hu, Yang Ming, Shuting Deng & Qianrong Tu

Research and application of new ecological slope protection technology for
highways in retention areas 388
Danxuan Xue, Xiaodong Zhu & Xingyu Zhang

Analysis of parking choice behavior in commercial areas based on hybrid
logit model 395
Shuting Deng, Yang Ming, Chunmei Hu & Qianrong Tu

Analysis of college students' bus travel behavior in the post-epidemic era 403
Congling Zheng & Xianghong Li

The design of revenue adjustment method for highway public-private-partnership
(PPP) projects under inflation 413
Zhenyao Wu

Impact analysis on dynamic response of asphalt pavement structure with
semi-rigid base effected by interlayer evolution properties under different
service conditions 420
YuBin Zhang, Xiangbiao Wang, Rui Zhang, Wenzhi Yuan & Qun Yang

Research on the ultimate protection ability of grade A W-beam barrier
based on the bus collision 431
Hao Wang, Shuai Gong, Hong Guo & Shuming Yan

Study on public transportation mode considering group heterogeneity 439
Jie Ji, Yutong Li & Ruiqi Dong

Traffic management with households under stochastic bottleneck
capacity when school is near home 445
Boyu Lin

Analysis of research articles and international standards related to wastewater 451
Li Yefu, Wang Rong, Hou Xiangyu & Zhang Shoujia

Structural optimization and practical engineering with the rise of modern
technology application 456
*Manli Boukari, Jianyong Pang, Monkam Ngameni Huguette Maeva &
Fatoumata Kir Kalissa*

Analysis of the influencing factors of personal credit for practitioners in the
field of engineering construction based on the fuzzy-DEMATEL model 467
Song Xue, Tong Su & Yi Zhou

Comparative analysis of carbon emissions of different asphalt pavement
structures 479
Yin Huai Ma, Cheng Gang Duan, Li Jiang & Yang Li

Low carbon strategies based on a steel prefabricated hotel in Shenzhen 486
Siyu Jiao, Hao Lu & Dongshan Ding

Research on the measurement of headlight downtilt 493
Zheng Tao

Experimental study on group root cooperative slope protection at normal
temperature in Xining area 501
Fanxing Meng, Hui Li, Ningshan Jiang, Chengkui Liu & Gencheng Liu

Modeling and analysis of vehicle infrastructure cooperative industry based
on tripartite evolutionary game theory 507
Yuan Yuan, Linheng Li, Xu Qu & Bin Ran

Author index 517

Water Conservancy and Civil Construction – Oke & Ahmad (Eds)
© 2024 The Editor(s), ISBN: 978-1-032-58618-2

Preface

The 2022 4th International Conference on Hydraulic, Civil and Construction Engineering (HCCE 2022) was held via virtual form in Harbin, China from December 16th–18th, 2022. Previous conferences of the past three years in this series were held in Guangzhou virtually or physically and it was agreed to hold the conference once a year. The purpose of this series of annual conferences is to establish and develop constant international collaboration.

In recent decades, interest in hydraulic and civil construction engineering problems has been flourishing all over the globe because of both the theoretical interest and practical requirements. Considering the trend, the fourth conference was organized in order to provide forums for developing research cooperation and to promote activities in the field of hydraulic and civil construction engineering.

Because solutions to hydraulic and civil construction engineering problems are needed in various applied fields, we entertained about 200 participants at the fourth conference and arranged various speeches and presentations which ranged from structural seismic resistance to smart city in the real world. Many researchers all over the world have contributed to the emerging technology of hydraulic and civil construction engineering. Assoc. Prof. Rohayu Che Omar from Universiti Tenaga Nasional, Malaysia addressed a keynote speech on Disaster Resilience Index and Indicators System for Managing Risks in Hazardous Terrain (DRIMS). She introduced to us that the methods for landslide mapping and landslide hazard assessment have experienced significant improvements during the last decade, but the requirements of the users have become more challenging, leading to improving landslide protection techniques to stabilize a landslide too costly in financial or environmental terms. This new context can only be managed with a better knowledge of the landslide mechanisms and their behavior. At the same time, different technical solutions, especially in the domain of risk mitigation, must be searched to guarantee the appropriate level of safety for the infrastructure and population.

HCCE 2022 has been endorsed by many international and national hydraulic civil engineering organizations and publishers. The papers collected and undergone peer review in the proceedings of HCCE 2022 are classified as follows: Engineering Structure, Intelligent Building, Smart City, Structural Seismic Resistance, Monitoring and Testing, Engineering Facility, etc.

Last but not the least is our gratitude. As editors we would like to express our sincere thanks to all the plenary and invited speakers, the members of the Program Committee and the Technical Committee for the success of the conference, which has given rise to this present volume of selected papers. We would also like to thank the CRC Press Balkema – Taylor & Francis Group for their effective work to make this volume published.

The Committee of HCCE 2022

Water Conservancy and Civil Construction – Oke & Ahmad (Eds)
© 2024 The Editor(s), ISBN: 978-1-032-58618-2

Committee Members

Conference Chairman
Prof. Fadi HAGE CHEHADE, ISBA TP, *Grande Ecole d'Ingénieurs de Spécialisation en Génie Civil, France*

Program Committee
Prof. Tetsuya Hiraishi, *Kyoto University, Japan*
A. Prof. Hazem Samih Mohame, *Southwest Petroleum University, Egypt*
A. Prof. Mohammad Arif Kamal, *Aligarh Muslim University, India*
A. Prof. Aeslina Abdul Kadir, *Universiti Tun Hussein Onn Malaysia, Malaysia*
Asst. Prof. Hamza Soualhi, *University of Laghouat, Algeria*
Senior Lecturer Mohammadreza Vafaei, *Universiti Teknologi Malaysia, Malaysia*
Senior Lecturer Au Yong Cheong Peng, *University of Malaya, Malaysia*
Senior Lecturer Nor Hasanah Binti Abdul Shukor Lim, *Universiti Teknologi Malaysia UTM, Malaysia*
Senior Lecturer Libriati Zardasti, *Universiti Teknologi Malaysia, Malaysia*
Ph. D. Dayang Zulaika Binti Abang Hasbollah, *Universiti Teknologi Malaysia, Malaysia*

Technical Committee
Prof. Dr. Mohammad Bin Ismail, *Universiti Teknologi Malaysia, Malaysia*
Prof. Ir. Dr. Hj. Ramli Nazir, *Universiti Teknologi Malaysia, Malaysia*
Prof. Dr. Muhd Zaimi Bin Abd Majid, *Universiti Teknologi Malaysia, Malaysia*
Prof. Lu, Jane Wei-Zhen, *City University of Hong Kong, Hong Kong, China*
Prof. Mingqiao Zhu, *Hunan University of Science and Technology, China*
Prof. QingXin Ren, *Shenyang Jianzhu University, China*
Prof. Bing Li, *Shenyang Jianzhu University, China*
Prof. Jianhui Yang, *Henan Polytechnic University, China*
Prof. Changfeng Yuan, *Qingdao University of Technology, China*
A. Prof. Bon-Gang HWANG, *National University of Singapore, Singapore*
A. Prof. Zhu Yuan, *Southeast University, China*
A. Prof. Chaofeng Zeng, *Hunan University of Science and Technology, China*
A. Prof. Weijun Cen, *Hohai University, China*
Asst. Professor Dr. Shah Kwok Wei, *National University of Singapore, Singapore*
Dr. Shaoyun Pu, *Southeast University, China*
Dr. Zhongzheng Lyu, *Dalian University of Technology, China*
Dr. Mohd Rosli Mohd Hasan, *Universiti Sains Malaysia, Malaysia*
Dr. Kim Hung Mo, *University of Malaya, Malaysia*
Dr. Yuen Choon Wah, *University of Malaya, Malaysia*
Dr. Huzaifa Bin Hashim, *University of Malaya, Malaysia*
Dr. Suhana Koting, *University of Malaya, Malaysia*
Dr. Sharifah Akmam Syed Zakaria, *Universiti Sains Malaysia, Malaysia*
Dr. Xian Zhang, *Southeast University, China*
Dr. Zhiming Chao, *University of Warwick, UK*
Dr. Jun Xie, *Central South University, China*
Dr. Derek Ma, *University of Warwick, England*
Dr. Ning Xu, *Shanghai Ershiye Construction CO., LTD., China*
Dr. Hongchao Shi, *Chengdu Technological University, China*
Dr. Li He, *Wuhan University of Science and Technology, China*

Intelligent building and automatic control system

Water Conservancy and Civil Construction – Oke & Ahmad (Eds)
© 2024 The Author(s), ISBN: 978-1-032-58618-2

Nano-indentation technique-based evaluation method of rock mechanical parameters for tight oil

Junwei Wang
LiaoHe Oil Company of PetroChina, Liaoning Panjin, China

Mingguang Che*
Research Institute of Petroleum Exploration and Development PetroChina, Beijing, China

Xiao Song
LiaoHe Oil Company of PetroChina, Liaoning Panjin, China

Xian Shi
China University of Petroleum (East China), Shandong Qingdao, China

Jian Su
LiaoHe Oil Company of PetroChina, Liaoning Panjin, China

ABSTRACT: Young's modulus and other key mechanical parameters are the basis for calculating the rock brittleness index (BI) of horizontal tight oil wells. Such elastic parameters are usually obtained in a laboratory by conducting rock mechanical tests using core plugs. However, the coring cost in horizontal wells is high, and sometimes high-quality cores cannot be obtained in reservoir sections, which makes it impossible to conduct mechanical tests indoors. To characterize mechanical properties, the nano-indentation technique is proposed, which effectively uses 1 mm × 1 mm small-scale broken core specimens or lithics to obtain key mechanical parameters. In addition, it also has the advantages of repeatable testing and simple operation. In using this method, the basic mineral facies of specimens should be analyzed first, and then the dot matrix indentation test should be designed and carried out. Besides, the convolution method should be used for scale upgrading so that the macroscopic mechanical parameters can be obtained. This method has been applied to the test of tight oil lithics from the LiaoHe Oilfield, and it is found that the error of the nano-indentation mechanical test results after scale upgrading is less than 10% compared with the results of the core plug test, which meets the need for BI calculation. Because irregularly broken core specimens or lithics can be used to obtain rock mechanical parameters, this method greatly widens the testing range of mechanical tests and provides a new idea and means for such engineering applications as the brittleness evaluation of tight oil wells.

1 INTRODUCTION

Rock mechanical parameters, such as Young's modulus, Poisson's ratio, and tensile strength, are of great significance to studying the evaluation of tight oil fracturing sweet spots as well as the formation and extension pattern of complex hydraulic fractures during

---

*Corresponding Author: chemg69@petrochina.com.cn

DOI: 10.1201/9781003450832-1

horizontal good fracturing [1–3]. Standard core plugs are mostly used in laboratories, and basic mechanical parameters are obtained by conventional uniaxial or triaxial rock mechanical tests [4–7]. Conventional rock mechanical tests not only have high requirements on rock dimensions and integrity but are also more difficult to carry out in horizontal good sections because of high coring costs and great difficulties. As a result, only the cores from the exploratory well or pilot well can usually be used to test the rock mechanical parameters of local good sections, and the test results are highly discrete, which has great limitations in engineering applications. Therefore, it is of great significance to try to apply a mechanical parameter testing method or technique using broken rock specimens, lithics, and irregular rock specimens.

The nano-indentation technique is an important technique that has rapidly developed for the testing of surface engineering mechanical properties in recent years [8–10], and many scholars have applied it to the testing of rock mechanical properties. It uses 1 mm × 1 mm small-scale broken core specimens or lithics to obtain mechanical parameters. Firstly, the basic mineral facies are analyzed, and then an indentation test is conducted. After that, the convolution method is used for scale upgrading to obtain the macro-mechanical parameters [11,12]. In this paper, HouHe tight sandstone in the LiaoHe Oilfield was studied. By conducting a nano-indentation test, the rock mechanical parameters of Young's modulus and Poisson's ratio were finally obtained. Compared with the rock mechanical test results of the core plug, the error of the indentation test results was less than 10%, which met the need for BI calculation. This method can use broken core specimens or lithics to obtain rock mechanical parameters, providing a new technical method to obtain testing parameters for the brittleness evaluation of tight oil/gas wells.

2 NANO-INDENTATION TEST

2.1 Testing method

The testing principle of the nano-indentation technique [13–16] has been introduced in detail in many pieces of literature, which will not be repeated in this paper. There are mainly two methods of testing for nano-indentation: the random indenting method and the matrix indenting method. In the case of random indenting, the rock surface is indented randomly, and indentation dots with greater depth are selected to acquire stable mechanical parameters. And then the averaging method is used to obtain mechanical parameters. Such results are highly discrete and can only reflect the mechanical properties at the nano-scale. In the case of matrix indenting, a mineral facies analysis is made first, and then the indentation test is carried out. After that, the convolution method is used for scale upgrading to obtain relevant mechanical parameters. This method provides more information on microscopic mechanical properties with higher accuracy, making it possible to comprehensively understand the mechanical properties and structural properties of all rock facies at the nano-scale, and explain the mechanical properties of rock at a larger scale by upgrading the physical model of rock. The indentation test in this paper was conducted using the matrix indentation method.

2.2 Testing process

Rock was processed into a sheet of 1 mm × 1 mm specimen, which was then subjected to a mechanical loading test using Hysitron TI Premier nano-indentation equipment, as shown in Figure 1. With a maximum load of 10,000 nN, a load resolution of 12 nN, and a maximum indentation depth of 80 um, this equipment can accurately test the mesoscopic mechanical property parameters of rocks in various formations, including elastic modulus, hardness, storage modulus, and loss modulus.

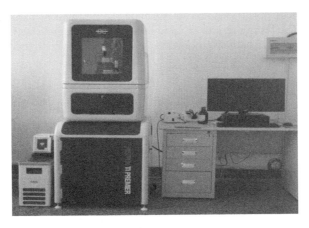

Figure 1. Testing equipment of nano-indentation.

At the moment, mechanical parameters can be tested using a dot matrix during an indentation test, and a gridded dot matrix can be used to better analyze the distribution characteristics of mechanical parameters corresponding to different minerals. The designed dot matrix had a size of 200 × 200 μm, and the matrix of indentation dots is shown in Figure 2.

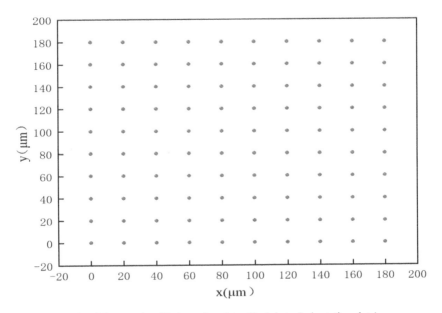

Figure 2. Schematic of the matrix of indentation dots (Red dots, Indentation dots).

With a maximum load of 10 mN, the nano-indenter approached the specimen surface at a speed of 20 nm/s. Both the load and its corresponding indenting depth were recorded automatically when the indenter touched the specimen surface. In order to eliminate the impact of load retention on mechanical parameters, no load retention time was set. A scanning electron microscope was applied to analyze the specimen surface and locate the mineral specimens at different positions so as to calculate the mechanical parameters of different indentation dots. The testing results of nano-indentation were subjected to scale upgrading to obtain macroscopic mechanical parameters, which were then compared with the results of the rock mechanical test.

2.3 *Method for scale upgrading*

What is obtained from an indentation test is the mechanical property of composite materials. However, when the indentation area is much smaller than the representative volume element, what is obtained from an indentation test is the mechanical property of a single component. Rock, as a multi-facies porous medium, is made up of particles of varying sizes and strengths.

The Mori-Tanaka mechanical model was used to upgrade the testing data from a small scale (micro-/nano-scale) to a macro-scale. Microscopically, rock was approximately regarded as a complex composed of quartz, feldspar, carbonate, clay, and porous media. The calculation equations of the Mori-Tanaka mechanical model are as follows:

$$K_{\text{hom}} = \left(\sum_{r=0} f_r \frac{k_r}{3k_r + 4\mu_0}\right) \Big/ \left(\sum_{r=0} \frac{f_s}{3k_s + 4\mu_0}\right) \tag{1}$$

$$G_{\text{hom}} = \frac{\sum_{r=0} f_r \frac{\mu_r}{\mu_0(9k_0+8\mu_0)+6\mu_r(k_0+2\mu_0)}}{\sum_{s=0} \frac{f_s}{\mu_0 9k_0+8\mu_0+6\mu_s(k_0+2\mu_0)}} \tag{2}$$

$$E_{\text{hom}} = \frac{9K_{\text{hom}} G_{\text{hom}}}{G_{\text{hom}} + 3K_{\text{hom}}} \tag{3}$$

$$v_{\text{hom}} = \frac{3K_{\text{hom}} - 2G_{\text{hom}}}{2G_{\text{hom}} + 6K_{\text{hom}}} \tag{4}$$

where K_{hom} is the shear modulus in Pa, f_r is the volume fraction of different mineral particles in %, f_s is the volume fraction of S facies in %, k_r is the volume modulus of different mineral particles in Pa, k_s is the proportion of clay minerals in %, μ_0 is the equivalent shear modulus of low-strength minerals in Pa, G_{hom} is the volume modulus in Pa, μ_r is the shear modulus of different mineral particles in Pa, k_0 is the equivalent volume modulus of low-strength minerals in Pa, μ_s is the shear modulus of clay minerals in Pa, E_{hom} is Young's modulus in Pa, and v_{hom} is the dimensionless Poisson's ratio.

where $r = 0$ represents dolomite and other carbonate minerals, $r = 1$ represents quartz, feldspar and other quartz minerals, and $r = 2$ represents clay minerals.

To accurately acquire the mechanical properties of all facies, a large number of nano-indentation tests are required. The indentation dots are arranged in a gridded matrix, with the spacing between any two adjacent grids being 1. Besides, it is required that 1 should be great enough to avoid interference between adjacent indentation dots. The volume fraction and mechanical properties of all facies can be obtained by fitting the frequency distribution curve of material mechanical properties and making a deconvolution analysis. The cumulative distribution functions of all facies were assumed to have a normal or Gaussian distribution during fitting, as expressed by:

$$P_J(x) = \frac{1}{\sqrt{2\pi s_J^2}} \exp(-\frac{(x - \mu_J)^2}{2s_J^2}) \tag{5}$$

where μ_J is the arithmetic average of the values of all terms.

3 TEST RESULTS

A total of 8 lithic specimens were tested, and core samples obtained from the same depth as the lithics were subjected to the conventional rock mechanical test.

3.1 Results of the nano-indentation test

Figure 3 shows the plane distribution of Young's modulus obtained from the nano-indentation test of lithic specimen 4. Because the rock has different mineral compositions at different indentation dots, Young's modulus is distributed in the range of 9.1–41.3 GPa, varying greatly planarly. The blue regions with low Young's moduli are caused by the locally high contents of clay or organic matter. Figure 4 shows the statistical distribution of Young's moduli obtained from the nano-indentation test of specimen 4. The Young's moduli are mainly distributed in the range of 18–37 GPa and peak at about 28 GPa. The test results of specimen 4 were subjected to scale upgrading to obtain macroscopic parameters, and the calculated Young's modulus was 25.7 GPa.

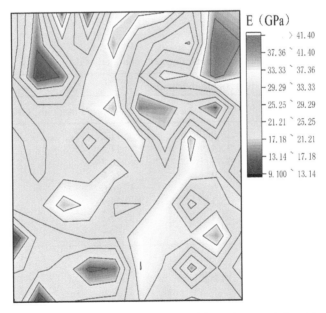

Figure 3. Plane distribution of Young's moduli obtained from the nano-indentation test of Specimen 4.

Table 1 shows the statistical results of 8 nano-indentation tests after scale upgrading. Young's modulus is in the range of 11.2–26.7 GPa, and the hardness is in the range of 2.2–6.8 GPa after scale upgrading.

3.2 Results of rock mechanical test

Mechanical parameters of the core plugs obtained from the depth corresponding to lithic specimens were tested using a tri-axial rock mechanical testing apparatus. The results of the rock mechanical test are shown in Table 2. Young's modulus ranges from 10.2 to 24.8 GPa, and Poisson's ratio ranges from 0.21 to 0.25.

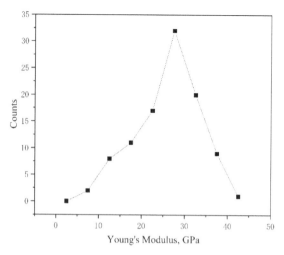

Figure 4. Statistical distribution of Young's moduli obtained from the nano-indentation test of Specimen.

Table 1. Calculation results of nano-indentation tests and after scale upgrading.

Specimen No.	Young's modulus (nano-indentation) Range, GPa	Peak value, GPa	Value after scale upgrading Young's modulus, GPa	Hardness, GPa
1	1.5–23.0	17.0	11.2	2.3
2	2.0–18.5	13.4	11.5	2.4
3	6.2–37.7	27.5	26.7	5.8
4	9.1–41.3	28.0	25.7	6.2
5	7.5–38.0	24.6	26.6	6.1
6	5.2–65.0	23.3	24.9	5.2
7	8.6–42.0	21.6	24.7	4.8
8	7.3–44.7	24.8	23.2	3.7

4 ANALYSIS AND APPLICATION

4.1 *Error of the results of nano-indentation rock mechanical tests and rock mechanical tests is less than 10%*

The Young's modulus after scale upgrading of the nano-indentation test results was slightly higher than the value obtained from the rock mechanical test. Table 2 compares the results of the tests, showing that the error of Young's modulus obtained from the two tests is 4–10%.

4.2 *What BI calculated based on the results of the nano-indentation test meets the requirements of fracturing design*

Figure 5 shows the comparison of BI's obtainment from log interpretation and BI's calculation based on the rock mechanical test and the nano-indentation test results for a tight oil well in Block Houhe of LiaoHe Oilfield, respectively. The BI's calculation was based on Poisson's ratio and Young's modulus [17].

Table 2. Comparison between the results of the nano-indentation test and the rock mechanical test.

No.	Comparison of Young's modulus (E)			Comparison of Poisson's ratio (v)		
	Indentation test GPa	Rock mechanical test GPa	Error %	Indentation test	Rock mechanical test	error %
1	11.2	10.2	9.80	0.25	0.24	4.17
2	11.5	10.7	7.48	0.23	0.24	−4.17
3	26.7	24.4	9.43	0.20	0.22	−9.09
4	25.7	24.1	6.64	0.22	0.22	0.00
5	26.6	25.2	5.56	0.22	0.23	−4.35
6	24.9	22.7	9.69	0.21	0.22	−4.55
7	24.7	22.7	8.81	0.23	0.23	0.00
8	23.2	22.3	4.04	0.23	0.22	4.55

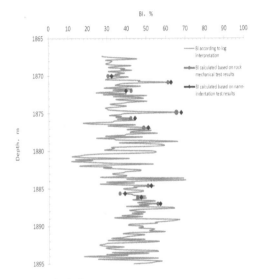

Figure 5. Comparison of BI's calculation.

The BI was 30.1–66.6 according to the log interpretation at the sampling points of cores and broken cores; what BI calculated was 30.5–65.3 when the calculation was based on the rock mechanical test results, and 32.3–67.8 when the calculation was based on the nano-indentation test results. Among the calculation results for samples obtained from 8 sampling points, the maximum error of what BI calculated based on the nano-indentation test results was 5.87% compared with what BI obtained from log interpretation, which met the requirements of fracturing design.

5 CONCLUSIONS

By applying the nano-indentation technique, 1 mm × 1 mm small-scale broken core specimens or lithics can be used to test the microscopic mechanical parameters, and the convolution method can be used for scale upgrading to obtain the macroscopic mechanical parameters.

Young's modulus obtained from the nano-indentation test after scale upgrading was slightly higher than that obtained from the rock mechanical test, and the error of the two tests was less than 10%.

The approach of brittleness evaluation based on Poisson's ratio and Young's modulus was applied for BI calculation. The maximum error of the calculation results of the nano-indentation test using the specimens of broken tight oil lithics from the LiaoHe Oilfield was 5.87% compared with the results of log interpretation, which met the requirements of fracturing design.

REFERENCES

[1] Block M D, Bonijoly D, Hofmann A, *et al. Shale Geomechanics: A Nano-indentation Application* [R]. Amsterdam RAI, The Netherlands: 76th EAGE Conference & Exhibition, 2014.

[2] Cała M, Cyran K, Kawa M, *et al.* Identification of Microstructural Properties of Shale by Combined use of X-ray Micro-CT and Nanoindentation Tests [J]. *Procedia Engineering*, 2017, 191: 735–743.

[3] Chen Yun, Jin Yan, Chen Mian. A Rock Brittleness Evaluation Method Based on Energy Dissipation [J]. *Chinese Journal of Theoretical and Applied Mechanics*, 2015, 47(06): 984–993.

[4] Hofmann A, Rigollet C, Portier E. *et al. Gas Shale Characterization: Results of the Mineralogical, Lithological and Geochemical Analysis of Cuttings Samples from Radioactive Silurian Shales of a Palaeozoic Basin, SW Algeria* [R]. Cairo: North Africa Technical Conference & Exhibition, 2013.

[5] Hou Zhenkun, Yang Chunhe, WEI Xiang, *et al.* Experimental Study on the Brittle Characteristics of Longmaxi Formation Shale [J]. *Journal of China Coal Society*, 2016, 41(05): 1188–1196.

[6] Kumar V, Curtis M E, Gupta N, *et al. Estimation of Elastic Properties of Organic Matter in Woodford Shale Through Nanoindentation Measurements* [R]. SPE 162778-MS, 2012.

[7] Li Qinghui, Chen Mian, Jin Yan, *et al.* Indoor Evaluation Method for Shale Brittleness and Improvement [J]. *Chinese Journal of Rock Mechanics and Engineering*, 2012, 31(08): 1680–1685.

[8] Liu K, Ostadhassan M, Bubach B. Applications of Nano-indentation Methods to Estimate Nanoscale Mechanical Properties of Shale Reservoir Rocks [J]. *Journal of Natural Gas Science and Engineering*, 2016, 35: 1310–1319.

[9] Ma Xinfang, Li Ning, Yin Congbin, *et al.* Hydraulic Fracture Propagation Geometry and Acoustic Emission Interpretation: A Case Study of Silurian Longmaxi Formation Shale in Sichuan Basin, SW China [J]. *Petroleum Exploration and Development*, 2017, 44(6): 974–981.

[10] Manjunath G.L, Jha B. Nanoscale Fracture Mechanics of Gondwana Coal [J]. *International Journal of Coal Geology*, 2019, 204: 102–112.

[11] Meng Junqing, Niu Jiaxing, Xia Junkai, *et al.* Study on Mechanical Properties and Failure Mechanisms of Coal at the Nanometer Scale [J]. *Chinese Journal of Rock Mechanics and Engineering*, 2020, 39(1): 84–92.

[12] Rick Rickman, Mike Mullen, Erik Peter, *et al.* A Practical Use of Shale Petrophysics for Stimulation Design Optimization: All Shale Plays Are Not Clones of the Barnett Shale [R]. SPE 115258-MS, 2008.

[13] Shi Xian, Cheng Yuanfang, Jiang Shu, *et al.* Experimental Study of Microstructure and Rock Properties of Shale Samples [J]. *Chinese Journal of Rock Mechanics and Engineering*, 2014,

[14] Shi Xian, Jiang Shu, Lu Shuangfang, *et al.* Investigation of Mechanical Properties of Bedded Shale by Nanoindentation Tests: A Case Study on Lower Silurian Longmaxi Formation of Youyang area in Southeast Chongqing, China [J]. *Petroleum Exploration and Development*, 2019, 46(1): 155–164.

[15] Sun Jianmeng, Han Zhilei, Qin Ruibao, *et al.* log Evaluation Method for Fracturing Performance in Tight Gas Reservoirs [J]. *Acta Petroleum Sinica*, 2015, 36(01): 74–80.

[16] Wang Sheng, Liu Bo, Fu Xiaofei, *et al.* Evaluation of the Brittleness and Fracture Characteristics for Tight Clastic Reservoirs [J]. *OIL&GAS GEOLOGY*, 2018, 39(06): 1270–1279.

[17] Yang Chao, Xiong Yongqiang, Wang Jianfeng, *et al.* Mechanical Characterization of Shale Matrix Minerals Using Phase-positioned Nanoindentation and Nano-dynamic Mechanical Analysis [J]. *International Journal of Coal Geology*, 2020, 229: 103571.

Overall design and key technologies of extra-wide deck space special-shaped arch bridge

Jie Liu* & Li-xin Wei*
Guangzhou Municipal Engineering Design and Research Institute Co. Ltd, Guangzhou, China

ABSTRACT: Dongguan Binhai Avenue Shachong Bridge is an urban landscape bridge. Considering the influence of flood control, structure stress, landscape function, economy, durability, and other factors comprehensively, the scheme conception and structure selection of the bridge were carried out. A novel and unique scheme of through arch bridges was proposed, aiming at the super-wide bridge deck with a width of 80 m, which adopted a new structure system of single span 120 m called an "open five-cable plane space special-shaped tied arch bridge". The key technologies such as main components, deck system scheme, economy, and durability were studied. The dynamic and static analysis of the bridge was carried out by using the spatial finite element software. The results showed that all the indexes of the bridge met the requirements of the code, and the stress of the bridge was reasonable and reliable. This kind of bridge has the advantages of a novel and unique structure, a beautiful appearance, reasonable stress, low engineering cost, and good durability. This paper will provide a new idea for an urban landscape bridge with a super-wide deck.

1 INTRODUCTION

Shachong Bridge is located on Binhai Avenue in Dongguan Binhai New Area. It runs east-west from Binhai Avenue in Humen Town in the west to Jiaoyiwan Avenue in the east. The total length of the route is about 5.5 km. The road is an urban trunk road, and the cross-section of the road is 8 lanes for the main road and 4 lanes for the auxiliary road, with sidewalks and non-motorized lanes on both sides. The standard width of the bridge is 80 m. Viewing platforms are set up on both sides of the upstream and downstream sides of the middle span. The section is partially widened to 88 m. At the bridge site, the designed tidal level of the Shachong River in 100 years is 4.63 m, the basic intensity of an earthquake is 7 degrees, and the peak acceleration of ground motion is 0.15 g. The geological structure of the site is stable. The bridge has a bored pile foundation with breezy granite as the pile end-bearing layer. Shachong Bridge is the cable-supported bridge with the widest deck under construction in the world at present. Based on the design concept of the bridge, key technologies such as bridge type selection, the structural stress system, the main structure of the bridge, deck selection, economy, and durability were studied.

---

*Corresponding Authors: 361728054@qq.com and szyhldj@163.com

2 OVERALL BRIDGE DESIGN

2.1 General layout

The plane of the Shachong Bridge is located in a straight line, and the longitudinal section of the bridge is limited by the elevation of the land parcels on both sides and the planned road intersection, so the bridge intersects with the planned road on both sides. On both sides of the bridge site, the planned elevation of the top of the flood embankment is 5.0 m, the tide level of the 100-year event is 3.46 m, and the minimum control elevation of the beam bottom is 4.2 m. The longitudinal slope of the bridge is 0.8% in both directions, and the radius of the vertical curve is 7500 m. According to the horizontal and vertical conditions, flood control requirements, and the elevation of the two sides of the bridge, the single-span 120-meter steel box tie arch bridge is adopted. The bridge layout is shown in Figure 1.

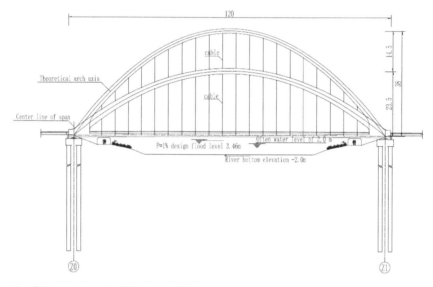

Figure 1. Elevation layout of Shachong Bridge.

The bridge adopts the whole layout with a standard width of 80 m for the section, including an 8.5-meter sidewalk and non-motorized lane, a 7.5-meter side road lane, a 5.5-meter side green belt, a 15.0-meter main road lane, an 8.0-meter green belt, a 15.0-meter main road lane, a 5.0-meter side green belt, a 7.5-meter side road lane, and an 8.5-meter sidewalk and non-motorized lane. It is partially widened to 88 m, and the bridge section arrangement is shown in Figure 2.

2.2 Arch rib structure

The arch rib structure includes the middle arch rib located in the vertical plane and the side arch ribs located on both sides of the middle arch rib. The arch feet at both ends of the arch ribs are consolidated with the main beam. The sagittal span ratio of the middle arch rib is 1/3.05, the design sagittal height is F = 38.00 m, the sagittal span ratio of the side arch rib is 1/4.936, and the vertical projection design sagittal height is F = 23.5 m. The golden ratio of arch height is between the middle arch and the side arch; there is no transverse support between the middle arch rib and the two side arch ribs. The middle arch rib is a vertical

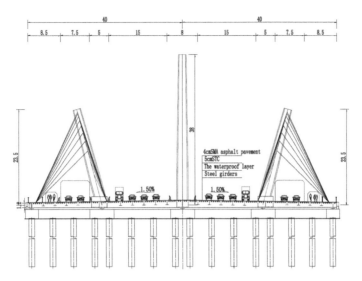

Figure 2. Layout of the cross-section of Shachong Bridge.

structure that is connected with the main beam through a single row of vertical cables. The side arch rib is an extrovert structure, and the angle between the side arch rib and the vertical plane is 15°. Each side arch is connected with the main beam by two rows of inclined cable. The vertical middle arch and two extroverted side arches constitute the spatial heterotypic arch bridge structure system of the "open space five cable plane inclined arch rib tied rod arch bridge". The shape of the system is not only like a spray but also like the bud of a flower, which is consistent with the design theme.

The arch ribs are all made of Q420qD structural steel, and the maximum thickness of the steel plate is 50 mm, which needs to meet the performance index of the steel plate in the thickness direction. The arch rib structure is made in sections in the factory. Each arch rib is made in 5 sections, considering the structural stress, transportation, lifting capacity, and other factors. The transverse section of the arch rib is shown in Figure 3.

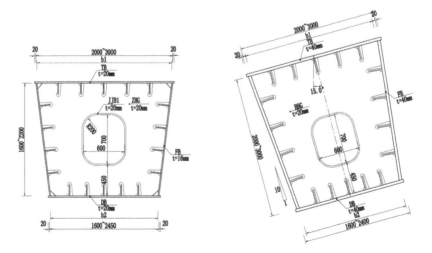

Figure 3. Cross-section of an arch rib (unit: mm).

2.3 Main beam construction

Due to the limit of road elevation on both sides of the bridge, the height of the main beam is relatively low; the minimum height of the main beam is 1.2 m, and the maximum height is 1.8 m. It is difficult to meet the stress requirement of the main beam with a concrete structure, while the main beam with a steel structure has the advantages of light dead weight and a large bearing capacity. Through calculation, the main beam adopts a steel structure, and the beam height meets the requirements.

The main beam structure adopts the whole wide-body steel box girder, which is composed of a top and bottom plate, longitudinal web, diaphragm, stiffening ribs, orthotropic steel deck, etc. The bottom plate of the box girder is horizontal, and the top plate is laterally inclined to form a two-way transverse slope of 1.5%. The width of the bridge is 80 m and there are 5 rows of cables in the cross-section. The force of the main beam is controlled by the cross-section, and the transverse force system of the bridge is a four-span continuous beam structure. After calculation and optimization, the main beam is optimized from the steel box girder to the integral vertical and horizontal beam system, with one longeron set every 2.0–3.5 m on the section, and the longeron corresponding to the arch rib and the cable adopts the box girder structure, and the bottom plane of the box girder is level and at the same height as the beam. The other position of the longeron has an I-shaped web height of 0.8–1.0 m and a bottom width of 0.4 m. The standard beam section of the steel main beam is provided with a diaphragm every 2.5 m, among which the diaphragm corresponding to the cable adopts the box section and the other diaphragms adopt the I-shaped section. The beam is 0.5 m high and 16 mm thick. Q345qD structural steel is adopted for the main beam except for the end beam. The cross-section of the main beam is shown in Figure 4.

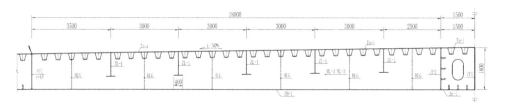

Figure 4. Cross-section of the main beam.

After the main beam is optimized from the integral steel box girder to the vertical and horizontal beam systems, the bridge cost is reduced from 22, 235 yuan per m^2 to 18, 700 yuan per m^2. The economic advantage of adopting the vertical and horizontal beam system for the main girder is obvious.

2.4 Cable design

The cable is one of the most important stress components in the arch bridge. Its reliability, durability, and adaptability are related to the safety and normal use of the bridge structure. This bridge adopts the flexible cable, and there are mainly two forms of parallel steel wire cable and steel strand cable, both of which are very mature. Considering the bridge cable force is not big if the steel strand cable is used, the clip is not easy to clamp, and the cable anchor head is easy to relax, so the cable adopts the parallel steel wire bundle scheme. The middle arch ribs and the main steel beam are connected by a vertical cable with a spacing of 5 m. The middle arch-cable adopts 73-φ5 high-strength and low-relaxation galvanized parallel steel wire bundles. The inner and outer sides of the side arch ribs are connected with the main steel beam by an oblique cable with a spacing of 5 m. To balance the overturning moment caused by the lateral road and sidewalk side cables on the arch ribs, eccentricity should be set between the cable on the side of the main roadway and the side arch plane. Side arch-cable with 55-φ5 parallel high strength, low relaxation galvanizing steel wire bundles, 127-φ7 parallel high strength low relaxation galvanizing steel wire bundle, and chill heading anchor rope body low-stress corrosion PES (FD), is used for cable. Super durable polyurea coating material to improve the durability of the cable, under normal maintenance conditions, only needs to replace the cable bridge's whole life cycle 2 to 3 times, which can save a lot of engineering cost.

3 RESEARCH ON KEY TECHNOLOGY OF STRUCTURE DESIGN

3.1 *Study of the bridge landscape scheme*

The selection of the Shachong bridge type comprehensively considers the regional cultural background, the theme of the bridge landscape, the surrounding environment of the region, the synergy of upstream and downstream related bridge landscapes, and other factors. Shachong Bridge is located in the Jiaoyiwan section of the Binhaiwan New Area. It is an important urban landscape bridge across the Shachong River and a key control project for the whole line. The Binhaiwan New Area is the core area connecting Dongguan with the Guangdong-Hong Kong-Macao Greater Bay Area and the Guangzhou-Shenzhen Science and Technology Innovation Corridor. In the future, it will be built into a new platform for the coordinated development of the Guangdong-Hong Kong-Macao Greater Bay Area and a new hub for the integrated development of the core area of the Pearl River Delta, which requires a high landscape function of the bridge.

Shachong Bridge, as an urban landscape bridge, has the following difficulties in bridge design: (1) The road width is 80 m and the bridge length is 120 m. Because of the overly wide road, the transverse and longitudinal ratios of the bridge are greater than those of a conventional bridge. A conventional arch bridge, such as a single arch or double arch, will make the whole bridge dull and bloated. (2) The height difference between the bridge deck and the river surge is small, so it is not easy to control the height and proportion of the arch aesthetically. To solve the above difficulties, the bridge design adopts a multi-arch configuration. Multi-arch has good mechanical properties and rich shape changes and can create a strong spatial feeling (Chen *et al.* 2005; Sheng 2009; Shao 2009; Song 2019; Yan *et al.* 2007).

Upstream of Shachong Bridge lies the Binhaiwan Bridge (Yulan Bridge), a space-twisted cable-stayed bridge with a single tower. The main tower is shaped like a Magnolia, the flower of Dongguan city, and the design concept is "Magnolia blossom, Silk Road light". Shachong Bridge adopts the shape of three arches, which resemble the tip of a Magnolia in full bloom. It is in line with the Binhaiwan Bridge (Yulan Bridge) and echoes from a distance, forming the sister bridge of the Binhaiwan New Area.

The scheme has three arch ribs; the middle arch is located in the central green belt of the bridge; and the cable is vertical and on a single cable plane. The two side arches are located in the side green belt between the secondary road and the main road and tilt outward at an angle of 75° with the road. The cable is located in the side green belt and on the outer side of the sidewalk, showing the motion of the spray. In terms of arch height, the middle arch has a golden ratio relationship with the two sides of the arch, and the scale relationship is appropriate, which is adapted to the surrounding environment. When completed, the Shachong Bridge will become another landmark of the Binhaiwan New Area, together with the Binhaiwan Bridge, forming a "sister flower" blossoming from the sea. After the completion of the bridge, the bridge effect is shown in Figure 5.

Figure 5. Real scene of Shachong Bridge.

3.2 *Research on the structural stress system*

Similar bridges have been built at home and abroad, such as the Dagu Bridge in Tianjin, the Nanning Bridge in Guangxi, the South Zhonghua Bridge in Shanxi, and the Aixi Lake Bridge in Nanchang. The arch ribs of these bridges are inclined outward, and the arch ribs on both sides can be designed as symmetrical or asymmetrical according to landscape requirements.

The bridge above is two pieces of an extroversion arch rib. Shachong Bridge's upper structure, according to the requirements of bridge deck width and modeling, is composed of three arch ribs: two side arch ribs are inclined and the middle arch ribs are upright. The vertical middle arch and two side arches constitute the open spatial special-shaped arch bridge structure system of the "steel box tie arch bridge under the space five-cable plane". The arch bridge is the first to use this type of structural system.

Compared with parallel arch bridges and introverted arch bridges, the mechanical properties of this bridge are more complex due to the outward inclination of the arch ribs. There is no transverse connection between the three arch ribs, resulting in a small transverse stiffness, which mainly depends on the transverse flexural stiffness and torsional stiffness of the arch ribs to ensure structural balance. The vertical middle arch is about 38 m high, using a single cable plane through the appropriate arch rib section to maintain the stability of the arch rib itself; the side arch adopts a double cable plane, which is not connected with the middle arch. The arch rib forms a triangular area through the multi-cable plane in space to maintain the lateral stability of the arch rib.

The length of the bridge is 120 m, the width is 80 m, and the ratio of width to span is large. The force of the main beam is controlled by the transverse bridge direction, and there are 5 cables in the transverse bridge section. The transverse force system of the bridge is a four-span continuous beam structure.

3.3 *Study on the joint section of arch and beam*

The bridge uses arch beam consolidation, arch rib and longitudinal beam transverse equal width, and arch rib web, and extends the roof of the longitudinal beam web butt, arch rib top, and bottom plates, with the main longitudinal beam roof welding, and installation in the corresponding position to strengthen the baffle. The side arch ribs and the side longitudinal beam roof are connected with high-strength bolts, and the arch foot is provided with a connecting flange, through which is the arch foot, with the side longitudinal beam roof bolt.

3.4 *Anchorage system design*

Commonly used cable-beam anchoring structures and cable-arch anchoring structures include anchor box types, anchor pipe types, ear plates, and anchor pull plates. The cable arch anchoring space of this bridge is limited. In order to facilitate tension and later maintenance, the upper end of the cable is anchored to the outer side of the arch box by ear plate anchoring. Since there is no tension condition in the box girder, the anchorage system can be fixed on the main beam with an ear plate or at the bottom of the main beam with an anchor box. Because the arch on the ear plate type is small and the late cable tension nut adjustable range is small, it is difficult to adjust the error, and the ear plate anchorage in the upper main girder and the bridge deck has the experience to influence the bridge landscape as a whole and the pedestrian thermal comfort. Therefore, the cable bottom with the steel anchor box at the bottom and the steel box girder anchor space beam body does not affect the bridge's beauty.

3.5 *Bridge deck design*

At present, there are two main types of orthotropic board pavement: flexible pavement and rigid pavement. Flexible pavement is asphalt material, such as stone mastic asphalt mixture (SMA) and epoxy asphalt mixture (EA). The rigid pavement material is ordinary concrete or ultra-high-performance concrete (UHPC). Large-span bridges generally adopt flexible pavement, while some medium- and small-span bridges adopt rigid pavement. The deck of the Shachong Bridge is 80 m wide. Due to the limitation of the height of the main beam, the dead load plays a key role in structural stress. According to the calculation, the key factor affecting the size of the arch rib structure is the second phase of the dead load of the bridge deck. Reducing the second phase dead load can optimize the section size of the arch rib and obtain a more graceful line shape. In order to reduce the second-phase dead load on the bridge, it is very important to adopt the pavement form of the bridge deck. This paper proposes three kinds of bridge deck pavement schemes, and the comparison of the three kinds of bridge deck pavement schemes is shown in Table 1.

Table 1. Comparison table of bridge deck pavement forms.

Option	Bridge Deck Pavement	Paving Option	Advantages and Disadvantages	Life Cycle Investment Cost
Scheme 1	Steel Deck Pavement	Epoxy Asphalt Concrete Pavement (double heterosexual pavement EA + SMA)	It has poor cracking and fatigue resistance, a short service life, high technical requirements, and a high cost. It is widely used in bridges.	Higher
Scheme 2	Combined Deck Pavement	150 mm Thick Steel Fiber Reinforced Concrete + Ordinary Asphalt Paving Layer	Although the bridge deck has high stiffness, high resistance to fatigue cracking, and a low cost, it is heavy and is commonly used in urban bridges.	Lower
Scheme 3	Light Combination Bridge Deck Pavement	50 mm Thick STC Layer + 40 mm Asphalt Concrete Wear Layer	It has strong fatigue cracking resistance, light dead weight of the structure, and high structural stiffness, making it popular but at a high cost.	A little higher

Table 1 shows that the dead weight of the steel deck pavement bridge in Scheme 1 of the three schemes is light, but the fatigue and durability of the orthotropic steel deck are prominent problems that require frequent maintenance and reinforcement, and the investment cost in the whole life cycle is high. In scheme 2, the steel-mixed composite bridge deck has large stiffness, good durability, a small maintenance workload, and low cost; however, the bridge deck has a large dead weight and poor crack resistance, and the concrete is easy to crack. The third scheme adopts an ultra-high toughness concrete (STC) lightweight composite deck, which can improve the crack resistance and durability of the bridge, reduce the dead weight, and solve the problems of the vulnerability of the orthotropic steel deck and the fatigue cracking of the steel structure. The investment cost over the whole life cycle is moderate, and the comprehensive economy is optimal. Based on the weight reduction and economic durability of the bridge, the bridge adopts Scheme 3—a steel and super toughness concrete (STC) lightweight composite bridge deck structure, which is composed of orthotropic steel bridge panels and STC, and is welded to the steel beam, and a 50-mm-thick STC layer is poured. Then a 40-mm-thick asphalt-concrete wear layer is spread on it. The lightweight composite structure has the advantages of large local stiffness, being lightweight, and having good durability, which comprehensively solves the inherent fatigue cracking and other typical disease problems of orthotropic steel bridge panels (DB 41/T 643-2010; Tian *et al.* 2016, 2017; Yang & Shi 2017) and can enhance the connection ability between the asphalt plane layer and the bridge panels and extend the service life. In the whole life cycle, the durability of the steel deck can be improved, the operation and maintenance costs of the steel bridge can be reduced, and the traffic interruption caused by the maintenance of the steel deck can be reduced.

3.6 *Structural durability design*

Shachong Bridge is located in a coastal environment, and the environmental effect level is III-C to III-F. In order to ensure the safety of the structure in the designed service life and meet the normal service function, a special durability design has been carried out for the concrete substructure, main beam, arch rib steel structure, cable, and deck system (Zheng & Xiao 2018), and a variety of comprehensive anti-corrosion strategies have been adopted. Several key technologies and structural measures for durability design are proposed based on research findings and practical experience in the field near the bridge site.

(1) For the cap and abutment and other concrete structures, the use of improved concrete strength grade, increased outermost net reinforcement protective layer thickness, strict control of crack width, and cap concrete mixed with steel rust inhibitor measures is required. The steel guard cylinder used for pile foundation construction shall be kept permanently, and the outer plane of the steel guard cylinder shall be coated with anti-corrosion.
(2) The warranty period of the anti-corrosion supporting scheme of the main beam and arch rib steel structure is more than 25 years, and the outer plane of the steel structure adopts the metal thermal spraying composite coating system for anti-corrosion. The inner plane of the steel structure adopts a heavy anti-corrosion coating system.
(3) The cable has a PES (FD) low-stress anti-corrosion cable body, and the outer layer of the cable sheath is protected by a new type of bridge cable with low wind resistance, impact resistance, and light resistance. The service life is more than two times that of the conventional HDPE casing.
(4) The bridge deck adopts a steel-STC ultra-high-toughness concrete lightweight composite bridge deck structure. The ultra-high-toughness concrete layer is a permanent structure layer that does not need to be replaced, and only the ordinary asphalt layer worn on it can be maintained regularly.

4 STRUCTURAL ANALYSIS AND CALCULATION

4.1 *Structural calculation model*

The space finite element software MIDAS Civil is used to establish the beam lattice model for the overall calculation and analysis of the bridge structure, and the main beam is simulated by the beam lattice model. The main beam and arch ribs are simulated by space beam elements, and the cable is simulated by truss elements. There are 8324 nodes and 10078 beam elements adopted in the whole bridge. The structural calculation model is shown in Figure 6.

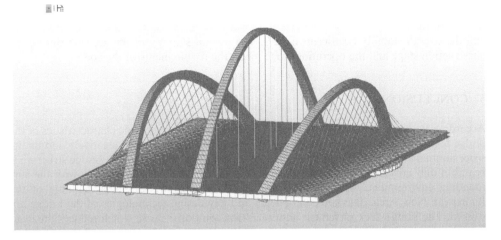

Figure 6. Finite element model of structural calculation.

4.2 *Structural static calculation and analysis*

(1) The main beam

Under the combination of basic loads, the maximum tensile stress of the upper side of the longeron is 129 MPa, and the maximum tensile stress of the lower side of the longeron is 122 MPa. The maximum tensile stress of the upper side of the beam is 109 MPa, and the maximum compressive stress of the lower side of the beam is −133 MPa. The strength of the vertical and horizontal beams meets the requirements.

(2) Arch ribs

Under the basic load combination, the maximum compressive stress of the upper side of the middle arch is −123 MPa near the arch foot, and the maximum compressive stress of the lower side of the middle arch is −90 MPa in the middle of the arch rib span. The maximum compressive stress of the upper side of the side arch is −213 MPa near the arch foot, and the maximum compressive stress of the lower side of the side arch is −178 MPa in the middle of the arch rib span. The strength of the arch ribs meets the requirements.

(3) Cable

The cable is made of galvanized steel wire with high strength and low relaxation, and its tensile strength is fpk = 1670 MPa. Table 2 shows that the safety factors of the cable meet the specification requirements and have a certain safety reserve.

Table 2. Maximum internal force table of cable.

The Cable Type	Type	Maximum Cable Force (kN)	Breaking the Completed (kN)	Safety Factor
Middle Cable	PES (FD) 5-73	711	2393	3.37
Side arch Inner Cable	PES (FD) 5-127	2491	8162	3.28
Side Arch Outer Cable	PES (FD) 5-55	456	1803	3.95

4.3 *Structural stability analysis*

The unfavorable conditions in the construction process are that the arch ribs have been installed, but no cable has been stretched. In addition, the wind speed is encountered once every hundred years, so the stability needs to be checked. According to the calculation results, the first is the outer instability of the middle arch, and the overall stability coefficient is 45.3. In the operation stage, the external instability of the middle arch plane is the first, and the overall stability coefficient is 12.0. The overall stability of the arch rib during the construction stage and the operation stage meets the requirements of the code.

5 CONCLUSION

According to the requirements of the urban landscape, the topographic characteristics of the bridge location, and the super wide deck structure, Shachong Bridge creatively adopts the space anomaly arch bridge with five cable planes and three arch ribs. The bridge structure is beautiful and novel; the structural stress is reasonable; and it is in harmony with the surrounding environment. After completion, it will become the landmark building of the Binhaiwan New Area. The key components and key design technologies of the bridge are studied. The bridge deck pavement adopts a light composite deck, which reduces the dead load in the second phase of the bridge, the size of arch ribs, and the amount of steel used, and solves the fatigue and durability problems of the steel deck. The dynamic and static analysis of the structure shows that all the components of the bridge meet the requirements of the code under the most unfavorable combination of loads, and the structural forces are reasonable and reliable. In the same span as a super wide arch bridge, this type of bridge is very competitive. The design outcome of this bridge can serve as a model for the design of a similar type of super wide deck landscape bridge. Shachong Bridge is the cable-supported bridge with the widest deck under construction in the world. Construction of the bridge started in October 2020, and it was completed and opened to traffic by the end of 2021.

REFERENCES

Chen Airong, Sheng Yong, Qian Feng. *Bridge Modeling* [M]. Beijing: People's Communications Publishing House, 2005.
DB 41/T 643-2010. *Technical Specification for Ultra-high Performance Lightweighted Composite Deck Structure* [S].
Shao Changyu. Technical Consideration and Features of Composite Structure Bridge of Jiubao Bridge in Hangzhou[J]. *Bridge Construction*, 2009, (6): 42–45. in Chinese
Sheng Hongfei. *Bridge Building Aesthetics* [M]. Beijing: People's Communications Press, 2009.
Song Shujun. Design of Through Continuous Arch Beam Composite System Irregular Shape Arch Bridge [J]. *Urban Roads Bridges & Flood Control*, 2019, (7): 79–82. in Chinese
Tian Qixian, Gao Liqiang, Zhou Shangmeng. Study of Mechanical Behavior of Composite Bridge Deck with Ultra High Performance Concrete and Orthotropic Steel Plate [J]. *Bridge Construction*, 2017, 47 (3): 13–18. in Chinese
Tian Qixian, Gao Liqiang, Du Xinxi. Study of Influences of Deck Plate Structural Design on Fatigue Performance of Orthotropic Steel Bridge Deck [J]. *Bridge Construction*, 2016, 46 (1): 18–23. in Chinese
Yan Donghuang, Liu Xuefeng, Tian Zhongchu, *et al.* A Summarized Account of Development and Application of Hybrid System Arch Bridges [J]. *World Bridges*, 2007, (2): 65–67. in Chinese
Yang Shili, Shi Zhou. Current Research of Fatigue Damage in Orthotropic Deck Plates of Long-Span Steel Box Girder Bridges in China [J]. *Bridge Construction*, 2017, 47 (4): 60–65. in Chinese
Zheng Qinggang, Xiao Haizhu. Design of Steel Box Girder for Main Girder of China-Maldives Friendship Bridge Aided to Maldives by China [J]. *Bridge Construction*, 2018, 48 (3): 95–99. in Chinese

Research on the algorithm of tooling position and attitude measurement based on a global camera system

Haifei Wang*, Shaozhong Cao & Wei Zhao
School of Information Engineering, Beijing Institute of Graphic Communication, Beijing, China

Zhenqing Gao
College of Mechanical and Electrical Engineering, Beijing Institute of Graphic Communication, China

ABSTRACT: Position detection of a six-degree-of-freedom parallel platform end-access probing tooling by the Global Camera's Thick Side Posture Measurement System is an important part of the unconstrained movement in the first stage of the assembly operation. The monocular camera is used to take the position picture on the right and upper sides of the penetration measurement tooling, determine the initial state posture of the penetration measurement tooling, arrive at the approximate position after unconstrained movement, and take the picture again, judge the current position and determine the close motion of the shaft hole in the second stage of the assembly operation, and record its posture and the movement state of the six-degree-of-freedom parallel platform. The system's proposal realizes the first step in the automation of assembly operations and lays a solid foundation for subsequent work.

1 INTRODUCTION

Assembly is an indispensable follow-up process in the production of modern industrial products. The application of robots to assembly is an inevitable requirement of modern industry and the most important step to achieving automatic production. The assembly of printing machine-bearing sleeves is a typical shaft-hole assembly operation and one of the most difficult problems in manufacturing printing equipment. Currently, manual assembly is mainly used in China, which is inefficient and harmful to workers' health and is one of the biggest bottlenecks restricting the automation of printing equipment manufacturing (Li 2016; Qian 2019).

The global camera coarse side poses measurement system is the first step to realizing assembly automation. Based on the initial pose and final pose of the probe measurement tooling that can be obtained by the system, the motion state of the 6-DOF parallel platform is recorded by the industrial computer. The subsequent assembly process only needs to control the 6-DOF parallel platform motion according to the data recorded by the industrial computer to reach the second stage shaft hole approaching motion state automatically. This paper combines the 6-DOF parallel platform, the probe-type measuring tooling, and the global camera to complete the automation of the assembly work's first stage of unconstrained motion (Jiang 2015; Wei 2021).

---

*Corresponding Author: 962775583@qq.com

DOI: 10.1201/9781003450832-3

2 CAMERA SYSTEM CONSTRUCTION

2.1 *Platform and control system*

The robot is based on a six-degree-of-freedom parallel platform with high precision and large bearing capacity and is equipped with high-power and high-performance electric cylinders. It has three measurements: a global camera rough position and orientation system, a three-dimensional laser measurement system, and a six-axis force sensor (Ren 2017). It can obtain the position and force information required for assembly in real-time, and differential can obtain the speed and other information. The measuring system has the advantages of fast speed, real-time performance, and no need for a slow surface hole searching process. It is very suitable for assembling complicated and severe printing press shaft sleeves. As a mixed product of multiple technologies, it is mainly based on computer and automatic control technology. It has many advantages over ordinary information-processing computers and is widely used in production and life. The main advantages are reliability, compatibility, real-time, etc. (Li 2016).

2.2 *Global camera rough pose measurement system*

It mainly includes two cameras mounted on the front side, the robot system's top, and the printing machine shell. The general pose of the through-hole axis relative to the edge of the printing machine shell is a known quantity. The current rough relative pose of the axial hole can be obtained through the relative pose of the printing machine shell and the robot in the global camera (Zhu 2019). The global camera layout is shown in Figure 1:

Figure 1. Camera layout.

2.2 *Workflow*

The pictures taken with the global camera need to be preprocessed by image, including image enhancement, median filtering, and binarization. Image enhancement aims to increase the definition of the pictures taken and reduce noise. The algorithm is gamma transform, mainly used for image correction and contrast enhancement. Median filtering is a typical nonlinear signal processing technology based on the ranking statistics theory that can effectively suppress noise. Binarization converts color pictures into grayscale images to distinguish the location of measuring tooling in pixels. Finally, the edge detection is carried out, and the position and orientation of the measuring tooling are calculated using the image obtained after the edge detection to obtain the rough position and orientation of the measuring tooling at this time (Wei 2022; Zou 2022; Zhang 2020).

3 ALGORITHM RESEARCH

3.1 *Solve angle*

After image processing, we get the contour and central axis of the target object and establish a spatial coordinate system around the central axis, as shown in Figure 2:

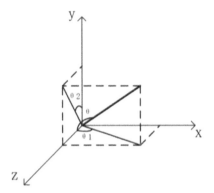

Figure 2. Diagram of 3D coordinate system.

It can be seen from the figure above:

$$\tan\theta = \frac{\sqrt{a^2\tan^2\theta_1 + \frac{a^2}{\tan^2\theta_2}}}{a^2} = \sqrt{\tan^2\theta_1 + \frac{1}{\tan^2\theta_2}} \qquad (1)$$

From the above formula, it can be further obtained that:

$$\theta = \arctan\sqrt{\tan^2\theta_1 + \frac{1}{\tan^2\theta_2}} \qquad (2)$$

At this time, the coordinates can be obtained:

$$\begin{aligned} k_x &= l\cos\theta\tan\theta_1 \\ k_y &= \frac{l\cos\theta}{\tan\theta_2} \\ k_z &= l\cos\theta \end{aligned} \qquad (3)$$

At this time, the equivalent shaft K is calculated:

$$K = \frac{(k_x, k_y, k_z) \times (0,0,1)}{|(k_x, k_y, k_z) \times (0,0,1)|} \qquad (4)$$

The rotation matrix of the central axis converted to the target attitude through the XYZ fixed angle coordinate system is:

$$R = \begin{bmatrix} c\beta c\gamma & -cas\gamma + sas\beta c\gamma & sas\gamma + cas\beta c\gamma \\ c\beta s\gamma & cac\gamma + sas\beta s\gamma & -sac\gamma + cas\beta s\gamma \\ -s\beta & sac\beta & cac\varphi \end{bmatrix} \qquad (5)$$

Two simultaneous equations can calculate the offset of the central axis:

$$\begin{aligned} \beta &= A\tan 2\left(-r_{31}, \sqrt{r_{11}^2 + r_{21}^2}\right) \\ \alpha &= A\tan 2\left(\frac{r_{21}}{c\beta}, \frac{r_{11}}{c\beta}\right) \\ \gamma &= A\tan 2\left(\frac{r_{32}}{c\beta}, \frac{r_{33}}{c\beta}\right) \end{aligned} \quad (6)$$

3.2 Solve displacement

In the image, the projection length of the central axis in the o-yz coordinate system and the o-xz coordinate system is:

$$\begin{aligned} M_1M_2 &= \sqrt{(y_1-y_2)^2 + (z_1-z_2)^2} \\ M_3M_4 &= \sqrt{(x_3-x_4)^2 + (z_3-z_4)^2} \end{aligned} \quad (7)$$

The actual length of M_1M_2 and M_3M_4 is:

$$\begin{aligned} M_1M_2 &= \frac{l\cos\theta}{\sin\theta_2} \\ M_3M_4 &= \frac{l\cos\theta}{\cos\theta_1} \end{aligned} \quad (8)$$

The ratio between the projected length and the actual length is:

$$\begin{aligned} p_1 &= \frac{l\cos\theta}{\sin\theta_2\sqrt{(y_1-y_2)^2 + (z_1-z_2)^2}} \\ p_2 &= \frac{l\cos\theta}{\cos\theta_1\sqrt{(x_3-x_4)^2 + (z_3-z_4)^2}} \end{aligned} \quad (9)$$

The final result is:

$$\vec{M_1P} = \left(x_0 - x_1', y_0 - y_1', z_0 - z_1'\right) \quad (10)$$

4 EXPERIMENTAL VERIFICATION

We set up the camera system as shown in Figure 3, adjust its position and angle to make the two optical axes orthogonal, and adjust the probe-type measuring tooling to any position, but we try to be close to the center of the images collected by the two cameras.

Figure 3. Camera system.

We take pictures of the probe-type measuring tooling at two angles and obtain Figure 4 through the above image preprocessing method. The red line represents the central axis of the measuring tooling.

Figure 4. Preprocess image.

The parameters used in the experiment and the calculated results are shown in Table 1. The calculated results are obtained by the calculation method mentioned above. After many tests, the offset and deflection at this time are input into the control software of the industrial computer to control the movement of the platform, and the probing measuring tools can reach the through-hole position, as shown in Figure 5.

Table 1. Input and output data.

Input					Output			
l	θ_1	θ_2	α	β	γ	x	y	z
140	1	2	0	−4.8893	−0.0337	−139.2827	0.0953	0.05512
140	3	1	0	−7.3557	−0.0514	−138.5674	0.0954	0.05513
140	4	2	0	−19.9651	−0.13502	−139.2844	0.0949	0.05514
140	3	3	0	−22.5485	−0.1491	−139.5223	0.0952	0.05509
140	2	5	0	−25.1173	−0.1597	−139.7129	0.0957	0.05511

Figure 5. Alignment model.

5 CONCLUSION

The algorithm's feasibility can be verified through experiments, and the unconstrained motion stage of the printing press sleeve assembly can be completed, which provides a basis for further realizing the assembly automation and reducing the problem of injury and illness.

In addition, this method can also be used in other projects to identify and measure the position and orientation of parts.

ACKNOWLEDGMENTS

1. Research on Key Technologies of Printing Machine Sleeve Assembly Robot (KM201910015003) of Beijing Municipal Education Commission in 2019.
2. Beijing Natural Science Foundation—Joint project of Beijing Municipal Education Commission (KZ202010015021).
3. 2021 Science and Technology Key Project—Research on Intelligent Recognition System for Surface Defects of Printing Rollers Based on Machine Vision (04190121007/002).

REFERENCES

Jiang (2015). Modeling and Control of a High-precision Tendon-based Magnetic Resonance Imaging–compatible Surgical Robot [J]. *Journal of Systems and Control Engineering*, 229(8): 711–727.

Li (2016). Multi-information Fusion Offset Press Precision Assembly Robot Calibration Method [J]. *Journal of Beijing Institute of Printing*, 24(06): 48–52.

Li (2016). *Research on Design, Modeling, and Control Methods of Printing Machine Sleeve Assembly Robot* [D]. Beijing Institute of Printing.

Qian (2019). *Application Research of Industrial Robot Based on Multi-vision* [D]. University of Electronic Science and Technology of China

Ren (2017). Orthogonal Binocular Vision Assembly End Pose Detection Method for Long Axis Parts [J]. *Journal of Harbin Institute of Technology*, 49(01): 60–65

Wei (2021). *Research on Some Problems of the 6-DOF Parallel Platform* [D]. Liaoning University of Technology.

Wei (2022). Research and Application of an Adaptive Image Preprocessing Method [J]. *Modern Electronic Technique.* 45(07): 53–57

Zhang (2020). Object Detection Algorithm Based on Sobel Operator Image Preprocessing [J]. *Electronic Technology and Software Engineering.* 151–153.

Zhu (2019). *Implementation of 3D Reconstruction Algorithm Based on Multi-view Image* [D]. Southeast University

Zou (2022). Research on Image Processing and Positioning Technology Based on vision [J]. *China's New Technology and New Products.* 46–48

Water Conservancy and Civil Construction – Oke & Ahmad (Eds)
© 2024 The Author(s), ISBN: 978-1-032-58618-2

Rigid-flexible coupling analysis of worm and gear reducer based on ANSYS and ADAMS

Yanting Yu*, Meili Song* & Xuejie Chen*
Nanjing University of Science and Technology, School of Mechanical Engineering, Nanjing, China

ABSTRACT: The 3D modeling software Soildworks established a brief 3D model of the worm gear and gear reducer. In ADAMS, the rigid dynamics model of the worm gear and worm gear reducer was established. The output speed and contact force of each part of the rigid dynamics model were simulated. The modal neutral file was output through the ANSYS Workbench module, and the modal neutral file was imported into the ADAMS rigid dynamics model. The flexible body was used to replace the original rigid body, and the rigid-flexible coupling was carried out. The coupled contact force curve of the worm gear and the worm was obtained by solving it and compared with the theoretical calculation value to verify the accuracy of the rigid-flexible coupling model. Furthermore, the stress-strain values were obtained by rigid-flexible coupling simulation, which provided a basis for the subsequent life calculation of reducer parts and theoretical support for the subsequent research on worm gear and worm reducer.

1 GENERAL INSTRUCTIONS

Worm gear and worm gear reducers are widely used. Still, the cost of gear and worm gear and worm gear processing and testing is high, so the stiffness strength analysis and simulation optimization of related parts in worm gear and worm gear reducer are of certain practical significance.

Many scholars have done a lot of research on the basic problems of worm gear and worm gear. Liu discovered that the motion response of the large structure machine is relatively large, and motion error varies along with the movement path in X, Y, and Z, which lays the foundation for the research of the control system. Wang established the rigid-flexible coupling virtual prototype model of the gear system, simulated the key components of the gear system, and verified the feasibility of the virtual prototype by comparing it with the test. Xu established a rigid-flexible coupling model of the rack and pinion and compared it with the rigidity analysis results to obtain the deviation of the bending stress of the tooth root under two different simulation conditions. Wang conducted a rigid-flexible coupling simulation on the contact pair of the worm gear and worm, verified the correctness of the simulation model, and provided a reference for product development and subsequent design. Meng derived the rigid-flexible coupling dynamics of a space robot system with flexible appendages and established a coupling model between the flexible base and the space manipulator, providing a theoretic basis for the system design, performance evaluation, trajectory planning, and control of such space robots.

The above research has carried out rigid-flexible coupling for gears or single worms and has not integrated multiple flexible bodies into a model. However, when multiple meshing

---

*Corresponding Authors: yyt1999_0825@163.com, 384667579@qq.com and 431341432@qq.com

DOI: 10.1201/9781003450832-4

places are engaged, more than one place often needs to be considered for flexibility. Therefore, based on the above research, this paper comprehensively considers the flexibility of multiple key components of the worm gear and worm reducer and analyzes relevant parameters such as the contact force. The stress-strain distribution of related parts is obtained, which provides theoretical support for the subsequent life prediction and fault diagnosis of the worm gear and worm reducer.

2 ESTABLISHMENT OF 3D MODEL OF WORM GEAR AND WORM REDUCER

The worm gear reducer is mainly composed of two pairs of gears a pair of the worm gear, a set of planetary gear reducers, and a motor. The rated torque of the known planetary gear reducer is 20 Nm, and the maximum torque is 60 Nm. The modulus of the two pairs of gears used is 2, and the number of teeth is 10 and 24, respectively.

Table 1.   Worm gear and worm parameters.

Basic Parameters	Parameter Values
Module	4
Number of Threads of Worm	1
Number of Worm Gear Teeth	15
Worm Gear Torque (Nm)	200
Center Distance (mm)	51
Worm Lead Angle (°)	3.49
Worm Lead (°)	12.56
Tooth Profile Angle (°)	20
Reference Diameter of Worm (mm)	42
Reference Diameter of Worm Gear (mm)	60

According to the above conditions, the axial force, circumferential force, and radial force between the worm gear and worm can be obtained as follows:

Worm circumferential force (snail gear axial force) is:

$$F_{t1} = -F_{x2} = \frac{2000T_1}{d_1} = 634.92N \qquad (1)$$

Worm axial force (worm gear circumferential force) is:

$$F_{x1} = -F_{t2} = \frac{2000T_2}{d_2} = 6666.67N \qquad (2)$$

The radial force of the worm (worm gear) is:

$$F_{r1} = -F_{r2} \approx -F_{t2} \tan \alpha = 2426.46N \qquad (3)$$

Then its resultant force is:

$$F = 7122.86N$$

where F_t = circumferential force; F_x = axial force; F_r = radial force; T = torque; d = diameter; α = tooth profile angle.

The simplified geometric model of the worm gear and worm reducer was established in Soildworks, as shown in Figure 1.

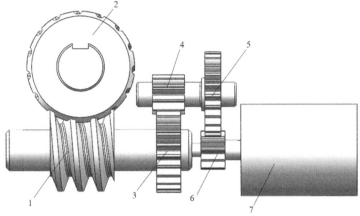

1-Worm; 2-Worm Gear; 3-Gear 1; 4-Gear 2; 5-Gear 3;
6-Gear 4; 7-Motor and Planetary Gear Reducer

Figure 1. Simplified model of worm gear and worm reducer.

3 RIGID BODY DYNAMICS ANALYSIS BY ADAMS

The simplified 3D model of the worm gear and worm reducer is saved in '.x_t' format and imported into ADAMS, and corresponding motion pairs are added to ADAMS to define the contact stress between the worm gear and gear. The specific parameters are shown in Table 2.

Table 2. Contact force parameters.

Contact Force Parameter	Numerical Value
Contact Stiffness (N/mm)	10000
Damping Factor (N*s/mm)	50
Force Exponent	1.5
Penetration Depth (mm)	0.1

We add 200000 Nmm load torque to the cochlear shaft, and the functional expression of drive gear speed is as follows: step (time, 0, 0, 0.3, 18000 d) + step (time, 0.3, 0, 1, 0) + step (time, 1, 0, 1.2, −18000 d) + step (time, 1.2, 0, 1.5, 0), drive gear speed function curve is shown in Figure 2. We set the simulation time to 1.5 s and the step length to 5000. The output speed of the cochlear shaft, the meshing force between each gear, and the meshing force of the parts on the worm gear and worm reducer can be obtained from the simulation results, as shown in Figures 3–5.

According to the speed function curve of the worm gear, when the input end runs stably, the output speed of the cochlear shaft fluctuates around 200 d/s. According to the meshing force curve of worm gear and worm reducer parts, the meshing force takes a big step in the initial instant due to the impact of load. When the motor runs statically, the meshing force between the worm gear and worm becomes a periodic stable cycle around 8000 N. When the drive is 0, the meshing force of the worm gear and the worm is provided by the load torque of 200000 Nmm. Its meshing force fluctuates around 8600 N. However, the meshing force

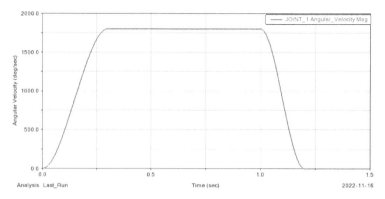

Figure 2. Speed function curve of driving gear.

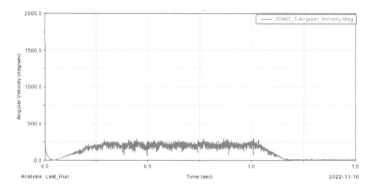

Figure 3. Speed function curve of cochlear shaft.

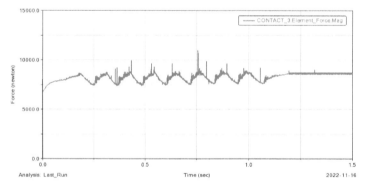

Figure 4. Meshing force curve of worm gear and worm.

obtained by theoretical calculation is 7122 N, and the simulation value greatly differs from the theoretical value. This is because the contact force fluctuation of the worm gear and worm is serious during meshing, and there are dynamic incentives, such as the impact of biting in and out.

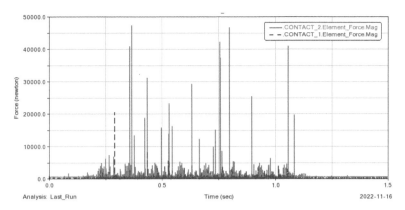

Figure 5. Meshing force curve of gear.

Moreover, when the meshing of two gears is running statically at the input end, there are big impacts, resulting in the actual meshing force being larger than the theoretical meshing force. The two gears of the worm gear reducer with a modulus of 2 and tooth number of 24 are flexible processing. Considering the larger torque of the worm gear, the worm gear is also flexible.

4 RIGID-FLEXIBLE COUPLING ANALYSIS BY ADAMS

4.1 *The establishment of modal neutral files*

Taking worm gear as an example, considering the larger torque borne by worm gear in the process of movement, the flexible treatment of worm gear is carried out. In the ANSYS Workbench module, the model module is used to import the worm gear Model, establish rigid in the worm gear center, grid it, insert Commands in the model, input Commands in Commands, solve the overall model, and then obtain the required '.mnf' file. The specific process is shown in Figure 6.

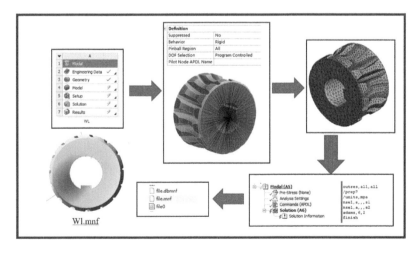

Figure 6. Process of modal neutral file establishment.

Similarly, the gear modal neutral file is obtained.

4.2 *Rigid-flexible coupling analysis by ADAMS*

Based on the above rigid body dynamics simulation, the contact between the worm gear and two gears of mold 2 tooth 24 was deleted, the modal neutral file replaced the corresponding rigid body, and the contact was redefined. We set the simulation time to 1.5 s and the step size to 5000. The simulation process is shown in Figure 7.

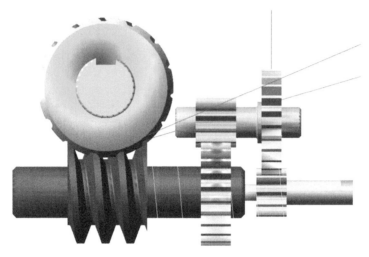

Figure 7. Rigid-flexible coupling simulation diagram.

The engagement force of the worm gear and the worm is shown in Figure 8.

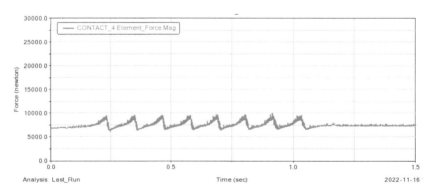

Figure 8. The rigid-flexible coupling of worm gear and worm meshing force curve.

By magnifying the curve from 1.2 s to 1.5 s, as shown in Figure 9, the meshing force fluctuates around 7300 N, and the theoretical value of 7122.86 N is close to it. The fluctuation is smaller than that in the simulation of rigid dynamics, and the sudden change of contact force is significantly reduced. Therefore, the rigid-flexible coupling dynamics simulation is more similar to the actual working condition of the reducer.

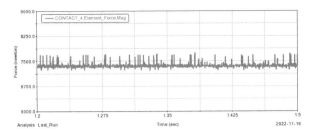

Figure 9. Engagement force curve of worm gear and worm at 1.2 sec to 1.5 sec.

5 CONCLUSIONS

A brief worm gear and gear reducer model was established based on the 3D modeling software Soildworks. ADAMS established the rigid dynamics model of the worm gear and gear reducer as a simulation platform. The output speed and contact force of each part of the rigid dynamics model were obtained by simulation. Through the ANSYS Workbench module, the modal neutral file was output and imported into the ADAMS rigid dynamics model instead of the original rigid file for rigid-flexible coupling. The coupled contact force curve of the worm gear and the worm was obtained, which was consistent with the theoretical calculation value, proving the accuracy of the rigid-flexible coupling model. The stress variation and load spectrum obtained by the rigid-flexible coupling simulation provide a basis for the follow-up life calculation of the reducer parts, prove the feasibility of fault diagnosis of the worm gear and worm gear reducer system through the virtual prototype, and provide theoretical support for the follow-up research on the worm gear and worm accelerator.

REFERENCES

He B, Xue H, Liu L, Pan Q, Tang W & Ostrosi E (2019). Rigid-flexible Coupling Virtual Prototyping-based Approach to the Failure Mode, Effects, and Criticality Analysis. *The International Journal of Advanced Manufacturing Technology*, 100, 1695–1717.

Li Y, Wang C, Huang W (2019). Dynamics Analysis of Planar Rigid-flexible Coupling Deployable Solar Array System with Multiple Revolute Clearance Joints. *Mechanical Systems and Signal Processing*, 117, 188–209.

Liu N, Zhang X, Lin Z, Shang D & Fan X (2018). Study on the Rigid-Flexible Coupling Dynamics of Welding Robot. *Wireless Personal Communications*, 102, 1683–1694.

Liu Z, Ye P, Guo X & Guo Y (2014). Rigid–flexible Coupling Dynamic Analysis on a Mass Attached to a Rotating Flexible Rod. *Applied Mathematical Modelling*, 28(21–22), 4985–4994.

Meng D, Wang X, Xu W & Liang B (2017). Space Robots with Flexible Appendages: Dynamic Modeling, Coupling Measurement, and Vibration Suppression. *Journal of Sound and Vibration*, 396, 30–50.

Sheng L, Li W, Wang Y, Yang X & Fan M (2019). Rigid-flexible Coupling Dynamic Model of a Flexible Planar Parallel Robot for Modal Characteristics Research. *Advances in Mechanical Engineering*, 11(1).

Wang M, Jing F & Wan C (2021). Research on Rigid-flexible Coupling Meshing of the Worm Gear and Worm Pair. *Coal Technology*, 42(05), 206–208.

Wang M, Wan C & Jing F (2021). Analysis of Contact Deformation of Gearbox Worm Pair. *Coal Technology*, 42(07), 202–205

Wang Y, Ma J & Meng G (2009). Modeling and Simulation of Rigid-flexible Coupling Dynamics of the Gear System. *Journal of Mechanical Transmission*, 33(04), 32–35.

Xu F, Wei X & Zhang M (2012). Rigid-flexible Coupling Meshing Analysis of Rack and Pinion Based on ADAMS. *Machinery Design & Manufacture*, (05), 200–201.

Research on the optimization of auto parts inventory cost based on an improved demand forecasting method

Weining Zhu & Cunrong Li*

Department of Industrial Engineering, School of Mechanical & Electronics Engineering, Wuhan University of Technology, Wuhan, P.R. China

ABSTRACT: An optimization method of auto parts inventory control based on a genetic-particle swarm (GA-PSO) hybrid algorithm optimized BP neural network is proposed for auto parts demand data characteristics. Firstly, the main factors affecting the auto parts demand are analyzed. Then a GA-PSO hybrid algorithm is introduced to optimize the initial weights and thresholds of the BP neural network for the defects of local optimum and slow convergence speed of the standard BP algorithm. Finally, based on the actual operation of the auto parts company and the parts demand forecasting results, an inventory cost model, including order lead time, storage space, and capital constraints, is established. The traditional differential evolutionary algorithm for solving the inventory cost model is improved to obtain the optimal decision strategy, and then the inventory cost is optimized.

1 INTRODUCTION

With the continuous changes in the automotive industry and the changing needs of the automotive market, the auto parts industry ushered in a variety of opportunities simultaneously but also faced a very serious test. In supply management, China's auto parts enterprises generally have high inventory, low availability, and a global supply and demand mismatch phenomenon. How to reduce inventory costs while ensuring normal and orderly production and operation activities through reasonable inventory control has become a new direction for more and more enterprises to study [1]. M Group is a typical parts manufacturing enterprise in the auto parts industry (e.g., oil coolers, cylinders, piston rods). The root causes of inventory control failure in M enterprises are identified through analysis: inaccurate demand forecasting, unreasonable inventory strategies, etc. Improvement measures are subsequently proposed for the existing problems.

Making an accurate forecast of the demand for automotive parts with the highest accuracy is a prerequisite for scientific inventory control and inventory cost optimization [2]. For the data characteristics of unstable demand for automotive parts and dramatic changes in demand, intermittent demand forecasting will be more accurate than continuous demand forecasting methods. Common intermittent demand forecasting methods include Croston's method, exponential smoothing method, neural network algorithm, and improvements of these methods. Although the above forecasting methods can be used to forecast auto parts materials, the error between the predicted demand quantity and the actual demand quantity is still unavoidable, and the data characteristics of dramatic changes and high instability of auto parts demand make a certain discrepancy between the demand obtained by forecasting and the actual demand [3]. Then, this paper will investigate how to reduce the error and

---

*Corresponding Author: cunrong_li@163.com

improve the accuracy of demand forecasting and how to build an inventory model that is more in line with the actual production operation of the company to reduce inventory costs based on the demand forecasting results obtained by the above method.

2 GA-PSO-BP NEURAL NETWORK ALGORITHM AND DEMAND FORECASTING MODEL

2.1 *Analysis of factors influencing demand forecast for automotive parts*

Demand is the result of a combination of factors. Before forecasting, companies need to identify the key factors important for forecasting and appropriately process the data information of these key factors. Before forecasting customer demand, some key factors must be considered, such as data and models. The data collected is cleaned, filtered, analyzed, and processed to influence the construction of prediction models and the selection of algorithms. The accuracy of model predictions depends on data quality and prediction methods.

2.2 *GA-PSO algorithm*

In optimizing the inventory cost of auto parts materials, accurate prediction of the demand for auto parts materials plays a crucial role. Still, the demand for auto parts is affected by the unit price of raw materials, the order quantity of individual SKU, delivery time, and other factors, which leads to a certain randomness and volatility of the demand for auto parts [4]. BP neural networks have been used in the demand forecasting field and are widely used and recognized. As the traditional BP algorithm converges slowly and easily falls into local optimum, intelligent optimization algorithms are used to replace the traditional algorithms for the optimization process of weights and thresholds in BP network training to improve the accuracy of network demand prediction. Among these optimization algorithms, the particle swarm algorithm (PSO) is simple and easy to use. Still, it has the disadvantages of low accuracy and easy to fall into local minima, which may not obtain the local optimum. To solve this problem, this paper proposes a genetic-particle swarm algorithm (GA-PSO) to improve the training efficiency of BP networks, which can be used in automotive parts demand forecasting. The GA-PSO improvement algorithm in this paper introduces the crossover and variation ideas of the genetic algorithm into the velocity update method of each particle in the traditional particle swarm algorithm, which effectively improves the global search ability of the algorithm and avoids the premature maturity of the population to a certain extent.

The velocity and position update equation of any particle in the GA-PSO improvement algorithm is as follows:

$$V_i(k+1) = \omega V_i(k) + c_1 r_1[pbest_i - x_i(k)] + c_2 r_2[gbest_i - x_i(k)] \tag{1}$$

$$X_i(k+1) = X_i(k) + V_i(k+1) \tag{2}$$

where k is the number of iterations. ω is the inertia weighting factor. c_1 and c_2 are the acceleration factor. r_1 and r_2 is a uniform random number between $[0, 1]$. *pbest* is the extreme individual value, i.e., the current optimal position of the particle. *gbest* is the global extremum, i.e., the current optimal position of the particle population. V_i is the velocity vector of the particle, $V_i \in [-v_{max}, v_{max}]$. v_{max} is the maximum speed of the particle set by the user and is a constant to prevent the particle from going too fast or moving too far during the update process.

In summary, the position update rate of the particle is expressed as follows:

$$X_i(k+1) = X_i(k) + \omega V_i(k) + c_1 r_1[pbest_i - x_i(k)] + c_2 r_2[gbest_i - x_i(k)] \tag{3}$$

2.3 GA-PSO optimized BP neural network algorithm

The GA-PSO algorithm is used to iteratively optimize the weights and thresholds of the BP neural network, which can effectively overcome the defects of the local optimum of the BP algorithm [5]. The algorithm flowchart of GA-PSO-BP is as follows.

(1) Initialization of the network structure. The number of nodes in the input, hidden, and output layers of the BP neural network are determined according to the vector dimension of the input quantity to be input and the output quantity to be obtained.
(2) Initialization of the population. The error function of the BP neural network ($E(\omega)$) is used as a method to evaluate the quality of each particle:

$$E(\omega) = \sum_{i=1}^{l} \|x_i - \sigma_i\|^2 \qquad (4)$$

GA-PSO iterative operation. The velocity and position of the updated particles are calculated according to Equation (3) to obtain the new population.

(3) Genetic manipulation of the particle population. The particles in the new population are successively crossed with extreme individual value and global extreme value particles with crossover probability. Then the particles with poor fitness values are randomly initialized with variation probability P_m.
(4) The optimal fitness function value of the updated particle, the optimal position, replaces the optimal particle of the previous generation of a genetic optimization.
(5) To determine whether the number of particle swarm iterations will be greater than the maximum number of iterations or whether the value of the fitness function is better than the previous generation, the optimal individual value is used as the neural network parameter.

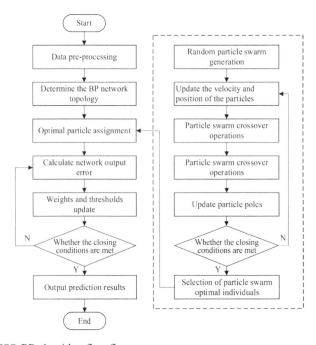

Figure 1. GA-PSO-BP algorithm flow figure.

3 CONSTRUCTION AND SOLUTION OF INVENTORY COST MODEL

3.1 *Conditions and assumptions for model building*

In inventory management, the company generally adopts the "maximum and minimum inventory control" strategy, in which the order cycle is 3 weeks. The order lead time is 3 days and checks the remaining quantity of materials in stock on the third Thursday of each inventory cycle to decide whether to initiate an order according to the current inventory level. This paper presents the necessary conditions and assumptions for constructing an inventory cost model:

(1) In an inventory cycle, the company orders multiple products simultaneously and initiates an order based on the current inventory level of each material. Each item is ordered independently of the other and the quantity ordered is not interchangeable between the various items.
(2) The warehouse has a 3-week lead time for all weekly material orders and a 3-day lead time for orders.
(3) It is assumed that component materials' consumption rate is uniform during each inventory cycle.
(4) When ordering parts and materials, the unit purchase cost and order cost of each material is fixed in each order cycle, and there is no quantity discount.
(5) The logistics cost of each material, i.e., the cost of delivering the material from the supplier to the company's warehouse, is consolidated into the acquisition cost of each material.
(6) The company needs to meet the needs of the customers of each vehicle factory as much as possible, with a high service level and customer satisfaction. When the auto parts material is out of stock, the material needs to be called from the field warehouse, so there will be a certain out-of-stock cost.
(7) When ordering parts and materials, there are constraints on capital and storage space available. The amount of capital available for each order varies depending on the company's business strategy.

3.2 *Establishment of the objective function*

In this paper, the inventory cost of the warehouse of an auto parts company is composed of ordering cost, purchasing cost, inventory holding cost, and out-of-stock cost. Among them, the ordering cost is not related to the number of current material orders but only to the number of orders; the purchasing cost (including logistics cost) is proportional to the unit price and quantity of materials purchased. The inventory holding cost is proportional to the storage cost per unit material in the unit storage cycle and the average inventory quantity per unit storage cycle. In the inventory cost model established in this paper, it is assumed that the consumption rate of each material is fixed within the unit inventory cycle, and the ordering cost, unit purchase cost, unit inventory holding cost, and unit inventory out-of-stock cost of each material are also fixed in each ordering cycle. The expressions for each component of inventory cost are as follows.

$$O_i = A_i Y_i \tag{5}$$

$$C_i = W_i Q_i \tag{6}$$

$$R_i = S_i H_i \tag{7}$$

$$L_i = B_i P_i \tag{8}$$

O_i is the order cost of the first material; A_i is the cost per order of the first material; C_i is the purchase cost of the first material; W_i is the unit purchase cost of the first material; Q_i is the order quantity of the first material; R_i is the inventory holding cost of the first material. S_i is the quantity of stock of the material in the next order cycle; H_i is the holding cost per unit of stock of the material; L_i is the stock-out cost of the material; B_i is the stock-out cost per unit of the material; P_i is the stock-out quantity of the material in the next order cycle; Y_i is the decision variable indicating whether the material is ordered or not.

$$Y_i = \begin{cases} 1 & Q_i > 0 \\ 0 & Q_i \leq 0 \end{cases} \tag{9}$$

The objective function of this inventory cost model is as follows:

$$\begin{aligned} TC &= \sum_{i=1}^{n}(O_i + C_i + R_i + L_i) \\ &= \sum_{i=1}^{n}(A_i Y_i + W_i Q_i + S_i H_i + P_i B_i) \end{aligned} \tag{10}$$

3.3 *Restraint condition*

The company's business strategy for certain auto parts varies over time. The capital budget for each type of part is limited and different when ordering in different order cycles. When ordering within the inventory cycle, the ordering cost of the company's warehouse and the funds provided to the warehouse must meet the following constraints:

$$\sum_{i=1}^{n}(A_i Y_i + W_i Q_i) \leq F \tag{11}$$

The storage space of auto parts and materials is limited. Therefore, when deciding whether to place an order in this order cycle, we should not only try to meet the planned demand for parts and materials, avoid loss of stock out, reduce customer satisfaction and meet the constraints of the number of funds provided to the warehouse by the company, but also consider the limitation of storage space, the relationship between the total space occupied by each part and the maximum total storage space:

$$\sum_{i=1}^{n} V_i S_i \leq V \tag{12}$$

V_i is the storage space occupied by the unit material of type i, and S_i is the inventory quantity of the unit material of type i in the next order cycle;

In the process of solving the inventory cost optimization model, to prevent the infeasible solution in the solution process, the following logical relationship constraints need to be met:

$$P_i = \frac{|D_i - S_i| + D_i - S_i}{2} \tag{13}$$

$$S_i = E_i + Q_i \tag{14}$$

$$E_i = \frac{|S_i' - D_i'| + S_i' - D_i'}{2} \tag{15}$$

P_i is the shortage quantity of the ith material, D_i is the actual demand quantity of the ith material, S_i is the inventory quantity of the ith material, and E_i is the shortage quantity of the $t - 1$ material in the th inventory cycle.

3.4 *Solving inventory cost model based on improved differential evolution algorithm*

3.4.1 *Improvement of differential evolution algorithm*

Differential Evolution Algorithm (DE) is a global optimization algorithm based on "greedy competition", which continuously evolves through mutation, crossover, and selection operations to guide the search to the optimal solution [6]. According to the optimization theorem, "There is no free lunch" [7]. If one learning algorithm is better than another in some respects, it will certainly be weaker than the other in some other aspects. A differential evolutionary algorithm is no exception. It has the advantages of simplicity, efficiency, and high optimization accuracy. At the same time, it will inevitably have premature convergence, search stagnation, and other problems during optimization, like other evolutionary algorithms [8].

Based on this, an improved differential evolution algorithm is proposed for the inventory cost model established in this paper, which introduces a new integer variable mutation strategy with neighborhood information and a new control parameter determination method with adaptive adjustment of cross probability A and scaling probability B, so that the differential evolution algorithm can better balance the relationship between "detection" and "development", solve the problem that its algorithm is prone to fall into local optimization or slow convergence, and improve the accuracy of the algorithm. The improved algorithm is used to optimize the inventory cost of the warehouse.

1) A new mutation strategy for integer variables

The performance of the DE algorithm largely depends on the formulation of the mutation strategy. Common mutation strategies include:

① "DE/rand/1":
$$V_i^G = X_r^G + F(X_{r2}^G - X_{r3}^G) \tag{16}$$

② "DE/best/1":
$$V_i^G = X_{best}^G + F(X_{r1}^G - X_{r2}^G) \tag{17}$$

③ "DE/rand/2":
$$V_i^G = X_{r1}^G + F(X_{r2}^G - X_{r3}^G) + F(X_{r4}^G - X_{r5}^G) \tag{18}$$

④ "DE/best/2":
$$V_i^G = X_{best}^G + F(X_{r1}^G - X_{r2}^G) + F(X_{r3}^G - X_{r4}^G) \tag{19}$$

⑤ "DE/current-to-best":
$$V_i^G = X_i^G + F(X_{best}^G - X_i^G) + F(X_{r1}^G - X_{r2}^G) \tag{20}$$

X_{r1}^G, X_{r2}^G, X_{r3}^G, X_{r4}^G and X_{r5}^G are individuals randomly selected from the population in a generation G, $r1 \neq r2 \neq r3 \neq r4 \neq r5 \neq i$. X_{best}^G is the best individual in the population and F is the scaling factor.

Taking the "DE/rand/2" mutation strategy as an example, its display in the two-dimensional plane is shown as follows:

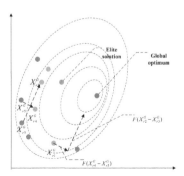

Figure 2. Two-dimensional plan of "DE/rand/2" mutation strategy.

The "DE/rand/1" mutation strategy has a good ability to search for elements. Still, it does not consider the information on the optimal solution for the population. Still, the development ability is insufficient, resulting in poor convergence performance in the later period. The "DE/best/1" mutation strategy introduces information on the optimal solution for the population and enhances the development ability of the algorithm. Still, premature convergence may occur and lead to local optimization [9]. Inspired by an improved particle swarm optimization algorithm (DNSPSO) proposed by Wang et al., which combines neighborhood search strategy and particle swarm optimization algorithm to achieve a balance between exploration and development capabilities, this section proposes a new integer mutation strategy (DE/best to neighbors/1) that introduces neighborhood search operations in mutation links and adaptively allocates the number of neighbors [10]. The variation formula is as follows:

$$V_i^G = X_{nbest}^G + F(X_{best}^G - X_i^G) + F(X_{r1}^G - X_{r2}^G) \tag{21}$$

X_{nbest}^G is the best individual among the N neighborhoods of X_i^G individual, X_{best}^G is the best individual in the whole population. When N is 8, the neighborhood search diagram is shown below.

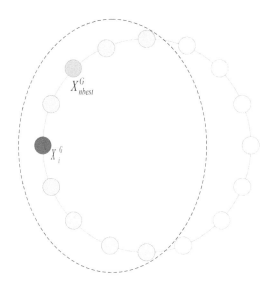

Figure 3. Schematic diagram of neighborhood search when a is 8.

In the process of mutation, if it V_i^G exceeds the search range, the mutation individual will be modified according to the following formula:

$$V_{i,j}^G = \begin{cases} 2X_{\sup} - V_{i,j}^G, & V_{i,j}^G > X_{\sup} \\ 2X_{\inf} - V_{i,j}^G, & V_{i,j}^G < X_{\inf} \end{cases} \tag{22}$$

The neighborhood's size greatly impacts the balance of algorithm development ability and exploration ability. Therefore, this paper proposes a method to dynamically adjust the size according to different search stages of the algorithm. The formula is derived as follows:

$$Variety(G) = \frac{1}{N_P} \sum_{i=1}^{N_P} \sqrt{\sum_{j=1}^{D} (X_{i,j}^G - X_{ave}^G)^2} \tag{23}$$

$$X_{ave}^G = \frac{\sum_{i=1}^{N_P} X_{i,j}^G}{N_P} \tag{24}$$

$$N = N_{\min} + (N_{\max} - N_{\min}) * Variety(G) \tag{25}$$

4 CASE VALIDATION OF INVENTORY COST OPTIMIZATION OF AUTO PARTS

To test the optimization effect of the inventory cost optimization model, this paper takes the inventory status of warehouse B in company M as an example, where the parts for oil cooler production are stored. It takes the order quantity and inventory control of an auto parts enterprise M within 24 inventory cycles from 2020 to 2021 as the historical data. The training sample is the data of the first 18 inventory cycles. The data of the last 6 inventory cycles are used as the experimental data of the performance of the prediction model and inventory cost optimization model. The input/output matrix of the model is 24 rows and 5 columns of normalized dimensionless processing of the original data, as shown in Table 1.

Table 1. The order quantity and inventory control of automobile parts enterprise M in 2020-2021.

Frequency	SKU order quantity/ piece	Market share	Brand recognition	Price fluctuation	Order satisfaction rate
1	260	0.401 2	0.857 3	0.656 7	0.848 7
2	265	0.465 6	0.804 2	0.622 3	0.826 5
3	276	0.496 5	0.791 5	0.648 9	0.815 7
4	308	0.612 1	0.846 4	0.702 7	0.796 2
5	296	0.402 2	0.790 6	0.608 6	0.745 2
6	296	0.541 3	0.857 3	0.632 1	0.812 1
7	284	0.508 3	0.747 1	0.699 7	0.764 2
8	352	0.595 2	0.857 5	0.645 6	0.839 6
9	288	0.642 3	0.786 3	0.689 7	0.768 8
10	368	0.665 3	0.847 4	0.604 2	0.801 3
11	280	0.622 4	0.816 1	0.639 1	0.799 6
12	300	0.612 6	0.766 8	0.585 6	0.755 2
13	268	0.452 2	0.775 9	0.589 8	0.795 1
14	260	0.405 5	0.769 5	0.608 4	0.802 2
15	280	0.483 7	0.775 3	0.625 9	0.824 6
16	324	0.665 9	0.910 6	0.727 6	0.790 4
17	276	0.425 7	0.847 3	0.690 2	0.702 1
18	300	0.586 2	0.862 6	0.609 3	0.772 1
19	240	0.456 3	0.706 2	0.734 5	0.765 3
20	364	0.599 4	0.890 6	0.647 2	0.839 7
21	332	0.656 2	0.776 8	0.650 5	0.791 3
22	249	0.713 6	0.826 5	0.607 1	0.823 6
23	348	0.656 2	0.876 9	0.660 5	0.821 1
24	320	0.586 7	0.779 7	0.667 4	0.757 0

The relationship between the number of iterations and the objective function value of the two methods was compared. As shown in Figure 4, the traditional evolutionary algorithm

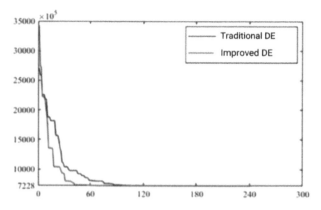

Figure 4. Inventory Cost Optimization figure.

converges to the objective value of 72,340.56 after 90 iterations, while the improved differential evolutionary algorithm converges to the objective value of 72,282.13 after 60 iterations. Therefore, the improved differential evolutionary algorithm proposed in this paper has better search results than the traditional differential evolutionary algorithm in terms of convergence speed and adaptation results.

5 CONCLUSION

This paper presents a method of inventory cost optimization based on the auto parts demand forecast. This method has the following advantages:

(1) A hybrid genetic-particle swarm (GA-PSO) algorithm is proposed to optimize the combined prediction model of the BP neural network for the characteristics of auto parts demand data. The crossover and variation operations in the genetic algorithm are introduced into the particle swarm algorithm to build a hybrid GA-PSO algorithm with global convergence of the genetic algorithm and fast convergence of the particle swarm algorithm to obtain the prediction results with higher accuracy.
(2) Based on the actual operation situation of the company, an inventory cost model, including order lead time, storage space, and capital constraints, is established. The proposed differential evolutionary algorithm for determining the probability of variation and crossover coefficients is used to solve the inventory cost model, to obtain the optimal decision strategy and, thus, optimize the inventory cost.

REFERENCES

[1] Brest J, Greiner S, Boskovic B, et al. Self-adapting Control Parameters in Differential Evolution: A Comparative Study on Numerical Benchmark Problems[J]. *IEEE Transactions on Evolutionary Computation*, 2006, 10(6): 646–657.
[2] David H. Wolpert, William G. Macready. No Free Lunch Theorems for Optimization. [J]. *IEEE Trans. Evolutionary Computation*, 1997, 1(1).
[3] Gong Qiguo, Huang Wenhui. A Review and Outlook of Centralized Inventory Research in Supply Chain Management[J]. *Management Review*, 2017, 29(11): 206–215.
[4] Hu, Cong, Xu, Min, Hong, Edward, Wang, Haixin, Liu, Cuiling, Xue, Xiaoru. Data-driven Demand Forecasting Based on Integrated LSTM model [J]. *Power Engineering Technology*, 2022, 41(06): 193–200.

[5] Li Ji, Sun Xiu-xia, Li Shi-bo. Improved PSO Based on Genetic Hybrid *Genes* [J]. *Computer Engineering*, 2008, 34(2): 181–183.

[6] Rainer Storn, Kenneth Price. Differential Evolution - A Simple and Efficient Heuristic for Global Optimization over Continuous Spaces. [J]. *Journal of Global Optimization*, 1997, 11(4).

[7] Tanabe R, Fukunaga A. *Success-history Based Parameter Adaptation for Differential Evolution* [C]// 2013 IEEE Congress on Evolutionary Computation. *IEEE*, 2013: 71–78.

[8] Wang H, Sun H, Li C, et al. Diversity Enhanced Particle Swarm Optimization with Neighborhood search [J]. *Information Sciences*, 2013, 223: 119–135.

[9] Zeng R. Current Status of Research on Demand Forecasting and Inventory Control of Automotive Aftermarket Parts[J]. *Automotive Parts*, 2020(11): 85–87.

[10] Zhou Yantao, Dai Jun, Yuan Huili, Lu Yue. Research on Demand Prediction and Planning Layout of Urban Electric Vehicle Charging Facilities[J]. *Power System Protection and Control*, 2021, 49(24): 177–187.

Water Conservancy and Civil Construction – Oke & Ahmad (Eds)
© 2024 The Author(s), ISBN: 978-1-032-58618-2

Design and identification management strategy of privately owned public space in NYC and its enlightenment on architecture public open space in China

Ying Shi, WanYing Wu & Ying Pan*

School of Architecture, South China University of Technology, Guangzhou, Guangdong, China

ABSTRACT: From the perspective of the design and identity management strategy of privately owned public space (POPS) in New York City (NYC), the article analyzed the problems in the development of POPS in NYC as well as their causes and solutions, with a focus on the identification approval procedure, design criteria, incentive policies, and supervision of POPS. Moreover, relevant experiences and lessons were explained. Finally, the impact of POPS in NYC on the architecture of public open spaces in China was discussed based on the current situation of urban development.

1 INTRODUCTION

With the rapid advancement of urbanization, problems such as lack of land resources, traffic congestion, and insufficient public spaces have become increasingly prominent in China. As the focus of urbanization shifts from the expansion of scale to the enhancement of spatial quality, urban public open space has attracted increasing attention, especially architectural public open space. However, the lack of design and identity management strategies for architecture in public open spaces in China often results in low construction enthusiasm and poor quality. Some scholars have studied the design elements and guidelines for the spaces, but no studies on the identification have been conducted. Therefore, it is necessary to develop an effective design and identification management strategy to guide the high-quality construction of architecture in public open spaces.

In 1961, the NYC Zoning Resolution formally proposed the policy of privately owned public space (POPS). It encouraged developers to build open space on privately owned space and make it open to the public, for which they could receive Floor Area Ratio (FAR) bonuses. For more than half a century, NYC has made several optimizations and adjustments to the management of POPS and has established a relatively complete management system. This article generalized the design and identity management system for NYC POPS and summarized the excellent experience in the formulation of the design and identification strategy, thus providing an important reference for the design and identity management strategy of architecture in public open spaces in China.

2 DESIGN AND IDENTIFICATION MANAGEMENT STRATEGY FOR POPS DEVELOPMENT HISTORY

The design and identity management strategy of POPS in NYC have undergone numerous modifications. It established a clear identification procedure, design standards, and supervision of POPS, which greatly boosted their construction and management.

---

*Corresponding Author: 13380056897@163.com

DOI: 10.1201/9781003450832-6

2.1 *Development of the design and identification procedure for POPS*

There are three types of approval procedures for POPS: as-of-right, certification, and discretionary procedures. The "as-of-right" procedure is relatively simple. The design of POPS only enters the architecture approval procedure as an appendix to the building design scheme to be approved by the Department of Buildings. The design scheme that conforms to the Zoning Regulation can obtain a construction permit and receive a FAR bonus.

Figure 1. "As-of-right" identification procedure.

The "certification" procedure is currently the most important approval procedure for POPS. The City Planning Commission must approve the POPS scheme and bonus area calculation. After the certification is passed, it will be submitted to the Department of Buildings along with the architectural scheme. Then, upon approval, a construction permit will be issued.

Figure 2. "Certification" identification procedure.

The "discretionary" procedure applies to POPS projects that are considered complicated and have a significant impact. The City Planning Commission will initiate the Uniform Land Use Review Procedure (ULURP), where POPS applications shall be approved by the City Planning Commission, Community Board, Borough Board, City Council, and Mayor. If there is no objection to the POPS design, it will be approved by the Department of Buildings, and the construction permit will be issued.

Figure 3. "Discretionary" identification procedure.

The 1961 Zoning Regulation in NYC specified two types of POPS: squares and arcades. Both types adopted the "as-of-right" approval procedure. POPS was widely welcomed by the real estate market due to its simple approval procedure and FAR bonuses. From 1968 to 1973, NYC added five types of POPS: raised plazas, cross-block arcades, wind and rain pedestrian spaces, sunken plazas, and outdoor plazas, which adopted the "discretionary" approval procedure.

In 1975, to solve the problem of low POPS quality, the NYC Zoning Regulation proposed three new POPS types: city squares, sidewalk widening, and residential squares, all of which

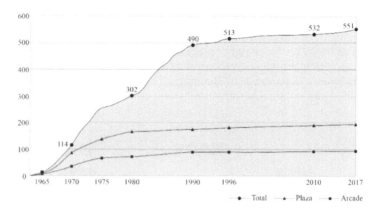

Figure 4. Numbers of POPS in NYC.

adopted the "certification" approval procedure. At the same time, after the POPS scheme was approved, the developer was required to pay a construction deposit and sign a legal contract with the City Planning Commission for operation and maintenance after completion.

To save administrative costs, the NYC Zoning Resolution adjusted the classification of POPS and integrated all types of POPS except arcades into the category of public squares. Since then, arcades have been identified following the "as-of-right" procedure, and public plazas have been identified following the "certification" procedure. For the complicated and impactful POPS projects, the "discretionary" procedure has been adopted to ensure fairness. All POPS identification follows one of the three procedures, which has improved the efficiency of administrative management.

2.2 Changes of the design standards for POPS in NYC

From 1961 to 1975, the NYC Zoning Resolution had only rough regulations on the size of the POPS, which led to poor POPS quality. Accordingly, the 1975 NYC Zoning Resolution mainly set detailed design standards for some plazas and added specific requirements for the configuration of service facilities. In June 2009, the Zoning Regulation refined 18 terms, including spatial form, location, orientation, visibility, walking path, elevation, steps, seating, vegetation, lighting, signage, etc. It has provided quantitative indicators for the design and identification of POPS.

Table 1. Quantitative design standards for POPS.

Element	Cutline	Element	Cutline
Plaza Configuration		Size Limit	
Minor Portion		Frontage	

Since 2016, the Zoning Regulation has included regulations on kiosks and open-air cafes in public plazas. Specifically, kiosks and open-air cafes shall be certified by the Chairperson of the City Planning Commission. At the same time, the layout and number of furniture such as tables and chairs, as well as storage locations during closures, were specified. In 2017, the Zoning Regulation added location restrictions on public plazas. In 2019, the Zoning Resolution added design standards for arcades, including the minimum depth, length, and area, and stipulated the range within which there should be no off-street parking spaces, passenger drop-offs, driveways, or off-street loading berths. Overall, the regulations were intended to ensure the quality and openness of POPS.

The design standards for POPS in NYC have evolved from rough regulations on POPS areas to regulations with comprehensive and precise quantitative indicators. In comparison, the current version pays great attention to space quality and problems in daily use. Moreover, the quantitative design standards facilitate identification, approval, and bonus granting. They also provide a useful reference for developers and designers to create POPS.

2.3 Incentive policies of POPS in NYC

Prior to 2014, the NYC Zoning Regulation primarily formulated incentive policies based on the POPS type. Since 2014, the policies have been optimized. The policies have taken into account not only the construction and maintenance costs of different POPS types but also the land costs due to land rents in different regions. Therefore, they have become more refined and reasonable.

Table 2. Incentive policies on public plazas.

District	Permitted Additional Square Feet of Floor Area per Square Feet	
	Only Commercial Uses	Both Commercial and Community Uses
C5-3 C5-5 C6-6 C6-7 C6-9	10 square feet	10 square feet
C4-7 C5-2 C5-4 C6-4 C6-5 C6-8	6 square feet	6 square feet
C4-6 C5-1	none	6 square feet
C6-1 C6-2	4 square feet	4 square feet
C6-3	4 square feet	none

2.4 Post-build management and re-identification of POPS in NYC

Initially, NYC developed a design and identity management strategy for POPS only at the approval stage. There was no effective management and supervision of the environment of POPS after completion. Since 2014, the Zoning Resolution has stipulated that the POPS owner shall submit a compliance report after identification. If the POPS fails to comply with the Zoning Resolution or the owner does not submit the report promptly, the Department of Building Commission of NYC has the right to revoke the building permit or the certification of occupancy.

2.5 Characteristics of the identification management strategy for POPS

To solve practical problems, the design and identity management strategy for POPS in NYC has constantly improved, ranging from comprehensive design standards and strict incentive policies to close supervision of spaces. The City Planning Commission has paid more attention to space quality and daily use and attached greater importance to enhancing people's experiences with urban spaces.

More importantly, management has become increasingly efficient. The City Planning Commission has set up a clear procedure for the application and identification of POPS, which saves communication costs between management personnel and developers and makes it convenient for management personnel to identify the scheme.

3 STATUS AND PROBLEMS OF DESIGN AND IDENTIFICATION OF ARCHITECTURE IN PUBLIC OPEN SPACES IN CHINA

There is no national-level legal document that regulates the construction of architecturally designed public open spaces in China. Meanwhile, based on the needs of urban development, some cities have put forward construction and reward policies for architecture public open spaces, which encourage developers to create architecture public open spaces and gain FAR bonuses.

There are some problems with the design and identification management of architecture in public open spaces on the Chinese mainland, such as unclear concepts and classification, ambiguous identification standards, a lack of special identification procedures, and imprecise incentive policies. In terms of the procedure for identification, the discretionary procedure is inefficient, which is likely to cause review bias. Also, the public lacks the right to approve. As for design standards, there is a lack of systematic guidelines for design elements and quantitative indicators for the construction of architecture in public open spaces. The spaces are prone to problems such as monotonous forms and functions, lack of facilities, and low openness. Furthermore, the ambiguity of design standards is detrimental to the formulation of incentive policies. As a result, the bonus does not take into account the difference from region to region in land prices and space operation costs.

In China, the construction of architectural public open spaces relies heavily on the outcome of urban design, so the motivation of designers and developers to design and build architectural public open spaces is low. Consequently, the spaces in China have low quantity and poor quality and are unable to play the role of connecting urban public spaces. Therefore, it is essential to refine the design and identification management of architectural public open space in China, which can encourage developers and designers to create these spaces and thus improve the environment of urban public space.

4 EXPERIENCE AND ENLIGHTENMENT OF POPS IN NYC

The current problems in the design and identity management strategy of architecture in public open spaces in China are similar to the problems in the POPS development in NYC. Accordingly, China can learn from the optimization of the NYC Zoning Resolution in the following aspects:

First, it is essential to clarify space classification and determine design elements. A clear definition is helpful in the identification and management of architecture in public open spaces for the department of city planning. In addition, the design element guidelines can provide a useful reference for developers and designers. Also, the standardized guidelines can reduce communication costs between the government and developers.

Secondly, it is important to set up quantitative design standards and standardized incentive policies. Currently, concerning quality and cost, the policy on architecture in public open spaces in China is relatively rough. Given the lessons of the poor quality of early POPS in NYC, cities in China can draw on the experience and establish systematic guidelines for design elements, quantitative design standards, and differentiated incentive policies to ensure the quality of the architecture of public open space. On the other hand, the incentive policy should fully reflect the difference in construction and maintenance costs for spaces of various types and functions to ensure precision.

Thirdly, it is necessary to optimize the approval system and increase public participation. Cities in China should learn from the experience of POPS in NYC and establish a special and multi-dimensional system of identification procedures for architecture in public open spaces. Moreover, public participation should be intensified, as the public also has the right to understand and supervise the construction of these spaces.

5 CONCLUSIONS

Policies about architecture and public open spaces in China are still in their initial stages. For further improvement, it is crucial to digest the experience and lessons of other countries during their development of public open space architecture and avoid similar problems. At the same time, it is vital to fully consider the national context when formulating a design and identity management strategy that applies to the architecture of public open spaces in China. By doing so, China will make steady progress in the construction of public open spaces and improve the environment of city public open spaces.

REFERENCES

Department of New York City Planning, (2014). *Uniform Land Use Review Procedure*. https://www1.nyc.gov/site/planning/applicants/applicant-portal/step5-ulurp-process.page.

Kayden J. (2000) *New York Department of City Planning, Municipal Art Society of New York. Privately Owned Public Space: The New York City Experience*. John Wiley, New York.

New York City Planning, (2019). *Zoning Resolution*. https://zr.planning.nyc.gov/.

Yang Y. (2017) *Study on Management and Control of Public Space of Building: Based on the New York City Experience of the POPS* (Guangzhou: University of Technology) p 42

Yu, Y. (2016) *Study on Guidelines of Public Space of Building: Based on the New York City Experience of the POPS. Urban Planning International*, 31 (02): 98–109

Summer thermal environment of building atrium in hot summer and warm winter areas

Jiating Chen* & Yan Liu*

Xiamen Institute of Technology, Fujian Xiamen, China

ABSTRACT: Under the influence of healthy building and energy saving and emission reduction policies, people have higher requirements for building indoor thermal comfort, a light environment, and building energy consumption. Lighted atriums are widely used in buildings, but the atrium thermal environment is poor, and energy consumption increases due to the presence of roof structure on the lighted roof. In a large number of public buildings, glass light roofs are often used as the structural form of the atrium roof. The glass roof not only increases the natural indoor lighting but also has a negative impact on the thermal environment of the atrium in summer. On the basis of meeting the lighting illumination, this paper discussed how to improve indoor thermal comfort and building energy consumption by integrating the design of glass roofs and shading became the focus of research. This paper took the building atrium in the hot summer and warm winter area in the hot summer area as the research object, analyzed the thermal environment characteristics of the lighting atrium and its influencing factors, studied and discussed the influence of the integrated design of glass lighting roof and shade under different shapes on the light and thermal environment of the atrium, and put forward the appropriate integrated optimization design scheme to achieve the goal of improving the thermal environment of the atrium.

1 INTRODUCTION

The number of modern tall space buildings is increasing, and the scale of buildings has grown tremendously. While tall space buildings are booming, energy problems such as high energy consumption and low utilization are becoming more prominent, and energy saving is becoming increasingly serious (Cheng 2022). A lighted atrium is a space that connects the interior to the outside world. For the light atrium, the main feature of its light and heat environment comes from the roof structure of the light roof. Due to the advantages of no shading, a good visual effect, a light load, and a short installation period, glass light roofs are often widely used as the roof structure of light atriums in libraries, shopping malls, and other public buildings (Cui 2021). The tall and light atrium space gradually shows the tendency to be huge and complex, which presents many problems in the control of the space thermal environment. The phenomenon of high temperatures in the upper layer and a large vertical temperature distribution gradient in the atrium space in summer due to the greenhouse effect makes it difficult to meet the requirements related to human thermal comfort without the regulation of air conditioning and ventilation, which will inevitably lead to huge energy consumption in order to meet the thermal comfort of human beings (Liu 2021).

---

*Corresponding Authors: 2907022993@qq.com and 164197619@qq.com

2 THERMAL ENVIRONMENT AND TYPICAL CHARACTERISTICS OF BUILDING ATRIUM

2.1 *Definition of the architectural atrium*

An architectural atrium is an interior hall in a building that penetrates multiple floors. In order to simplify the discussion and make the study more focused, the atrium form of the research object is narrowed down to the core atrium with typical representatives (Li 2020). The light-transmitting panels of this research object are limited to the most commonly used glass material for architectural light roofs, and the rest of the materials do not belong to the scope of this research.

2.2 *Thermal environment characteristics of building atrium*

The construction of the building atrium determines both the chimney effect and the greenhouse effect, resulting in a significant vertical temperature gradient as hot air collects in the upper part of the atrium and cooler temperatures in the active areas at the bottom (Ren 2021).

The chimney effect, also known as the "thermal pressure effect," is a difference in pressure caused by the difference between indoor and outdoor temperatures. Air is drilled into and out of the atrium through doors, windows, or other gaps, resulting in the flow of indoor air.

The greenhouse effect is the result of the sunlight shining on the atrium's light roof. Short-wave heat radiation enters the interior through the glass light roof or curtain wall, and its energy is absorbed through the wall surface, floor slab, or other objects, causing the temperature of the atrium to rise. At the same time, the internally heated surface emits longer-wavelength infrared radiation, which cannot be transmitted through the glass, thus making the temperature rise (Zhang 2021).

2.3 *Analysis of thermal environment impact factors*

There are many factors affecting the thermal environment, such as geographical location, outdoor meteorological parameters, heat transfer performance of the building envelope, airflow organization, and so on. The significant difference between the lighted atrium and other buildings is the presence of tall space and a lit interface in the lighted atrium; therefore, when studying the influencing factors of the thermal environment, we should focus on analyzing these two parts. The influencing factors of the indoor thermal environment of the building lighting atrium are shown in Figure 1.

Through the methods of review and simulation, the researchers analyzed the influence of the physical environment composition factors affecting the atrium building, mainly the plan size, plan shape, section size and shape, skylight form, and enclosure method of the light atrium on the light environment, sound environment, and thermal environment. With the goal of energy consumption control, scholars qualitatively studied the influencing factors of the spatial composition of the light atrium from four basic perspectives: spatial form, outdoor environment, physical building, and light interface (Zhang 2022). The spatial composition of the light atrium was classified into five factors: type, plan characteristics, vertical characteristics, scale, and combination. The purpose of this paper was to study the design of light roofs and tall atrium spaces, especially the influence of atrium geometry on the indoor thermal environment, in order to further optimize them under the condition of ensuring light. Therefore, the research object was limited as follows: the shape of the atrium section is vertical, the basic enclosure of the atrium is of the corridor type, and the influence of the atrium in terms of spatial proportion, sun shading, skylight design, and roof ventilation was mainly studied in the preliminary research, experiment, and later simulation. Figure 2 depicts the factors that influence the spatial composition of the lighted atrium.

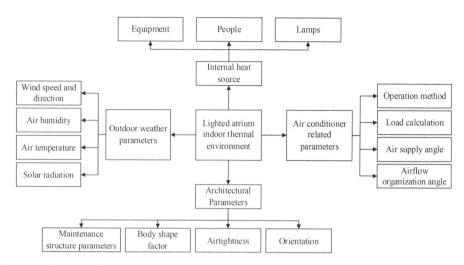

Figure 1. Factors influencing the indoor thermal environment of the building's lighted atrium.

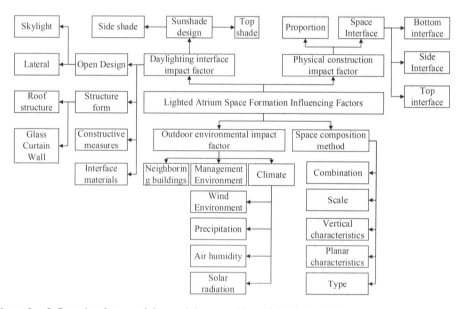

Figure 2. Influencing factors of the spatial composition of the lighted atrium.

3 ANALYSIS OF THE SUMMER THERMAL ENVIRONMENT OF THE BUILDING'S LIGHTED ATRIUM

3.1 *Selection of the architectural atrium*

Space scale and proportion (light roof area, atrium height), interface shading situation (shading form), light form (atrium form, light roof form, light roof shape), and roof ventilation situation were the main elements of atrium space that we focused on the indoor thermal environment. For this purpose, 10 buildings with tall spaces and light roofs were selected for research in Xiamen, mainly for field investigation of light roof form, sun-shading form, light roof area, and space height (Xu 2020).

SAR is the geometric characteristic of the vertical direction of a lighted atrium. It is defined as the ratio of the height of the atrium (H) to the width of the light roof (w) in the north–south direction. The higher the SAR value is, the less the ground is influenced by solar radiation.

$$SAR = \frac{H}{w} \tag{1}$$

To ensure that the experimental and simulation results were comparable and representative, this paper counted a total of 55 atrium building cases published in Architectural Technology and HVAC from January 1, 2012, to January 1, 2022, as well as 10 building atria from the field research. Due to the irregular shape of the atria, the dimensionless section aspect ratio (SAR*) value was used instead of SAR to reflect the relationship between atrium height and atrium area. When the height-to-width ratio of the atrium SAR* was less than 4, 95% of the buildings in the study area were satisfied. The form of the light roof was mostly flat, and the glass material of the light roof was mostly hollow laminated glass.

$$SAR^* = \frac{Atrium\ height}{\sqrt{Atrium\ area}} \tag{2}$$

According to the above analysis, Xiamen A Hospital with SAR* = 1.385, a recently completed sawtooth light roof, and Xiamen B Library with SAR* = 0.627 were chosen.

The Xiamen B Library with a flat roof and SAR* = 0.627 was selected for analysis, and the requirements of no personnel disturbance and non-air-conditioning conditions could be met during the test period.

3.2 *Xiamen a hospital thermal environment experiment*

The building had two atriums, each 24.2 m long and 7.8 m wide: Atrium 1, which ran from the second floor to the fourth floor, was 18.5 m high, and Atrium 2, which ran from the second floor to the fourth floor, was 14.2 m high.

The test time was April 28, 2022, a sunny day with no air conditioning; the actual measurement time was 9:00–16:00, testing the temperature change law under the atrium of the lighting roof during the day. The specific idea in the vertical direction of the lanyard method measurement point arrangement was to use the corridor railings to fix the steel plastic rope, and then the instrument was fixed on the steel plastic rope. The measurement point arrangement was tested continuously up until the end of the test period. In the fixed fourth-floor corridor of the 5-meter-tall vertical uprights, from the top of the arrangement of five measurement points, the uprights were arranged under the glass. The indoor corridor area, first-floor atrium area, and outdoor temperature and humidity were tested by using a temperature and humidity recorder T every 15 minutes, and other parameters were tested every 30 minutes by using handheld equipment.

3.3 *Xiamen b library thermal environment experiment*

The library building was a 6-story frame structure with one basement level and five floors above ground; each floor height was 4.5 meters, covering a total area of 19, 500 square meters and a total building area of 27, 907 square meters, with two glass-lit atriums. The scope of this study was the atrium area, including the two light roofs. The building structure type was mainly glass curtain walls and a light roof. A total of 6 mm tempered light green clear glass was used for the open-frame glass curtain wall, 15 mm tempered clear glass was used for the ribbed point-supported curtain wall surface, 8 + 1.52 PVB + 8 mm tempered clear glass was used for the ribbed glass, and 10 + 1.52 PVB + 8 mm tempered laminated clear glass was used for the light roof. The larger the shading area was, the less direct solar radiation entered the room, and the roof ventilation device could exhaust the hot air accumulated on top under a clear sky with few clouds. Therefore, during the actual measurement, the influence of two factors, namely air conditioning condition and shading area on the indoor environment, were considered. In this test, the ventilation volume was 47819 m^3/h, which was equivalent to 4.57 air changes per hour in the atrium.

In order to exclude the influence of solar radiation on the test apparatus, tin foil was applied to the outer surface of the temperature and humidity sensors and the temperature autometer, as well as the vertical measuring rod, to reduce the influence of direct solar radiation on the experiment.

3.4 *Heat and humidity environment evaluation index*

The difference between the lighted atrium and other building forms lies in the influence of solar radiation on the indoor thermal environment. Solar radiation raises the temperature of the ground, the light surface, and the glass of the light roof, radiating heat exchange with the human body, while the air converts heat exchange with the human body. In the non-artificial environment, human thermal comfort in the lighted atrium is mainly influenced by the radiation temperature and the air-dry bulb temperature. Therefore, this paper took the body temperature as the evaluation index of the thermal environment.

$$t_{op} = A \cdot t_a + (1 - A) \cdot t_r \tag{3}$$

where t_a denotes the air temperature, t_r denotes the average radiation temperature, and A denotes the coefficient, usually chosen as 0.5.

In engineering practice, the average radiation temperature was calculated as shown in (4).

$$T_r = \frac{T_1 A_1 + T_2 A_2 + \cdots + T_N A_N}{A_1 + A_2 + \cdots + A_N} \tag{4}$$

where T_N denotes the temperature of the Nth surface and A_N denotes the area of the Nth surface.

3.5 *Library thermal and humid environment analysis*

The difference between the body temperature and the limit value was then calculated. Finally, the difference between the sensory temperature and the evaluation limit was compared in the four working conditions, and the smaller the difference was, the better the thermal environment was.

The difference between the sensory temperature and the evaluation limit value for different working conditions of the library is shown in Figure 3.

The effect of shading was compared: With ventilation, the difference ΔT is 2.28°C lower than without shading. In the absence of ventilation, the difference ΔT is 1.78°C lower than without shading. The effect of ventilation was compared: when there is no shade, the difference $\triangle T$ can be reduced by 0.31°C compared with no ventilation; when there is shade, the difference $\triangle T$ can be reduced by 0.81°C compared with no ventilation.

It is clear from the above that both shading and ventilation can optimize the atrium thermal environment, and shading is more effective than ventilation for optimization.

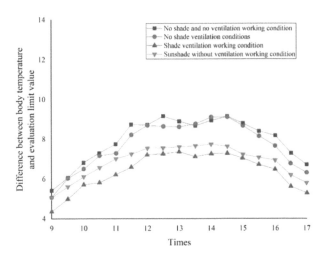

Figure 3. Difference between body temperature and evaluation limit values for different working conditions in the library.

3.6 *Comparison of thermal and humid environments in experimental buildings*

The hospital's experimental working conditions were without shade and ventilation, and the thermal and humid environment was compared to the library's corresponding working conditions. The differences between hospital and library light atrium spaces were (1) SAR* = 1.385 in the hospital and SAR* = 0.627 in the library; (2) the form of the light roof in the hospital atrium was serrated and had an inclination angle, while the light atrium in the library was flat; and (3) the light roof glass in the hospital was 6 + 1.52 PVB + 6 + 13A + 6LOW-E tempered laminated insulating glass. The above three points might optimize the indoor thermal environment of the atrium. The difference between the physical temperature of the atrium of the experimental building and the evaluation limit is shown in Figure 4.

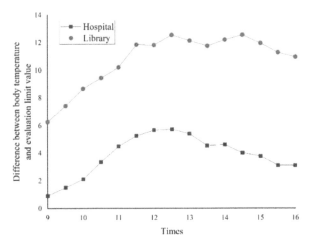

Figure 4. Difference between atrium temperature and the evaluation limit value of an experimental building.

4 CONCLUSION

The four conditions of the library atrium were compared horizontally, and it was found that both shading and ventilation could optimize the thermal environment of the atrium under the summer condition with 4.57 times as much ventilation, and the shading was more effective than the ventilation. When comparing the thermal and humid environments of the library and hospital atriums under summer conditions, the hospital outperformed the library. In order to have a better indoor thermal environment in the core-type lighted atrium, the design of the core-type lighted atrium should consider increasing the connecting area of the lighted roof, reducing the area of the lighted roof, increasing the height of the atrium, and increasing the inclination angle of the lighted roof as much as possible under the premise of meeting the lighting requirements. The physical environment of the atrium in this study was a light and heat environment. In future research work, more diversified atrium optimization models can be established according to other design requirements and physical environments (e.g., sound environment). The single-factor analysis was used in the dimensionless factor optimization process, and the cross-influence of multiple factors was not analyzed in depth. The interaction between different spatial elements can be studied in depth in subsequent research.

ACKNOWLEDGEMENT

This paper was supported by the following foundation project fund: the 2021 Fujian Province Young and Middle-aged Teachers Educational Research Project Grant (JAT210566).

REFERENCES

Cheng, J.J. & Ye, L.J. & Li, C.X. *et al* (2022). Research on Summer Thermal Environment Testing in Residential Areas of Guilin. *Energy Conservation*, 41 (07): 15–18.

Cui, Y.Q. & Lai, Z.Z. & Liu, C. (2021). A Review of Research on Optimization of Indoor Light and Heat Environment in Glass-roofed Atrium Buildings. *Building energy efficiency (in English)*, 49 (09): 51–56.

Liu, H.J. & Li, M.X. & Zhang, L. *et al* (2021). Research on the Application of Reflective Thermal Insulation Coatings on Building Facades in Hot Summer and Warm Winter Areas in Guangxi. *Energy conservation in buildings (in English)*, 49 (08): 109–115.

Li, Z.W & Li, X.Y & Li, N. *et al* (2020). Research on Architectural Design Strategies for Hot Summer and Warm Winter Regions based on Energy Saving and Indoor Environment Quality Improvement – Taking Nanning Garden Expo Horticultural Pavilion as an example. *Journal of Architecture*, 95–99.

Ren, C.X. & Pang, G.Y. & Fang, C. *et al.* (2021). Numerical Simulation of the Summer Thermal Environment of a Building Atrium in Shenzhen. *Building Thermal Ventilation and Air Conditioning*, 40 (06): 68–70 + 63.

Xu, M.K. & Zhao, S.J. (2020). Research on Summer Thermal Environment and Thermal Comfort of Suzhou Metro Station. *Journal of Suzhou University of Science and Technology (Engineering Technology Edition)*, 2020, 33 (04): 74–80.

Zhang, L.N. (2021). Optimization Analysis of Summer Thermal Environment in the Atrium of a High-rise Green Office Building in Shanghai. *Green Building*,13 (04): 58–62.

Zhang, M.Y. & Wang, C. & Fan, Q. *et al.* (2022). Study on the Optimization of Physical Environment Design of Atrium in Public Buildings in Beijing-Tianjin-Hebei region. *Western Journal of Habitat Environment*, 2022, 37 (04): 93–100.

Water Conservancy and Civil Construction – Oke & Ahmad (Eds)
© 2024 The Author(s), ISBN: 978-1-032-58618-2

Surface settlement prediction of composite stratum shield construction based on e-GRA-LSTM

Lili Feng*, Changming Hu & Liang Li
School of Civil Engineering, Xi'an University of Architecture and Technology, Xi'an, Shaanxi, China

ABSTRACT: To accurately predict the surface settlement of shield tunneling under composite stratum conditions, the e-GRA-LSTM surface settlement prediction model is proposed. Firstly, to distinguish the influence degree of different tunneling parameters and improve the prediction accuracy of the LSTM model, entropy-gray relationship analysis (e-GRA) was used to calculate the correlation degree of each tunneling parameter on surface settlement. The parameters with greater influence on surface settlement were taken as input parameters for LSTM, and they were reasonably weighted. Secondly, considering the engineering geological parameters, geometric parameters, and shield tunneling parameters, the long short-term memory (LSTM) model was constructed. Finally, relying on a section project of Qingdao Metro Line 4, the input parameters and the existing tunnel settlement monitoring values were trained and tested, and the prediction results were compared with those of the BP neural network model. The results showed that the e-GRA-LSTM model constructed in this paper had a qualified rate of 93.33% for the prediction of surface settlement, with a root mean square error ($RMSE$) of 0.7355, a correlation coefficient ($R2$) of 0.9318, and an average absolute percentage error (MAPE) of 4.61%, indicating that the model was highly reliable. In terms of $RMSE$ and R^2, the prediction accuracy and fitting ability had been greatly improved compared with the BP model, so the model had good reliability and practicability and could provide a theoretical basis for the prediction and control of the surface settlement of shield tunneling under composite stratum conditions.

1 INTRODUCTION

Shield tunnel construction has become one of the most important methods of rail transit construction in recent years (Zhang 2018), but it will inevitably disturb the soil surrounding the stratum, resulting in the consolidation of the gap between the surrounding rock and the support, the surrounding rock and the lining due to the decline of the groundwater level, resulting in the subsidence of the support in the weak surrounding rock and causing surface settlement (Li 2021). Therefore, it is of great significance to predict and control the surface settlement to ensure the smooth progress of the shield's construction. At present, the research on surface subsidence prediction mainly focuses on the empirical formula method (Hu 2018), the numerical simulation method (Huang 2020), and the artificial intelligence method (Sun 2020). The empirical formula method and numerical simulation method have great limitations in practical application, while the artificial intelligence method has broad application prospects in surface deformation prediction.

Xu et al. (2022) used the BP neural network method to build a prediction model for the total thrust of the composite stratum and the torque of the cutter head, which controls the

---

*Corresponding Author: 1378762271@qq.com

DOI: 10.1201/9781003450832-8

main parameters of the shield machine operation. Pourtaghi et al. (2012) proposed a prediction method for maximum ground subsidence by combining wavelet theory with an artificial neural network, or wavelet network. Chen et al. (2019) compared the prediction results of six algorithms for tunnel surface settlement and found that the prediction results of the GRNN and RF algorithms were the most accurate. Li et al. (2020) compared the applicability of recurrent neural networks (RNN) and BP neural networks in surface settlement prediction, and the results showed that RNNs had a better prediction effect on tunnel settlement.

Most of the existing studies have not considered the time sequence of surface subsidence (K 1999). Li's work (2020) showed that RNN integrating time series could better predict land subsidence. RNN is an artificial neural network that has a tree-like hierarchical structure, and the network nodes recurse the input information according to their connection order. It can mine the temporal information and semantic information in data, so it is very effective for processing data with sequential characteristics. Traditional RNN, on the other hand, is prone to gradient disappearance or explosion. The improved variant of RNN, short and long-term memory network (LSTM), has better performance in long sequence prediction (Wang 2018). LSTM can selectively read or forget information by adding a gate mechanism, which can well alleviate the problem of gradient disappearance when processing long sequences. Based on the above analysis, this paper adopted entropy-weighted grey correlation analysis (e-GRA) to analyze the impact of various tunneling parameters on the surface settlement based on existing research. Based on the analysis results, the input data of LSTM were reasonably weighted. Considering engineering geological parameters, geometric parameters, and shield tunneling parameters, the e-GRA-LSTM model for surface settlement prediction under composite stratum conditions was established. Based on an interval project for Qingdao Metro Line 4, the ground settlement was predicted and compared with the prediction results of the BP neural network model to verify the reliability of the model.

2 PRINCIPLES OF THE E-GRA-LSTM SURFACE SETTLEMENT PREDICTION MODEL

2.1 *Principle of the entropy-grey correlation analysis method*

Entropy is often used for objective weighting. The smaller the entropy value is, the higher the importance of the index is. By calculating the gray correlation degree, the grey correlation analysis method describes the tightness and order of the relationship between different objects (Zhou 2019). The correlation between the comparison sequence and the reference sequence can be divided into the following three grades: $0 < \xi_i(j) \leq 0.35$, indicating weak correlation; $0.35 < \xi_i(j) \leq 0.65$, indicating medium correlation; $0.65 < \xi_i(j) \leq 1$, indicating strong correlation.

2.2 *Principles of the LSTM neural network*

The long-short-term memory neural network (LSTM) (Qin 2021) is an improved variant of the traditional recurrent neural network (RNN). Time series are used in LSTM. By adding three control units, namely the forgetting gate, input gate, and output gate, it can effectively solve the limitations of RNN in dealing with long-term dependence.

2.3 *E-GRA-LSTM surface settlement prediction model*

The input parameters of the LSTM model will have a direct impact on the final prediction results. Due to the obvious difference in the influence degree of each tunneling parameter on the surface settlement, to distinguish the influence degree of different tunneling parameters and improve the prediction accuracy of the LSTM model, the entropy weight gray correlation method (e-GRA) was first used to calculate the correlation degree of each tunneling parameter on the surface settlement. The parameters with greater influence on the surface settlement were then reasonably weighted, and together with geometric and geological parameters, they formed the input data of the LSTM model. The structural parameters will have a significant impact on the learning and generalization abilities of the model. To find the super parameter combination with the minimum prediction error, the k-fold cross-validation method was used to optimize the number of hidden layers (ls), hidden nodes (N), learning rate (lr), and iteration times (iter) of LSTM. After the network structure was determined, the model was trained, tested, and compared with the measured data to verify the reliability of the model. The flow of the e-GRA-LSTM surface subsidence prediction model is shown in Figure 1.

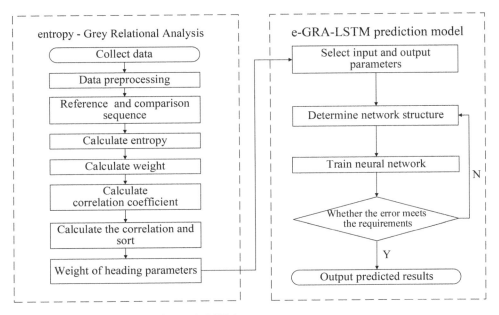

Figure 1. Prediction process of e-GRA-LSTM.

2.3.1 *Data preprocessing*
The shield machine will inevitably produce abnormal tunneling parameters during tunneling. The box and line diagram method was used to identify and eliminate abnormal values (Mahalanobis 1936), and the missing values were supplemented by linear interpolation.

As the dimensions of each parameter were different, in order to facilitate comparison and analysis, normalization shall be conducted according to the maximum and minimum value method, and the sample data shall be divided into the training set and test set according to the 8:2 ratio.

2.3.2 *Entropy-grey correlation analysis*
The steps of entropy-weight gray correlation analysis were as follows: (1) selecting reference series and comparison series; (2) computing information entropy; (3) calculating weight; (4) calculating the correlation coefficient; (5) calculating the degree of correlation.

2.3.3 *Selection of input and output parameters of LSTM*

The input parameter was selected as follows: (1) The geometric parameter (namely tunnel size and burial depth H): as the tunnel specifications remain unchanged in this project, the depth span ratio H/D of the tunnel section was selected as the geometric input parameter (Zhang 2020). (2) The geological parameters: based on the research in the literature on the relationship between cohesion, internal friction angle, formation weight, and deformation modulus, the deformation modulus E_0, coefficient of static lateral pressure K, and proportion of composite formation rock λ were selected as input parameters. (3) Driving parameters: according to the results of the entropy-gray correlation analysis, the tunneling parameters that had a greater impact on the surface settlement were selected as the input parameters of the LSTM prediction model, and they were reasonably weighted. (4) Surface subsidence: the surface settlement data was obtained from the construction site monitoring, and its value was taken from the value after the deformation of the surface settlement monitoring point was stabilized.

2.3.4 *Determine the structure of the LSTM model*

The LSTM model structure includes an input layer, an output layer, and a hidden layer. The number of LSTM hidden layers (ls), the number of hidden nodes (N), the learning rate (lr), and the number of iterations ($iter$) will have a significant impact on the learning and generalization abilities of the model. The k-fold cross-validation method was used to find the optimal super parameter combination (Hochreiter 1997).

2.3.5 *Evaluation indicators of the LSTM model*

Before evaluating the settlement prediction effect of the LSTM prediction model, it is necessary to reverse normalize the prediction results. Root mean square error ($RMSE$), mean absolute error (MAE), mean absolute percentage error ($MAPE$), and goodness of fit (R^2) were used to evaluate the model.

2.3.6 *Models for experimental comparison*

In order to verify the superiority of the LSTM model in surface settlement prediction of composite strata, the results were compared with those of the BP neural network model. The input and output parameters and training times of the BP neural network were consistent with those of the LSTM model, and other super parameters were determined by k-fold cross-validation.

3 ENGINEERING EXAMPLES

3.1 *Project overview*

This study was based on an interval project of Qingdao Metro Line 4, with a total length of 1072.00 m, a tunnel vault buried depth of 6.20–15.20 m, and a floor buried depth of 13.5–17.0 m. The stratum crossing the section from top to bottom in a vertical direction was the main medium coarse sand layer, silty clay layer, gravel soil layer, and locally strong, medium, and slightly weathered tuff, which belonged to the typical soft upper and hard lower strata; strata with different lithology and weathering degree were distributed successively in the horizontal direction. Therefore, the geological conditions of the section were composite strata.

3.2 *Entropy-grey correlation analysis*

A series of corresponding surface settlement values Y_0 above the tunnel axis of the section were selected as the reference sequence, and eight tunneling parameters, including the propulsion speed X_1, total thrust X_2, cutter torque X_3, cutter speed X_4, soil bin pressure X_5, auger speed X_6, synchronous grouting amount X_7, and soil yield X_8, were taken as the comparison sequence to calculate the correlation between tunneling parameters and surface settlement. The calculation results are shown in Table 1.

Table 1. Results of entropy weight-grey correlation analysis.

Tunnelling Parameters	Weight Value	Correlation Value	Relevance Ranking	Degree of Relevance
Total Thrust	4.30%	0.2647	3	Weak Correlation
Propulsion Speed	40.30%	0.7690	1	Strong Correlation
Cutter Head Torque	9.10%	0.1845	4	Weak Correlation
Cutter Head Speed	6.00%	0.1176	5	Weak Correlation
Screw Machine Speed	36.00%	0.7395	2	Strong Correlation
Silo Pressure	3.70%	0.0690	6	Weak Correlation
Grouting Volume	3.00%	0.0063	7	Weak Correlation
Unearthed Volume	2.00%	0.0041	8	Weak Correlation

The above calculation results show that on the one hand, the parameters that had a greater impact on the surface settlement were respectively the propulsion speed, the rotation speed of the screw, the total thrust, the rotation speed of the cutter head, the torque of the cutter head, and the soil bin pressure, which should be paid more attention to in the later stage of the surface settlement control as the input parameters of the LSTM model. On the other hand, due to the great difference in the influence of various tunneling parameters on the surface settlement, when building the LSTM model, the input sample data should be reasonably weighted based on the results of gray correlation analysis to improve the prediction accuracy of the model.

3.3 LSTM shield surface settlement prediction model

3.3.1 Selection of LSTM input and output parameters

The geometric parameters, geological parameters, tunneling parameters, and surface settlement monitoring data of 70 surface settlement monitoring sections on the left line of an interval of Qingdao Metro Line 4, including upper soft and lower hard geology, were selected as samples of the LSTM prediction model. The output parameter was the surface settlement value of the current settlement monitoring point; input parameters included geometric parameters, geological parameters, shield tunneling parameters, and the surface settlement of the previous settlement monitoring point.

3.3.2 Construction of the E-GRA-LSTM model

(1) Structure of the E-GRA-LSTM Model

The LSTM model structure was set as 1 input layer (13 input nodes), 1 output layer (1 output node), and a hidden layer. Based on the LSTM layer of the first layer, the RMSE of the settlement amount of the training sample was taken as the objective function for super parameter optimization. The value range and optimal solution of the hyperparameter are shown in Table 2. The Adam optimizer was used to continuously update the weight parameters.

Table 2. Hyperparameter optimization results of LSTM model.

Name	ls	N	lr	$iter$
Value Range	1, 2, 3, 4, 5	16, 32, 64, 128, 256	0.001, 0.01, 0.05, 0.1	100, 200, 250, 300, 350, 400
Optimal Solution	1	128	0.05	350

The final LSTM model consisted of a hidden layer (128 nodes), an input layer (13 nodes), an output layer (1 node), and a full connection layer. The loss function during training was set as the RMSE function. The initial learning rate was set to 0.05, and 350 rounds of training were conducted.

BP neural network prediction model was used as the comparison model. The input, output layer, and training times of the BP neural network were consistent with those of the LSTM model, and their values were 13, 1, and 350 respectively. Other super parameters were searched and determined by the k-fold cross validation method. The average root mean square error ($RMSE$) of each validation set was used as the objective function in the optimization process, and the value of the final super parameter was determined as $[ls, N, lr] = [1, 10, 0.05]$.

3.4 *Analysis of prediction results for surface settlement*

After the structure of BP and LSTM was determined, the training set was used to train the model iteratively to obtain the final model. The prediction effect of test set data was used, and the prediction error and fitting accuracy of different models were compared. The prediction results and relative errors of surface subsidence are shown in Figure 2, Table 3, and Figure 3, respectively.

Figure 2 shows that the predicted value curve and measured value curve of the BP neural network model and LSTM model can fit well, which indicates that both models can deeply dig out the nonlinear mapping relationship between geometric, geological, and tunneling parameters and surface settlement in the iterative process of the training set and have good adaptability to the prediction of surface settlement in shield tunneling. However, the LSTM model is more accurate in predicting surface subsidence.

By analyzing the data in Table 3 and Figure 3, it can be seen that each evaluation index of the LSTM model is better than that of the BP model. As far as R2 is concerned, the goodness of fit of the LSTM model is 0.9381, which is closer to 1 than that of the BP model, indicating that the model has stronger goodness of fit. As far as the prediction error is concerned, all kinds of errors in the LSTM model are smaller than those in the BP model. The RMSE, MAE, and MAPE of the LSTM model are 0.7355, 0.1638, and 4.61%, respectively, which are 12.03%, 28.50%, and 68.12% lower than those of the BP model; the maximum relative error is 56.03% lower, and the prediction qualification rate is 27.27% higher.

In conclusion, the LSTM model was obviously superior to the BP model in terms of both model fitting accuracy and prediction error control, which showed that the LSTM model had stronger fitting and generalization ability for the problem of surface settlement prediction in this paper.

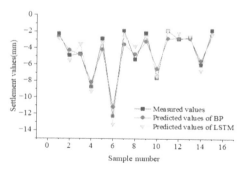

Figure 2. Settlement prediction values of different models.

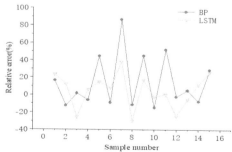

Figure 3. Prediction relative errors of different models.

Table 3. Comparison of prediction errors between different models.

Model Name	RMSE	MAE	R^2	MAPE (%)	Max MAPE (%)	Qualified Rate* (%)
LSTM	0.7355	0.1638	0.9381	4.61	38.09	93.33
BP	0.8361	0.2291	0.9199	14.46	86.63	73.33

*Sample data with a relative error of less than 30% between the predicted value and the measured value of surface subsidence is considered qualified.

4 CONCLUSIONS

Based on the actual project data, this paper used the entropy-gray correlation method to analyze the influence of tunneling parameters on the surface settlement. On the basis of geometric and geological parameters, an LSTM neural network model was built to predict the surface settlement of shield tunneling. The prediction results were compared with the BP model, and the following conclusions were drawn:

(1) According to the results of the entropy-gray correlation analysis, among the various tunneling parameters that affected the surface settlement, the propulsion speed, auger speed, and total thrust had the greatest impact, while the influence of the amount of excavated soil and grouting was small. The analysis results can provide a theoretical basis for the classification of the importance of the factors affecting the surface settlement of shield tunneling and the establishment of the settlement prediction model.
(2) The cross-validation method was used to optimize the super parameters of the LSTM model. The results showed that the more hidden layers in the model were not better. Increasing the number of layers did not necessarily improve the nonlinear fitting ability of the model. The prediction accuracy of single-layer BP and LSTM models was better than that of the multi-layer model.
(3) The prediction pass rate of the time series model, LSTM, reached 93.33%, while that of the non-time series model, BP, was less than 75%, demonstrating that the time series model was more suitable for the prediction of surface settlement during shield tunneling. The $RMSE$, MAE, and R^2 of the $LSTM$ model for surface settlement prediction were 0.7355, 0.1638, and 0.9381, respectively, and MAPE was only 4.61%, which has higher prediction accuracy and nonlinear fitting ability than the BP model.

The e-GRA-LSTM model constructed in this paper could accurately predict the surface settlement caused by shield tunneling in composite strata and had good practicability and reliability. In the future study, it is proposed to use the settlement prediction results obtained by the model to optimize and control the tunneling parameters in a timely and effective manner during the shield construction process, so as to ensure the safety of the construction and improve the construction efficiency.

REFERENCES

Chen, R. P., Zhang, P. & Wang, H. N. (2019). Prediction of Shield Tunneling-induced Ground Settlement Using Machine Learning Techniques. *J. Frontiers of Structural and Civil Engineering*. 13 (6).
Hu, C. M., Feng, C. & Mei, Y. (2018). Improvement of Peck Settlement Prediction Formula for Shield Tunneling in Xi'an Water Rich Sand Layer. *J. Journal of Underground Space and Engineering*. 14, 176.
Huang, Z., Zhang, C. & Fu, H. (2020). Numerical Study on the Disturbance Effect of Short-Distance Parallel Shield Tunnelling Undercrossing Existing Tunnels. *J. Advances in Civil Engineering*.

Hochreiter, S. & Schmidhuber, J. (1997), Long Short-term Memory. *J. Neural Computation.* 9 (8).

Lee K.M., Ji H.W. & Shen C.K. (1999). Ground Response to the Construction of Shanghai Metro Tunnel-Line 2. *J. Soils and Foundations.* 39 (3).

Li, S., Li, P. & Zhang, M. (2021). Analysis of Additional Stress for a Curved Shield Tunnel. *J. Tunnelling and Underground Space Technology Incorporating Trenchless Technology Research.* 107.

Li, L. B., Gong, X. N. & Gan, X. L. (2020). Prediction of Maximum Ground Settlement Caused by Shield Tunnel Based on Cyclic Neural Network. *J. Journal of Civil Engineering.* 53 (S1), 13–19.

Mahalanobis, P. C. (1936). On the Generalised Distance in Statistics. *J. Proceedings of the National Institute of Science of India.* 2, 49–55.

Pourtaghi, A. & M, A. L. (2012). Wavenet Ability Assessment in Comparison to ANN for Predicting the Maximum Surface Settlement Caused by Tunneling. *J. Tunnelling and Underground Space Technology incorporating Trenchless Technology Research.* 28.

Qin, C., Shi, G. & Tao, J. (2021). Precise Cutterhead Torque Prediction for Shield Tunneling Machines Using a Novel Hybrid Deep Neural Network. *J. Mechanical Systems and Signal Processing.* 151.

Sun, J. & Wen, H. Y. (2020) Application and Practice of Artificial Intelligence Science in Deformation Prediction and Control of Soft Soil Underground Engineering Construction. *J. Tunnel Construction (Chinese and English).* 40 (01), 1–8.

Wang, X., Wu, J. & Liu, C. (2018). Prediction of Fault Time Series Based on LSTM Recurrent Neural Network. *J. Journal of Beijing University of Aeronautics and Astronautics.* 44 (04), 772–784.

Xu, Y. F., Wang, S. M.& He, C. (2022) Prediction of Shield Tunneling Parameters in Composite Strata Based on BP Neural Network. *J. Railway Standard Design.* 66 (07), 120–125.

Zhang, N., Li, J. & Jing, L. J. (2018). Research and Application of Intelligent Control System for TBM Tunneling Parameters. *J. Tunnel Construction (Chinese and English).* 38 (10), 1734–40.

Zhou, J. F., Hu, S. Y. & An, J. K. (2019). Research on Key Indicators of Power Supply Level of Medium Voltage Distribution Network Based on Entropy Weight Method and Grey Correlation Segree. *J. Science and Technology Bulletin.* 35 (12), 101–104.

Zhang, P., Wang, H. N. & Cheng, R. P. (2020). Hybrid Meta-heuristic and Machine Learning Algorithms for Tunneling-induced Settlement Prediction: A Comparative Study. *J. Tunnelling and Underground Space Technology incorporating Trenchless Technology Research.* 99 (C).

Numerical analysis of wall shedding based on thermal coupling

Hongli Zhang
Ande College, Xi'an University of Architecture and Technology, Xi'an, China

Changming Hu*
Xi'an University of Architecture and Technology, Xi'an, China

ABSTRACT: To determine the building exterior insulation structure has the degree of danger of breakage in without repair case by using the finite element software ABAQUS, the typical three kinds of both the damage state of retaining structures and construct a three-dimensional transient thermal structure coupling model is analyzed, calculating the heating layer under the condition of different functions in the summer the temperature field, stress field, and displacement field of distribution. The calculation results show that EPS strip cavitation is more dangerous in the heating process. In this state, the displacement of the ceramic tile layer is the largest and the safety is low. The larger the heating rate is, the larger the tile layer displacement is. In the same temperature rise interval, the smaller the heating rate is and the longer the time is, the larger the tile layer displacement is; in the condition of the EPS plate strip empty drum, the smaller the EPS plate thickness, the greater the stress of tile layer, the more dangerous; the temperature difference between the inner and outer layers of ceramic tile caused by the strip empty drum of ceramic tile is greater than that of other damaged ceramic tile. The strip empty drum of the EPS plate will cause the temperature difference between the inner and outer layers of the EPS plate to be larger than that of other existing damaged states, reaching 38.96°C, which is more likely to cause the falling off of ceramic tile and injury.

1 INTRODUCTION

In the early 1980s, our country began to implement building energy saving (Zhang 2019). The energy consumption generated during the use of the house is up to 30% of the total energy consumption of the building, and this part of consumption mainly comes from the external wall which accounts for 74% of the outer surface of the building maintenance structure (Zhang 2021). Related research shows that there are many quality problems in buildings with existing external insulation systems. The falling off of the building's external insulation system causes falling objects to hurt people and smash cars, causing a series of infringement disputes and resulting in serious economic losses, casualties, and adverse social impact. The falling objects often block veneer tiles. The veneer layer is on the outside of the building, and the phenomenon of cracking, local, or large area hollow drum appears. When it is not treated in time, it continues to deteriorate under the action of environmental thermal expansion and contraction stress and rain erosion. Therefore, the analysis and study of the risk degree of the existing damage of external insulation systems under the action of the environment are conducive to property management and other relevant departments to improve the safety management and emergency management measures (Shi 2021; Tong 2021), which is conducive to the government to further carry out targeted prevention and control.

---

*Corresponding Author: 13609161448@163.com

DOI: 10.1201/9781003450832-9

For the current situation of exterior wall insulation systems, related scholars have carried out corresponding research. Ling Hongwei (Ling 2018) investigated the main defects of the exterior insulation system of 73 civil residential buildings in the Jiaxing area and found that the defects of the exterior insulation system were more likely to occur in the east and south, and the building gables were more likely to fall off. The scholar conducted a bond strength test on the bonding mode between the EPS board and base wall and found that the failure started from the part with a small bonding area, then developed to the part with a large connected cavity, and gradually fell off, which was consistent with the falling off the process of ceramic tile in reality.

The phenomenon of ceramic tile falling off has caused more scholars to think about and study the causes of the damage to the external insulation systems. Hu Changming (Hu 2018) simulated the temperature field and heat flux field of the external walls of buildings through ANSYS and found that the large temperature difference in winter would lead to adverse phenomena such as thermal bridges and condensation. Le Duong Hung Anh (Le A 2021) discussed the factors affecting the thermal conductivity of thermal insulation materials and proposed that temperature and moisture content were the main reasons for the failure of the external thermal insulation structure. Cheng Jie (Cheng 2018) studied the weather-ability of the external wall insulation system, calculated the temperature field, thermal stress field, and displacement distribution of different functional layers, and found that the thermal insulation mortar layer had a large displacement, and the interface mortar layer was subjected to a large stress. Gu Bin (Gu 2018) established a finite element model of the temperature effect of EPS external insulation system through ABAQUS and found that the temperature inside the insulation layer presented a linear distribution, and the range of variation was about equal to the temperature difference between the inner and outer surfaces of the wall.

It can be seen that the falling off of ceramic tiles is a thermodynamic coupling problem, but there are no studies on the damaged envelope structure to carry out further research. For urban high-rise buildings, due to the funding problem, the governance order should be formulated to deal with it in turn.

Based on this, this paper will conduct a thermodynamic coupling numerical simulation of the existing damage status of the external wall insulation structure, and study the damage of different degrees and levels of the envelope structure, to provide a reference for determining the risk degree, treatment sequence, and targeted treatment measures of the existing damaged wall.

2 HEAT TRANSFER THEORY

According to the second law of thermodynamics, in the absence of external work input, heat is always spontaneously transferred from the high-temperature region to the low-temperature region, and the heat transfer is caused by the temperature difference within or between objects. The three heat transfer pathways in heat transfer theory do not exist independently. In this study, heat conduction and radiation were considered for simulation.

In the natural environment, the heat exchange situation is considered as follows: the basic wall and the indoor temperature are kept constant, which is set as a temperature boundary constant, while the outer surface of the wall mainly considers the convective heat transfer, heat conduction, and radiation with the external natural environment (Tomás 2017). The three-dimensional temperature field of the insulation wall is simplified into a one-dimensional temperature field along the thickness of the wall, and the process is considered as one-dimensional transient heat conduction without an internal heat source whose temperature field changes with time. The differential equation is:

$$\rho c \frac{\partial T}{\partial t} = \lambda \frac{\partial T^2}{\partial^2 x}$$

Where T is the instantaneous temperature of the wall, unit °C; X is the coordinate along the wall thickness; t is time, unit h; ρ is density, unit kg / m^3; C is specific heat capacity, unit J / (kg °C); λ is thermal conductivity, unit W / (m °C)

3 BASIC ASSUMPTIONS AND THE COMPUTATIONAL MODEL

Based on the research of literature, this paper assumes that the enclosures are all infinite plates, and assumes isotropic, completely elastic, and object continuity. And according to the "Code for Thermal Design of Civil Buildings" (Liu 2019) in our country, boundary conditions for convection heat transfer on internal and external surfaces are determined. In the boundary conditions, according to the actual situation in a cold region, in summer, the convective heat transfer coefficient between outdoor air and wall surface is βo = 19 W/ (m^2·°C), and the set temperature between the wall inner surface and the indoor surface is 26°C constant, without considering the convective heat transfer between indoor air and wall inner surface. The external thermal insulation wall model of the external wall adopts a three-dimensional thermal coupled solid element, which is a hexahedron with 8 nodes. In the simulation conducted in this paper, constraints of the wall are considered, and fixed constraints are set around the model (Liu 2016).

In this paper, the gable wall with the size of 6 m * 3 m was simulated, and the existing damage of three kinds of walls with the ceramic tile layer running through the hollow drum, the EPS board running through the hollow drum, and the ceramic tile layer falling off locally were simulated. The initial defect was the tile hollow drum 4 cm, the EPS board hollow drum 4 cm, the initial defect of the ceramic tile falling off the hollow drum model was 0.6 cm*0.6 cm, and the hollow drum was set around the part of the ceramic tile falling off. The hollow drum is raised outwards 0.2 cm, in which the EPS board has three thickness sizes of 70 mm, 60 mm, and 50 mm. Considering that the most adverse load on the outer surface of the exterior wall is 70°C, the wall damage further develops under the summer environment, which increases from 26°C to 70°C after eight hours.

Applying the above assumptions and values, a finite element model in line with the actual engineering construction is established for the external insulation wall of the external wall. The wall thickness direction model is shown in Figure 1. In Figure 1, ①-the base wall, ②-the interface mortar layer, ③-the EPS board, ④-the anti-crack mortar layer, and ⑤-the ceramic tile. The finite element model of each wall is shown in Figure 2. The material parameters of each layer of the wall are shown in Table 1.

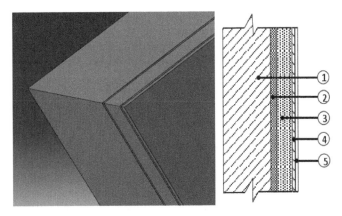

Figure 1. Wall thickness direction model.

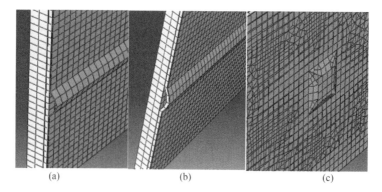

(a) (b) (c)

Figure 2. Finite element model.

Table 1. Physical properties of each layer material.

	Wall	Inter-facial mortar	EPS plate	Anti-crack mortar	Tile
Thickness (mm)	240	5	50/60/70	5	10
Density (Kg·m^{-3})	800	1800	30	1500	2100
Heat conductivity (W·(m·k)$^{-1}$)	0.54	0.93	0.042	0.85	0.1
Thermal expansivity (e^{-6} m·k^{-1})	8	12	60	12	10
Specific heat (J·(Kg·k)$^{-1}$)	1050	1050	1380	1050	876
Poisson ratio	0.2	0.28	0.1	0.25	0.25
Elasticity modulus (e^9 Pa)	8.8	4.9	0.0091	4.9	3.5

Note: The density, thermal conductivity, and specific heat capacity indexes of various materials are cited from Appendix B for the calculation parameters of the thermal-physical properties of building materials commonly used in document (Liu 2019).

4 MODEL CALCULATION AND OUTCOME ANALYSIS

4.1 Displacement analysis

The four working conditions of tile strip hollow drum, EPS plate strip hollow drum, tile falling hollow drum, and intact wall were simulated. The 8-hour heating process and 6-hour cooling process were set. The heating process was from 26°C to 70°C, and the cooling process was from 70°C to 20°C, and the displacement of the tile layer changed, as shown in Figure 3.

As can be seen from Figure 3, the maximum value of the wall displacement occurs in the heating up to 70°C, and the subsequent 8-hour warming process will be taken as the main research object.

4.1.1 Analysis of displacement results

The model as shown in Figure 2 was established to simulate four working conditions of tile strip hollow drum, EPS plate strip hollow drum, tile falling hollow drum, and intact wall. The summer heating process of external surface temperature rising from 26°C to 70°C for 8 hours was set, and the displacement changes of different functional layers were compared, as shown in Figure 4.

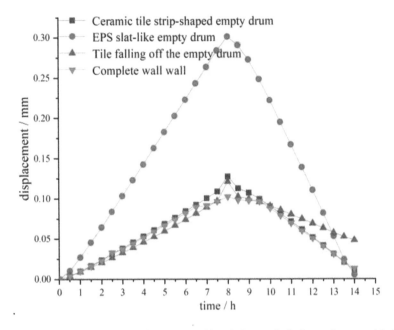

Figure 3. The curve of displacement of the external insulation wall tile layer changes with time.

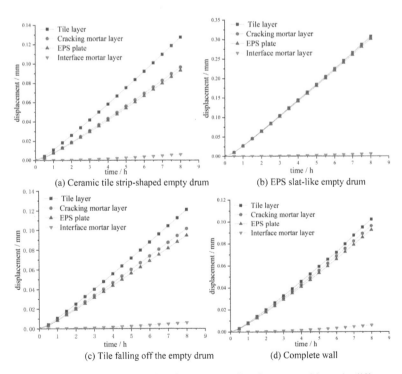

Figure 4. Displacement changes of the heating process of each structural layer in different states.

As shown in Figure 2(a), a strip hollow drum occurs in the tile layer. Figure 4(a) shows the displacement changes of each structural layer over time when the temperature rises in this state. This is in line with the actual life, and the tile structure located at the outside of the structure is the most prone to damage.

As shown in Figure 2(b), the EPS board of the enclosure structure has an empty drum, and the associated anti-crack mortar layer and tile layer have the same initial empty drum state. The simulation results are shown in Figure 4(b). The maximum displacement of ceramic tile reaches 0.307mm, and the maximum displacement of the anti-crack layer and EPS board is 0.304mm and 0.301mm respectively.

As shown in Figure 2(c), the ceramic tile layer partially falls off in the middle of the wall, and the ceramic tile layer surrounding the falling-off position is empty. The simulation results are shown in Figure 4(c). The maximum displacement of the ceramic tile layer reaches 0.121mm, and the maximum displacement of the anti-crack mortar layer and EPS board is 0.101mm and 0.095mm.

The simulation results of the complete wall are shown in Figure 4(d). The maximum displacement of the tile layer reaches 0.102 mm, and the maximum displacement of the crack-resistant layer and EPS board is 0.096 mm and 0.092 mm respectively.

The tile layer in the EPS board drum state will cause greater displacement, reaching 0.307 mm in the state of the tile drum, the maximum displacement of the tile layer is 0.127 mm in the state of the tile drum, and the maximum displacement of the tile layer is 0.121mm.

4.1.2 *Displacement analysis of different warming rates*
In different dimensions, the heating rate of the summer environment is different. To study the displacement changes of ceramic tile layer under different heating rates, 8 hours, 5.5 hours, and 4 hours (26°C to 70°C) are set respectively.

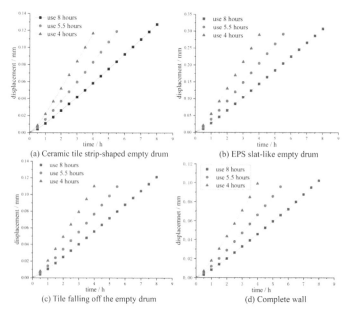

Figure 5. Tile layer displacement changes of different heating rates in the four situations. (a) Ceramic tile strip-shaped empty drum (b) EPS slat-like empty drum (c) Tile falling off the empty drum (d) Complete wall.

As shown in Figure 5, the ceramic tile hollow drum, EPS plate strip hollow drum, ceramic tile falling hollow drum, and the original envelope were compared and analyzed. It was found that the EPS hollow drum took 4 hours to realize the maximum slope of the heating process from 26°C to 70°C, which was twice the slope of the heating process when it took 8 hours. However, the final displacement of the ceramic tile layer after 8 hours of heating is greater than that after 4 hours of heating, which indicates that under the same heating conditions, the longer the heating time, the larger the final displacement of the ceramic tile layer will be. Under the empty state of the EPS board, the displacement of the ceramic tile layer after 8 hours of heating reaches 0.307mm.

4.2 *Temperature field and stress field analysis*

To study whether different damage states will affect the temperature difference of different structural layers, the ceramic tile hollow drum, EPS plate strip hollow drum, ceramic tile falling hollow drum, and the original enclosure structure are heated from 26°C to 70°C for 8 hours, and the temperature distribution along the wall section is analyzed. The foundation wall is 0, the thickness of the anti-crack mortar layer is 5 mm, and the thickness of the EPS plate is 50 mm. The thickness of the interface mortar layer is 5 mm and the thickness of the tile layer is 10 mm. The temperature distribution is shown in Table 2 below.

Table 2. Distribution of wall temperature along the section.

	Tile empty drum (°C)	EPS empty drum (°C)	Fall off and empty drum (°C)	Master mold (°C)
0 mm	30.62	29.93	27.37	30.64
5 mm	30.63	30.02	29.43	30.80
55 mm	66.33	68.98	66.72	66.65
60 mm	66.55	69.02	66.99	66.87
70 mm	69.41	69.21	67.42	67.39

As can be seen from the table, under the empty drum state of the EPS board, the outer temperature of the EPS board reaches 68.98°C, the inner temperature reaches 30.02°C, and the temperature difference is 38.96°C. The temperature difference between the empty drum state of ceramic tile, the empty drum state of ceramic tile, and the original model is 35.70°C, 37.29°C, and 35.85°C, respectively. When the tile layer is empty, the temperature of the outer layer is 69.41°C, the temperature of the inner layer is 66.55°C, and the temperature difference between the inside and outside of the tile layer is 2.86°C. The temperature difference between the empty EPS board, the empty tile falling off, and the tile layer of the original model is 0.19°C, 0.43°C, and 0.52°C respectively.

The analysis shows that when the EPS board is empty, the temperature difference between the inside and outside of the EPS board is the largest in this damaged state, and the temperature difference between the inside and outside of the EPS board is 38.96°C. When the ceramic tile layer is empty, the temperature difference between the inside and outside of the ceramic tile layer is the largest in the damaged state, and the temperature difference is 2.86°C.

In these damaged states, the displacement of the tile layer reached 0.307mm under the condition of an EPS strip hollow drum. To study the influence of EPS plate thickness on the tile layer, the change of EPS plate thickness under the condition of an EPS strip hollow drum was simulated, and the 8-hour heating process was set, from 26°C to 70°C.

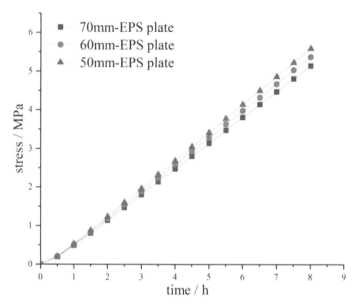

Figure 6. Stress distribution of ceramic tile layer with different plate thicknesses.

When the thickness of the EPS plate is 70 mm, the maximum stress at the damaged part of the ceramic tile layer is 5.14 MPa; when the thickness of the EPS plate is 60 mm and 50 mm, the maximum stress is 5.37 MPa and 5.58 MPa. It can be seen that under the condition of an EPS plate strip empty drum, the smaller the thickness of the EPS plate is, the higher the stress value of the tile layer, the more dangerous it is.

5 CONCLUSIONS

By simulating the heating process of the intact wall, wall tile strip hollow drum, EPS plate strip hollow drum, and tile falling hollow drum in 8 hours, the temperature rise process of 26°C to 70°C was achieved, and the different displacements of each functional layer under uniform heating and different heating rates were compared. At the same time, the temperature simulation of 50 mm, 60 mm, and 70 mm thick EPS plates under the condition of a strip empty drum of EPS plate was carried out, and the temperature field, stress field, and displacement field were analyzed, and the following conclusions were drawn:

(1) The original strip hollow drum defect of the EPS board will cause a large displacement of the tile layer. When the temperature reaches 70°C, the tile layer displacement reaches 0.307mm, which is much larger than the tile layer displacement under the condition of tile strip hollow drum and falling hollow drum, and the risk is greater;
(2) The higher the heating rate is, the greater the final displacement of the ceramic tile layer will be. The longer the heating time will be, the larger the final displacement of the ceramic tile layer will be. Under the empty drum condition of the EPS board, the tile layer displacement reaches the maximum of 0.307mm after heating for 8 hours.
(3) Under the initial condition of the strip hollow of EPS board, the stress of the tile layer of 50 mm thick board is 5.58 MPa, which is greater than that of 60 mm and 70 mm thick EPS board. When the initial state is a strip hollow of an EPS board, the smaller the thickness of the EPS board, the greater the stress value of the tile layer, and the more dangerous.

(4) The temperature analysis shows that the temperature difference between the inner and outer layers of the ceramic tile caused by the strip hollow drum is 2.86°C, which is larger than the temperature difference between the inner and outer layers of the ceramic tile caused by the strip hollow drum; it is larger than the temperature difference between the inner and outer layers of the ceramic tile caused by the strip hollow drum and the outer layer of the ceramic tile; it is larger than the temperature difference between the ceramic tile caused by the strip hollow drum and the inner and outer layer of the ceramic tile; it is larger than the temperature difference between the ceramic tile and the EPS board under the existing damaged state, which is 38.96°C. This will result in bad heat transfer, eventually, leading to ceramic tile layer damage and fall-off.

REFERENCES

Cheng Jie, Xu Xinhua, Zhang Xiaoyingi. Numerical Simulation of the Weathering Performance of an Exterior Wall External Insulation System Under Heating-cooling Cycles. [J]. *Journal of Engineering Science*, 2018, 40(06):754–759.

Gu Bin, Xue Tao, Xie Fuzhe. Study on the Temperature Field and Temperature Stress of the EPS External Thermal Insulation System[J]. *Sichuan Building Science Research*, 2018, 44(01):102–107.

Hu Changming, LI Xiang, HE Xiaoshan. Finite Element Analysis of the Temperature Field of the Insulated Wall Panel[J]. *Building Science*, 2018, 34(02).

Ling Hongjie, Xiong Houren, Yu Anni. Statistical Survey of the Degradation of Finishing Coating for External Thermal Insulation Composite Systems in Walls[J]. *New Type Building Material*, 2018, 45 (12):114–118.

Le Duong, Hung Anh, Zoltán Pásztory. An Overview of Factors Influencing Thermal Conductivity of Building Insulation Materials, *Journal of Building Engineering*, Volume 44, 2021

Liu Siqi. *Research on Reinforcement Technology of Polystyrene Board Thin Plastered Exterior Insulation System[D]*. Shandong Jianzhu University, 2019.

Liu Yongjian, LI Nan, Wang Qiang. Simulation Study on Temperature Stress of XPS Board Exterior Insulation System in the Cold Region [J]. *Building Energy Efficiency*, 2016, 44(9):38–41.

Shi Juan, ZHENG Peng, CHANG Dingyi. Governance of Urban Public Safety in the Context of Big Data: Blockchain Technology Enablement [J]. *China Safety Science Journal*, 2021, 31(02): 24–32.

Tong Ruipeng, Wang Lulu, LI Hongwei. Discussion on the Relationship Between Safety Management, Risk Management, and Emergency Management Based on the Macro-security View[J]. *China Safety Science Journal*, 2021, 31(05): 36–44.

Tomás Lourenço, Luís Matias, Paulina Faria, Anomalies Detection in Adhesive Wall Tiling Systems by Infrared Thermography, *Construction and Building Materials*, Volume 148, 2017, Pages 419–428

Zhang Chunli. *Study on the Applicability of Building Wall Insulation Structure in Zhejiang Province [D]*. Zhejiang University of Technology, 2019.

Zhang Taiyong. Relationship between Energy Saving and Ecological Environment of New Urban Residential Building[J]. *Environmental Science and Management*, 2021, 46(07):35–39.

Water Conservancy and Civil Construction – Oke & Ahmad (Eds)
© 2024 The Author(s), ISBN: 978-1-032-58618-2

Study on properties of permeable brick with recycled aggregate from construction waste

Xiaoya Huang, Junxia Sun, Ting Hu, Yan Zhang & Limei Wang
Chongqing College of Architecture and Technology, Chongqing, China

ABSTRACT: In this paper, a new type of permeable brick with recycled aggregate from construction waste is prepared with recycled aggregate, cement, and fly ash as raw materials. By simplifying the preparation process, a reliable reference is provided for the optimal design of permeable bricks. We select an orthogonal test to systematically study the effects of recycled aggregate gradation, the content of fly ash, and aggregate-binder ratio on the compressive strength and permeability coefficient of permeable bricks. The research results showed that the performance of permeable bricks prepared with recycled aggregate from construction waste can meet the requirements of Floor Tiles and Permeable Bricks with Recycled Aggregate (JC/T 400-2012), and have excellent mechanical properties and water permeability. The compressive strength of permeable brick decreases with the increase of recycled coarse aggregate content, fly ash content, and aggregate-binder ratio. The permeability coefficient increases with the increase of recycled coarse aggregate content, decreases with the increase of fly ash content, and increases with the increase of aggregate-binder ratio. This research will be helpful to the reuse of recycled aggregate and the popularization of permeable bricks.

1 INTRODUCTION

In China, the construction waste produced in urban and rural construction has seriously endangered the sustainable development of society and the ecological environment every year. The construction waste mainly consists of concrete and bricks. If the construction waste can be broken into recycled aggregates (RA), it can then be made into a new pavement material — permeable bricks with recycled aggregate. On the one hand, it can reduce not only the amount of natural aggregate in the preparation of permeable brick which is a new pavement material but also the amount of cement, solving the environmental problems caused by construction waste to a certain extent, and reducing the production cost of permeable brick. On the other hand, it can not only realize the construction of a sponge city and beautiful countryside but also vigorously promote the resource utilization of construction waste.

Permeable brick as a pavement component has many gaps inside of it, which makes them have excellent functions of air permeability and drainage and leads to a significant reduction in the strength and durability of permeable bricks. Therefore, the focus of scientific researchers' study was always to ensure the permeability of permeable bricks, improve their strength and durability, and reduce the complexity of their preparation process and production costs.

At present, many research achievements have been made on RA and permeable brick. Li W (Li 2007) used 0.63~1.25 mm, 1.25~2.5 mm, and 2.5~5 mm aggregates to prepare permeable bricks. The research found that with the increase in particle size, the strength of permeable bricks first increased and then decreased. Yang G et al (Yang 2012) prepared permeable bricks by mixing 0.63~2.36mm and 2.36~4.75mm aggregates, and studied their properties. Wang Z J et al (2019) made permeable bricks with 4.75~9.5 mm recycled coarse

74

DOI: 10.1201/9781003450832-10

aggregate (RCA) as raw materials. Through relevant experiments, it was found that with the increase of target porosity, the permeability coefficient of permeable bricks increased, but the flexural and compressive strength showed a significant downward trend. Yan et al (2019) prepared permeable bricks with fly ash and steel slag as the main raw materials. They found that the strength of permeable bricks decreased with the increase in the content of fly ash, but the permeability coefficient increased significantly. When the steel slag content was 6%, the content of fly ash was 30%, and the porosity design goal was 20%, the compressive strength and permeability coefficient of permeable bricks could reach a good level. Ye W S et al (2018) made a detailed study on the water-retention property of permeable bricks and found that the water-retention property of permeable bricks increased with the increase of coarse aggregate. Zhao X et al (2016) used waste concrete to prepare permeable bricks. Research shows that with the increase in the content of fly ash, the strength increases, and the permeability coefficient decreases. Yuan (2018) carried out a life cycle impact assessment and life cycle cost calculation for permeable bricks and found that the proportion of the proportioning process of permeable bricks in the cost was up to 78.13%. Reducing cement consumption and improving the recovery rate of basalt powder and crushed stone can effectively reduce the environmental impact and production cost of the whole process. When making permeable bricks, the consumption of cement and crushed stone is reduced by 5%, the overall environmental impact is reduced by about 2.21% and 1.79%, and the production cost is reduced by 1.02 yuan and 0.15 yuan respectively. In addition, the use of additives to reduce cement consumption is expected to reduce economic costs and environmental impact. Li Z et al (2018) explored the effects of high carbon fly ash content, forming pressure, and sintering temperature on the permeability and mechanical properties of permeable bricks through experiments. The results showed that the increase of high carbon fly ash content would increase the permeability and water absorption of permeable bricks, and reduce the density and splitting tensile strength. Through experiments, Zhang Z Q (Zhang & Liu 2017) found that the preparation of pervious concrete by adding an appropriate amount of recycled clay brick aggregate into the natural aggregate has little effect on its permeability. But when the volume percentage of recycled clay brick aggregate in the total aggregate exceeds 40%, it will greatly influence the mechanical properties and durability of previous concrete. Zuo F Y et al (Zuo & Sun 2008) believed that the optimal forming pressure of permeable bricks was 4 MPa. When the pressure holding time reached 90 seconds, the strength was maintained at a certain value.

At present, many achievements have been made by the research on permeable bricks, but traditional permeable bricks have large porosity, and small cohesion between aggregate particles, and the recycled aggregate itself has certain physical and mechanical performance defects, under the premise of meeting the requirements of permeability, which makes the permeable bricks with recycled aggregate have low strength and poor durability. It is difficult to meet the road requirements. Aiming at a series of existing problems, we use a recycled aggregate of 1.18~2.36 mm, 2.36~4.75 mm, and 4.75~9.50 mm with certain aggregate grading, the aggregate-binder ratio of 2.8, 3.0, and 3.2, and fly ash content of 25%, 35%, and 45%, and the static pressure forming process to prepare permeable brick with recycled aggregate from construction waste. Under the condition of a 100% replacement rate of recycled aggregate, the effects of recycled aggregate gradation, the content of fly ash, and aggregate-binder ratio on the basic performance of permeable brick is studied, which will help the reuse of recycled aggregate and the popularization of permeable brick.

2 THE RAW MATERIALS

(1) Cement. P·O 42.5 cement is produced by Chongqing Lafarge Cement Plant, which is tested according to General Portland Cement (GB175-2007). Table 1 is the test results of the basic performance of cement.

Table 1. The basic performance of cement.

Cement type	Cement density (g/cm^3)	Initial setting time (min)	Final setting time (min)	Bending strength (MPa) 3d	Bending strength (MPa) 28d	Compressive strength (MPa) 3d	Compressive strength (MPa) 28d	Volume stability
P·O 42.5	3.13	165	255	4.8	7.2	19.4	46.5	Qualified

(2) Aggregate. Table 2 lists the test results of the performance of recycled aggregate.
(3) Fly ash. Fly ash with Grade II.
(4) Water reducer. Naphthalene series superplasticizer, which water reduction rate is 25%.
(5) Water. Tap water.

Table 2. Performance test of recycled aggregate.

Recycled aggregate size (mm)	Water absorption (%)	Water content (%)	Bulk density (kg/m^3)	Apparent density (kg/m^3)
4.75 to 9.5	3.5	2.32	1390	2380
2.36 to 4.75	3.8	2.64	1280	2310
1.18 to 2.36	3.6	2.53	1340	2340

3 MIX DESIGN

3.1 Preparation process

During the preparation of permeable brick, we used the slurry coating method similar to the concrete mixing process of the cement sand gravel method, which is used to evenly wrap the cementitious materials on the surface of recycled aggregate. The specific preparation process is shown in Figure 1.

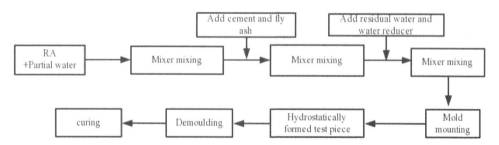

Figure 1. Specific preparation process.

3.2 Mix proportion design of permeable brick test

Based on the existing research, the recycled aggregate gradation, the content of fly ash, and aggregate-binder ratio are selected as orthogonal test variables, and we designed an orthogonal test with three factors and three levels to study the specific effects of each factor on compressive strength and permeability coefficient of permeable bricks. When forming the piece of permeable brick with recycled aggregate, we select 4.0MPa forming pressure, take 1% water reducer dosage, adjust the mixing water consumption according to the workability requirements of the mixture, and steam curing for 10h. The level of each factor is shown in Table 3, and the design of the test mix proportion is shown in Table 4.

Table 3. Horizontal factors.

Horizontal	Recycled aggregate grading (4.75~9.5mm: 2.36~4.75mm: 1.18~2.36mm)	Content of fly ash (%)	Aggregate-binder ratio
I	8:2:0	25	2.8
II	6:3:1	35	3.0
III	4:4:2	45	3.2

Table 4. Test mix design.

Number	Recycled aggregate grading (4.75~9.5mm: 2.36~4.75mm: 1.18~2.36mm)	Content of fly ash (%)	Aggregate-binder ratio
A	8:2:0	45	2.8
B	8:2:0	25	3.0
C	8:2:0	35	3.2
D	6:3:1	35	2.8
E	6:3:1	45	3.0
F	6:3:1	25	3.2
G	4:4:2	25	2.8
H	4:4:2	35	3.0
I	4:4:2	45	3.2

4 TEST RESULTS AND ANALYSIS

4.1 *Effects of different factors on compressive strength of permeable brick with recycled aggregate*

It can be seen from Figure 2 that the effects of recycled aggregate grading on the compressive strength of permeable brick are very significant. With the increase of recycled coarse aggregate, the compressive strength of permeable brick decreases gradually. When the recycled coarse aggregate content increases from 40% to 60%, the compressive strength of permeable brick with recycled aggregate decreases to a certain extent, but the compressive strength of permeable brick with recycled aggregate is greater than 30MPa. When the recycled coarse aggregate content is increased to 80%, the compressive strength of permeable brick will be reduced to below 30MPa. As the internal structure of permeable brick is loose, its strength formation mainly depends on the cohesive force between recycled aggregates. The recycled coarse aggregate mainly plays the role of the skeleton in permeable bricks, while the recycled fine aggregate mainly plays the role of filling, transferring, and dispersing external forces. The recycled fine aggregate can fill the gap between the recycled coarse aggregate and increase the bonding points between the aggregates, thus showing good mechanical properties. However, with the increase of recycled coarse aggregate content, the internal pores of permeable bricks increase, and the contact points between recycled coarse

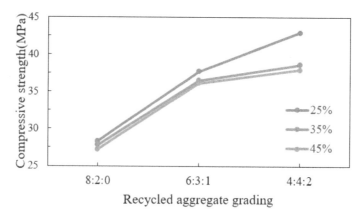

Figure 2. Effects of recycled aggregate grading on compressive strength of permeable brick.

aggregates decrease, which cannot transfer external forces better. Therefore, as the recycled coarse aggregate content continues to increase, its compressive strength decreases significantly.

It can be seen from Figure 3 that the compressive strength of permeable brick decreases slightly with the increase of the content of fly ash. Under the same aggregate gradation, the content of fly ash has little effect on the compressive strength of permeable brick. The content of fly ash increases from 25% to 45%, and the compressive strength of permeable brick decreases only by 1 to 5 MPa. This is mainly due to the low activity of fly ash, which mainly reacts with the hydration product $Ca(OH)_2$ of cement. However, with the increase of the content of fly ash, the cement content decreases, and the later hydration products decrease, which will lead to the strength decrease.

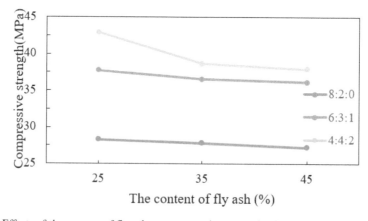

Figure 3. Effects of the content of fly ash on compressive strength of permeable brick.

From Figure 4, we can see that the compressive strength of permeable brick with recycled aggregate decreases with the increase in aggregate-binder ratio. This is because the aggregate-binder ratio will directly affect the thickness of the slurry coated on the surface of recycled aggregate. The aggregate-binder ratio will directly affect the slurry coating thickness on the surface of recycled aggregate. When the aggregate-binder ratio is small, there are

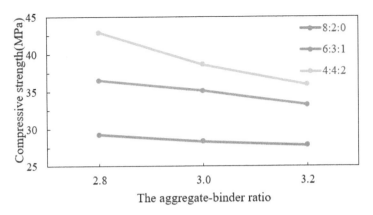

Figure 4. Effects of aggregate-binder ratio on compressive strength of permeable brick.

more cementitious materials, the thickness of the slurry wrapped with recycled aggregate is large, and the bonding strength between the recycled aggregates is high. However, with the increase of aggregate-binder ratio, the amount of cementitious material decreases, the thickness of slurry wrapped with recycled aggregate decreases, the bonding strength between recycled aggregates decreases, the resistance to external forces decreases, the pores increase, and the compressive strength decreases.

4.2 *Effects of different factors on the permeability coefficient of recycled aggregate permeable brick*

It can be seen from Figure 5 that under the effects of recycled aggregate grading, the permeability coefficient gradually increases with the increase of recycled coarse aggregate content, and the permeability coefficients of permeable bricks with recycled aggregate are greater than 1×10^{-2}cm/s. When the amount of recycled coarse aggregate is small, a large amount of recycled fine aggregate can fill the gap between recycled coarse aggregates, thus reducing the formation of internal connecting pores of permeable bricks, forming a dense structure inside the permeable bricks, and resulting in a low permeability coefficient. With the increase of recycled coarse aggregate, the content of recycled fine aggregate decreases, and the internal pores of permeable bricks increase. When the recycled coarse aggregate reaches 80%, there is a lack of sufficient recycled fine aggregate to fill the internal pores, and

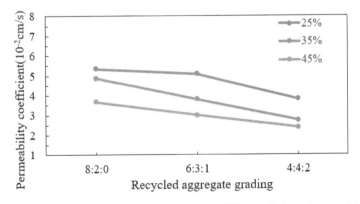

Figure 5. Effects of recycled aggregate grading on permeability coefficient of permeable brick.

the gap between the coarse aggregates further increases. The internal pores of permeable bricks are large and many, and it is easier to form connecting pores, which shows an increase in the permeability coefficient.

It can be seen from Figure 6 that with the increase of the content of fly ash, the permeability coefficient gradually decreases, and the decline range of the permeability coefficient gradually increases. Because the density of fly ash is lower than cement, when it replaces cement with equal quality, the volume of cementitious material increases, and the thickness of mortar coating on the aggregate surface increases, then preventing the formation of connected pores. Due to the ball effect of fly ash, the friction of the mixture can be reduced, and the fluidity of the slurry can be increased. The recycled fine aggregate and slurry are easier to fill the pores in the permeable brick, making the permeable brick more densely, then the permeability coefficient of the permeable brick decreases.

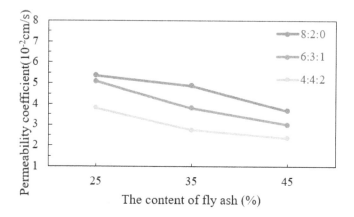

Figure 6. Effects of the content of fly ash on the permeability coefficient of permeable brick.

It can be seen from Figure 7 that the permeability coefficient increases with the increase of the aggregate-binder ratio. During the preparation of permeable brick by static pressure forming, the recycled aggregate is in close contact, and the cementitious material wrapped with recycled aggregate is pressed into the internal pores of permeable brick under the effect of extrusion force. When the aggregate-binder ratio is small, there are more cementitious materials, the thickness of the slurry wrapped with recycled aggregate is thick, and a dense

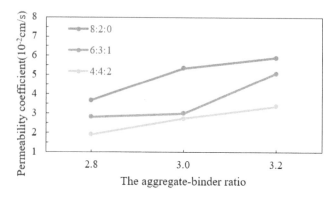

Figure 7. Effects of aggregate-binder ratio on permeability coefficient of permeable brick.

structure is formed inside of the permeable brick, so the permeability coefficient is low. With the increase of aggregate-binder ratio, the amount of cementitious material decreases, the thickness of the slurry wrapped with recycled aggregate is thin, the amount of cementitious material in the pores of permeable bricks decreases, and the friction between recycled aggregates increases. It is not easy to move, and larger pores can be formed, which increases the permeability coefficient.

5 CONCLUSION

This study shows that the performance of permeable bricks prepared with recycled aggregate from construction waste can meet the requirements of Floor Tiles and Permeable Bricks with Recycled Aggregate (JC/T 400-2012). The compressive strength of permeable brick decreases with the increase of recycled coarse aggregate content, fly ash content, and aggregate-binder ratio. The permeability coefficient increases with the increase of recycled coarse aggregate content, decreases with the increase of fly ash content, and increases with the increase of aggregate-binder ratio.

Recycled brick aggregate is used as the main raw material to form permeable bricks, which makes full use of the characteristics of large porosity and strong water absorption of brick aggregate, so the permeable bricks with recycled aggregate have excellent water permeability and corresponding mechanical properties, meet the needs of practical projects, have strong practicability, and have great ecological significance.

ACKNOWLEDGMENTS

Source: Science and Technology Research Program of Chongqing Municipal Education Commission in 2022 (KJQN202205204), and Chongqing Municipal Education Commission - Research Project on Teaching Reform of Higher Vocational Education in 2021(Z212061).

REFERENCES

Li W. (2007) Development of High-strength Concrete Permeable Brick. *Building Blocks and Block Building*, (2): 25–27.

Li Z., Wang N.J., Zhao L., *et al.* (2018) Preparation and Performance Study of High Carbon Fly Ash Permeable brick. *Concrete*, (03): 118–120.

Wang Z.J., Yang C., Weng D.Q., *et al.* (2019) Study on the Effects of Target Porosity on the Performance of Recycled Permeable Bricks. *New Building Materials*, 46(4): 51–54.

Yan L., Wei T., Rajendra P.S. (2019) Study on Compressive Strength and Water Permeability of Steel Slag -Fly Ash Mixed Permeable Brick. *Applied Sciences*, 9(8):1542–1552.

Yang G. (2012) Experimental Study on the Effect of Mix Proportion and Admixture on the Performance of Permeable brick. *Brick and Tile*, (2): 17–18.

Ye W.S., Zhang R.J., Nie L.W. (2018) Study on Preparation and Properties of Recycled Aggregate Water Retaining Permeable Bricks for Sponge Cities. Proceedings of 2018 International Green Building and Building Energy Efficiency Conference. 699–702.

Yuan X.L., Tang Y.Z., Li.Y., *et al.* (2018) Environmental and Economic Impacts Assessment of Concrete Pavement Brick and Permeable Brick Production Process - A Case Study in China.

Zhang Z.Q., Liu Y.X. (2017) The Effects of Recycled Clay Brick Aggregate Content on the Performance of Pervious Concrete. *Concrete*, (8): 123–127.

Zhao X., Liu H.S., Li G.Z. (2016) Research on Technology and Performance of Construction Waste Permeable Brick. *Brick and Tile*, (10): 32–34.

Zuo F.Y., Sun K.W. (2008) Research on the Technology of Preparing Permeable Bricks from Waste Concrete. *Brick and Tile World*, (2): 39–40.

Water Conservancy and Civil Construction – Oke & Ahmad (Eds)
© 2024 The Author(s), ISBN: 978-1-032-58614-4

Analysis of ships vibration causes and application of active vibration control technology

Xiaohai Qi
CGN Research Institute Co., Ltd, Guangdong Shenzhen, China

Jincai Huang*
Sinomach Intelligence Technology Co., Ltd, Guangdong Guangzhou, China

Hao Wang
Sinomach Academy of Science and Technology Co., Ltd, Beijing, China

Xun Zhao
Suzhou Nuclear Power Research Institute, Jiangsu Suzhou, China

Wangxuan Zhang
CGN Research Institute Co., Ltd, Guangdong Shenzhen, China

ABSTRACT: With the continuous development of maritime traffic, the research on the stability and comfort of various types of ships at sea has been deepened. When large ships are sailing at sea, they are not only affected by the vibration of their internal equipment but also may encounter the problem of hull shaking and vibration caused by natural environmental factors such as sea waves. Therefore, targeted vibration solutions are required. By analyzing the vibration causes of ships, the problems and hazards that the ships may encounter in the design and navigation stages are considered. After comparing the principle of different types of vibration reduction technologies, characteristics of vibration reduction devices, and vibration reduction effects, ship stabilization technology based on air-floating active control devices is proposed, and the development trend of ship damping devices in the future is predicted.

1 INSTRUCTION

With the vigorous development of maritime traffic and the continuous exploitation of Marine resources, the design level, and navigation technology of large ships are also developed and increasingly mature. Vibration is a common and critical problem for ships. Vibration has many adverse effects on the normal navigation of ships. And the research on vibration control related to ships has become a hot spot in the field of ships. Zou et al. (2003) determined the main factors affecting the vibration of ship structure through numerical analysis of ship structure vibration, providing a new way to consider vibration prediction in the design stage of ships. Yang (2017) studied the ship vibration and noise induced by the fluctuating pressure of the propeller on the ship and summarized the vibration characteristics of the dynamic influence of the propeller. Wang et al. (2021) sorted out the difficulties in the application of active control technology in the suppression of low-frequency vibration and

---

*Corresponding Author: acwanghao@163.com

DOI: 10.1201/9781003450832-11

noise of ships. Hong & Chen (2012) summarized the status and development trend of the overall ship anti-roll technology.

The beginning of the paper clarifies the various factors that are easy to cause ship vibration, analyzes the vibration characteristics and hazards of various vibration sources, then sorts out the commonly used anti-rolling devices and vibration control methods for ships. Finally, a new technology for ship stability based on the active control device of air flotation is proposed.

2 CAUSES AND HAZARDS OF SHIP VIBRATION

2.1 *Vibration problems in naval ship design*

In the phase of ship design, the influences of the following factors on ship vibration should be comprehensively considered:

(1) Hull shape and the vibration frequency and vibration form in free vibration state;
(2) The excitation force causes hull vibration, to determine the excitation size and frequency of the ship's power equipment;
(3) Free vibration frequency, dynamic displacement, and dynamic load stress of main structures in the hull;
(4) Vibration shape curve and vibration acceleration in free vibration state.

2.2 *Vibration of ships during navigation*

Ships may encounter violent vibrations or even a rocking state when sailing on the sea, and the main reason is that the ship will be affected by internal vibration sources and external vibration sources in the process of moving (Zhao 2001). The internal vibration source refers to the vibration and noise generated during the operation of power equipment in the ship, such as the vibration generated during the operation of the Marine gearbox, generator, air compressor, and other power machinery. The external vibration source refers to the impact of the external natural environment factors on the whole ship during the voyage, mainly the impact of sea wind and sea waves, especially when the ship is attacked by huge waves. The enormous force causes the hull to wobble with multiple degrees of freedom (Ma 2006).

2.3 *The internal vibration sources*

2.3.1 *High power and heavy-duty gearbox*
The high-power and heavy-duty gearbox mainly play the functions of clutch, deceleration, and propeller thrust during operation. Its vibration and noise are mainly generated by high-speed dynamic meshing when the gear pair is running. The main transmission path is as follows. The driving force drives the gear meshing to generate exciting force to make the gear pair vibration itself, and then transmits to the bearing and bearing support through the drive shaft, and finally to the gearbox and its base position to cause violent vibration. The dynamic impact generated inside the heavy-duty gearbox will seriously affect the working performance of the transmission system, mainly including the system stability and the normal working environment of the equipment, and easily shorten the service life of the components inside the gearbox (Cheng 2018).

2.3.2 *Marine diesel generator set*
The diesel generator set equipment itself has a complex structure, more vibration source points, a wide frequency band, and various forms. The vibration is generally divided into three categories. The first type is the whole machine vibration, which is mainly generated by the inertia force, inertia moment, and overturning moment of the crank connecting rod. The

second type is structural elastic vibration, mainly aerodynamic and inertial force on the piston, connecting rod, and other elastic structural parts of the role. The third type is the torsional vibration of the shafting of the generator set. Excessive vibration of the Marine diesel generator set will lead to fatigue damage of hull and mechanical equipment, not only damaging the structural strength of the diesel set itself but also directly affecting the service performance of the equipment on board, such as reducing the measurement accuracy of instruments and sensitive devices and causing the malfunction of mechanical and electrical equipment in the ship (Shen 2015).

2.3.3 *Marine air compressor*

The air compressor provides the air source power for the power equipment and converts the mechanical energy into gas pressure energy. During the operation of the air compressor, due to the influence of the eccentric mass and the reciprocating movement of the piston, the air compressor produces an unbalanced inertia force, which is transmitted to the main body of the equipment through the main bearing and causes strong vibration. However, the unbalanced moment generated by dynamic eccentricity makes the equipment tend to topple side to side (Tian et al. 2001). The Marine air compressor has a high speed, and the vibration exceeding the standard is easy to cause fluctuation in the working pressure of the air compressor, leading to unstable working performance and even damage to the instrument and electrical parts (Liu 2015).

2.4 *The external vibration sources*

The characteristics of ship vibration and shaking caused by external vibration sources, such as sea wind and waves, are random and nonlinear. The randomness is mainly reflected in the randomness of the Marine natural environment. The vast majority of sea waves are irregular waves. The main reasons that affect the nonlinear characteristics of ship roll are as follows: nonlinear recovery moment, nonlinear damping moment, the nonlinear additional moment of inertia, and nonlinear wave disturbance moment (Li 2007).

Violent roll will not only cause physical discomfort to the crew on board, such as excessive fatigue or seasickness but also affect the normal operation of the important equipment ships and the fixing of articles ships. When the swing amplitude of the ship exceeds the equilibrium point it can bear, the ship will be damaged as a whole, and even capsize and sink in serious cases.

3 APPLICATION OF SHIP VIBRATION CONTROL TECHNOLOGY

3.1 *Vibration control of ship internal power equipment*

The configuration of the power equipment inside the ship is complicated, and the equipment is composed of relatively much more moving parts. Therefore, the vibration and noise sources are large in number and intensity, and the spectrum components of vibration signals are complex. Given vibration hazards of ship internal power equipment, common vibration and noise reduction measures mainly include:

3.1.1 *Single- and double-layer vibration isolation devices*

Single-layer vibration isolation device refers to the use of a group of vibration isolators as support, the target power equipment, and the ship base is separated, using the vibration isolation function of the vibration isolator to reduce the excitation force transmitted by the power equipment to the hull base. R. Plunkett (1998) found through tests that the single-layer vibration isolation device has a good effect in the middle and low-frequency segments, but a poor effect in the high-frequency segment. And its energy loss to the power equipment is generally about 20 dB, which is the most basic form of vibration isolation device.

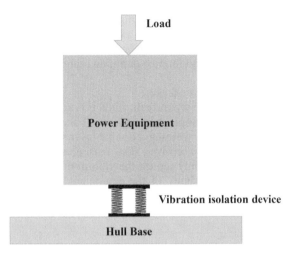

Figure 1. Structure diagram of single-layer vibration isolation device.

The double-layer vibration isolation device is to first fix the power equipment to a middle base through the upper vibration isolator, and then connect the middle base with the bottom vibration isolator to form the upper and lower double vibration isolation layer. The double-layer vibration isolation device has a more significant vibration isolation effect than the single-layer vibration isolation device. According to the experimental results, the effect of the double-layer vibration isolation device is 10 dB~20 dB better than that of the single-layer vibration isolation device. Among these, when the mass of the middle base is 40%~100% of the top power equipment, it generally can achieve a better vibration isolation effect. The double-layer vibration isolation device is mainly used for generator sets, ventilator sets, pump equipment, etc. (Wei et al. 2004).

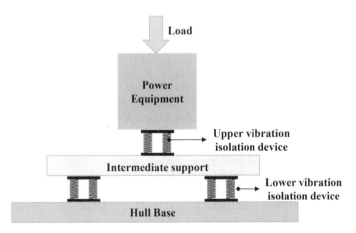

Figure 2. Structure diagram of double-layer vibration isolation device.

3.1.2 *Damping vibration isolation device*
A damping vibration isolation device is by using the hysteretic energy dissipation characteristics of viscoelastic materials to consume the energy transmitted through vibration, reduce vibration and noise, and reduce the dynamic response of the structure. For the power

equipment on the ship, the damping vibration isolation device is usually installed on the base of the supporting equipment. According to the experimental data, in the middle and high-frequency bands, the equivalent acceleration stage can be reduced by 10~20 dB for the base with both vertical and horizontal and interior damping materials.

3.1.3 *Floating raft vibration isolation device*

Floating raft vibration isolation is to separate the main power equipment and other auxiliary equipment on the ship, so that multiple host equipment shares an intermediate platform, namely the floating raft frame, thus forming a double-layer vibration isolation system with multiple equipment and multiple incentives. The reasonable design of the floating raft vibration isolation device can not only reduce the vibration and noise of the power equipment and improve the working stability of the equipment but also centrally arrange the equipment to save space. Experiments show that the floating raft vibration isolation device can reduce the vibration acceleration level of the power equipment base by 20~40 dB.

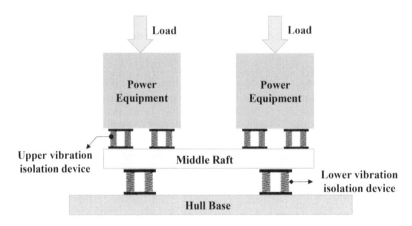

Figure 3. Structure diagram of floating raft vibration isolation device.

3.2 *Vibration control of the whole ship*

For the overall ship structure, to find measures to reduce hull rocking, researchers have applied more than 20 kinds of rocking reduction devices in practice. The most important and widely used technologies include bilge keel, stabilizer fin, damping tank, and rudder anti-roll, most of which are passive anti-roll technologies (Yang et al. 2002).

3.2.1 *Bilge keel*

The bilge keel is a longitudinal member mounted on the outside of the bilge column and perpendicular to the outside of the bilge plate along the captain's direction. The structure mainly consists of a bilge keel body and a transition part connected with the hull. The bilge keel can change the distribution of water around the ship when it rolls and reduces the roll by increasing the additional mass moment of inertia of the roll. To obtain relatively large damping during the swing, the optimal position of the bilge keel should be the oblique line from the intersection of the hull half-width line and the baseline to the ship's center of gravity point.

The bilge keel has the features of simple structure, convenient installation, and low cost, and can work effectively under any circumstances, with a rolling reduction efficiency of about 20%-25%. Bilge keels have been installed in almost all types of ships, and have become indispensable parts of the ship. (Dong et al. 2008)

3.2.2 *Stabilizer fin*

A fin stabilizer, also known as a side rudder, is an active fin stabilizer mounted on both bilges underwater in the hull. The structure mainly includes three parts: the wing fin, the rotating fin transmission device, and the control system. The working principle of the stabilizer fin is that when the ship is rolling in the process of sailing, the control mechanism rotates the stabilizer fin to reverse deflect it and generate the angle of attack in the opposite direction. At the same time, the water body generates a pair of reverse lift forces on the fin to form a stable stabilizer moment to reduce the rolling moment for the ship.

The anti-roll fin has obvious advantages, which can greatly improve the overall safety performance of the ship, optimize the seaworthiness conditions, ensure the stable speed of the ship, and improve the working conditions of the crew. The damping efficiency of the fin can reach 90% in the ideal circumstance.

3.2.3 *Damping tank*

The working principle of the damping tank is to use the water level difference on both sides of the tank to generate a rolling recovery moment to offset the external interference force. The frequency of the damping tank is the same as the natural frequency of the ship roll in this design. When resonance occurs, the tank and the ship move synchronously, and the water movement in the tank lags the roll angle by 90°. At this time, the natural frequency of the ship roll is equal to the frequency of wave disturbance, and the roll angle of the ship lags the external interference force by 90°. Therefore, the movement of water in the tank lags behind the external interference force by 180°, and the direction of the recovery moment generated by gravity in the damping tank is exactly opposite to that of the interference force, thus reducing the hull roll. The damping efficiency of the tank is up to 70%.

The damping tank structure is relatively simple, the installation cost is low, the driving frequency is only 5% of the damping fin, the energy consumption is very low, and almost not affected by the speed. With the development of the computational model and control theory, the control method and response speed of the tank are constantly optimized. (Holder et al. 2011)

3.2.4 *Rudder rolls damping*

The automatic rudder and related control system equipped with the ship can not only control the course of the ship but also control the roll reduction. When the ship is sailing, there is a certain height difference between the action center of wave disturbance power on the rudder and the center of gravity for the ship. During rudder steering, not only the yaw moment to change the course will be generated, but also the roll moment. Moreover, the mass moment of inertia of the hull around the roll axis is tens of times larger than that around the roll axis, so the roll period is smaller than the roll period. The roll period of a ship is generally between 7s and 15s, while the yaw is between 20 s and 40 s. The main working principle of rudder roll reduction is to control the rudder angle by relying on the difference of the rudder's response to the roll and yaw cycle and to effectively use the roll moment generated by the rudder to partially offset the roll moment generated by the sea wave, to achieve the roll reduction control.

Compared with other roll reduction devices, rudder roll reduction has the advantages of low cost, small space occupation, convenient operation, and maintenance, and ease of installation and transform the ship. Studies have shown that the rocking reduction effect of the rudder is up to 70% (Sun 2009).

3.3 *Ship stabilization technology based on the air-floating active control device*

Based on the vibration control theory, the vibration control characteristics of the active air flotation control device are given full play. The fixed load on the ship is used to generate the reverse rocking moment of the ship. With the help of the accelerometer and controlled damper, shaking and vibration can be suppressed and eliminated to reduce the possibility of capsizing the ship. The main technical process is shown in Figure 4.

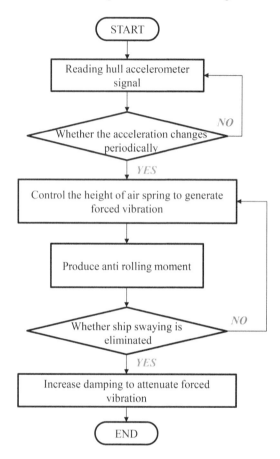

Figure 4. Technical flow chart.

The main technical characteristics are:

Adaptive ship rolling stability technology. With the help of the accelerometer installed on the left and right sides of the hull, the control system in the active air buoyancy control device can automatically identify the current vibration state, record the characteristic parameters such as frequency, amplitude, and phase, and generate the relevant vibration curve as shown in Figure 5, A-B-C curve. Accordingly, the control system of the active control device can cause the ship to produce forced vibration. Furthermore, the frequency and phase of the forced vibration are further adjusted to make it equal to the frequency of the ship roll, with a

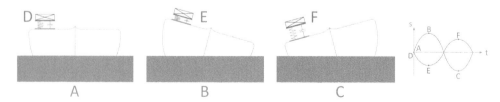

Figure 5. Diagram of adaptive stability control technology for ship roll.

phase difference of a half cycle, as shown in the D-E-F curve in Figure 5, to form a rocking torque and ensure the stability of the control position.

The air flotation active control device is arranged in three ways on the ship: one side, two sides, and across. The air flotation active control device is arranged in an appropriate position close to the ship side according to different actual conditions, to achieve the best rocking effect. For large ships, the air flotation active control device can be arranged separately or in multiple sets scattered on both the left and right sides of the ship. Active air flotation controls may grow and span the entire deck of a ship as shown in Figure 6.

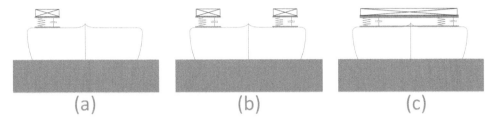

Figure 6. Three types of layouts.

The active air flotation control device mainly consists of the following parts: ① ship fixed load, ② load platform, ③ air spring, ④ vent valve, ⑤ inflatable valve, ⑥ controlled damper, ⑦ platform accelerometer, ⑧ Control system, and ⑨ hull accelerometer (Figure 7).

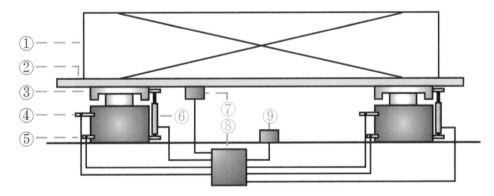

Figure 7. Schematic diagram of air floating active control device.

The technology is realized with the help of the active air flotation control device and its internal control system. Through the design of the actual situation of the ship, one or more active air flotation control devices are arranged on both sides of the ship, so that the ship's fixed load produces the forced vibration which is equal to the ship's roll frequency and the phase difference is half a cycle. The forced vibration produces the roll reduction moment, to reduce the ship's roll. Compared with traditional passive damping devices, it has a better damping effect.

4 DEVELOPMENT TREND OF SHIP VIBRATION CONTROL TECHNOLOGY

Active control technology has been continuously developed and applied with a higher degree of intelligence. By adopting an active control device of aid float type, high-precision vibration control is carried out for the specific area. At the same time, with the modern control theory active control algorithm and the actual vibration observation record, the active optimal control force is achieved as far as possible, and the optimal control effect in the region is obtained. An intelligent roll reduction device will also be able to automatically identify changes in the environment, hull, and other parameters and make active corrections to ensure the optimal rolling reduction effect in the whole process. (Hong & Chen 2012)

Full-speed roll reduction technology has been applied more widely with faster development. To keep the ship always in good running condition, the research on the whole speed damping device is more and more thorough.

The development of multi-degree-of-freedom integrated roll reduction technology can make the stability performance of ships more reliable during navigation. Ship roll in the actual course of sailing is generally the cross-coupling of roll, bow roll, and other motion states. A single rolling reducer has certain limitations in adapting to the speed and effect of rolling reduction. The most reasonable and effective way of rolling reduction is determined by optimizing the configuration of the comprehensive rolling reducer.

Accurate perception of ship attitude and advanced prediction technology will be the key research objectives. There is a time difference between the response of the hull to the external excitation and the complete attitude change of the ship, so the response of the damping device is also relatively lagging. By constructing a very short-term and high-precision forecast model and cooperating with the ship attitude control system, the active control anti-roll device can respond quickly and maximize its advantages.

5 CONCLUSION

(1) The main causes of ship vibration or rocking are reciprocating force, aerodynamic force, unbalanced torque, and other exciting forces generated during the operation of various types of power equipment units inside the ship, which lead to vibration transmission inside the ship and the ship as a whole is affected by the external environment.

(2) For the vibration reduction methods of power equipment in the ship, the main methods include single and double-layer vibration isolation devices, damping vibration isolation devices, floating raft vibration isolation devices, and other passive vibration isolation devices. For the whole ship roll reduction, it is generally adopted to attach bilge keel, fin, and tank or rudder roll reduction technology. These methods can achieve certain vibration reduction or vibration reduction effects.

(3) An active control device based on air buoyancy is proposed, which makes adaptive stability control of abnormal vibration of the ship at any time through the accelerometer, air spring, controlled damping, and other modules, and makes reasonable arrangements in different forms according to the actual situation of the ship.

REFERENCES

Chao Hong & Ying-Xia Chen. (2012). Current Situation and Tendency of Development of Ship Stabilizer Technique. *J. Ship Engineering*. 2, 236–244.

Cheng-En Yang, *et al.* (2002). The Retrospect and Prospect of Rudder Roll Damping Technology. *J. World Shipping*, 4, 4–6.

Christian Holder, Tristan Perez & Thor I. Fossen. (2011). A Lagrangian approach to nonlinear modeling of anti-roll tanks on the development of ship anti-roll tanks. *J. Ocean Engineering*. 38, 341–359.

Chun-ping Zou, Duan-shi Chen & Hong-xing Hua. (2003). Study on Structural Vibration Characteristics of Ship. *J. Journal of Ship Mechanics*. 7, 103–115.

Gengxian Zhao. (2001). Ship Vibration is One of the Key Design Factors for Ship and Offshore Structures. *J. Ship&Boat*. 5, 35–41.

Hong-fang Sun. (2009). The Research of Oblique Rudder Ship Comprehensive Stabilization Control and Prediction Method. D. 16–23.

Jian-Hua Shen. (2015). The Vibration Isolation Design and Analysis of a 5L21/31 Marine Diesel Generator Set. D. 143–178.

Ji-de Li. (2007). Ship Seakeeping. M. 18–20.

Jie Ma. (2006). Research on Ship Movement Prediction and Anti-roll Control Method. D. 132–211.

Mei-hua Dong, *et al.* (2008). Research Progress of Ship Anti-rolling Technology. *J. Journal of the University of Jinan.* 2, 183–187.

Plunkett R. (1998). Interaction between a Vibratory Machine and its Foundation. *J. Noise Control*, 4.

Qian Cheng. (2018). Vibration Characteristic Analysis and Vibration Reduction Technology Research of Marine Gearbox. D. 13–30.

Qiang Wei, *et al.* (2004). The Power Flow Numerical Analysis of Double—layer Vibration Isolation System on Ship Base. *J. Ship Science and Technology*. 6.

Qi-xue Zhou, *et al.* (2019). Application of Measures of Controlling Vibration and Noise in Science-research Ship's Mechanical Equipment. *J. Ship & Ocean Engineering.*48.

Tao Tian & Ren-de Wang. (2001). A Research of the Mechanical Features of V-Type Compressors. *J. Mechanical and Electrical Equipment*, 6.

Tao Yang. (2017). Study on Pressure Fluctuation of Propeller and Induced Hull Vibration and Cabin Noise. D. 16–30.

Ying-chun Wang, *et al.* (2021). Application of Active Control Technology on Ship Vibration and Noise. *J. Journal of the Naval University of Engineering*. 33, 57–63.

Ze-lin Liu. (2015). The Vibration Control of a Certain Type of High-Pressure Air Compressor on the Ship. D. 87–110.

Hysteretic behavior analysis of new energy dissipation fuse system

Dingyu Chen & Xiaotong Peng*
School of Architecture and Civil Engineering, University of Jinan, Jinan, China

Chen Lin
School of Architecture and Landscape Design, Shandong University of Art & Design, Jinan, China

ABSTRACT: To solve the problem of decentralized energy dissipation in the eccentric frame structure, a new energy dissipation substructure called a fuse system is proposed. Based on this, the non-linear finite element analysis on a fuse system is conducted to investigate the failure mode and seismic behavior. Four series of parametric analyses are performed, in which four main influential factors are considered, including length of reduced beams, distance from the reduced starting point to the end plate, reduced length, and reduced depth. The results of the analysis indicate that decreasing the reduced distance and length, increasing the length of reduced beams, and selecting moderately reduced depth can improve the performance of the whole structure. Finally, design commendations based on the analysis are presented.

1 GENERAL INSTRUCTIONS

The traditional eccentrically braced steel frame structure is prone to buckling instability under earthquakes, and the energy dissipation area is not concentrated enough. To solve the problems, the energy dissipation beam is separated from the main load-bearing components and arranged centrally in the form of a substructure, which is convenient for centralized replacement after earthquakes, thus reducing the maintenance cost. Based on the above ideas, a new fused frame structure is proposed, which is composed of a bending resisting frame and a fused system. The substructure is usually set on the side span of the structure, consisting of centrally placed energy-dissipating beams with the Reduced Beam Section (RBS) and double support columns. The RBS beam dissipates energy through shear deformation, and other components remain elastic under the design load level. It can achieve the concentration of plastic deformation and the unified replacement of energy-dissipating beam segments.

At present, the research on fuse systems is still in its initial stage at home and abroad. Balut first proposed the innovative idea of the replaceable energy-dissipation beam (Nicolae et al. 2003). The bolts are used to separate and connect the energy-dissipation beam and the frame beam in the flexural frame, and the reduced beam is applied as the replaceable energy-dissipation beam, which makes the energy-dissipation beam easier to enter the plasticity phase. It can ensure that other components remain in the elastic stage and replace the damaged energy-dissipation beam after earthquakes, which provides a new idea for the seismic design of the movement of the plastic hinge outward. Dougka proposed the concept of the FUESE system for the first time. The systems with an energy-dissipating beam of different cross-section sizes are tested to verify the feasibility of the FUESE system (Georgia 2014). Avgerinou concluded that the high-strength steel energy-dissipated beam in the fuse

---

*Corresponding Author: pengxito@163.com

system can greatly improve lateral resistance and have good seismic performance (Stella 2020). The fuse system is applied to the reinforcement project of reinforced concrete structure by Tsarpalis through finite element analysis. It is concluded that reinforced concrete structures with a fuse system can significantly improve the seismic performance of reinforced concrete structures (Panagiotis 2021).

The research on the new substructure is not deep enough, mainly because the test data of the structure is less, and the structural parameters of the energy dissipation beam on the overall performance of the substructure cannot be summarized. In this paper, 10 models with 4 series are established by using ABAQUS finite element software to analyze the hysteresis performance of the fuse substructure and to make design recommendations.

2 FINITE ELEMENT SIMULATION OVERVIEW

2.1 *Structural parameters*

The prototype structure of the model is a 6-storey and 3-span fused frame, which is designed by using finite element software SAP2000 and referring to relevant regulations (GB50011, 2010, GB 50017, 2017, JGJ99, 2015), as shown in Figure 1. The fuse substructure on the third floor of the frame is selected as the BASE model for finite element analysis, as shown in Figure 2. The support column is made of wide flange H-shaped steel with a specification of $400\times400\times13\times21$, and the steel is Q345. The energy dissipation beam section is made of thin-walled H-shaped steel with the specification of $175\times90\times4\times6$, and the steel is Q235B. The height of the structural floor is 3400 mm, the length of the beam is 760 mm, the spacing between the upper and lower beam is 850 mm, and the end plate of $315\times100\times20$ and 16 mm diameter high-strength bolts with 10.9 level are used to connect the support column and the energy dissipation beam. The reduced parameters of the energy dissipation beam should meet the requirements (FEMA350 2000), as shown in Figure 3, l is beam length, a is the distance from the reduced starting point to the end plate, b is reduced length, and c is reduced depth.

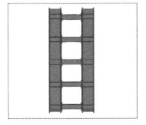
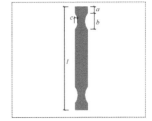

Figure 1.　Fuse frame structure.　Figure 2.　Fuse system model.　Figure 3.　Reduced parameters.

2.2 *Simulation parameters*

The constitutive relationship of Q235 is a trilinear model considering the strengthening and descending sections. The constitutive relationship of high-strength bolts and Q345 is a bilinear model considering the strengthening section. Under cyclic loading, a kinematic strengthening model is adopted and the Mises yield criterion and corresponding flow rule are adopted.

Solid elements are used in all models. To simplify the meshing in the thickness direction of each component, the non-conforming integral element C3D8I is used. The support column and energy dissipation beam are connected by end plates and high-strength bolts. To simulate the actual working conditions of high-strength bolts, it is necessary to apply pre-tightening force on high-strength bolts. Three analysis steps are used to apply the bolt preload. The first analysis step applies a small preload on the bolt to establish a contact constraint between the nut and the end plate (column). The second analysis step applies a complete 100 kN preload. The third analysis step fixes the length of the bolt rod.

To prevent the entire substructure from displacement outside the plane, the displacement in the z-direction of the support column is constrained. Constant amplitude loading is performed at multiples of the yield displacement during cyclic loading. Each of the first two stages is cycled once, and then each stage is cycled twice. Since ABAQUS does not provide a fixed failure criterion, the specimen is considered damaged when the capacity of the specimen exceeds the ultimate load and drops to 0.85 times the ultimate load (Qian 2001).

2.3 *Model validation*

To verify the feasibility of the modeling method, the test data of the M1 specimen in Reference [2] are selected for verification. The hysteresis curves of the finite element and the test are basically consistent. The comparison between the experimental value and the simulated value is shown in Table 1. The load and lateral stiffness in the negative direction of the test and simulation are greater than those in the positive direction. The simulated value of the lateral stiffness in the positive direction is 32.86 kN / mm, and the experimental value is 30.66k N/mm, which is about 93% of the simulated value. The simulated value of the ultimate load in the positive direction is 352.45 kN, and the experimental value is 334.02 kN, which is about 95% of the simulated value. In summary, the test results are basically consistent with the simulated results. The above modeling method can be used for the analysis of the failure modes and the calculation of ultimate capacity.

Table 1. Comparison between test value and simulation value.

Mode	Lateral stiffness (kN/mm)		Ultimate capacity (kN)	
	Positive direction	Negative direction	Positive direction	Negative direction
Test	30.66	33.59	334.02	365.53
Simulation	32.86	32.88	352.45	359.65
Ratio	0.93	1.02	0.95	0.98

3 FINITE ELEMENT ANALYSIS

3.1 *Hysteresis performance of the BASE model*

The hysteretic curve obtained by the cyclic loading test on the BASE model is shown in Figure 4. The overall curve is relatively full, the hysteretic performance is stable, and the elastic section, plastic section, and strengthening section are clear. As the area of the hysteresis loop gradually increases, the energy dissipation connection begins to yield slowly after the model enters the strengthening section from the elastic section, and the hysteresis curve gradually bends, indicating that the replaceable energy dissipation connection has good plastic deformation capacity. Under the cyclic loading, the stiffness of the model deteriorates, and the capacity of the compression time-delay loop curve decreases, resulting in softening.

When the energy-dissipating beam segment begins to yield, the stress is mainly concentrated in the flange end of the beam segment and the weakened flange. With the increase in the cyclic load at each stage, the plastic deformation first appears in the flange end and the weakened part, and then spreads to the area between the flange end and the weakened part. After the plastic hinge appears at the flange, the inelastic deformation of the web begins to appear. When the loading is completed, only the energy dissipation beam segment has plastic deformation, and other parts are in the elastic stage, as shown in Figures 5 and 6. The inelastic deformation of the energy-dissipating beam segment mainly occurs at the

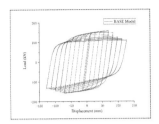 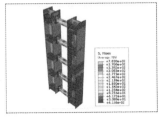 

Figure 4. Hysteresis curve.   Figure 5. Global stress.   Figure 6. Beam strain.

weakening of the flange of the energy-dissipating beam segment, which can ensure that the middle area of the energy-dissipating beam segment keeps a large stiffness and prevent the collapse of the structure.

3.2 *Parameter analysis*

Based on the BASE model, a total of 10 models are designed for cyclic loading analysis considering different parameters affecting the dissipative performance of the fuse substructure: the length of the energy dissipating beam (BL series), the distance from the reduced starting point of the energy dissipating beam to the end plate (RS series), the reduced length of the energy dissipating beam segment (RL series), and the reduced depth of the energy dissipating beam (RD series). The changed parameters are following the provisions of FEMA-356. The specific structure and analysis results are shown in Table 2. The initial stiffness and ultimate capacity can reflect the strength of the substructure. The ductility can be expressed as the ratio of ultimate displacement to yield displacement.

3.2.1 *BL series*

The series studies the influence of the length of the energy dissipation beam on the seismic performance of the fuse system (Table 2). With the increase in the length of the energy dissipation beam segment, the stiffness and ultimate capacity of the substructure decrease all the time, and the ductility first increases and then decreases. In the range of 660 mm to 810 mm, the ultimate capacity decreased by 8.51%, 5.52%, and 5.23% for every 50 mm increase in length. The stiffness decreased by 1.84%, 2.45%, and 3.18%, respectively. The ductility coefficient first increased by 3.79%, 0.59%, and then decreased by 2.16%. The rate

Table 2. Analysis results.

Model	Beam length l (mm)	Reduced distance a (mm)	Reduced length b (mm)	Reduced depth c (mm)	Initial stiffness K (kN/mm)	Ultimate capacity F (kN)	Ductility coefficient u
BASE	760	60	120	20	12.57	158.36	6.23
BL660	660	60	120	20	14.54	165.39	6.05
BL710	710	60	120	20	13.30	162.35	6.28
BL810	810	60	120	20	11.91	153.32	6.19
RS50	760	50	120	20	12.50	154.73	6.34
RS70	760	70	120	20	12.65	161.73	5.74
RL130	760	60	130	20	12.56	158.99	5.98
RL140	760	60	140	20	12.54	158.99	5.79
RD18	760	60	120	18	12.74	162.00	6.09
RD22	760	60	120	22	12.40	153.91	6.06

of increase in ductility coefficient is not obvious compared with the rate of decrease in capacity. Therefore, increasing the length of the energy dissipation beam segment is a significant way to improve the overall performance of the structure, but this method also increases the engineering cost of the structure.

3.2.2 *RS series*

This series explores the influence of the distance from the reduced starting point of the energy dissipation beam system to the end plate on the seismic performance of the fuse substructure (Table 2). The further the weakening distance is, the greater the initial stiffness and ultimate capacity of the structure will be. For every 10 mm increase, the increase in each parameter will be less than 3%. However, the longer the weakening distance is, the greater the reduction rate will be: in the range of 50-70 mm, the ductility decreases by 1.77% and 7.86% for every 10 mm increase. If the weakening length is too small, the plastic hinge will appear in the beam-column structure, which will easily cause brittle failure and lose the control role of the plastic hinge in the weakening type energy dissipation beam segment. Therefore, the weakening length should not be too small or too large, and it is recommended to take a smaller value within the specification range.

3.2.3 *RL series*

In this series, the influence of the reduced length of the energy dissipation beam on the seismic performance of the fuse system is studied (Table 2). The influence of the weakening length of the energy dissipation beam on the initial stiffness and ultimate capacity of the fuse substructure is not obvious, and the change is less than 0.4%. However, the increase in the weakened length of the energy dissipation beam segment will lead to a significant reduction in the ductility coefficient: within the range of 120 mm-140mm, the ductility coefficient decreases by 3.93% and 3.21% for every 10 mm increase in the length. Therefore, within the value range specified in the specification, the design of weakening length should take a smaller value.

3.2.4 *RD series*

The influence of the weakening depth of the energy dissipation beam segment on the seismic performance of the fuse substructure is studied (Table 2). With the increase in the weakening depth, the initial stiffness and ultimate capacity of the energy dissipation beam segment gradually decrease, while the ductility coefficient first increases and then decreases: within the depth range of 18-22 mm, the initial stiffness decreases 1.31% and 1.36% for every 2 mm increase; The ultimate capacity decreases by 2.25% and 2.81% respectively; The ductility coefficient first increases by 2.17% and then decreases by 2.66%. Therefore, the moderate weakening depth of the energy dissipation beam segment cannot only ensure the strength of the structure but also be conducive to the energy dissipation of the structure.

3.2.5 *Significance analysis*

To better accurately and intuitively analyze the significance of various influencing factors on the results of finite element analysis, the variance analysis method is used to analyze the initial stiffness K, ultimate capacity F, and ductility coefficient u of all models. Using Origin software for data inspection, the above parameters of the model are significantly from the normal distribution population at the 0.05 level, and the overall variance is not significantly

Table 3. Variance analysis.

K variance analysis			F variance analysis			u variance analysis		
Variance Source	Value	Significance	Variance Source	Value	Significance	Variance Source	Value	Significance
RL	14.07	-	RL	0.36	-	RD	0.09	-
RS	71.69	*	RS	3.50	*	BL	0.10	*
RD	168.7	**	RD	4.05	**	RL	0.22	**
BL	1000.1	***	BL	5.22	***	RS	0.32	***

different. Therefore, the influence of each parameter on the ultimate moment capacity and initial rotational stiffness of the model can be analyzed by the variance analysis method, and the analysis results are shown in Table 3. The results show that the influence degree of each influencing factor on the initial stiffness and ultimate capacity of fuse substructure is arranged from large to small as follows: BL, RD, RS, and RL. The order of influence on the ductility coefficient of the results is RS, RL, BL, and RD.

4 CONCLUSION

(1) The fuse substructure has excellent energy dissipation capacity and high ductility. The stiffness of the structural support column is much larger than that of the energy dissipation beam section. The flange weakening energy dissipation beam section is adopted so that the plastic deformation of the structure is easier to concentrate on the energy dissipation beam section.
(2) The length of the link beam will significantly affect the initial stiffness and ultimate capacity of the structure. Properly lengthening the energy-dissipation beam section within the scope specified in the specification will effectively improve the overall performance of the structure, but at the same time will greatly increase the cost of the structure, and should not be the primary choice to improve the overall performance of the structure.
(3) The decrease in length and distance from the end of the reduced section will improve the ductility of the structure, but the value is too small to make the plastic hinge appear at the beam-column joint of the structure and cause brittle failure. Therefore, it is recommended that these two values be taken as small values within the specification range.
(4) The decrease in the reduced depth will increase the initial stiffness and ultimate capacity of the substructure. The ductility increases first and then decreases, but the ductility does not change much compared with other parameters.

ACKNOWLEDGMENTS

The work was sponsored by the Natural Science Foundation of Shandong Province (ZR2019MEE009), the Ministry of Education University-Industry Collaborative Education Program (201802276002; 201902204001; 201902204002), Shandong Postgraduate Education and Teaching Reform Research Project (SDYJG19039), Science and Technology Project of Housing and Urban-Rural Development of Shandong Province (2020-K2-3). The writers gratefully acknowledge all the support provided.

REFERENCES

Avgerinou, S. & Vayas, I. (2020). High-strength Steel on Dissipative Elements in Seismic Resistant Systems: Tests and Simulations. *Journal of Constructional Steel Research*. 172, 1–16
Balut, N. & Gioncu, V. (2003). Suggestion for an Improved "dog-bone" Solution. *Stessa*. 2003, 129–134
Dimakogianni, D. & Dougka, G. (2012) Innovative Seismic-resistant Steel Frames (FUSEIS1-2) Experimental *Analysis. Steel Constructional Design and Research* 5(4), 212–21
Dougka, G. & Dimakogianni, D. (2014). Innovative Energy Dissipation Systems (FUSEIS1-1) Experimental Analysis. *Journal Constructional Steel Research*. 96(5), 69–80
Feng, Q. & Ping, C. (2001) Civil Engineering Structure Test. *China Architecture Press.*
FEMA350 (2010). *Recommended Seismic Design Criteria for New Steel Moment-frame Buildings*. 3.38–3.42
GB50011 (2010). *Code for Seismic Design of Buildings*. 8.1–8.5
GB50017 (2017). *The Standard for the Design of Steel Structures*. 3.1–3.5
JGJ99 (2015). *Technical Specification for Steel Structure of the Tall Building*. 3.1–3.9
Tsarpalis, P. & Vayas, I. (2021). Rehabilitation of Reinforced Concrete Buildings Using the Fuse is a Beam-link System. *Structures*. 34, 3300–3314.

Water Conservancy and Civil Construction – Oke & Ahmad (Eds)
© 2024 The Author(s), ISBN: 978-1-032-58618-2

SVR algorithm-based prediction of scraper wear for super-large-diameter slurry shield

Bowen Tao*, Lei Zhang, Jian Li & Zhipeng Shi
CCCC Tunnel Engineering Co., Ltd, China

ABSTRACT: Given the difficulty in predicting the scraper wear of super-large-diameter slurry shields and combining the Yangtze River Tunnel project in Jingjiang, Jiangyin, the parameters of the support vector machine (SVM) and the influencing factors of scraper wear are selected to establish an SVV-based prediction model for the scraper wear of a slurry shield. In addition, the result calculated by this model is comparatively analyzed with the current empirical formula and the actual wear degree. The results show that (1) it is reasonable and feasible to take the cutter head speed, tunneling speed, cutter head torque, cutter head thrust, weathering index, penetration depth, and the length of the cutter's cutting trajectory as the prediction parameters for the cutter's wear extent. (2) The SVR-based prediction model for the scraper wear of the slurry shield can be used as the judgment criterion for scraper replacement because its accuracy is higher than that of the empirical formula. (3) It is recommended that 3 cm be used as the threshold value for the degree of cutter wear and as a judgment criterion for cutter replacement during subsequent shield tunneling.

1 INTRODUCTION

As major technical equipment with high technical content and high added value, shield machines have been extensively applied to subways, highways, railways, municipal works, hydroelectric tunneling works, etc. [1]. The tunneling condition and stress state of shield machines are extremely complicated during construction [2]. The wear of shield cutters is one of the important difficulties faced by safe and fast shield tunneling. Too large cutter wear can seriously affect the construction efficiency and frequent cutter replacement can easily trigger the instability of the excavation face, which even worse can lead to the collapse of the tunnel face, thus increasing construction risks and costs. Therefore, it is very essential to predict the cutter's wear extent of shield machines during tunneling through tunneling parameters, reduce the frequency of opening-chamber cutter inspection, set the threshold for the wear extent, and grasp the appropriate cutter replacement time.

In essence, cutter wear is the result of complex interactions during soil cutting, which involves complicated physical-mechanical mechanisms and is influenced by various factors. Super-large-diameter shield machines (cutter head diameter > 14 m) are subjected to more complicated stress conditions due to the larger cutter head-stratum contract area. In recent years, the prediction of shield cutter wear has been extensively investigated by many domestic (Chinese) and foreign scholars. Wu *et al.* [3] put forward a prediction formula for cutter wear by studying the wear mechanism of shield cutters. Yang *et al.* [4] used the Support Vector Machine (SVM) and Genetic Algorithm (GA) to predict the wear of shield cutters. Han *et al.* [5] optimized the BP neural network model through the GA algorithm to

---

*Corresponding Author: 1284584654@qq.com

98   DOI: 10.1201/9781003450832-13

analyze cutter wear. Wang et al. [6] obtained the scraper wear coefficient applicable to unevenly hard-soft composite strata by using scraper wear extent and shield tunneling parameters. Zuo et al. [7] studied the relationship between torque and cutter wear state. Zhou et al. [8] innovatively measured the scraper wear degree by constructing a sand tray indoors to simulate scraper tunneling, calculated the wear coefficient by 3D printing technology, and obtained a prediction formula for the wear degree of shield knives in sand and pebble strata. Yuan [8] and Zhang [10] et al. analyzed multiple groups of cutter wear data under different shield tunneling conditions through the Partial Least Squares Regression (PLSR) method. Niu [11] and [12] et al. predicted the wear extent of shield scrapers through orthogonal experiments and the Elman neural network. In summary, the wear extent of scrapers has been predicted mainly from three aspects, numerical prediction models, indoor simulation experiments, and subitem fitting of theoretical formulas. In addition, the prediction of shield scrapers' wear extent is strata-specific.

In this study, an SVM-based prediction model for the scraper wear of slurry shields is proposed. This model simultaneously considers the influences of formation conditions, parameters of the shield machine, and the length of cutting trajectory on cutter wear. Under the Yangtze River Tunnel Project background in Jingjiang, Jiangyin, the data on influencing parameters and cutter wear extent acquired on the field are extracted as the test set to acquire a prediction model. In addition, this prediction model for the scraper's wear extent is tested using cutter wear data of the shield machine in the subsequent tunneling stage.

2 PROJECT PROFILE

The Yangtze River Tunnel in Jingjiang, Jiangyin, which is located between Jiangyin Bridge and Taizhou Bridge, connects Jingjiang in the north and Jiangyin in the south. The main tunnel starts at about 300 m from the north side of S356 in Jingjiang, joins in the Yangtze River after underpassing S356 and Binjiang No.1 Road in the south, extends to Jiangyin City, underpasses Jiangfeng Road, Binjiang Road, and Tongfu Road, and connects the ground at about 270 m from the south side of Tongfu Road, with a total length of 6,445 m. In the river-crossing section of this tunnel, a shield machine with a diameter of 16.09 m is used for tunneling construction, the total length of the shield section is 4,937.19 m, and this section passes through silty clay, silty soil, silt, silty fine sand, and medium-coarse sand formations, with sandbox stones locally exposed. The average hydraulic pressure is high in this tunnel and the maximum hydraulic pressure reaches 0.82 MPa (the central section of the Yangtze River). The planar graph of the Yangtze River Tunnel Project in Jingjiang, Jiangyin, is displayed in Figure 1.

According to the geological conditions of the Yangtze River Tunnel Project in Jingjiang, Jiangyin, the super-large-diameter slurry balance shield machine developed by CCCC

Figure 1. Route plan of Yangtze river Tunnel in Jingjiang, Jiangyin.

Tianhe Machinery & Equipment Manufacturing Co., Ltd is used with a maximum cutter head excavation diameter of 16.19 m and an aperture rate of 41%. A six-spoke normal-pressure cutter head is used and equipped with cutter wear detection devices at 6 positions, accompanied by 4 cutter head distributions. In addition, 360 scrapers and 18 edge protection scrapers are arranged for the cutter head. The cutter layout plan for the Yangtze River Tunnel Project in Jingjiang, Jiangyin, is shown in Figure 2.

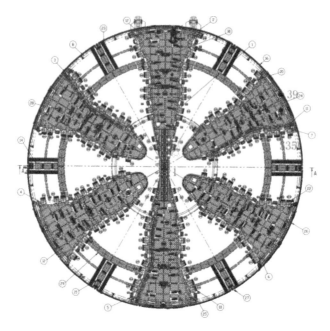

Figure 2. Cutter layout plan for the Yangtze River Tunnel project in Jingjiang, Jiangyin.

3 DETERMINATION OF PREDICTION MODEL FUNCTION

The regression-type Support Vector Machine (SVR) is a statistical theory created by Vapnik and aims at the minimization of structural risks. Compared with the traditional BP neural network algorithm, SVR can acquire the global optimal solution with a few samples. No iterative trials are required to determine the network structure and this algorithm can shorten the calculational time. Vapnik *et al.* [13] expressed its regression equation as follows:

$$
\begin{aligned}
f(x) &= \omega^* \Phi(x) + b^* = \sum_{i=1}^{l} (a_i - a_i^*) \Phi(x_i) \Phi(x) + b^* \\
&= \sum_{i=1}^{l} (a_i - a_i^*) K(x_i, x) + b^*
\end{aligned}
\tag{1}
$$

Where ω^* and b^* parameters are hyperplane parameters and $\phi(x)$ is nonlinear mapping functions, the corresponding sample x_i is the support vector in question. $(a_i - a_i^*)$ is the optimal solution vector of the hyperplane obtained by the dual transformation of the Lagrange where only some parameters are non-zero.

$K(x_i, x)$ is the kernel function, which is an extension of the vector's inner product space and transforms the nonlinear regression problem into an approximately linear regression problem through the kernel function. In this study, the Radial Basis Function (RBF) kernel is chosen in the compilation process. RBF (Radial Basis Function) is a monotone function that represents the Euclidean distance from any point x to a center x_c in space. The most commonly used RBF kernel is the Gaussian kernel function. It can be expressed as follows.

$$K(x_i, x) = \exp\left(-\frac{\|x - x_i\|^2}{2\sigma^2}\right) \quad (2)$$

Where xi is the kernel function center and σ is the width parameter of the function. If x is close to x_i, the kernel function equals 1. If x is far from x_i, the kernel function equals 0. We bring Formula (2) into Formula (1) and the final prediction model is expressed below.

$$T = \mathrm{sgn}\left(\sum_{i=1}^{n} W_i \exp\left(-\frac{\|x - xi\|^2}{2\sigma^2}\right)\right) + b \quad (3)$$

Where n denotes the number of support vectors, sgn represents a step function, and x_i stands for the post-training support vector. Therefore, the data to be solved through the SVR algorithm include the correction coefficient b and support vector coefficient Wi.

4 SELECTION OF PREDICTION DATA FOR SHIELD CUTTER WEAR

The wear of shield scrapers is decided by multiple factors during stratal tunneling, so it is very important to select appropriate influencing parameters of cutter wear. Too many parameters and calculation volume can lead to a poor fitting effect and insufficient parameters can result in the poor prediction effect of the prediction equation.

4.1 Selection of shield cutter wear data

During the shield tunneling process of the Yangtze River Tunnel Project in Jingjiang, Jiangyin, 9 open bunker inspections are conducted and 731 cutters are inspected each time. To ensure the representativeness of scraper wear data, the cutter wear extent on the cutter head is measured during each inspection, and samples are picked out to establish this

Figure 3. Field chamber-opening cutter inspection.

prediction model. The field inspection of some representative cutters is shown in the following pictures (from Field Chamber-opening Cutter Inspection Yangtze River Tunnel Project in Jingjiang, Jiangyin).

- The alloy wear extent of the 1# cutter is 2 mm, no obvious wear is observed on the cutter matrix, and there is no muck residue on the cutter surface;
- The alloy wear extent of the 5# cutter is 5 mm, no obvious wear is observed on the cutter matrix, and there is no muck residue or mud cake on the cutter surface;
- The alloy wear extent of the 35# cutter is 7 mm, no obvious wear is observed on the cutter matrix, and the wear extent on its counter-soil surface is 1 cm;
- The wear extent of the 39# cutter is 1 cm and no obvious wear or eccentric wear is observed on the cutter matrix. The alloy wear extent of the 35# cutter is 7 mm.
- Samples are selected according to the following principles:
- The wear laws of scrapers are investigated, so the relevant wear data of edge protection scrapers are eliminated.
- The wear extent is influenced greatly by different scraper morphologies. The scraper morphologies of the shield cutter head in this project can be divided into three types, normal-pressure replaceable scrapers (58), advanced scrapers (144), and bolt-type scrapers (158), which are distributed at different heights on the cutter head. Therein, advanced scrapers cut into the rock mass first under more complicated stress conditions. Bolt-type scrapers are mainly distributed at spoke edges and their replacement is very cumbersome. For the practicability of cutter wear extent prediction and the convenience of field operation, only the wear data of normal-pressure replaceable scrapers are selected while the wear data of the other two types are excluded.
- Some replaced cutters are subjected to special circumstances, such as the breakage of cutter rings, the pressure loss of cutter shafts, and eccentric wear, their wear data cannot serve as sampled data, so the wear data of replaced cutters are excluded, while only the wear data of cutters that can be normally used on the cutter head are adopted.
- The wear position varies with the unfixed contact surface between the scraper and the formation during shield tunneling. For unity, only the wear extent of alloys on scrapers is investigated in this study.

A total of 58 normal-pressure replaceable scrapers arranged on the spoke are included as the study objects through the initial screening of scraper wear data. The measured data in each chamber-opening inspection are imported, including 731 pieces of data in 9 groups. 630 pieces of data are reserved after 101 pieces of abnormal data are excluded. Meanwhile, it can

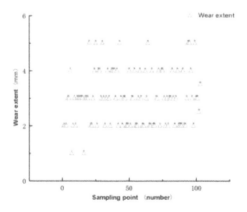

Figure 4. Point Cloud Image of Cutter Wear Extent.

be seen from Figure 2 that the cutters are symmetrically arranged on this cutter head. To simplify the model, the influence of the posture change on cutter wear during shield tunneling is not considered. Therefore, the cutters on 1/6 major-minor spokes of the whole cutter head are studied, including 105 pieces of data. The point cloud image of cutter wear extent in this project is exhibited in Figure 4.

4.2 *Selection of influencing factors for shield cutter wear*

In general, the factors influencing shield scraper wear include the shield thrust, the length of the scraper cutting trajectory, cutter head torque, shield tunneling speed, cutter head speed, the pressure of the slurry chamber, penetration depth, and formation weathering index. The above 8 parameters within the chamber-opening scraper replacement section are acquired from the host control system of the shield machine for the correlation analysis. All data participating in the correlation analysis are the mean values of data acquired during the one-time chamber-opening scraper replacement. Before this, the data should be normalized through the following formula.

$$\xi = \frac{e - e_{\min}}{e_{\max} - e_{\min}} \qquad (4)$$

Where e stands for the average value of the parameters obtained from the shield machine host control system for the scraper replacement section and $e_{\max}$ and $e_{\min}$ represent the maximum value and minimum value acquired in the section of one-time cutter inspection, respectively.

The normalized data are imported into SPSS software for correlation analysis. Then, the correlations of shield parameters in this project are obtained as shown in Figure 5.

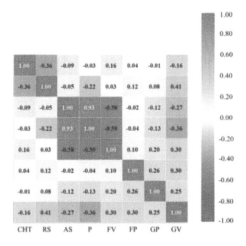

Figure 5. Correlation analysis of shield tunneling parameters.

It can be seen from the above figure that among the eight parameters selected, the slurry chamber pressure and the shield machine thrust are significantly positively correlated, with the correlation coefficient reaching 0.93. Therefore, the parameters are subjected to dimension reduction, and only the influences of the other 7 parameters, except for the pressure of the slurry chamber, on the wear extent of shield scrapers are considered. The values of parameters, such as the thrust of shield machines, cutter head torque, shield tunneling speed, cutter head speed, and penetration depth, have been selected in relevant domestic and foreign literature, so this study does not describe such details, but only two parameters, scraper cutting trajectory length and stratigraphic weathering index, are further described.

4.2.1 *Length of scraper cutting trajectory*
When cutting the rock mass, as the shield advances, the scraper moves in a circular motion around the blade. Without considering the scraper autorotation, its moving trajectory is a helical line. For the Yangtze River Tunnel Project in Jingjiang, Jiangyin, the cutter spacing of the scraper on major-minor spokes is equal (20 cm), as shown in Figure 6.

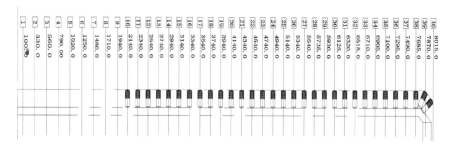

Figure 6. Schematic diagram of scraper cutter spacing.

Therefore, the scraper trajectory can be approximated as a helical line rotating around the cylinder, which is expressed by the following formula.

$$L = \sqrt{S^2 + (\pi D)^2} \qquad (5)$$

Where S denotes the shielding distance and D represents the diameter of the circle in which the scraper rotates around the cutter head center. The relevant data are imported in this formula to obtain the length of the cutting trajectory of each scraper during the tunneling. The point cloud image of the scraper-cutting trajectory is displayed in Figure 7.

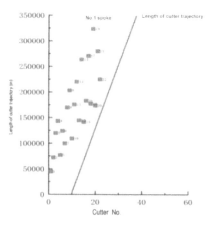

Figure 7. Point cloud image of scraper cutting trajectory.

It can be seen that during the tunneling process, the scrapers that are farther from the center of the cutter disk travel along a longer trajectory. Meanwhile, only one scraper is arranged on the same spoke considering the helical line-type cutter arrangement rule on the cutter head. Therefore, the trajectory length of each cutter should correspond to the specific scraper on the cutter disc.

4.2.2 *Formation weathering index*

So far, the project has passed through clay, silt, and sandstone strata in the tunneling process. Sandstone strata with different degrees of weathering are passed during the 3-5 cave cutter inspections. Given this, the weather index KI is used to evaluate the factors influencing formation weathering, where KI = 0.6 indicates moderately weathered sandstone strata and KI = 0.4 indicates strongly weathered sandstone strata. A schematic diagram of strata passed by the Yangtze River Tunnel in Jingjiang, Jiangyin, is shown in Figure 8.

Figure 8. The schematic diagram of strata passed by the Yangtze River Tunnel in Jingjiang, Jiangyin.

In summary, in this paper, tunneling speed, thrust force[14], cutter head torque, cutter head speed, weathering index, and penetration depth are selected as the Prediction Parameters of Cutter Wear. The relevant parameters of shield tunneling are shown in Figure 9.

5 WEAR DATA PREDICTION AND EVALUATION

In this study, the SVR model is programmed by MATLAB software to build the SVR model and evaluate its performance. The program flowchart is displayed in Figure 10.

In the SVR model, the data fitting quality of the final model is influenced by the penalty function C of the model, the insensitive loss ε, the cross-validation frequency n, the kernel function, the number of support vectors, and the variance g in the kernel function. In this study, cross-validation is used to automatically find the best variance of the penalty function and kernel function to prevent overfitting of the prediction model, shorten calculation time, and mitigate the model complexity. The cross-validation frequency is 10 and the insensitive loss function is 0.1.

5.1 *Establishment of prediction formula for shield cutter wear data*

The shield tunneling parameters extracted in the previous section and SVR algorithm parameters are imported into the program, where 80 groups of tunneling parameters and shield cutter wear extent data are served as the training set. The effects of training and prediction models are depicted in Figure 10, in which the abscissa displays the sample mark number and the ordinate denotes the wear extent of shield cutters. The red color displays the field-measured cutter wear extent and the blue color represents the cutter wear extent predicted through the SVR algorithm.

It can be seen from Figure 10 that the value obtained through the trained model matches well with the true value, the mean square error (MSE) is 0.0344, and the confidence coefficient R^2 is 0.85. The trained prediction model is expressed by the following formula:

$$T = \text{sgn}\left(\sum_{i=1}^{80} W_i \exp(-[g]\|SVs - x\|^2)\right) + 0.2454 \qquad (6)$$

Where Wi, [g], and SVs form an 80*80 matrix generated from the training set.

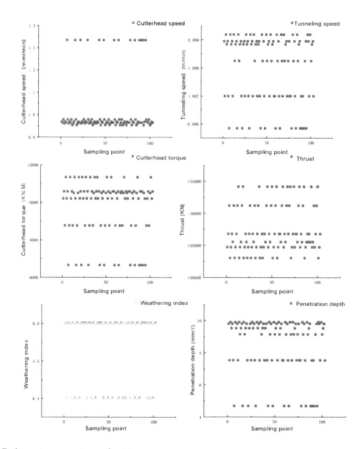

Figure 9. Relevant parameters of cutter wear.

5.2 *Verification of prediction formula*

The remaining 25 groups of samples' tunneling parameter data are input into the acquired prediction model to gain the prediction curve of scraper cutter wear data for the slurry shield. The prediction curve is compared with the actual scraper wear extent, as shown in Figure 12.

It can be known from the above figure that by comparing the predicted value with the actual value, the MSE is 0.019 and the confidence coefficient R^2 is 0.2832. The MSE is smaller than 2%, indicating a relatively accurate prediction result. The model established through the SVR algorithm can predict the wear extent of shield cutters very well. However, the confidence coefficient is small (0.2832), which reflects the ordinary generalization ability of this model. In the future study, the model will be further trained using richer and more accurate data.

5.3 *Comparison with the empirical formula*

The widely used empirical formula [15] for the wear extent of shield cutters at present is as follows.

$$\delta = \frac{K * \pi * D * N * L}{V} \quad (7)$$

Where δ is the scraper wear extent; K stands for the wear coefficient; D denotes the diameter of the shield cutter head; N is the cutter head speed; L is the shielding distance; V represents

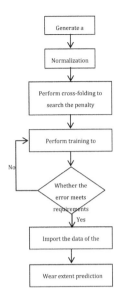

Figure 10. Schematic diagram of SVR algorithm flow.

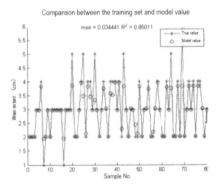

Figure 11. Model acquisition of scraper wear training set.

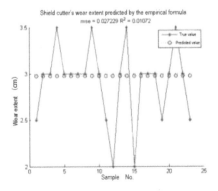

Figure 12. Prediction of scraper cutter wear extent.

the tunneling speed. Cao, L. [16] *et al*. stated that the K-value can be approximately taken as 0.0025 in the study of similar strata. The relevant measured data are imported into the empirical formula to compare the predicted wear extent of shield scrapers and the measured value, as shown in Figure 13.

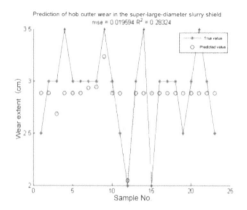

Figure 13. Shield scraper's wear extent predicted by the empirical formula.

It can be seen from the above figure that the value predicted by the empirical formula tends to be flat (3 cm). Compared with the measured value, the Mean Square Error (MSE) is 0.027, while the confidence coefficient (R^2) is only 0.01, indicating that the prediction conclusion is incredible. Therefore, the degree of cutter headwear of the mud shield can be solved through a linear prediction formula under the complex multi-factor comprehensive influences.

5.4 *Scraper replacement time*

In practice, if the shield tunneling parameters are abnormal, the gate chamber should be opened for inspection, and if the degree of scraper wear is found to be excessive, the scraper needs to be replaced. It can be known from the preceding part of the paper that except for the excessive degree of wear at individual points, in this project, once the degree of scraper wear reaches about 3 cm, the scraper needs to be replaced. Therefore, a threshold can be assigned to shield cutter wear in the subsequent tunneling process of this project. Chamber-opening inspection and scraper replacement can be considered when the predicted data reaches this threshold.

6 CONCLUSIONS AND SUGGESTIONS

In this study, an SVR algorithm-based prediction method for the scraper wear of a super-large-diameter slurry shield is proposed.
- It is feasible to take the cutter head speed, tunneling speed, cutter head torque, cutter head thrust, weathering index, penetration depth, and the length of cutting trajectory as influencing factors for the wear level of mudguard scrapers.
- The final prediction model is obtained by testing 100 groups of selected data and used to predict the remaining 25 groups of data. The acquired prediction result fits well with the

measured data, so the prediction model is suitable for predicting the degree of blade wear during the subsequent shield tunneling process.

- This prediction method can reduce the chamber opening frequency. It is recommended to set a wear level threshold for the blade in normal tunneling conditions. Chamber-opening and scraper replacement can be considered if the predicted wear extent exceeds this threshold.
- This method proposes high requirements for data accuracy. Therefore, high-precision measurement of cutter wear extent can be further implemented with the expectation of obtaining a more accurate prediction formula.

REFERENCES

Han, B. Y., Yuan, D. J., Jin, D. L., Li, D. S. Wear Analysis and Prediction of Shield Cutters in Composite strata [J]. *China Civil Engineering Journal*, 2020, 53 (S1): 137.

Li, X., Su, X. J. A New Method for Forecasting Shield's Disc-cutters Wearing Based on Elman Neural Network [J]. *Journal of Liaoning Technical University*, 2010, 29 (6): 1122.

Li, X., Su, X. J. Forecast of Wear Shield's Disc Cutters Based on Multivariate Nonlinear Regression [J]. *Journal of Liaoning Technical University*, 2009, 28 (2): 281.

Lv, R. H., Wang, G. H. Analysis of Research Status on Shield Cutters' Wear Laws and Antifriction & Anti-wear Measures [J]. *Tunnel Construction*, 2012, 32 (S2): 41.

Niu, J. C., Chen, L., Liu, J. Z., Guo, J. B. Simulation Analysis and Optimization for 19-inch disc Cutter Based on LS-DYNA [J]. *Journal of Shijiazhuang Tiedao University*, 2015, 28 (2): 73.

Vapnik, V. N. An Overview of Statistical Learning Theory. *IEEE Trans. Neural Netw.* 1999, 988:999.

Wang, H. J., Wang, Z. Y., Zhou, Z., Zhang, X. P., Jiang, H. L., Tang, S. H., Li, F. Y. Prediction and Snalysis on the Wear Law and Cutting Life of Disc Cutter of Earth Pressure Balanced Shield TBM in Composite Strata. *Collected Papers From the 2019 National Academic Annual Conference on Engineering Geology*.

Wu, J., Yuan, D. J., Li, X. G., Jin, D. L., Shen, X. Wear Mechanism and Prediction Analysis of Shield Cutters [J]. *China Journal of Highway and Transport*, 2017, 30 (8): 109.

Xu, L. M., Yang, Y. D., Zhou, J. J., Chen, K., Cai, G. Y. Prediction of Shield Scraper Wear and Replacement for the Sea-crossing Section of Xiamen Rail Transit Line 2 [J]. *Tunnel Construction*, 2016, 36 (11): 1379.

Yang, J. Z., Qiu, C. Y. Prediction of Cutting Tool Wear Rate [J]. *Construction Technology*, 2016, 45 (22): 25.

Yuan, D. J., Wu, J. Analysis and Prediction of Composite Shield Cutter Wear in Extremely Hard Rock Strata in Dalian [J]. *China Civil Engineering Journal*, 2015, 48 (S1): 250.

Zhang, M. F., Yuan, D. J., Huang, Q. F., Huang, C. B., Zhang, F. Analysis of Dynamic Abrasion of Shield Cutters in Sandy Cobble Stratum [J]. *Chinese Journal of Rock Mechanics and Engineering*, 2008, 27 (2): 397.

Zhao, Q., Duan, J. C., Yang, T., Duan, Z. H., Yan, Q. X. Research on Service Life Prediction After Abrasion of Shield Hob in Weathered Granite Formation [J]. *Subgrade Engineering*, 2016, 2: 94.

Zhou, J. J., Song, J. P., Tan, Z. S. Research on Disc Scraper Abrasiveness of Subway Shields in Sandy Cobble Strata [J]. *China Civil Engineering Journal*, 2017, 50 (S1): 31.

Zuo, C. F., Tang, D. G., Rong, X. L., Liao, B. State Recognition of Cutter Wear Based on Frictional Torque Parameters of Cutter Head [J]. *Tunnel Construction*, 2016, 36 (3): 344.

Water Conservancy and Civil Construction – Oke & Ahmad (Eds)
© 2024 The Author(s), ISBN: 978-1-032-58618-2

Study on dynamic stability of geogrid-reinforced cracked slope based on finite element limit analysis

Cheng Li & Llidan Li
School of Architectural Engineering, Kaili University, Guizhou, China

Xingqian Xu*
College of Water Conservancy, Yunnan Agricultural University, Kunming, China

Xin Qu*
School of Civil and Architecture Engineering, Anyang Institute of Technology, Anyang, China

Haijun Wang
College of Water Conservancy, Yunnan Agricultural University, Kunming, China

Junduo Yang
Power Supply Bureau, Rongjiang, Kaili, Guizhou, China

ABSTRACT: Earthquakes lead to slope damage, especially on slopes with water-filled cracks. Although the influence of water-filled cracks is seldom considered in the study of slope dynamic stability, these water-filled cracks must be considered in slope reinforcement design. In this study, the Mohr-Coulomb strength criterion, the finite element limit analysis method, and the quasi-static method are used to discuss the influence of water-filled cracks on the dynamic stability of a reinforced homogeneous soil slope. This study only considers the impact of the horizontal seismic load and the seismic loads are 0.1 g, 0.2 g, and 0.3 g. A series of stability charts for slope inclinations of 1:1 ($\beta = 45°$), 4:7 ($\beta = 60°$), and 2:7 ($\beta = 75°$) (vertical to horizontal), internal friction angles of $20°$, $25°$, $30°$, and $35°$; and cohesions of 20 kPa, 25 kPa, 30 kPa, and 35 kPa are presented. These charts show the influence of water-filled cracks on the safety factor of a slope. In addition, this study further reveals the difference between a water-filled cracked slope and a dry cracked slope in terms of damage modes and axial force forms. The results can provide support for the stability evaluation of a water-filled cracked slope subjected to an earthquake.

1 INTRODUCTION

Tensile stress (such as seismic load or external static load) or drying-wetting cycles can cause cracks in a slope. The cracks can reduce the stability of the slope. Furthermore, when the slope is subjected to earthquakes, the preexisting cracks will lead to a significant decrease in slope stability (Senior 1981; Michalowski 2013; Utili 2013;). The extent to which vertical cracks reduce the safety factor of the slope mainly depends on their location and depth (Utili & Abd 2016). Cracks are not only a potential part of sliding surfaces but also channels

---

*Corresponding Authors: xuxingqian_123@163.com and xqu1987@163.com

110

DOI: 10.1201/9781003450832-14

through which rainwater can easily flow. This reduces the effective stress of the soil and imposes lateral stress, leading to failure when the cracks are filled with water.

Soil has a certain compressive and shear strength but the tensile strength is very low. After laying an appropriate amount of soil nails in the soil, the tensile strength of the soil can be increased and its deformation characteristics can be improved. Adding soil nails, such as geogrids, to slopes to form reinforced soil structures can improve the stability of slopes with vertical cracks affected by earthquakes and prevent slope slides. Therefore, reinforced retaining structures have attracted extensive attention from researchers in slope support engineering (Collin *et al.* 1992; Juran & Christopher 1989; Sandri 1997; Tatsuoka *et al.* 1997; 2007).

At present, numerical analysis methods, such as the limit equilibrium method, the finite difference method, the finite element method, and the limit analysis method, are often used to evaluate the seismic stability of reinforced soil structures. Considering the existence of cracks, scholars have used the limit equilibrium method to evaluate the stability of unreinforced and reinforced slopes by modifying the failure surface geometry (Chowdhury & Zhang 1991;Baker Leshchinsky 2001, 2003). Mendonca used the finite difference method to analyze the dynamic response characteristics and structural deformation performance of a geogrid-reinforced embankment subjected to an earthquake (Mendonça & Lopes 2011). Soltani used the finite element method to study the seismic response of a geogrid-reinforced slope top strip foundation and compared the response to that of an unreinforced strip foundation. The results showed that the acceleration response spectrum of the geogrid reinforced slope was completely different from that of the unreinforced slope (Soltani 2021). Ausilio *et al.* applied the kinematic theorems of limit analysis to derive expressions for different failure modes. This allows engineers to easily calculate the reinforcement force required to prevent slope failure and the yield acceleration required to withstand a seismic load (Ausilio *et al.* 2000). Porbaha *et al.* used the limit analysis method to study the stability of numerical models for reinforced and unreinforced slopes with centrifugal weightlessness (Porbaha *et al.* 2000). Abd and Utili used the limit analysis method to derive the functions of uniform cohesion and friction angle, tensile strength, shear angle, and slope inclination, and obtained design diagrams of the required reinforcement strength and embedment depth for slope stability conditions (Abd & Utili 2017). Chehade *et al.* used a discrete technique and a limit analysis method to analyze the internal stability of reinforced soil retaining walls under seismic loads. They discussed the influence of cracks, seismic load, soil properties, heterogeneity, and layered soil on structural stability (Chehade *et al.* 2022).

In summary, the analytical theory of reinforced cracked slopes subjected to an earthquake has not been fully studied. At present, the following challenges exist in the study of the seismic performance of reinforced cracked slopes: (1) there is a lack of quantitative research on the dynamic stability of dry cracked and water-filled cracked reinforced slopes. (2) The changing trend of the soil nail axial force and strain in a reinforced cracked slope subjected to earthquake needs to be further clarified. Given this, this study uses the finite element limit analysis method to study the changing trend of the safety factor, axial force, failure mode, and strain difference of a geogrid-reinforced cracked slope subjected to an earthquake.

2 NUMERICAL MODELING

2.1 *Finite element limit analysis*

The commonly used slope stability analysis methods include the limit equilibrium method, the finite element method, and the limit analysis method. Among these methods, the limit equilibrium method has been widely used in geotechnical engineering because it has a simple calculation process and a clear physical concept. However, each of these methods has its disadvantages. For example, the limit equilibrium method cannot strictly consider the stress equilibrium equation of the soil. The calculation process of the finite element method is complex, the calculation results are affected by many factors, and the results often require subjective experience judgment. It is very difficult to construct a suitable velocity field and stress field with the limit analysis method. Therefore, the application of the limit analysis method is greatly limited. Compared with other calculation methods, the finite element limit analysis method has the advantages of low calculation costs, high calculation accuracy, and strong applicability. The finite element limit analysis method has been widely utilized in geotechnical engineering and this method has been successfully applied to various research projects in geotechnical engineering (Lyamin & Sloan 2002a,b; Krabbenhoft & Lyamin 2015;Li et al. 2019). Therefore, the finite element limit analysis method is used for calculations in this study.

2.2 Calculation conditions

The slope model in this research comes from the study of Abd and Utili (2017) with a height of 60 m. The slope geometry, crack distribution, and soil nailing grid spacing distribution are shown in Figure 1. The crack is a vertical crack with a length of 3.28 m and a horizontal distance of 2.2 m from the top of the slope. The spacing of the geogrid is equidistantly distributed. A total of six layers of soil nails are laid with an interval distance of 1.67 m. The length of a single steel bar is 10.2 m, and the distance between the top and bottom steel bars from the top and bottom of the slope is 0.825 m. The model is subjected to horizontal seismic acceleration and the quasi-static method is used to consider the influence of an earthquake. For simplicity, the computational model conforms to the following assumptions. ① The influence of strain is not considered; ② The influence of strength loss of rock and soil under earthquake action is not considered; ③ The influence of dynamic pore water pressure is not considered; ④ The influence of vertical seismic action is not considered; ⑤ The influence of elevation amplification effect is not considered; ⑥ The influence of geomorphic factor is not considered; ⑦ The soil property is uniform and isotropic.

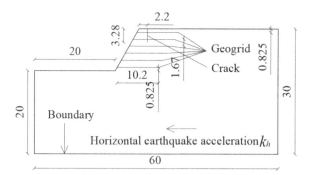

Figure 1. Slope geometry (unit: m).

The numerical model is generated using an adaptive grid method. The number of adaptive iterations is 3, the initial element number is 1000, and the maximum element number is 2000. The control variable generated by the adaptive mesh is shear dissipation. The finite element discretization results and boundary conditions of the upper bound solution of the Mohr-Coulomb failure criterion are shown in Figure 2 without considering the effects of earthquakes and fractures. The slope safety factor is obtained with the strength reduction method. In the model of this study, horizontal constraints are set on the side boundary and fixed constraints are set on the bottom boundary.

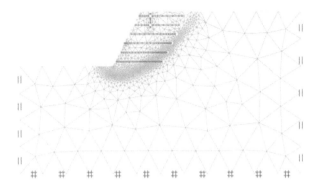

Figure 2. Finite element discretization results and boundary conditions.

The geogrid of the reinforced soil is simulated by a geogrid unit. The design stiffness EA is 450 kN/m and the yield force np is 45 kN/m. The slope model is homogeneous soil and the soil material adopts the Mohr-Coulomb constitutive model. The unit weight (γ) is 20 kN/m^3, the elastic modulus is 30 MPa, and the Poisson's ratio is 0.25. In addition, the mechanical parameters of the fracture are the cohesion of 0 kPa and internal friction angle of 15.0°. The calculation conditions are shown in Table 1. The cracks are divided into water-filled cracks and dry cracks. According to the orthogonal test design principle, a total of 384 calculations are performed.

Table 1. Calculation conditions.

c (kPa)	k_h (g)	φ (deg)	β (deg)
20	0	20	45
25	0.1	25	60
30	0.2	30	75
35	0.3	35	

3 RESULTS ANALYSIS

3.1 Slope safety factor

Figures 3–5 show the changing trend of the safety factor of the water-filled cracked slope and the dry cracked slope subjected to an earthquake when β = 45°, 60°, and 75°. In general, the slope safety factor decreases with the increase in the earthquake and slope inclination, showing a negative correlation, and increases with the increase in the soil shear strength, showing a positive correlation. Whether the cracks are filled with water affects the safety coefficient of slopes. The safety factor for water-filled cracked slopes is lower than for dry cracked slopes. When β = 45° and φ = 20°, the maximum safety factor and the minimum safety factor of the dry cracked slope are 1.96 and 1.235, respectively. The maximum safety factor and the minimum safety factor of the water-filled cracked slope are 1.863 and 0.891, respectively (the slope safety factor is less than 1, indicating that the slope is unstable). When β = 60° and φ = 20°, the maximum safety factor and the minimum safety factor of the dry cracked slope are 1.882 and 0.918, respectively. The maximum safety factor and the minimum safety factor of the water-filled cracked slope are 1.784 and 0.868, respectively. When β = 75° and φ = 20°, the maximum safety factor and the minimum safety factor of the dry cracked slope are 1.762 and 0.856, respectively. The maximum safety factor and the minimum safety factor of the water-filled cracked slope are 1.703 and 0.85, respectively.

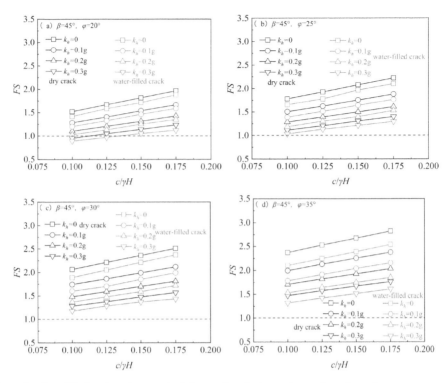

Figure 3. β = 45°, the changing trend of slope safety factor under earthquake action.

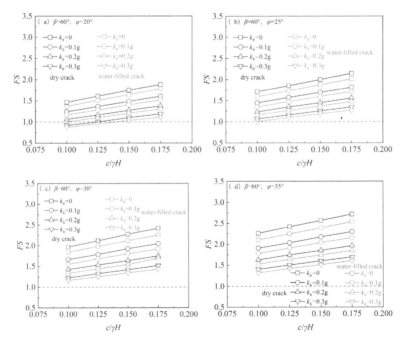

Figure 4. β = 60°, the changing trend of slope safety factor under earthquake action.

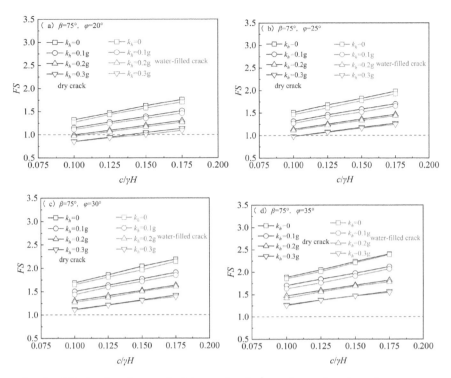

Figure 5. β = 75°, the changing trend of slope safety factor under earthquake action.

The influence of water-filled cracks on the slope stability subjected to earthquakes is further analyzed, as shown in Figure 6. Point A in Figure 6a shows the calculation process. The horizontal axis △FS1-△FS4 indicate the calculation results using cohesion c = 20kPa, 25kPa, 30kPa, and 35kPa. The ordinate is the slope safety factor when kh = 0.1 g, 0.2 g, or 0.3 g. Figure 6 shows that when k_h is low (0.1–0.2 g), the water-filled crack has little influence on the slope safety factor subjected to an earthquake. When k_h = 0.3 g, whether the water-filled crack has a significant influence on the slope safety factor subjected to earthquakes. Specifically, for Point A in Figure 6a, when the crack is filled with water, the slope safety factor change rate is 0.666. For the dry crack, the slope safety factor change rate is 0.587.

In summary, when the slope with water-filled cracks is subjected to a strong earthquake, the shear strength of the soil is reduced, thus the slope safety factor will be greatly reduced. Therefore, in practical engineering, taking engineering measures to limit the water content of the internal cracks of slopes is the key to slope support design.

3.2 Failure mode

Figures 7–10 show the shear dissipation changes of the dry cracked slope and the water-filled cracked slope subjected to earthquakes when β = 45° and 60°, c = 20 kPa, and φ = 20°. The different colors represent different shear dissipation and the size of the shear dissipation increases with the change of color. The shear dissipation represented by the colors is a relative value, not an absolute value. The possible potential sliding surfaces are observed with color comparison (Peng et al. 2022). The combined analysis indicates the presence of shear dissipation penetration on the slope. The area starts from the toe of the slope and extends upward along the arc to the top of the slope. The potential sliding surface slides in a

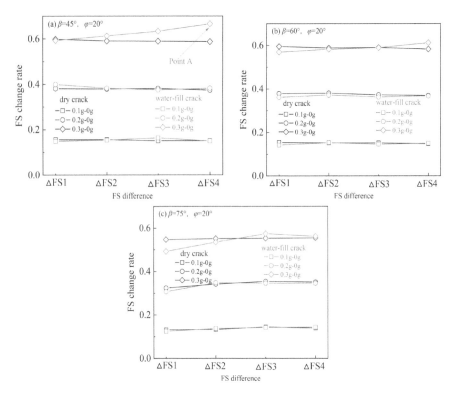

Figure 6. $\varphi = 20°$, the change rate of slope safety factor under earthquake.

Figure 7. $\beta = 45°$, the shear dissipation changes of dry fracture slope.

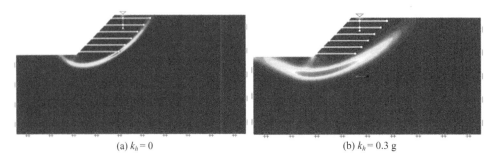

Figure 8. $\beta = 45°$, the change of shear dissipation of water-filled cracked slope.

Figure 9. β = 60°, the change of shear dissipation of dry crack slope.

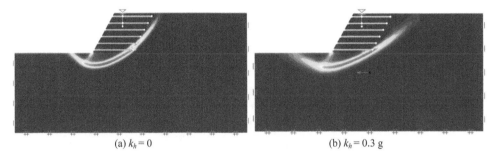

Figure 10. β = 60°, shear dissipation changes of water-filled cracked slope.

steep arc. The shear dissipations outside the potential sliding surface are very small and some are close to zero. These areas usually do not undergo large damage and deformation. In addition, in the absence of seismic loading (k_h = 0), the position of the potential sliding surface of the slope is almost unchanged, showing an arc shape that is basically along the end of the soil nailing.

Compared with the potential sliding surface of the slope without the seismic load, the position of the potential sliding surface of the slope under a seismic load of 0.3 g changes significantly. The upper half of the sliding surface is separated from the end of the soil nail. In particular, when the crack is filled with water, the upper half of the sliding surface is more obvious, and the rate of shear dissipation near the toe of the slope is very high. This means that shear failure will occur near the foot of the slope. At this time, the sliding force provided by the soil in the water-filled cracked slope is greater than the anti-sliding force of the slope. This is because the water reduces the shear strength of the fracture and the upper part of the slope is pulled, weakening the integrity. The safety factor obtained using numerical calculation is 0.868, indicating that the water-filled cracked slope is in an unstable state. The water-filled crack can significantly reduce the dynamic stability of the slope.

Figures 11–14 show the horizontal displacement changes of the dry cracked slope and the water-filled cracked slope subjected to an earthquake when β = 45° and 60°, c = 20 kPa, and φ = 20°. The comprehensive analysis shows that horizontal displacement near the toe of the slope is larger than that at the top of the slope. This is because the toe of the slope is pushed laterally and the embankment slope slides greatly along the horizontal direction, which is consistent with the situation revealed by the shear dissipation diagram. The positive area of horizontal displacement is mainly distributed beyond the potential sliding surface of the slope. The region produces a small displacement along the opposite direction of the seismic load. Its value is negligible and this displacement is opposite to the sliding direction of the slope. Therefore, the displacement value is expressed as a positive number. The maximum

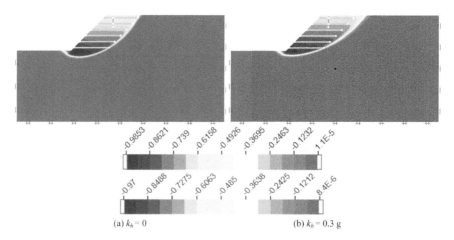

Figure 11. β = 45°, horizontal displacement changes of dry cracked slope.

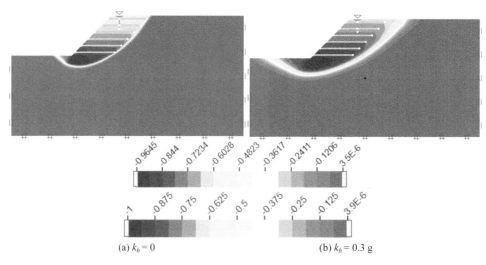

Figure 12. β = 45°, horizontal displacement changes of water-filled cracked slope.

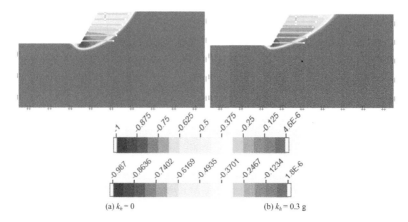

Figure 13. β = 60°, the horizontal displacement variation of the dry cracked slope.

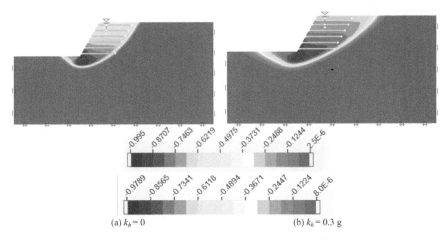

Figure 14. β = 60°, the horizontal displacement change of the water-filled cracked slope.

horizontal displacement values of the water-filled cracked slope and dry cracked slope are close to -1 but the range of the maximum horizontal displacement of the water-filled cracked slope is significantly larger than that of the dry cracked slope. It indicates that water-filled cracks contribute to the development of the horizontal displacement of slopes. This is mainly due to the redistribution of the soil stress caused by the water-filled cracks under seismic loads, which weakens the integrity of the slope.

3.3 *Axial force*

Figures 15 and 16 show the changes in the axial force of the soil nails for the dry cracked slope and the water-filled cracked slope at different slope inclinations without considering an earthquake (c = 20 kPa, φ = 20°). The comprehensive analysis shows that the top soil nailing force is the smallest and increases gradually from the top to the bottom. When the slope inclination is 45° or 60°, the soil nails in the 2nd-6th layers are generally in the form of large in the middle and small in the two ends. The soil nails in the first layer are not sufficiently stressed, showing that the first half of the force is small and the second half is large, and the axial force is far from the yield strength of the soil nails (45 kN/m). When the slope inclination increases to 75°, the soil nails of each layer are fully stressed. The form of the force is large in the middle and small at the ends. In addition, for slopes with an inclination of 45°, the axial force in the third layer of soil nails grows fastest. For a slope inclination of 60°, the axial force growth rate of soil nails in the second layer is the fastest. For a slope inclination of 75°, the axial force growth rate of soil nails in the first layer is the fastest. The slope inclination has an obvious influence on the location of the rapid growth of the axial force of soil nails.

Comparing Figure 15 and Figure 16, it can be seen that in the 4th-6th layers of soil nails, whether the cracks are water-filled or not has little effect on the axial force of soil nails. When the slope inclination is 45°, the growth rate of the axial force of the soil nails in the dry cracked slope is greater than that in the water-filled cracked slope in the 1st-3rd layers of soil nails. The growth rate of the axial force of the soil nails gradually decreases with the increase in slope inclination. This indicates that when the cracks are filled with water, the soil nails quickly perform their performance to further increase the integrity and anti-sliding force of the slope. Therefore, it is necessary to check the location of cracks and achieve as good drainage as possible in practical projects. This is done to reduce the effect of water on soil nail erosion and to avoid water reducing the shear strength of the slope soil.

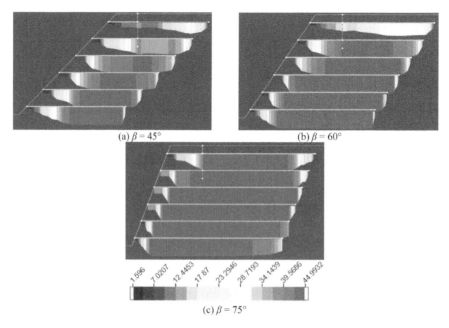

Figure 15. $k_h = 0$, the axial force changes of soil nails in a dry cracked slope.

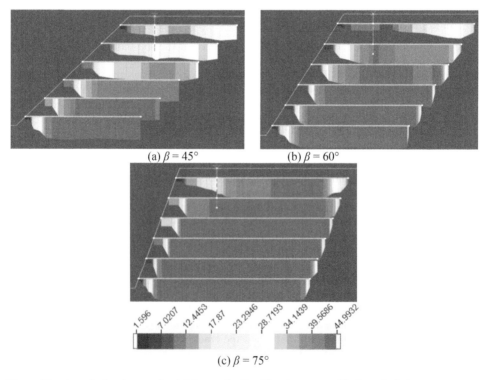

Figure 16. $k_h = 0$, the change of axial force of soil nailing in a water-filled cracked slope.

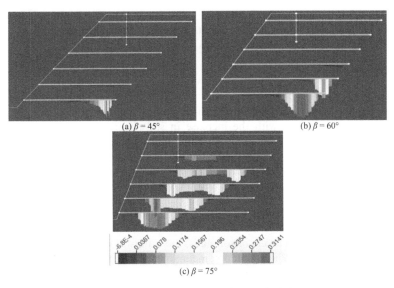

Figure 17. $k_h = 0$, the strain change of soil nailing in a dry cracked slope.

Figure 17 and Figure 18 show the strain changes of soil nails on the dry cracked slope and the water-filled cracked slope with different slope inclinations when the earthquake is not considered (c = 20 kPa, φ = 20°). It can be seen from the comprehensive analysis that when the slope inclination is 45° or 60°, the soil nails in the 1st-3rd layers hardly produce any strain. The strains of soil nails are gathered at the end and show a "peak shape", while the soil nails in layers 4th-6th show a strain trend that gradually increases from top to bottom. The soil nail strain gradually increases from the top to the bottom with the increase in the soil depth when the slope inclination increases to 75°. The maximum value of the soil nail

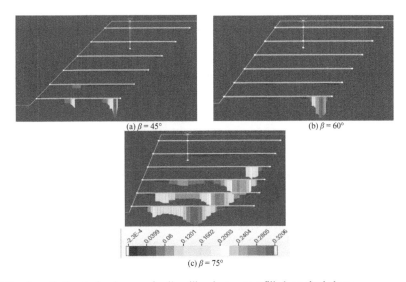

Figure 18. $k_h = 0$, the strain change of soil nailing in a water-filled cracked slope.

strain moves from the tail to the front. The maximum value of the soil nail strain on the water-filled cracked slope is greater than that on the dry cracked slope.

4 CONCLUSIONS

1. The slope safety factor decreases with the increase in slope inclination, showing a negative correlation, and increases with the increase in soil shear strength, showing a positive correlation. Water-filled cracks affect the safety factor of the slope and the safety factor of water-filled cracked slopes is lower than that of the dry cracked slope.
2. Compared with the dry cracked slope, the shear dissipation degree near the foot of the water-filled cracked slope is very high and the "detachment" phenomenon in the half of the sliding surface is more obvious. The range of the maximum horizontal displacement of the water-filled cracked slope is significantly larger than that of the dry cracked slope.
3. The slope inclination significantly affects the position of the rapid growth of the soil nail axial force. In the 4th–6th layers of soil nails, the water-filled crack has little effect on the axial force of soil nails. When the slope inclination is 45°, the axial force growth rate of soil nails in the dry cracked slope is greater than that in the water-filled cracked slope in the 1st-3rd layers. The axial force growth rate of soil nails decreases with the increase in slope inclination.
4. When the slope inclination is large ($\beta = 75°$), the soil nail strain increases gradually from top to bottom. With the increase in the soil depth, the maximum value of the soil nail strain moves from the tail to the front and this value is greater in the water-filled cracked slope compared to the dry cracked slope.
5. The lateral deformation of the soil is produced under self-weight and the seismic load and the reinforced slope can transfer the lateral deformation of the soil to the nails. The lateral deformation of the soil is limited due to the large tensile modulus of the soil nails. Therefore, the next research direction of this study is to optimize the flexible reinforcement scheme in the presence of water-filled cracks in the actual design process.

ACKNOWLEDGMENTS

This study has been financially supported by the National Natural Science Foundation of China (Grant No. 41867040), National Natural Science Foundation of China (Grant No.42107195), Qiandongnan Science and technology plan project ((2022) No.51), and the National Natural Science Foundation of China (Grant No. 52069007).

REFERENCES

Utili, S. 2013. An Investigation by Limit Analysis on the Stability of Slopes with Cracks *Geotechnique.* 63(2) 140–154.

Michalowski, R. L. 2013. Stability Assessment of Slopes with Cracks Using Limit Analysis *Canadian Geotechnical Journal.* 50(10) 1011–1021.

Senior, R. B. 1981. Tensile Strength, Tension Cracks, and Stability of Slopes *Soils and Foundation.* 21(2) 1–17.

Utili S and Abd A H 2016 On the Stability of Fissured Slopes Subject to Seismic Action. *International Journal for Numerical and Analytical Methods in Geomechanics.* 40(5) 785–806

Juran, I. and Christopher, B. 1989. Laboratory Model Study on Geosynthetic Reinforced Soil Retaining Walls *Journal of Geotechnical Engineering.* 115(7) 905–926.

Collin, J. G., Chouery-Curtis, V. E., and Berg, R. R. 1992. Field Observations of Reinforced Soil Structures Under Seismic Loading *International symposium on earth reinforcement practice.* (pp.223–228).

Sandri, D. 1997. A Performance Summary of Reinforced Soil Structures in the Greater Los Angeles Area After the Northridge Earthquake *Geotextiles and Geomembranes.* 15(4-6) 235–253.

Tatsuoka, F., Tateyama, M., Uchimura, T. 1997. Geosynthetic-reinforced Soil Retaining Walls as Important Permanent Structures 1996-1997 mercer Lecture *Geosynthetics International.* 4(2) 81–136.

Tatsuoka, F., Tateyama, M., Mohri, Y., Matsushima, K. 2007. Remedial Treatment of Soil Structures Using Geosynthetic-reinforcing Technology *Geotextiles and Geomembranes.* 25(4-5) 204–220.

Chowdhury, R. N., and Zhang, S. 1991. Tension Cracks and Slope Failure. In: Proceedings of the International Conference: Slope Stability Engineering, Developments, and Applications *J. London: Thomas Telford.* (pp.27–32).

Baker, R. and Leshchinsky, D. 2001. Spatial Distribution of Safety Factors *Journal of Geotechnical and Geoenvironmental Engineering.* 127(2) 135–145.

Baker, R. and Leshchinsky, D. 2003. Spatial Distribution of Safety Factors: Cohesive Vertical cut *International Journal for Numerical and Analytical Methods in Geomechanics.* 27(12)1057–1078.

Mendonça, A. and Lopes, M. L. 2011. Role of the Damping Ratio of Reinforcement on the Behavior of Geogrids-reinforced Systems *Geotechnical and Geological Engineering.* 29(3) 375–388.

Soltani, N. 2021. Seismic Response Evaluation of Strip Footing on Geogrid-reinforced Slope *Innovative Infrastructure Solutions.* 6(4)1–9.

Ausilio, E., Conte, E., and Dente, G. 2000. Seismic Stability Analysis of Reinforced Slopes *Soil Dynamics and Earthquake Engineering.* 19(3)159–172.

Porbaha, A., Zhao, A., and Kishida, T. 2000. Upper Bound Estimate of Scaled Reinforced Soil Retaining Walls *Geotextiles and Geomembranes.* 18(6) 403–413.

Abd, A. H., and Utili, S. 2017. Design of Geosynthetic-reinforced Slopes in Cohesive Backfills *Geotextiles and Geomembranes.* 45(6) 627–641.

Chehade, H. A., Dias, D., Sadek, M. Jenck, O., and Chehade, F. H. 2022. Seismic Internal Stability of Saturated Reinforced Soil Retaining Walls Using the Upper Bound Theorem of Limit Analysis *Soil Dynamics and Earthquake Engineering.* 155(107180).

Lyamin, A. V. and Sloan, S. W. 2002a. A Lower Bound Limit Analysis Using Nonlinear Programming *International Journal for Numerical Methods in Engineering.* 55(5)573–611.

Lyamin, A. V. and Sloan, S. W. 2002b. Upper Bound Limit Analysis Using Linear Finite Elements and Non-linear Programming *International Journal for Numerical and Analytical Methods in Geomechanics.* 26(2) 181–216.

Krabbenhoft, K. and Lyamin, A. V. 2015. Strength Reduction Finite Element Limit Analysis *Géotechnique Letters.* 5(4) 250–253.

Li, A. J., Qian, Z., Jiang, J. C., and Lyamin, A. 2019. Seismic Slope Stability Evaluation Considering Rock Mass Disturbance Varying in the Slope *KSCE Journal of Civil Engineering.* 23 (3)1043–1054.

Peng, W. Z., Zhao, M. H., Zhao, H. 2022. Seismic Stability of the Slope Containing a Laterally Loaded Pile by Finite-Element Limit Analysis *International Journal of Geomechanics.* 22(1) 06021033.

Water Conservancy and Civil Construction – Oke & Ahmad (Eds)
© 2024 The Author(s), ISBN: 978-1-032-58618-2

Research on the real-time scheduling model of pumping station group based on differential evolution algorithm – take the Jin'an river basin as an example

Yang Chen
School of Resources & Civil Engineering, Northeastern University, Shenyang, China

Chao Wang*
State Key Laboratory of Simulation and Regulation of Water Cycle in River Basin, China Institute of Water Resources and Hydropower Research, Beijing, China

Shuanglin Jiang
Research Center of Fluid Machinery Engineering and Technology, Jiangsu University, Zhenjiang, China

Chen Ji
College of Water Resource and Hydropower, Sichuan University, Chengdu, China

Dianchang Wang
Yangtze Ecology and Environment Co., Ltd, Wuhan, China
Chongqing Three Gorges Eco-Environmental Technology Innovation Center Co., Ltd, Chongqing, China

Ruozhu Shen
Beijing Capital Eco-Environment Protection Group Co., Ltd. Beijing, China

Haocheng Huang
State Key Laboratory of Simulation and Regulation of Water Cycle in River Basin, China Institute of Water Resources and Hydropower Research, Beijing, China

ABSTRACT: Given the frequent urban flooding events caused by extreme rainstorms, and considering the actual engineering problems faced by the Jin'an river basin in Fuzhou and the influence of tidal level changes on the joint scheduling of pumping stations, this paper constructs a real-time scheduling model for a cluster of terrace drainage pumping stations considering the influence of tidal level, and obtains a joint scheduling scheme for the terrace pumping stations under different tidal levels and simultaneous rainstorms in the joint optimal scheduling study of terrace drainage pumping stations. The simulation results show that the optimal dispatching scheme has a better flood control effect, cost, and ecological water level restoration effect than the traditional dispatching rules, which can provide technical support for the development of the optimal dispatching scheme of drainage pumping stations in the watershed and improve the ability of the city to cope with the rainstorm.

1 INTRODUCTION

In recent years, the probability and intensity of extreme water storm events in coastal cities have increased significantly due to the greenhouse effect (Kirk-Davidoff 2018; Guo etal. 2017). For coastal towns with developed water systems, rapid urbanization has led to

---

*Corresponding Author: wangchao@iwhr.com

124   DOI: 10.1201/9781003450832-15

changes in land use, resulting in a decrease in infiltration and an increase in runoff (Suriya & Mudgal 2012), leading to frequent flood disasters, heavy economic losses, and an impact on everyday work, production, and life. Therefore, how to alleviate the economic loss caused by urban waterlogging has attracted widespread attention among the general public (Huang et al. 2008).

The large-scale drainage pumping station is one of the vital flood drainage and flood prevention facilities related to the safety of Fuzhou city. How to improve the economy and security of the operation of the cascade pumping station while ensuring the drainage effect and how to quickly restore the ecological water level of the river after the flood disaster are the focuses of this paper. Therefore, based on the flood drainage and waterlogging prevention work in Fuzhou City, the river basin, which has the main drainage task, is selected as the study area, and a real-time scheduling model based on penalty function and differential evolutionary algorithm (Chiou et al. 2004; Liu 2005; Piotrowski & Adam 2014; Storn & Price 1997; Spadoni & Stefanini 2009) for a cluster of stepped drainage pumping stations considering the influence of tide level is established, combined with the study of optimal operation of the pumping station cluster to improve the basin with flood control capacity.

2 CONSTRUCTION OF REAL-TIME DISPATCHING MODEL FOR CASCADE DRAINAGE PUMPING STATION

2.1 *Objective function*

(1) The peak water level of Qinting Lake is the lowest.

Reducing the peak water level of Qinting Lake as much as possible to ensure the flood control safety of the Qinting Lake area is an important dispatching objective of the joint optimization operation of the pumping station groups. The specific objective function is expressed as follows.

$$\min F_1 = \min\{\max Z_{1,t}\} \quad t = 1, 2 \ldots T \tag{1}$$

Where $Z_{1,t}$ represents the water level of Qinting Lake at time t, and T is the total number of scheduling periods.

(2) The peak water level of Guangminggang is the lowest.

Reducing the water level of Guangminggang to ensure the upstream essential flood area flood control safety is another joint optimal dispatching goal of the must-go pumping station group. Considering that there is no water conservancy project at both ends of Qingting Lake to Guangminggang, the water level of Guangminggang is used to reflect the regional flood control safety and the specific objective function is expressed as follows.

$$\min F_2 = \min\{\max Z_{2,t}\} \quad t = 1, 2, \ldots T \tag{2}$$

Where $Z_{2,t}$ represents the water level of Guangminggang at time t.

(3) The pumping station has the lowest power consumption.

On the premise of ensuring the flood control safety of all regions, reducing the power consumption of pumping stations as much as possible is another dispatching objective that should be considered in the joint optimal dispatching of pumping station groups. The objective function is expressed as follows.

$$\min F_3 = \min \sum_{t=1}^{T} \sum_{n=1}^{N} \frac{\rho g Q_{n,t} H_{n,t}}{1000\eta} \Delta T + K \cdot \Delta T \cdot a \quad t = 1, 2, \ldots T \quad n = 1, 2 \tag{3}$$

Where N represents the number of pumps; n = 1 represents the Qinting Lake pumping station; n = 2 represents the Kuiqi drainage station; ρ is the density of water; g is gravitational acceleration: $Q_{n,t}$ denotes the flow rate of the pumping station n at time t; $H_{n,t}$ denotes the head of the pumping station n at time t; ΔT represents the length of a single period; K indicates the operating cost in addition to energy consumption (million yuan/hour*unit); a is the number of starting pumps.

(4) The difference between the water level at the end of the operation period and the ecological water level of Qinting Lake is the smallest.

$$\min F_4 = \min\{Z_{1,t} - h_1\} \quad t = 24 \tag{4}$$

(5) The difference between the water level at the end of the operation period and the ecological water level of Guangminggang is the smallest.

$$\min F_5 = \min\{Z_{2,t} - h_2\} \quad t = 24 \tag{5}$$

2.2 Constraints

The following constraints need to be considered for the joint optimal dispatching model of the pumping station group in the Jin'an River Basin.

(1) Water level constraint:

$$Z_n^{\min} \leq Z_{n,t} \leq Z_n^{\max} \quad n = 1,2 \tag{6}$$

Where $Z_n^{\min}$ and $Z_n^{\max}$ are the upper and lower upstream water level constraints for pumping station n, respectively.

(2) Flow constraint:

$$Q_n^{\min} \leq Q_{n,t} \leq Q_n^{\max} \quad n = 1,2 \tag{7}$$

$$Q_n^{\min} = 0m^3/s, \quad Q_1^{\max} = 25m^3/s, \quad Q_2^{\max} = 120m^3/s \tag{8}$$

Where $Q_n^{\min}$ and $Q_n^{\max}$ are the upper and lower limits of drainage capacity of pump station n, respectively.

3 MODEL SOLUTION

The real-time scheduling model of the cascade drainage station is a multi-objective optimization scheduling problem. For this kind of optimization problem, this paper transforms the multi-objective problem into a single-objective problem by constructing a penalty function and solves it using a differential evolutionary algorithm. The specific processing methods are as follows.

$$\min F = \min\left\{\left(F_1 - F_1^{obj}\right) \cdot \alpha_1 + \left(F_2 - F_2^{obj}\right) \cdot \alpha_2 + F_3 + F_4 + F_5\right\} \tag{9}$$

$$\alpha_1 = \begin{cases} 0, & F_1 - F_1^{obj} \leq 0 \\ 1 \times 10^{10}, & F_1 - F_1^{obj} > 0 \end{cases} \quad \alpha_2 = \begin{cases} 0, & F_2 - F_2^{obj} \leq 0 \\ 1 \times 10^{9}, & F_2 - F_2^{obj} > 0 \end{cases} \tag{10}$$

Where α_1 and α_2 represent the penalty factor of scheduling objectives 1 and 2, respectively.

4 CASE STUDY

4.1 Study area

The study area covers an area of 73.52 square kilometers and its location is shown in Figure 1. Jin'an River is the longest urban inland river in Fuzhou, spanning the north and south of the Jin'an district, with a total length of 6,682.7 meters and a width of 24-54 meters.

Qinting Lake is located in the upper reaches of the Jin'an river and the water area is 19 hectares. The lake receives waterlogging water from Xindian Stream, Masa stream, Xiafang Stream, and Jiefang Stream, with a catchment area of 43 km^2. The water level of Qinting Lake is generally kept at about 4.50 m. The ground elevation around Qinting Lake is above 7.80 m. When the lake level is below 7.80 m, there is no waterlogging around the lake.

Guangminggang is the downstream water system of the Jin'an River. The incoming water from the Jin'an River needs to be discharged into the Minjiang River through Guangminggang. The river is about 6,500 m long and its endpoint is the Kuiqi drainage station.

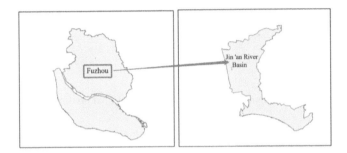

Figure 1. Study area.

4.2 Project overview

Qinting Lake is located at the upper end of the Jin'an River with a storage capacity of 1.11 million m^3. Qinting Lake pumping station is located at the outlet of Qinting Lake, with 5 pumps and a total design flow of 25 m^3/s. Guiqi drainage station, situated at the end of Guangminggang, is responsible for the main drainage task. There are 3 pumps with a total design flow of 120 m^3/s. When the tide level is higher than the water level of the inland river, it is necessary to close the Kuiqi sluice to prevent the tide from topping off. When the river level rises to the warning water level, the Kuiqi drainage station is opened to drain the water. When the water level falls to the warning water level, the Kuiqi pumping station unit is

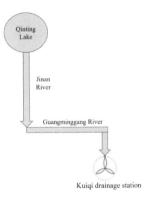

Figure 2. River basin generalization map.

closed. When the tide level of the Minjiang River is lower than the water level of the inland river, the Kuiqi sluice for drainage is opened without opening the Kuiqi pump station.

4.3 Data input

To verify the dispatching effect of the real-time dispatching model of the cascade drainage pumping station group, the once-in-a-century flood is selected and the simulation results of two scenarios of flood peak meeting high tide level and flood peak meeting low tide level are analyzed, respectively. Figure 3 shows the flooding process and tide level process.

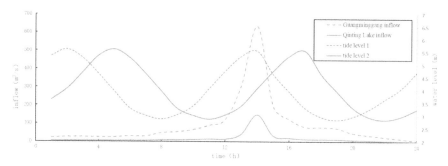

Figure 3. Qinting Lake, Guangminggang inflow process, and tidal process.

5 RESULTS AND DISCUSSION

Taking the typical flood scenario as an example, the joint optimal dispatching model of the Qinting Lake and Kuiqi cascade pumping station is simulated. The changes in water level and outflow during the dispatching period are shown in Figures 4, 5, 6, and 7. The comparison of the dispatching effect is shown in Table 1.

It can be seen from the simulation results that in the case of a once-in-a-century design flood:

(1) When the flood peak meets the low tide level

In terms of the flood prevention effect, the peak water levels of Qinting Lake and Guangminggang are 4.74 m and 5.17 m, respectively, which do not exceed the warning level. The water levels are reduced by 33.0% and 4.7%, respectively, compared with the regular scheduling plan. In terms of the cost, the energy consumption is 58,700 kWh and the number

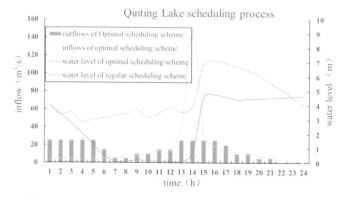

Figure 4. Qinting lake scheduling process (flood peak meets low tide level).

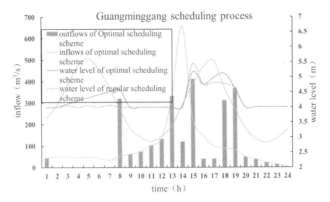

Figure 5. Guangminggang scheduling process (flood peak meets low tide level).

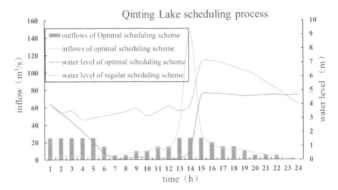

Figure 6. Qinting lake scheduling process (flood peak meets high tide level).

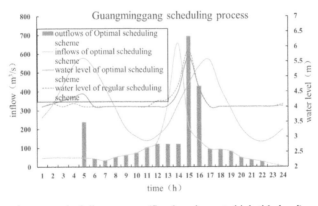

Figure 7. Guangminggang scheduling process (flood peak meets high tide level).

of times the pumping station is opened and closed is 26, which is reduced by 37% and 37%, respectively, compared with the conventional scheduling scheme. In terms of ecological water level restoration, the water levels of Qinting Lake and Guangminggang are 4.66 m and 4 m at the end of the dispatching period, both of which are restored to the landscape water level.

Table 1. Effect comparison of conventional and optimized scheduling schemes for cascade pumping stations.

	Optimization Goals	Regular Scheduling	Optimized Scheduling
Low tide level	Power consumption (kWh)	92800	58700
	Opening and closing times	41	26
	Peak water level of Qinting Lake (m)	7.10	4.74
	Peak water level of Guangminggang (m)	5.43	5.17
High tide level	Power consumption (kWh)	122800	100700
	Opening and closing times	41	24
	Peak water level of Qinting Lake (m)	7.10	4.74
	Peak water level of Guangminggang (m)	5.87	5.65

(2) When the flood peak meets the high tide level

In terms of the flood prevention effect, the peak water levels of Qinting Lake and Guangminggang are 4.74 m and 5.65 m, respectively, which do not exceed the warning level. The water levels are reduced by 33.0% and 3.7%, respectively, compared with the regular scheduling plan. In terms of the cost, the energy consumption is 100,700 kWh and the number of times the pumping station is opened and closed is 24, which is reduced by 37% and 41% compared with the conventional scheduling scheme. In terms of ecological water level restoration, the water levels of Qinting Lake and Guangminggang are 4.67 m and 4 m at the end of the dispatching period, both of which are restored to the landscape water level.

6 CONCLUSIONS

The influence of extreme rainstorms has led to an increase in the number of flooding incidents in the urban areas of Fuzhou year by year. Therefore, building the cascade drainage pumping station group of scheduling model considering the influences of tide level can generate a scheduling plan and provide a scientific basis for scheduling decision-making. Using typical flood scenarios as input conditions, the flood control effect, cost, and ecological water level recovery of the dispatching scheme are compared. The simulation results showed that (1) the optimized scheduling plan can reduce the flood peak and prevent the river from overflowing compared with the regular scheduling plan. (2) With a similar effect of optimized dispatching and conventional dispatching scheme, the optimized dispatching scheme consumes less energy and starts and closes pumps less often, which can effectively reduce cost and save energy. (3) For Guangminggang, the optimized scheduling scheme and the conventional scheduling scheme have the same effect on restoring the landscape water level, while for Qinting Lake, the optimized scheduling scheme has a better ability to restore the landscape water level.

ACKNOWLEDGMENTS

The authors thank the Editor and the anonymous reviewers for their valuable suggestions, which helped improve the quality of the paper. This study was supported by the National Key R&D Program of China (2021YFC3001405), Key Program of the National Natural Science Foundation of China (U2240203).

REFERENCES

Chiou, J. P., Chang, C. F., Su, C. T. Ant Direction Hybrid Differential Evolution for Solving Large Capacitor Placement Problems [J]. *IEEE Transactions on Power Systems*, 2004, 19(4):1794–1800.

Guo, X., Zhengfang, W. U., Haibo, D. U., *et al.* Spatio-temporal Change Characteristics and Circulation Causes of Extreme Precipitation in Fujian Province [J]. *Resources Science*, 2017.

Huang, X., Yu, J., Hua, W. The Characteristics, Causes and Prevention Countermeasure of Waterlogging Disaster in the Region of Jianghuai [J]. *Chinese Agricultural Science Bulletin*, 2008.

Kirk-Davidoff, D. The Greenhouse Effect, Aerosols, and Climate Change [J]. *Green Chemistry*, 2018:211–234.

Liu, M. G. Differential Evolution Algorithms and Modification [J]. *Systems Engineering*, 2005.

Piotrowski, Adam , P. Differential Evolution Algorithms Applied to Neural Network Training Suffer from Stagnation [J]. *Applied Soft Computing*, 2014, 21:382–406.

Suriya, S., Mudgal, B. V. Impact of Urbanization on Flooding: The Thirusoolam sub-watershed – A Case Study [J]. *Journal of Hydrology*, 2012, 412-413; 210–219.

Storn, R., Price, K. Differential Evolution–A Simple and Efficient Heuristic for Global Optimization Over Continuous Spaces [J]. *Journal of Global Optimization*, 1997, 11(4):341–359.

Spadoni, M., Stefanini, L. Handling Box, *Linear and Quadratic-Convex Constraints for Boundary Optimization with Differential Evolution Algorithms [J]*. 2009.

Water Conservancy and Civil Construction – Oke & Ahmad (Eds)
© 2024 The Author(s), ISBN: 978-1-032-58618-2

Study on the alignment characteristics of high bridge-tunnel ratio road sections based on Catmull-Rom spline curve—take the section of Baima-Longtan expressway as an example

Yudan Liu* & Xianghong Li*
School of Energy Science and Engineering, Henan Polytechnic University, Jiaozuo, Henan Province, China

ABSTRACT: The high bridge-tunnel ratio road section has the characteristics of frequent section switching and sudden changes in the longitudinal driving environment, which are unfavorable to drivers. To effectively analyze the geometric characteristics of the alignment of the high bridge-tunnel ratio road section in China, a road alignment is fitted based on the CatMull-Rom spline curve using Matlab program code while extracting the characteristic parameters. The results of this paper are as follows. Based on the visual geometric alignment model, it is used to analyze the causes of road accident-prone sections, find the design causes of the poor visual road environment, propose improvement measures, and optimize the design of the visual road environment.

1 INTRODUCTION

Since most of the highways in the central and western parts of the country pass through the mountains, there are a large number of high bridge-tunnel ratios and high-risk routes. The bridge-tunnel ratio is defined as the ratio of the sum of bridge mileage and tunnel mileage to the total road mileage [Cao 2022]. Mountain highway alignment is complex and has special structures. Due to the constraints of topography, geology, geomorphology, and other factors, there are usually a large number of curves, ramps, tunnels, bridges, cliffs, and high-drop sections in the design of different lines and their combinations. At the same time, there are often multiple tunnel connections, multiple bridge connections, or multiple tunnel and bridge intervals connected [Li 2015]. Road alignment has a great impact on the driver's driving status [Zhang 2022]. Research and evaluation of road alignment can effectively prevent traffic accidents. With the increase in the proportion of dense bridge-tunnel sections in the construction of highways, the current research results should also be in line with the development characteristics of mountainous highways. How to better analyze and evaluate the advantages and disadvantages of road alignment is an important question. It is a better method to load out the road alignment and extract and process the data information contained in it, which is more commonly used at home and abroad.

J Son et al. proposed a lane recognition algorithm for various illumination conditions, such as harsh climates and nighttime [J 2015]. Hur J et al. considered the lane boundary as a collection of small line segments, proposed an improved Hough transform to detect the small line segments, and clustered the small line segments based on similarity measures to achieve lane line recognition [J 2013]. Wang J et al. proposed a lane detection algorithm based on inverse perspective mapping [J 2014]. Rahmdel, P. S. et al. proposed a lane line detection method combining multilayer fractional Fourier transform (MLFRFT) and a state-of-the-

---

*Corresponding Authors: 212202020097@home.hpu.edu.cn, lixianghong001@126.com, 212202020097@home.hpu.edu.cn and lixianghong001@126.com

132   DOI: 10.1201/9781003450832-16

art lane detector [Rahmdel 2013]. For the study of linear geometric characteristics of road perspective, He Sihong et al. established a visual lane model to describe the information of road surface characteristics in the driver's eyes, rationally evaluate the road design, and analyze the principle that the road surface linear environment affects the driver's behavior [He 2013]. Chen Yuren et al. divided the traditional road into three identical functional partitions according to the perspective drawing method and proposed the basic concepts of shape parameters, such as apparent curve length, apparent curvature, and the corresponding calculation methods[Chen 2018].

The spline curves and parabolas commonly used to fit lines are Bézier curves and B spline curves, etc. The CatMull-Rom curve differs from the first two types of spline curves in that the fitted curve obtained by interpolation can pass through all control points and has strong adaptability, which makes the curve important in the control of the shape of the fitted line. At the same time, the CatMull-Rom curve is also locally adjustable, and changing a control point does not affect other control points. Therefore, the CatMull-Rom curve is more suitable for fitting road alignment than the other two types of curves, both from the perspective of control point location and curve fitting[Chen 2015]. And this paper constructs a road alignment fitting model based on the Catmull-Rom spline curve to study the geometric characteristics of the alignment of mountainous highway sections with high bridge-tunnel ratio, which can provide a theoretical basis for the prevention of traffic accidents on bridge-tunnel ratio sections.

2 TYPICAL ROAD SELECTION

The section investigated in this paper is the section from Baima to Longtan of Yuxiang Expressway, with a total length of about 254.12 km and four lanes in both directions, crossing the mountainous and heavy hilly area along the route, and the bridges and tunnels in the section are generally characterized as "many, long, and dense", accounting for 61.85% of the total, which is a typical mountainous high bridge-tunnel ratio. The whole section is divided into 14 basic sections by interchanges, and the basic road traffic information of each section is shown in Table 1.

Table 1. Basic road traffic information table.

Serial number	Road section	Length (km)	Tunnel ratio (%)	Bridge ratio (%)	Bridge-Tunnel Ratio (%)	Average daily traffic volume (Vehicles/day)	Number of Accidents (times)
1	Baima-Wulong	26.28	78.78	11.12	89.90	17236	303
2	Wulong-Huangcao	17.80	47.21	24.79	71.99	14839	81
3	Huangcao-Pengshui West	20.72	62.90	22.84	85.74	15571	68
4	Pengshui West - Pengshui East	10.32	58.26	20.43	78.70	12136	55
5	Pengshui East - Baojia	10.07	9.05	23.13	32.18	13643	45
6	Baojia-Qianjiang West	41.31	39.39	24.35	63.74	13854	180
7	Qianjiang West - Qianjiang South	13.20	70.01	13.88	83.89	12592	50
8	Qianjiang South - Maushui	15.82	11.82	21.87	33.69	13861	90
9	Zhuoshui-Yujiagou	10.30	4.59	28.75	33.33	13372	81
10	Yujiagou-Heishui	14.18	28.97	26.76	55.73	13102	106
11	Heishui-Taohuayuan	25.48	22.46	14.66	37.11	13121	209
12	Taohuayuan- Youyang South	12.78	22.35	32.56	54.91	12013	110
13	Youyang South-Banxi	4.77	0.00	20.12	20.12	12673	27
14	Banxi-Longtan	20.92	41.57	24.47	66.04	12636	104
15	Total	243.95	40.30	21.55	61.85	13922	1509

3 LINE SHAPE FITTING AND GEOMETRIC FEATURE ANALYSIS

3.1 *Mathematical basis of the curve*

Catmull-Rom spline curves are a series of cubic interpolation spline curves, which are formulated by computing the tangent line at each point p_i using the previous and next points of the spline curve $\tau(p_{i+1} - p_{i-1})$, and the geometric matrix is derived from Equation (1) [Twigg 2003]:

$$P(s) = \begin{bmatrix} 1 & u & u^2 & u^3 \end{bmatrix} \begin{bmatrix} 0 & 1 & 0 & 0 \\ -\tau & 0 & \tau & 0 \\ 2\tau & \tau - 3 & 3 - 2\tau & -\tau \\ -\tau & 2 - \tau & \tau - 2 & \tau \end{bmatrix} \begin{bmatrix} p_{i-2} \\ p_{i-1} \\ p_i \\ p_{i+1} \end{bmatrix} \quad (1)$$

3.2 *Lane centerline fitting and geometric feature parameter extraction*

After the image processing of the typical road section, the final road lane line and road median fitting effect map can be obtained by running the code, as shown in Figure 1. The control point spacing and directional angle of the control points are calculated based on the control points in the fitted road median and the extracted control point directional angle and control point spacing are shown in Table 2, followed by the feature analysis.

Figure 1. Final fitting effect.

3.3 *Road alignment characteristics analysis*

The model divides the lanes in the field of view into three regions. near view, middle view, and far view, and selects the length of the visual curve as the shape parameter of the model for each region. Therefore, three shape parameters are extracted for each lane line, where d12, d23, and d34 are the lengths of the visual curves in the near view, middle view, and far view, respectively, and Equation (2) for the length of the apparent curve is as follows.

$$d_{i(i+1)} = d_{i+1} - d_i \# \quad (2)$$

Table 2. Extracted control point orientation angles and control point spacing (partial).

Image serial number	q_1	q_2	q_3	q_4	d_{12}	d_{23}	d_{34}
1	101.23	108.64	117.74	115.4	100.93	38.77	14.18
2	109.59	102.59	100.48	101.34	145.85	42.05	14.43
3	101.01	123.77	130.62	127.89	136.27	34.86	13.91
4	132.46	138.8	142.04	141.07	172.07	61.23	22.79
5	85.66	73.64	62.63	66.13	119.56	35.75	13.2
6	92.3	218.72	210.64	214.33	99.79	33.02	13.08
7	82.03	45.26	14.8	22.51	119.31	31.95	13.41
8	104.69	252.19	248.5	250.01	114.96	30.17	11.07
9	74.48	50.09	20.98	25.62	129.23	42.19	18.77
10	74.34	83.6	78.93	80.05	90.03	31.74	11.58
11	89.21	89.5	88.02	88.77	100.46	35.37	12.7
12	68.57	18.92	9.86	13.09	126.98	47.13	16.47
13	88.27	58.73	25.87	31.74	110.37	38.15	17.02
14	79.79	49.69	41.91	44.68	102.07	36.43	12.71
15	66.93	61.12	66.25	65.72	71.39	50.03	19.25
16	74.21	89.54	94.1	93.49	75.38	27.3	10.06
17	87.54	96.48	117.45	118.87	98.15	26.99	10.39
18	70.41	89.21	79.73	87.95	94.54	29.19	11.27
19	86.26	82.23	74.24	75.09	71.01	46.91	19.69

q_i indicates the directional angle of the ith control point, i = 1, 2, 3, 4.
d_{12} indicates the distance between control points 1, 2; d_{23} indicates the distance between control points 2, 3; d_{34} indicates the distance between control points 3, 4.

Where di is the control point, which also indicates the accumulated curve length of the lane edge at the control point di, and i is taken 1, 2, and 3. The results of the correlation test of the shape parameters using SPSS software are shown in Table 3. It can be seen that d_{12}, d_{23}, and d_{34} correlate significantly with each other, as well as q_1, q_2, q_3, and q_4, but

Table 3. Correlation test of characteristic parameters.

		d_{12}	d_{23}	d_{34}	q_1	q_2	q_3	q_4
d_{12}	Pearson Correlation	1	-.385[**]	-.462[**]	.092	.133	.133	.128
	Sig.(two-tailed)		.000	.000	.386	.210	.212	.230
d_{23}	Pearson Correlation	-.385[**]	1	.966[**]	.068	.059	.050	.053
	Sig.(two-tailed)	.000		.000	.527	.579	.637	.621
d_{34}	Pearson Correlation	-.462[**]	.966[**]	1	.052	.063	.049	.054
	Sig.(two-tailed)	.000	.000		.628	.556	.648	.616
q_1	Pearson Correlation	.092	.068	.052	1	.538[**]	.632[**]	.622[**]
	Sig.(two-tailed)	.386	.527	.628		.000	.000	.000
q_2	Pearson Correlation	.133	.059	.063	.538[**]	1	.821[**]	.818[**]
	Sig.(two-tailed)	.210	.579	.556	.000		.000	.000
q_3	Pearson Correlation	.133	.050	.049	.632[**]	.821[**]	1	.999[**]
	Sig.(two-tailed)	.212	.637	.648	.000	.000		.000
q_4	Pearson Correlation	.128	.053	.054	.622[**]	.818[**]	.999[**]	1
	Sig.(two-tailed)	.230	.621	.616	.000	.000	.000	

Notes
[**]The correlation was significant at the 0.01 level (two-tailed).

Table 4. Normality test.

	Kolmogorov-Sminov (V)[a]			Shapiro Wilke		
	Statistics	Degree of freedom	Significance	Statistics	Degree of freedom	Significance
d_{12}	.072	90	.200[*]	.986	90	.479
d_{23}	.100	90	.027	.927	90	.000
d_{34}	.118	90	.004	.864	90	.000
q_1	.061	90	.200[*]	.983	90	.305
q_2	.177	90	.000	.797	90	.000
q_3	.135	90	.000	.912	90	.000
q_4	.138	90	.000	.897	90	.000

Notes
[*]This is the lower limit of true saliency.
[a]Riley's Significance Correction.

$d_{i(i+1)}(i=1,2,3)$ and $q_n(n=1,2,3,4)$ do not correlate significantly with each other. The correlation between $d_{i(i+1)}(i=1,2,3)$ and $q_n(n=1,2,3,4)$ is not significant. $d_{i(i+1)}(i=1,2,3)$ and $q_n(n=1,2,3,4)$ are tested for normality in a single set of data and Table 4 is obtained. It can be seen that only the data of d_{12} and q_1 are significantly correlated under the Kolmogorov-Sminov normality test, while the correlation is not significant in the Shapiro-Wilk normality test.

According to the available data analysis, from d_{12}, d_{23}, d_{34}, the correlation between each other is significant, it can be concluded that although driving on a high bridge-tunnel ratio road, the driver is still a gradual process of receiving near, medium, and far views when processing the information from the road ahead, the driver can see the road ahead of the field of view is limited. In the section of the road that can be seen, the driver has enough reaction time to make the corresponding operation. From q_1, q_2, q_3, q_4, significant linear correlation between each other can be derived, in the high bridge-tunnel ratio road section, the road centerline curve change is a gentle transition process. In the driver's field of vision, there is generally no sudden change in the line.

The data collected are sorted, the data with large errors are eliminated and imported into SPSS for analysis, and the results are calculated for the total distance between the extracted control points, as shown in Figure 2. It is assumed that this value is the normal sight distance

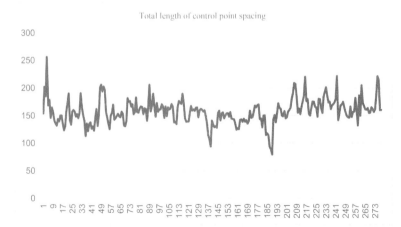

Figure 2. The total length of control point spacing.

within the driver's field of vision, samples above this value show open sight distance, while samples below this value show poor sight distance. Poor sight distance is largely caused by poor alignment, so we should focus on the analysis of the poor sight distance section, compare it with the actual road map, find the reasons for the poor alignment of the visual road environment, propose improvement measures, and optimize the design of the visual road environment. At the same time, it can be seen that the data fluctuates more often, indirectly indicating that the road alignment changes in the eyes of drivers are complex, which is also in line with the road characteristics of mountain roads with a high bridge-tunnel ratio.

4   CONCLUSION

(1)   Based on the previous research, a lane line identification model based on the Cat Mullrom spline curve is established, the alignment of the road section from Baima to Longtan is fitted as an example, and the feature parameters are extracted from the fitted alignment and analyzed.

(2)   SPSS (Statistical Product Service Solutions) is used to perform correlation analysis on the extracted control point spacing length and control point direction angle and the alignment characteristics of the road section are analyzed based on the correlation.

(3)   The analysis results show that the entire high-speed line of the section from Baima to Longtan is relatively good, the overall change in road alignment is continuous, the driver does not feel the abruptness of the line on the way, and some sections have poor sight distance (i.e., the total distance between control points is lower than the average line), which is likely to cause traffic accidents in the actual driving process and should be improved.

(4)   One of the purposes of establishing the model is to evaluate the safety of existing roads, predict the number of possible accidents, analyze the causes of accidents to identify dangerous sections, and carry out safety management to provide a basis for traffic accident prevention, thus avoiding traffic accidents.

(5)   The research results of this paper can identify the bad alignment sections of the road and take corresponding measures to improve them, which has some reference value for the comprehensive management of similar sections. However, its applicability needs further verification and investigation. At the same time, there are some limitations in the sample collection, model construction, and data processing of this paper, and the study of the geometrical characteristics of road alignment should be combined with the factors of people, vehicles, roads, and environment for comprehensive analysis.

REFERENCES

Cao Xuejuan, Huang Mingxuan, Wu Bowen, Yang Xiaoyu. BP-Garson Algorithm for Traffic Accident Prediction on High Bridge-tunnel Ratio Road Sections [J]. *Journal of Chongqing University of Technology (Natural Sciences)*, 2022, 36(03):119–125.

Chen Yuren, Fu Yuntian,Wang Fan. Establishment and Application of Support Vector Regression-based Sight Distance Calculation Model [J]. *Chinese Journal of Highways*, 2018, 31(04):105–113. DOI:10.19721/j.cnki.1001-7372.2018.04.013.

Chen Yuren, Wang Ruiyun, Dong, Y. J., He Sihong. Research on the Method of Applying the CatMull-Rom Spline Curve to Describe the Centerline Feature of Road Perspective view [J]. *Journal of Chongqing Jiaotong University (Natural Science Edition)*, 2015, 34(04):45–51+112.

He Sihong, Chen Yuren. Research on Visual Lane Model Based on Road Feature Information [J]. *Journal of Beijing University of Technology, Nature*, 2013, 33(S1): 156–159.

Hur, J., Kang, S. N., Seo, S. W. Multi-lane Detection in Urban Driving Environments Using Conditional Random Fields [C] 2013 IEEE Intelligent Vehicles Symposium (IV). IEEE, 2013:1297-1302.

Li Qiaoling. Research on the Linkage of Safety Emergency and Early Warning Methods for Mountain Highways [D]. Chongqing Jiaotong University, 2015.

Rahmdel, P. S., Shi, D., Comley, R. Lane Detection Using Fourier-based Line Detector [C] 2013 IEEE 56th International Midwest Symposium on Circuits and Systems (MWSCAS). IEEE, 2013:1282-1285.

Son, J., Yoo, H., Kim, S., *et al.* Real-time Illumination Invariant Lane Detection for Lane Departure Warning System [J]. *Expert Systems with Applications*, 2015, 42(4): 1816–1824.

Twigg, C. Catmull-rom Splines [J]. *Computer*, 2003, 41(6): 4–6.

Wang, J., Mei, T., Kong, B., *et al.* An Approach of Lane Detection Based on Inverse Perspective Mapping [C] 17th International IEEE Conference on Intelligent Transportation Systems (ITSC). IEEE, 2014:35-38.

Zhang Zou, Guo Xin. Correlation Analysis of Speed Limit Control and Route Index of Mountain Highway [J]. *Western Transportation Science and Technology*, 2022 0(05):20–23. DOI:10.13282/j.cnki. wccst.2022.05.007.

Water Conservancy and Civil Construction – Oke & Ahmad (Eds)
© 2024 The Author(s), ISBN: 978-1-032-58618-2

Study on smart aggregate for monitoring asphalt pavement health system

Shengchao Cui
Field Engineering College, Army Engineering University of PLA, Nanjing, China

Yanxiang Zhang
Logistics command department, Army Logistics Academy, Chongqing, China

Guangzheng Wang, Qingna Ma & Feng Sun*
Field Engineering College, Army Engineering University of PLA, Nanjing, China

ABSTRACT: Asphalt pavement diseases are often caused by excessive stress and strain on the pavement structure. In recent years, the rapid development of sensor technology has made great breakthroughs in monitoring the internal stress and strain of the pavement. However, the drawbacks of both traditional strain/stress gauges and current advanced fiber-optic sensors are also very obvious, such as poor interference immunity and low sensitivity. Therefore, this paper introduces the concept of piezoelectric smart aggregate into the field of asphalt pavement structure information monitoring, which can obtain real-time and long-term monitoring of the internal stress and strain of asphalt pavement. This work describes a detailed analysis of this sensing technology to understand the state-of-the-art progress and required future works for health monitoring technologies in road engineering. The development, principles, applications, existing problems, and future research suggestions of piezoelectric smart aggregate are elaborated on in detail. This paper provides a reference for road construction and road disease monitoring, as well as building the road health monitoring system.

1 INTRODUCTION

Sensors can detect and sense external signals, physical conditions (such as light, heat, and humidity), or chemical composition (such as smoke), and transmit the detected information to other devices (EB/OL 2011), so they are widely used in medical, agriculture, transportation, aviation, shipbuilding, and other fields. In the engineering field, structures such as houses, highways, and bridges are often designed to have a service life of decades or even centuries. To ensure service safety, it is necessary to apply sensor technology to monitor and perceive structures' health problems, such as stress–strain behaviors during their service period. For highway engineering, the road is directly exposed to the natural environment during its service time. The working loads include man-made loads (vehicles, etc.), natural environmental loads (ice, snow, rain, etc.), corrosion effects, material aging, fatigue effects, etc. As a result, damage will inevitably occur inside the structure, which can affect the durability, applicability, and safety of the structure itself. Therefore, it is necessary to monitor the health of pavement structures. In China, by the end of 2021, the total road mileage is 5.2 million km, of which 95% is asphalt pavement. During the asphalt pavement service period, it can experience millions of vehicle loads. It is the excessive stresses and strains that lead to pavement failure. If the internal stress and strain of the asphalt pavement are

_____
*Corresponding Author: sunfengnjjs@163.com

DOI: 10.1201/9781003450832-17

139

monitored, the reasons and patterns of the damage can be analyzed. Meanwhile, the strain–stress information can evaluate road performance and damage conditions, which can provide maintenance information in time to prevent road damage, realizing road health monitoring. The development of structural health monitoring (SHM) technology is based on advanced sensing elements. SHM refers to a detection technology that uses on-site non-destructive sensing technology and analyzes the characteristics of the structural system to monitor structural damage or degradation (Sun 2014). SHM involves the long-term observation of a structure or mechanical system using measurement data at periodic intervals. These data are usually dynamic responses but many alternative measures, such as temperature, image, or sound data, can be used. Damage-sensitive features are extracted from these data and the analysis of these features can be used to determine the current state of system health (Charles & Farrar 2012).

For a long time, the strain on asphalt pavements has been mainly tested by external strain gauges. In actual application, the strain gauges are mainly pasted on the surface of the structure. Many pieces of literature have introduced strain gauges to monitor stress and strain (Bose et al. 2020; Iriarte, Aginaga & Gainza 2021; Kikumoto & Togashi 2022;). The use of strain gauges also has many disadvantages. For example, it is difficult to bond well with the structural layer, it has poor moisture resistance, the solder joints of the strain gauges are prone to falling off, the obtained test data results are relatively discrete, and when the transmission distance exceeds 100 m, it is susceptible to 50 Hz industrial frequency electrical interference.

In recent years, optical fiber sensor technology has developed rapidly and it can be used for stress and strain measurements. For example, the European accelerated loading test program AFD40 and the U.S. full-scale accelerated loading test program COST374 have developed corresponding road optical fiber grating strain sensors. The literature (Churin, Nemov & Skvortsov 2021; Kuang et al. 2018; Liu et al. 2021; Liu et al. 2020; Qu et al. 2018; Xiang & Wang 2018; Zhang et al. 2022) shows strain and stress sensing based on Fiber Brag Grating (FBG) sensors. Fiber grating uses the grating wavelength as the sensing medium and senses change through wavelength drift. To widen the measurement range, the broadband light source should be used. On the contrary, to improve the resolution, the reflection line width should be narrowed, which can greatly reduce the power utilization of the broadband light source. Therefore, in the fiber grating sensing application system, a broadband high-power light source should be used to improve its signal-to-noise ratio and achieve reliable signal detection. In addition, due to wavelength drift, a high-performance monochromator or spectrometer should be used to enhance detection sensitivity and resolution. As a result, it is bound to increase the cost of the entire system and reduce its practicability. In summary, fiber grating sensors require a complex focusing optical system and precise displacement movement technology.

Recently, piezoelectric materials are widely studied because they can be used as sensors and actuators. It is a functional material with electromechanical conversion characteristics. Piezoelectric materials have a wide frequency response range, fast response speed, simple structure, low power consumption, and low cost (Wegert et al. 2022). It is widely used in various high-tech fields, such as national defense, communications, transportation, medical treatment, aerospace, aviation, etc. Piezoelectric materials can be divided into many types, such as piezoelectric ceramics, piezoelectric polymer materials, etc.

With the rapid development of science and technology, the development and exploration driven by application requirements have given new impetus to the research of piezoelectric materials. Due to the unremitting efforts of scientific and technical personnel in basic research and production process improvement, new types of piezoelectric materials have continuously emerged in the past ten years. Nowadays, piezoelectric materials are developing toward the trends of lead-free, high-performance, and thin-film, which promotes the corresponding application device research (Meng 2013). Therefore, piezoelectric sensors based on piezoelectric smart materials have also been rapidly developed.

The development of piezoelectric sensors is significant for pavement performance detection, safety, and maintenance during road engineering. For example: (1) the direct and reverse piezoelectric effects of piezoelectric materials can be used to intelligently monitor the speed and distance of vehicles; (2) the internal stresses and strains of structures can be monitored; (3) the health status of structures can also be assessed. Therefore, the design and manufacture of piezoelectric smart sensors are of great value for studying the true stress and strain level inside the road and for improving pavement service life.

This paper aims to introduce the new trends and developments in the research on piezoelectric materials technology at home and abroad in recent years and the research on SHM based on piezoelectric sensors. It also looks forward to the direction of future research work and puts forward some suggestions.

2 PIEZOELECTRIC MATERIALS

In 1880, Curie, P. and Curie, J. first discovered the piezoelectric effect of quartz crystals while studying the pyroelectric effect and crystal symmetry (Damjanovic 2009). In 1954, American Jaffe, B. discovered lead zirconate titanate (PZT) piezoelectric ceramics (Quan & He 1997), which greatly expanded piezoelectric ceramics application fields. In 1969, Japanese scholars discovered the piezoelectric effect of PVDF for the first time (Zhang et al. 2008). According to statistics (Hakkiyal et al. 1984; Meng 2016; Sencadas et al. 2009), thousands of research papers on PZT and PVDF piezoelectricity were published in the next 10 years. Piezoelectric ceramic materials were widely used and developed in the fields of science and technology. In the 1980s, Hakkiyal, A. first prepared glass ceramics (Santos et al. 2021). Compared with ceramic materials, there was no inherent aging and polarization in glass ceramics. Moreover, the sensors made of glass ceramics can work under high temperatures. Due to the shortcomings of brittleness and fragility of ceramics, the research on piezoelectric ceramic-polymer composite materials has been developed rapidly. Newnham, who came from the American Materials Laboratory of Pennsylvania State University, devoted himself to the investigation of composite materials. The literature (Banerjee et al. 2021; Cortes et al. 2021; Li et al. 2022; Mawassy et al. 2021; Zhang et al. 2022) investigated the properties of modified piezoelectric composites that have good electromechanical coupling coefficients, piezoelectric constants, mechanical quality factors, and Curie temperature and stability. The development of piezoelectric ceramic materials was from binary system to ternary system and multi-system. As the research on piezoelectric ceramics gradually slowed down, the development of piezoelectric materials also slowed down. Nowadays, due to the increasing awareness of environmental protection, there is more and more research on lead-free piezoelectric materials. Wang (2016), Ju (2012), and Ge (2011) studied lead-free piezoelectric materials and found that the piezoelectric constant and dielectric constant were higher than those of lead-containing piezoelectric ceramics. Das Mahapatra et al (Chen 2021; Das Mahapatra et al. 2021; Rafiee 2022; Wegert 2022) investigated the functional collections of piezoelectric materials, including energy harvesting and sensing applications. Therefore, the current development direction of piezoelectric ceramics is not only to improve the piezoelectric performance of piezoelectric ceramics but also to be more environmentally friendly and multifunctional.

2.1 *Piezoelectric effect*

Piezoelectric materials have a piezoelectric effect. The piezoelectric effect is divided into the direct piezoelectric effect and the inverse piezoelectric effect. Figure 1 shows the direct piezoelectric effect. When the pressure F parallel to the polarization direction is applied to the PZT element, the piezoelectric effect is caused by the compressive deformation of the PZT element.

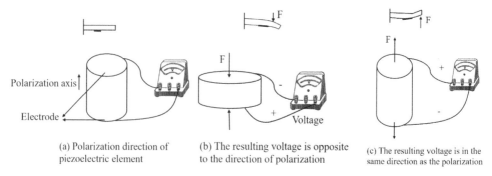

Figure 1. The direct piezoelectric effect.

Figure 2 shows the inverse piezoelectric effect. When an electric field in the same direction as the polarization is applied to the PZT element, the piezoelectric effect is caused by the increase in the polarization strength of the PZT.

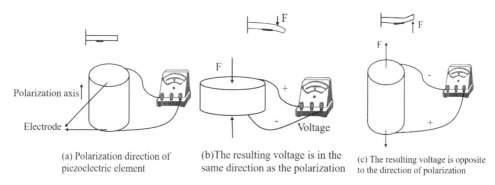

Figure 2. The inverse piezoelectric effect.

The direct piezoelectric effect how efficiently the piezoelectric material converts mechanical energy into electrical energy, while the inverse piezoelectric effect shows how efficiently the piezoelectric material converts electrical energy into mechanical energy. Information about the deformation of a piezoelectric element or structure can be obtained by detecting the change of charge on the piezoelectric element. Therefore, sensors can be produced based on the direct piezoelectric effect of piezoelectric materials. On the other hand, if the voltage is applied to the piezoelectric element, displacement will be displaced by the positive and negative charges due to the electric field, resulting in the deformation of the piezoelectric element. Thus, the actuator can be made based on the inverse piezoelectric effect of piezoelectric materials. The piezoelectric effect shows the transformation relationship between the mechanical and electrical quantities of the piezoelectric materials.

2.2 *Piezoelectric equation*

The piezoelectric equation indicates the comprehensive relationship between piezoelectric crystal mechanical quantities (i.e., stress tensor σ, strain tensor x) and electrical quantities (i.e., electric field strength E, electric displacement D). According to the application status and test conditions of piezoelectric ceramics, it can be divided into the following two situations.

(1) Mechanical boundary conditions: These include mechanical free boundary conditions and mechanical clamping boundary conditions. The mechanical free boundary condition means that the boundary of the piezoelectric shaker is free to deform when the excitation signal frequency is low or the center of the piezoelectric shaker is clamped. At this time, the stress on the boundary is zero or a constant value but the strain is not zero. And if the frequency of the excitation signal is much higher than the resonant frequency, the piezoelectric shaker is not free to deform. At this time, the strain is zero or a fixed value but the stress is not zero.
(2) Electrical boundary conditions: These include electrical open-circuit boundary conditions and electrical short-circuit boundary conditions. The electrical open circuit boundary condition means that when the resistance value of the external circuit between the two electrodes of the piezoelectric vibrator is much higher than that of the internal resistance, the external circuit is an open circuit. At this time, the charge on the electrode surface of the piezoelectric vibrator remains unchanged and the electric displacement vector is zero. However, the electric field intensity in the piezoelectric vibrator is not zero. The electrical short-circuit boundary condition means that when the resistance value of the external circuit between the two electrodes of the piezoelectric vibrator is much lower than that of the internal resistance, the external circuit is in a short circuit. At this time, the internal electric field intensity of the piezoelectric vibrator is zero but the electric displacement is not zero.

Although the current research on piezoelectric materials has been advanced, there are still some existing shortcomings. For example, (1) the piezoelectric effect can decay. Piezoelectric materials will depolarize under hundreds of thousands of cyclic loads, that is, the piezoelectric effect may disappear. At this time, the output voltage value of the piezoelectric material will significantly decrease or even disappear. (2) The mechanical properties are poor. Although the mechanical properties of piezoelectric composites are significantly higher than those of ordinary piezoelectric ceramics, they still cannot withstand heavy loads. (3) The high-temperature resistance is poor. The current commercial piezoelectric materials are mainly PZT, whose Curie temperature is about 150–360 °C. The low Curie temperature value can limit the application of piezoelectric ceramics in high-temperature environments. Many scientists have studied piezoelectric ceramic materials of perovskite-type or bismuth layered oxide, which can increase the Curie temperature to 900 °C. However, there are disadvantages, such as low piezoelectric activity and high coercive electric field. (4) Environmental protection is poor. At present, almost all commercial piezoelectric materials contain the lead element. With the increasing awareness of environmental protection, the focus should be on lead-free piezoelectric materials. These problems severely restrict the development of piezoelectric materials and need to be studied urgently.

3 THE SENSING PRINCIPLE OF PIEZOELECTRIC SENSORS

3.1 *Active monitoring method of piezoelectric smart aggregate*

Figure 3 shows the structural diagram of the piezoelectric smart aggregate sensor. The piezoelectric smart aggregate sensor is composed of silicone waterproof layers, PZT sheets, shielded wire, and concrete. It can be made into cubes, cylinders, or other shapes based on the mold. The

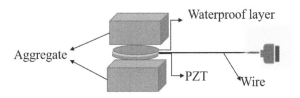

Figure 3. The structural diagram of the piezoelectric smart aggregate sensor.

PZT sheet is located in the center and is the core of the sensor. One end of the shielded wire is connected to the PZT sheet, the other end is connected with the shielding connector and the shielding connector is connected to the signal transmitting and collecting channel.

The basic principle of the wave-based method is to attach the piezoelectric sensors to the surface of the structure or embed them inside the structure so that the piezoelectric sensors and the monitored structure form a piezoelectric intelligent structure system, as shown in Figure 4.

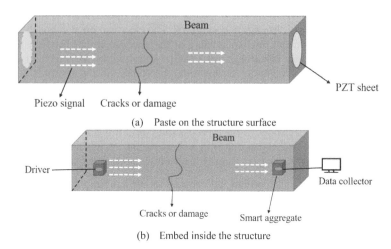

Figure 4. The wave-based principle of piezoelectric sensors.

Based on the direct and inverse piezoelectric effect, the sensors emit the stress wave and the actuators receive the stress wave. The damage and defects of the structure can be identified by analyzing the signal difference between the sensors and the actuators. For example, the damage can cause attenuation of signal amplitude, modal changes, and delay of propagation time. The wave method is widely used in the field of damage detection of metal structures and composite structures. The Lamb waves method is the most widely used (Le et al. 2021; Li et al. 2021; Zhang et al. 2021; Zhang et al. 2022). There are three stages of development of smart aggregates. First, the development of new materials has promoted the properties of piezoelectric materials. Second, the wave method is applied to the field of health monitoring of civil engineering structures. Third, piezoelectric patches are used as both sensors and receivers to simulate structure damage by pasting on both ends of the structure body. These methods are simple, practical, and accurate. On this basis, a wireless signal transmission and reception system applied to the online detection of cable tension has also been developed. Some scholars attached PZT patches at different positions of the steel bars to detect the defects of the reinforced concrete, such as the debonding of the steel bars, the concrete matrix, and the yield of the steel bars. The test shows that the amplitude of the received wave increases with the increase in the debonding length. The wave arrival time increases when the steel bar yields. When the concrete beam is bent, the amplitude of the received wave increases with the expansion of the crack.

3.2 Passive monitoring method of piezoelectric smart aggregate

The high impedance characteristics of piezoelectric materials make piezoelectric smart sensors suitable for the measurement of dynamic mechanical quantities. The high sensitivity of piezoelectric materials to strain changes is suitable for the passive monitoring of structural states. The piezoelectric passive monitoring technology directly uses sensors to monitor the real-time state of the structure and determines the health of the structure based on the characteristics of the monitoring parameters, combined with mechanical modeling analysis, prior knowledge, and related instruments (acoustic emission meters, etc.). This method requires only sensors, no actuators, and a relatively simple monitoring system. It is mostly used in structural vibration status, impact load identification and positioning, and monitoring based on acoustic emission technology.

4 THE DEVELOPMENT OF PIEZOELECTRIC SENSORS

In 1916, to study seabed geomorphology, Langzhiwan made sensors with quartz crystals that could transmit and receive signals in the water. In 1921, quartz resonators and filters are successfully developed. In 1967, V. I. Popkov (1967) designed a piezoelectric transducer to measure the dynamic stress exerted by a mechanism on the foundation. It is found that its sensitivity characteristic is linear in the range of 40–50 kHz and the error does not exceed 20%. In 1971, Kertzman (1971) designed a piezoelectric sensor and used the sensor for pollution detectors, meteorology monitors, and research instruments. In 1980, the SYC piezoelectric quartz pressure sensor was successfully developed. With the development of piezoelectric materials, piezoelectric sensors in the fields of aviation and medicine have developed rapidly. In the engineering field, piezoelectric sensors can be used for SHM or damage diagnosis. There are two application methods. It can be either pasted on the surface of the structure or embedded in the structure based on the actual engineering.

4.1 *Pasted application*

The pasting method is to paste the piezoelectric elements or piezoelectric sensors on the surface of the structure. The specific steps are as follows. First, we polish the structure surface. Second, we use acetone or absolute ethanol to wipe the piezoelectric sensor surface and the paste surface. Finally, we use 502 glue or epoxy resin to paste the sensor on the structure surface. The advantages of this method are simple operation, no negative impact on the structure, and simple force analysis of the piezoelectric element. However, the piezoelectric elements or sensors are exposed to the external environment in this method and are affected by the ambient temperature and humidity, which in turn affects their service life. At present, the paste method is mostly used in the health monitoring and damage identification of concrete composite panels and metal structures based on PZT impedance technology and Lamb wave technology.

Tan Ping studied a theoretical active control method for the analysis of traveling waves propagating from long thin beams using a pasted piezoelectric ceramic actuator. The modeling on the affixed piezoceramic actuator was put up. He also deduced the force-volt relationship between piezoceramic slices on account of the affection of glue layer deformation. Yan Shi studied the application principle of piezoelectric actuators and established the mechanical model of pasted piezoelectric ceramic actuators based on the inverse piezoelectric effect. The piezoelectric actuators solve the problem of driving control of building structure detection. The results obtained are that the output signal of the piezoelectric ceramic is linearly related to the voltage under the alternating voltage, and the thickness of the bonding layer and the elastic shear modulus affects the piezoelectric ceramic. The thinner the thickness of the bonding layer is, the greater the force transmitted. The higher the shear modulus of the bonding layer, the higher the force transmitted. Li used piezoelectric patches to achieve active vibration control of conical shells and found that the optimal placement scheme of the piezoelectric patches obtained by the genetic algorithm was efficient in the active vibration control of the conical shell.

4.2 *Embedded application*

The embedded method application is to embed the piezoelectric element or the piezoelectric sensor in the monitored structure. This method can effectively avoid the shortcomings of the paste-type method's susceptibility to the environment and human factors and ensure the service life of the piezoelectric elements or the piezoelectric sensors. However, due to the fragility of piezoelectric materials, it is not conducive to construction. To effectively protect piezoelectric components, Song proposed the concept of "smart aggregate" (Gu 2006; Song 2007). Song monitored the early stress change pattern of concrete by encapsulating piezoelectric ceramics into aggregates, embedding them in the concrete structure, and building a fuzzy logic system to calculate the early stress of concrete. Wen-I Liao (2013) investigated SHM of RC structures under seismic loads using smart aggregate. The smart sensing system is based on using one smart aggregate as an exciter to generate excitation waves. Other distributed smart aggregates are used as sensors to detect the response. It is found that this method can predict the health status of the tested frames through the proposed damage index matrix. Chang (Roh 1999; Wang 2000) at Stanford University monitored the health of the material by embedding piezoelectric ceramics in the composite material based on the changes in the stress wave generated inside the structure. Shen et al. (1996) monitored the composite curing condition and impact location using the embedded PZT as actor and receiver. Kawiecki (1998) studied the damage to the concrete beam by placing mass blocks on the beam. In this method, he placed piezoelectric ceramics at both ends of the beam as sensors and actuators. The results show that there is a one-to-one correspondence among frequency with the magnitude, position, and amplitude of the simulated damage. In addition, it has the advantages of high sensitivity and good repeatability. Li et al. (2001, 2002) proposed 2-2 cement-based piezoelectric composites in 2001. The basic application principle is similar to that of "smart aggregate". The difference is that it connects multiple piezoelectric ceramic sheets in pairs at the same level and embedded them in cement slurry at equal intervals to form piezoelectric composites. Later, Yang Xiaoming (2006) and Li Lifei (2011) improved the manufacturing method of the 2-2 cement-based piezoelectric sensor and designed a mold to make the manufacturing easier.

Gao (2021) pre-embedded piezoelectric ceramic actuators and sensors in concrete specimens. He studied the damage evaluation index based on the maximum amplitude of the signal. Yang (2020) studied the damage status of concrete structures by pasting PZT sensors on the structure in detail. It can be found that the conductivity curves measured by each paste PZT show a significant downward shift at the resonance peak as the damage increases. Moreover, Lee (2016) investigated the EMI resonance peak magnitude and frequency to determine solidification time. Ghafari (2018) found that EMI can be used as a non-destructive testing (NDT) method to enable in-situ measurement of the strength gain process of cement paste. Therefore, SHM can be achieved regardless of whether the PZT sensor is pasted or embedded. Based on the peak change, the actual damage location can be determined. The damage degree also can be judged based on the peak increment.

At present, the research and application in the field of intelligent aggregate engineering are mainly concentrated in concrete structures, such as beams and columns. However, there are few relevant studies in road engineering. Hou Shuang (2021) investigated vehicle load monitoring for asphalt concrete pavement based on smart aggregates. When asphalt concrete pavement is embedded with smart aggregates, the sensitivity of smart aggregates under dynamic vehicle load is linear. The change of the elastic modulus of pavement at room temperature has a negligible effect on the sensitivity of the smart aggregate. Lei Jinfang (2013) used the piezoelectric intelligent pavement embedded with piezoelectric smart aggregate to measure the driving vehicles on-site, realizing the dynamic weighing of the road vehicles by the piezoelectric sensor. For road engineering with such a huge mileage, monitoring the internal stress and strain state of the pavement can realize road information mastery, which is significant to the maintenance of road damage. Therefore, the following problems need to be solved.

(1) At present, when new high-grade roads are built, they are mainly asphalt concrete pavements. The size of the piezoelectric sensor affects the performance of the asphalt mixture, such as Void Volume (*VV*), Voids in Mineral Aggregate (*VMA*), Voids Filled Asphalt (*VFA*), Marshall Stability (*MS*), Flow Value (*FL*), high-temperature stability (Dynamic stability), low-temperature resistance to splitting, water stability, etc.
(2) When vehicles are driving on asphalt concrete pavement, the periodic alternating stress can be simplified as the sine wave load (Zhang 2006). According to the *Chinese Road Design Specifications* (2017), the tire ground pressure of the standard axle load (BZZ-100) is specified as 0.7Mpa. Therefore, the mechanical properties and sensitivity performance of the piezoelectric sensor should be studied.
(3) Although the embedded application method can effectively improve the survival rate of the piezoelectric patches, it puts forward high requirements for construction.

5 CONCLUSION

This study provides a comprehensive review of piezoelectric technology, including piezoelectric materials, piezoelectric sensors, and the application of smart aggregate. On this basis, a new method for monitoring the health condition of asphalt pavements is proposed. The conclusions are drawn as follows:

(1) Piezoelectric materials have direct and inverse piezoelectric effect properties, which can be used as sensors and actuators. Piezoelectric sensors can be attached to surfaces as well as embedded in structures to monitor changes in stress and strain.
(2) The design and fabrication of piezoelectric smart aggregates need to be coordinated with the actual engineering structure.
(3) Our state-of-the-art pavement monitoring health system is proposed in the paper, which is expected to solve the pavement damage problem as well as improve pavement maintenance issues.

This paper introduces the sensing principle of PZT sensors and forms smart aggregates to monitor the structure damage. The authors expect that smart aggregates can inspire future researchers to come out with more novel ideas and propose new research directions toward the development of pavement health monitoring. For example, (1) the principle of piezoelectric ceramic power generation can be used to provide electrical energy for data transmission. (2) At present, the obtained data are transmitted by wires. In the future, wireless data transmission can be investigated.

REFERENCES

Banerjee, S., Bairagi, S., Ali, S. W. A Critical Review on Lead-free Hybrid Materials for Next Generation Piezoelectric Energy Harvesting and Conversion. *Ceramics International*, 2021, 47(12): 16402–16421.

Bose, A. K., Zhang, X. Z., Maddipatla, D., *et al.* Screen-Printed Strain Gauge for Micro-Strain Detection Applications, *IEEE Sensors Journal*, 2020: 12652–12660

Charles, R., Farrar, K. W. *Structural Health Monitoring: A Machine Learning Perspective.* John Wiley & Sons, 2012.

Chen, Y., Zhang, D. L., Ji, X. R. Review of Research on the Rare-Earth Doped Piezoelectric Materials. *Frontiers in Materials*, 2021, 8: 67–75.

Churin, A. E., Nemov, I. N., Skvortsov, M. I. Glue Independent FBG Sensor with Enhanced Stability. *Optical Fiber Technology*, 2021, 64: 1–5.

Cortes, L. Q, Sanches, L., Bessaguet, C., *et al.* Improving Damping Capabilities of Composites Structures by Electroactive Films Containing Piezoelectric and Conductive Fillers. *Smart Materials and Structures*, 2021, 30(8): 1–12.

Damjanovic, D. Comments on Origins of Enhanced Piezoelectric Properties in Ferroelectrics [J]. *IEEE Transactions on Ultrasonics, Ferroelectrics, and Frequency Control*, 2009,56(8):1574–1585.

Das Mahapatra, S., Mohapatra, P. C., Thakur, V. K. Piezoelectric Materials for Energy Harvesting and Sensing Applications: Roadmap for Future Smart Materials. *Advanced Science*, 2021, 8(17): 1–73.

Gao, Z. H. *Damage Identification and Monitoring of Concrete Based on PZT*. Taiyuan University of Technology, 2021.

Ge, H., Xia, Y. Preparation and Piezoelectricity of NaNbO3, High-Density Ceramics by Molten Salt Synthesis [J]. *Journal of the American Ceramic Society*, 2011, 94(12):4329–4334.

Ghafari, E., Yuan, Y., Wu, C., *et al.* Evaluation of the Compressive Strength of the Cement Paste Blended with Supplementary Cementitious Materials Using a Piezoelectric-based Sensor. *Construction and Building Materials*, 2018, 171: 504–510.

Gu, H., Song, G., Dhonde, H., *et al.* Concrete Early-age Strength Monitoring Using Embedded Piezoelectric Transducers. *Smart Materials and Structures*, 2006, 15(2):1837–1845.

Hakkiyal, A., Safari, A., Bhalla, A. S. Grainoriented Glass-ceramics for Piezoelectric Devices [J]. *Journal of American Ceramic Society*, 1984, 67(5):331–335.

Iriarte, X., Aginaga, J., Gainza, G., *et al.* Optimal Strain-gauge Placement for Mechanical Load Estimation in Circular Cross-section Shafts, *Measurement*, 2021: 108938–108938.

Ju, S. K., Kim, J. P., Chang, W. A. Ferroelectric and Piezoelectric Properties of Lead-free (K, Na, Li) (Nb, Ta, Sb)O3 Ceramics Fabricated by Using a Citric-acid-assisted Modified sol-gel Method [J]. *Journal of the Korean Physical Society*, 2012, 61(6):966–970.

Kawiecki, G. Feasibility of Applying Distributed Piezo Transducers to Structural Damage Detection. *Journal of Intelligent Material Systems and Structures*, 1998, 2(3):189–197.

Kertzman, J. Piezoelectric Sensors for Use as Pollution Detectors, Meteorology Monitors and Research Instruments: 25th Annual Symposium on Frequency Control, Atlantic City, NJ, USA, USA, 1971.

Kikumoto, M., Togashi, Y. Method for Measuring Three-Dimensional Strain Tensor of Rock Specimen Using Strain Gauges, *Rock Mechanics and Rock Engineering*, 2022.

Kuang, Y., Guo, Y., Xiong, L., *et al.* Packaging and Temperature Compensation of Fiber Bragg Grating for Strain Sensing: A Survey [J]. *Photonic Sensors*, 2018,8(4):320–331.

Le, H. V., Kim, M. K., Kim, S. U., *et al.* Enhancing Self-stress Sensing Ability of Smart Ultra-high-performance Concretes Under Compression by Using Nano Functional Fillers. *Journal of Building Engineering*, 2021, 44: 17–28.

Lee, J. C., Shin, S. W., Kim, W. J., *et al.* Electro-mechanical Impedance-based Monitoring for the Setting of Cement Paste Using Piezoelectricity Sensor. *Smart Structures and Systems*, 2016, 17: 123–134.

Lei, J. F. *Research on Vehicle Dynamic Weighing Technology Based on the Piezoelectric Intelligent Pavement*. Dalian University of Technology, 2013.

Li, L. F. *Research on Stress Measurement and Damage Identification of Concrete Components Based on Piezoelectric Ceramics*. Hunan University, 2011.

Li, P. F., Jiang, W., Lu, R., *et al.* Design and Durability of PZT/PVDF Composites Based on Pavement Perception. *Construction and Building Materials*, 2022, 323: 1–11.

Li, Y., Ma, Y. L., Hu, X. B. Early-age Strength Monitoring of the Recycled Aggregate Concrete Using the EMI Method. *Smart Materials and Structures*, 2021, 30(5): 1–15.

Li, Z. J., Zhang, D. Cement-based Smart Material-cement Matrix Piezoelectric *Composites. Concrete Science and Engineering*, 2001, 3(10):116–120.

Li, Z. J., Zhang, D., Wu, K. R. Cement-based 0-3 Piezoelectric Composites [J]. *Journal of the American Ceramic Society*, 2002,85(2):305–313.

Liao, W. C. H. L., Hwang, J. S., *et al.* Seismic Health Monitoring of RC Frame Structures Using Smart Aggregates. *Earthquake Engineering and Engineering Vibration*, 2013(12):25–32.

Liu, S. C., Lin, Z. C., Lu, M., *et al.* Design of Storage and Management Device of Packaged FBG Sensors Based on RFID, 2021, Journal of Physics: Conference Series, 1840: 012012–012015.

Liu, Z., Liu, X., Zhu, S. P., *et al.* Reliability Assessment of Measurement Accuracy for FBG Sensors Used in Structural Tests of the Wind Turbine Blades Based on Strain Transfer Laws. *Engineering Failure Analysis*, 2020, 112: 104506–104520.

Mawassy, N., Reda, H., Ganghoffer, J. F., *et al.* A Variational Approach of Homogenization of Piezoelectric Composites Towards Piezoelectric and Flexoelectric Effective Media, *International Journal of Engineering Science*, 2021, 158: 800–805.

Meng, Y. Y. *Mechanical Model and Experimental Study of Piezoelectric Intelligent Aggregate*. Dalian University of Technology, 2013.

Meng, Y. Y. *Piezoelectric Intelligent Sensing-Actuator Mechanical Properties and Its Application*. Wuhan University Press, 2016.

Ministry of Communications of China. *JTG D50-2017 Specification for Design of Highway Asphalt Pavement.* 2017.

Popkov, V. I. Piezoelectric Transducer for Measuring Dynamic Stress Exerted by a Mechanism on its Foundation [J]. *Measurement Techniques*, 1967, 10(9):1067–1069.

Qu, Y., Hong, L., Jiang, X., *et al.* Experimental Study of Dynamic Strain for Gear Tooth Using Fiber Bragg Gratings and Piezoelectric Strain Sensors [J]. *Proceedings of the Institution of Mechanical Engineers Part C-journal of Mechanical Engineering Science*, 2018, 232(21): 3992–4003.

Quan, X. P., He, F. C. Application of PVDF Sensor in Robot Intelligence Technology. *Mechanical and Electrical Engineering*, 1997(4):54–56.

Rafiee, M., Granier, F., Therriault, D. Multi-Material, Multi-Process, Planar, and Nonplanar Additive Manufacturing of Piezoelectric Devices. *Advanced Engineering Materials*, 2022: 2200294–2200308.

Roh, Y. S. *Built-in Diagnostics for Identifying an Anomaly in Plates Using Wave Scattering.* Stanford University, 1999.

Santos, J. A., Sanches, A. O., Akasaki, J. L., *et al.* Influence of PZT Insertion on Portland Cement Curing Process and Piezoelectric Properties of 0-3 Cement-based Composites by Impedance Spectroscopy. *Construction and Building Materials*, 2021, 238: 334–344.

Sencadas, V., Gregorio, J. R., Lanceros-Mendez, S. α to β Phase Transformation and Microstructural Changes of PVDF Films Induced by Uniaxial Stretch[J]. *Journal of Macromolecular Science*, 2009, 48 (3):514–525.

Shen, B. S., Tracy, M., Roh, Y. S, *et al.* Built-In Piezoelectrics for Processing and Health Monitoring of Composite Structures: AIAA/ASME/AHS Adaptive Structures Forum, Salt Lack City, UT, 1996.Apr. 18–19.

Singh, D., Alam, J., Alam, S., *et al.* Performance Analysis of Footstep Power Generation Using Piezoelectric Sensors. 2021 International Conference on Intelligent Technologies.2021: 25–27.

Song, G., Gu, H., Mo, Y. L., *et al.* Concrete Structural Health Monitoring Using Embedded Piezoelectric Transducers. *Smart Materials and Structures*, 2007, 16(4):959–968.

Sun, H. P. *Research on Signal Stability of Piezoelectric Smart Aggregate.* Shenyang Jianzhu University, 2014.

Wang, C. S., Chang, F. K. *Diagnosis of Impact Damage in Composite Structures with Built-in Piezoelectrics Network: Smart Structures and Materials 2000: Smart Electronics and MEMS, Stanford*, 2000.

Wang, Z., Wang, J., Chao, X. Synthesis, Structure, Dielectric, Piezoelectric, and Energy Storage Performance of (Ba0.85Ca0.15) (Ti0.9Zr0.1) O_3 Ceramics Prepared by Different Methods [J]. *Journal of Materials Science: Materials in Electronics*, 2016, 27(5):5047–5058.

Wegert, Z. J., Roberts, A. P., Challis, V. J. Multi-objective Structural Optimization of Piezoelectric Materials, *International Journal of Solids and Structures*, 2022, 248: 1–12.

Wegert, Z. J., Roberts, A. P., Challis, V. J. Multi-objective Structural Optimization of Piezoelectric Materials. *International Journal of Solids and Structures*, 2022, 248: 111666–111677.

Xiang, P., Wang, H. Optical Fiber-based Sensors for Distributed Strain Monitoring of Asphalt Pavements [J]. *International Journal of Pavement Engineering*, 2018, 19(9): 842–850.

Xu, S. Y., Yeh, Y. W., Poirier, G. Flexible Piezoelectric PMN-PT Nanowire-based Nanocomposite and Device [J]. *Nano Lett*, 2013, 9(4):393–397.

Yang, X. M. *Research on Performance Monitoring System and Damage Identification Method of Civil Engineering Structure.* Tianjin University, 2006.

Yang, Z. L. *Research of Concrete Damage Detection Based on PZT Sensor Under Controlled Temperature Conditions.* Huazhong University of Science and Technology, 2020.

Zhang, J. Z., Tan, T., Duan, C., *et al.* Measurement of Residual Stress Based on a Ring FBG Array. *IEEE Transactions on Instrumentation and Measurement*, 2022, 71: 1–7.

Zhang, S. W., Zhai, C. P., Liu, K. Y., *et al.* Quantitative Evaluation of Energy Harvesting Capabilities on Flexoelectric and Piezoelectric Materials. *Journal of Applied Physics*, 2022, 131(6): 101–108.

Zhang, X. J., Huang, C., Gu, D. W., *et al.* Privacy-preserving Statistical Analysis Over Multi-dimensional Aggregated Data in Edge Computing-based Smart Grid Systems. *Journal of Systems Architecture*, 2022, 127: 102508–102516.

Zhang, X. J., Huang, C., Xu, C. X., *et al.* Key-Leakage Resilient Encrypted Data Aggregation With Lightweight Verification in Fog-Assisted Smart Grids. *IEEE Internet of Things Journal*, 2021, 8(10): 8234–8245.

Zhang, X. L. *The Viscoelastic Principle, and Application of Asphalt and Asphalt Mixture.* People's Communications Press. Beijing, 2006.

Zhang, Z. Z., Song, H. J., Men, X. H. Effect of Carbon Fibers Surface Treatment on the Tribological Performance of Polyurethane (PU) Composite Coating [J]. *Wear*, 2008, 264(7):599–605.

Water Conservancy and Civil Construction – Oke & Ahmad (Eds)
© 2024 The Author(s), ISBN: 978-1-032-58618-2

Process analysis for the demolition work of mid-partition walls in complex foundation pits in super high-rise buildings

Jianping Zhu & Yalong Liu*
Digital Construction College, Shanghai Urban Construction Vocational College, Shanghai, China

Jiao Zhang
School of Municipal and Ecological Engineering, Shanghai Urban Construction Vocational College

ABSTRACT: According to the requirements of the construction process, the mid-partition walls set up for a complex foundation pit are generally removed after the top slab of the basement is formed and its toughness has reached the design level. Therefore, the demolition of the wall is difficult. In this paper, a building-intensive project with 54 floors above ground and 4 floors below ground in Shanghai urban area is taken as an example to analyze the reconstruction of the basement and the removal of the middle slab in an attempt to provide a potential reference for the removal of specific walls. The results show that (1) the demolition of the mid-partition wall is the key point for the basement reconstruction schedule. (2) When the mid-partition wall is demolished, the stress released from the wall can sudden changes in the maximum bearing capacity of the pit. To avoid overturning the mid-partition wall, we divide the mid-partition wall into horizontal blocks and vertical layers and demolish the wall by parts in a certain order. (3) Static-cutting technology is more suitable for complex environments.

1 INTRODUCTION

Some super high-rise buildings have complex foundation pits. According to the requirements of construction procedures, demolition is usually required after the basic construction is completed. The narrowness in the working area increases the difficulty of hoisting, so the finished building structure needs to be protected. The workload of demolition and concrete clinker transportation is high. There are many uncertainties in the demolition process, which can easily cause a redistribution of stresses in the existing structural members of the basement (Chen 2005). Therefore, the construction technology that minimizes the negative impact on the structural stability of the building and the surrounding environment should be carefully selected and the environment and characteristics of the construction process should be considered before the removal procedure of the partition wall in the foundation pit (Lv 2019).

The mid-partition wall in the complex foundation pit of a high-rise building in the urban area of Jing'an District, Shanghai, has to be demolished after finishing the foundation pit construction to complete the structural connection. Considering the demands of construction and targets in safety, economy, efficiency, noise control, and environmental protection, the project adopts alternative bay construction methods and static concrete-cutting techniques during the demolition of the mid-partition wall, which assists well in completing the target (Li 2017). The corresponding construction techniques and experiences are summarized to provide references for similar projects in the future.

---

*Corresponding Author: 1246987280@qq.com

150

DOI: 10.1201/9781003450832-18

2 PROJECT OVERVIEW

The project is a building-compact construction project in the urban area of Shanghai with 54 floors overground and 4 floors underground. The frame-core wall system is a mixed-use building. The foundation pit depth is 25 meters. The eastern part of the site is a historic building under the government's preservation program and the western part has 2 high-rise residential buildings. To accommodate the construction demands of the underground station, the foundation pit area of the tower has to be constructed first. The footings of the podium can be constructed at a later stage according to the schedule to allow for further adjustments in the design. Therefore, a mid-partition wall is required to divide the whole foundation pit into two independent parts.

3 PROJECT FEATURE AND SCHEME SELECTION

The construction site is aligned east-west and a mid-partition wall around the K-axis divides the foundation pit into two north and south regions, which are currently being modified and constructed by two separate construction companies. After the structure construction of the skirt building foundation pit basement is completed and the $\pm$ 0.000 basement top-plate is done, the mid-partition wall needs to be demolished and the foundation-pit concrete-structural system on both sides of the mid-partition wall needs to be connected.

Due to the limited construction area, efforts are needed to protect the completed building structure during the demolition process.

The workload for the demolition of the mid-partition wall is heavy. To maximize the speed of construction, some cutting equipment should be used and lifted at multiple crane points simultaneously.

The demolition area is located in an urban district with demanding regulations on dust pollution. Measures to prevent dust pollution should be closely monitored.

There is a relatively large variety of workers involved in the construction, including scaffold work, cutting work, and vertical transportation work, which makes it particularly important that all technicians coordinate smoothly with others. This feature places some extra demands on the organizational capability safety management of the demolition team.

According to the environment and key features of the construction project, the static cutting technique for concrete stands out for its safety, economy, efficiency, environmental friendliness, and relatively low noise level (Qiu 2022).

4 DEMOLITION OF MID-PARTITION WALL

4.1 *Demolition processes*

After the completion of the basement renovation in the north-south pit (formation of the basement roof slab) and after reaching the established toughness, the demolition of the mid-partition wall should be started. This wall is 1 m thick, about 25.5 m high, and 68 m long. Before the construction of the top slab, the middle partition wall must be removed. (Liu 2022) The vertical cutting area ranges from the bottom of the basement to the top slab of the basement.

The detailed technical processes are listed below:

We prepare as many parts and lengths of equipment and concrete as necessary according to the construction drawings.

The project management team and others who might be concerned should attend the design clarification meeting before the beginning of construction to get themselves familiar with the construction drawings and the actual arrangement of the site.

We build the infrastructure according to the blueprints for the cutting site of the mid-partition wall, e.g., roads (using the construction roads arranged by the general contractor), water, electricity, necessary equipment, etc.

We set up and check the machinery before construction work and make sure water, power supply, and lighting at night are in place.

The construction site is small in area. We reasonably arrange all machinery and equipment, toolboxes, the container of the cutting machine, etc.

We also need to pay attention to the construction water. There should be no less than 10 6.67 cm taps in the pit and 5 water separation tanks for cooling the cutting machine.

4.2 *The cutting schemes of the mid-partition wall*

4.2.1 *Selection of cutting equipment*

The main equipment used is chainsaws and drills both from a certain brand.

4.2.2 *Construction techniques of chainsaw-cutting*

(a) We set up the power supply and water supply system on the site.
(b) We divide the mid-partition wall with cutting guidelines and make sure that the mass of each part divided by these lines does not exceed 1.5t and the size is less than 0.6m $\times$ 1m.
(c) We drill a connection hole of not less than 108mm in diameter at the connection of the cutting lines of the central partition wall with an electric drill for winding the chain.
(d) We install the cutting equipment on the septum wall and connect it to the power supply and water source.
(e) We unscrew the chain and connect the connector correctly with a hydraulic clamping tool.
(f) We open the water pipe, adjust the velocity of the water flow, and tighten the chain with the controller.
(g) We fix the size of the pin on the drilling stand. The distance between the cutting line and the middle of the pin should be 42 cm.
(h) j. In principle, we cut horizontally before vertically and tighten with an electrical chain block.

4.2.3 *The lighterage and hoist-out of the demolished concrete blocks*

The following measures are required due to the cutting technique used here.

(a) During the cutting process, we hold up the target concrete block with an electrical chain block to prevent it from collapsing.
(b) We determine the size of the target concrete block with the transport capability of forklifts.
(c) The cut concrete blocks should be placed gently onto the flatbed truck right away. The worker on the truck should stake up the hoisted concrete blocks.
(d) The concrete blocks should be arranged for track-loading immediately after being transported out of the pit.

4.2.4 *Machinery selection*

Due to the restriction from the construction site, the mid-partition wall is cut into parts and hoisted out (Jiang 2021). We equip it with a 25T sling van based on the current stage of construction and the supporting weight of the concrete and use 2 cranes if necessary.

4.2.5 *The transport of cutting debris*

The cut wall pieces are lifted by the 25T crane onto a flatbed truck and transported to the concrete yard waiting to be crushed.

4.3 *The horizontal demolition procedure of mid-partition wall*

The horizontal forces of the foundation pit mainly come from the horizontal structural support of the basement and the conduction of the mid-partition wall (Mao 2020). When the mid-partition wall is demolished, the stress released from the wall can cause sudden changes in the maximum bearing capacity of the foundation pit. To avoid the problem above, we divide the mid-partition wall into horizontal blocks and vertical layers and demolish the wall by parts in a certain order.

The thickness of the horizontal layer of the diaphragm wall should not exceed 6 meters. We demolish the wall from the middle outward to both sides. At the same time, we skip the adjacent part and demolish the next part, as shown in Figure 1.

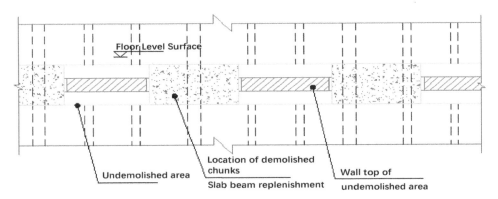

Figure 1. Partition wall demolition plan.

The arrangement for plan separation is shown in Figure 2. We demolish each part in ascending order of number.

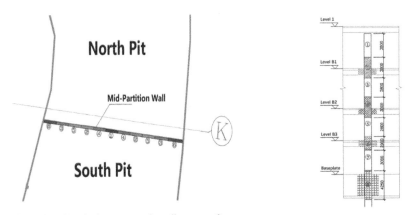

Figure 2. Vertical segmentation diagram of partition wall demolition.

Figure 3. Horizontal segmentation diagram of partition wall demolition.

Step 1: We demolish the blocks numbered ①, ②, and ③ in order.
Step 2: We carry out the structural replenishment of slab beams for the demolished parts in ①, ②, and ③.

Step 3: After the strength of the replenished slab beams in ①, ②, and ③ which reaches 75% of the designed toughness, we demolish blocks ④, ⑤, and ⑥ in order.

Step 4: We carry out the structural replenishment of slab beams for the demolished parts in blocks ④, ⑤, and ⑥.

Step 5: We repeat steps 1 to 4 to demolish the middle partition wall from top to bottom.

4.4 *The vertical demolition procedure of mid-partition wall*

The mid-partition wall is divided into 8 parts in the vertical direction, which are demolished from top to bottom part by part. Each structural level contains 2 demolished sub-levels and the height of these sub-levels is shown in Figure 3. When demolishing the 2nd, 4th, and 6th levels, the floor beams were partially removed together with the joints of the central partition walls.

4.5 *The demolition of single-layer wall body*

The static cutting technique is adopted for demolition (Cai 2021). The lifting holes are first drilled with a drilling tool. It is fixed with electrical chain blocks during the cutting process. When the cutting process finishes, it is carried to the ground with a forklift and lifted by a crane onto a transport truck that is transported out of the construction site.

We set up operating platforms on both sides of the mid-partition wall. It should be noted that these platforms are also used as protective platforms.

We place wooden boards above the floor slabs in the demolition area around the septum wall to prevent the floor slabs from being damaged by the concrete debris falling from the wall. These concrete fragments should be slid onto the bottom floor slab through the recesses set. It is forbidden to fall to the ground under gravity or just throw them away.

To ensure ventilation in the basement during the demolition process, several blowers are set up around the central partition wall. The electricity comes from the electrical box of that basement level.

4.6 *The structural replenishment of the mid-partition wall*

The replenishment of the floor slab structure should be started after the cutting process of the mid-partition wall on that level is done.

We prepare construction joints on both sides of the partition wall and on the wall itself at a distance of 800 mm from the beam by attaching relays to the reinforcement on the beam.

During the concrete casting, the floor slabs on both sides of the mid-partition wall are at the very edge of the wall. After the walls are removed, the vertical reinforcement within the floor slab is stripped out and the two ends are welded together 600 mm apart. The seam is poured after welding.

We prepare construction joints at a distance of 1000 mm from the two bottom surfaces of the middle partition wall and the wall itself; attach the relays to the reinforcement in the bottom slab.

We pick the reinforcement out of the joint with a relay and fill the joint with concrete.

5 CONCLUSION

The demolition of the mid-partition wall in a complex foundation pit has many factors that are highly likely to bring stress redistribution to the structure and sometimes even cause serious damage. Therefore, this type of work has high requirements for construction techniques (Gong 2018). For the demolition work, all necessary precautions should be taken and the corresponding demolition guidelines should be followed to ensure the quality and efficiency of demolition.

In this paper, the demolition techniques of diaphragm walls in footings are discussed as an example for super high-rise buildings in an attempt to provide a potential reference for the demolition of specific walls. The analysis of the demolition technology of the partition wall in the foundation pit, such as the main methods, process flow, and precautions, is beneficial to the progress of the whole project (Lu 2015).

REFERENCES

Guihong, Chen, Yuwen, Li, Yuguang, Zhao. (2005) Mechanical Behavior of Middle Wall in Continuous Arch Tunnel. *J. China Railway Science*. Vol.26. 20–26.

Guodong, Lv. (2019). The Influence of Temporary Support on the Mechanical Behavior of Middle Wall Construction in a Continuous Arch Tunnel. *J. Journal of Guangdong Communication Polytechnic*. Vol.18. 47–52.

Jianyu, Li. Jianhui, Yang. Zhenxing, W. (2017). Study on Selection of Structural Type Selection and Stability for Mid-partition. *J. China Civil Engineering Journal*. Vol 50, 236–242.

Jingwei, Qiu. (2022). Optimization of a Zoned Excavation of a Foundation Pit Adjacent to the Subway. *J. Soil Eng. and Foundation*. Vol.36.187–190.

Jiajie, Liu. (2022) Research on Excavation Technology of Utility Tunnel with Shared Partition Wall between Adjacent Chambers. *J. Municipal Engineering Technology*. Vol. 40. 165–170

Shichao, Jiang. (2021) Advanced Reinforcement Construction Technology of Shallowly Buried and Underground Excavation Comprehensive Pipe Gallery Under Urban Trunk Road. *J. Construction Mechanization*. Vol.42. 30–32.

Junqiang, Mao. (2020). Construction Technology of Middle Partition Wall Demolition in Deep Foundation Pit. *J. Building Construction*. 42.1381–1385.

Wenjie, Cai. (2021). Research on Removal Technology of Middle Partition Wall Under Overlapping Condition Between Middle Partition Walls and Frame Columns. *J. Building Construction*. Vol.43 1009–1011.

Yin, Gong. (2018). Construction Technology for Closure of Two Side Structure of Mid-Partition Wall in Basement. *J. Building Construction*. 40.1731–1733.

Xianrong, Lu. (2015). Construction Technology for Changing of Compartment Wall Support and for Wall Removal in Deep Foundation Pit. *J. Building Construction*. Vol. 37. 1254–1255.

Water Conservancy and Civil Construction – Oke & Ahmad (Eds)
© 2024 The Author(s), ISBN: 978-1-032-58618-2

Study on the technology of lime expanded pile foundation treatment for airport runway in high-temperature permafrost regions

Danze Wu* & Youshi Hao
Northeast Branch Company of China Airport Planning & Design Institute Co., Ltd., Shenyang, China

Guoguang Liu & Li Feng
College of Transportation Science and Engineering, Civil Aviation University of China, Tianjin, China

Yifan Wu
Northeast Branch Company of China Airport Planning & Design Institute Co., Ltd., Shenyang, China

ABSTRACT: Airport runways have a high standard for foundation settlement and uneven settlement, and there is a limited experience in airport runway construction in permafrost areas at home and abroad. In recent years, how the frozen soil foundation treatment measures adopted in the construction of iron (public) roads in frozen soil areas can be applied to the construction of airport runways still needs experimental research. According to the geological conditions of the proposed extension section of the runway of Mohe Airport, the indoor model test and field test of lime squeeze pile foundation treatment in permafrost were carried out. Through the water absorption and heat reaction of lime piles, the permafrost foundation was completely thawed and the soil moisture content was reduced. Data analysis proves that under the effect of lime pile compaction, the compactness of the foundation soil is improved, thereby increasing the bearing capacity and compression modulus of the composite foundation. Using the method of the numerical simulation analysis for the foundation temperature field, it is proved that the introduction of lime piles as a heat source into the foundation can achieve the purpose of thawing the permafrost, and the ground temperature stabilizes the positive temperature state after the rise-fall changes. This article describes the application effect of the lime squeeze pile method in the treatment of deep-buried, high-temperature, and unstable permafrost foundations, so the foundation conditions can meet the construction standards for runway settlement and uneven settlement. It has accumulated certain theoretical and experimental support and perfected the key technologies for the construction of airport runways in permafrost regions in China.

1 INTRODUCTION

Permafrost in China has an area of about 21.5×10^5 km^2, accounting for 22.3% of the country's land area. It is mainly located in the small and large Xing'an Mountains, the Qinghai-Tibet Plateau, and the high mountains in the west [Wang 2011]. After the permafrost thaws, the strength and shear strength are significantly reduced, which is likely to cause uneven settlement of the foundation [GB50324-2014 2014]. With the development of civil aviation and general aviation, a large number of airports will be built in the permafrost regions in the future. It is inevitable to solve the problem of permafrost foundations [Liu *et al.* 2015]. There are a number of airports located in the cold zone, such as the Alaska region of the United States, the Russian

---

*Corresponding Author: 9432287@qq.com

156   DOI: 10.1201/9781003450832-19

Far East, and the Nordic region. Most of these airports are very small and mainly used for military purposes. The construction and maintenance costs are extremely high, so the guiding significance is limited for the construction of civilian airports in permafrost regions. A series of methods for permafrost foundation treatment, such as the replacement method, the hot rod roadbed, and the thermal insulation roadbed, have been derived from the theories of "maintaining frozen state design", "gradually thawing state design" and "pre-thawing state design" in highway, railway, and airport engineering fields at home and abroad [Dou & Hu 2016; Fang et al. 2016; Oldenborger & LeBlanc 2015; Short et al. 2014]. At present, there are limited cases of airport runway construction in permafrost areas in China. The first runway of Mohe Airport built in permafrost areas in China is treated with sand and gravel material replacement method. The local maximum replacement depth is 8 m. The runway has been opened for navigation in 2008 and the operation is in a good condition.

Generally speaking, high-temperature frozen soil refers to frozen soil with a temperature between 0.0 and $-1.5°C$. Due to the existence of a large amount of unfrozen water, its physical and mechanical properties tend to thaw soil [Liu & Zhang 2012]. Affected by climate warming and human activities, the permafrost degradation in the Mohe area is significant, which is manifested by the decrease of the upper limit of permafrost, the increase in temperature, and the thinning of the thickness of permafrost [He et al. 2009]. According to the survey data, the runway extension section of Mohe Airport is a high-temperature and unstable permafrost distribution area, and the buried depth is mostly between 8 and 15 m. The replacement method has certain limitations. Drawing on the experience of foundation treatment of permafrost squeezed piles on the Qinghai-Tibet Railway [Ma et al. 2005] and the research results of the Zhengxi high-speed rail collapsible loess lime soil compacted the pile test section [Wang et al. 2020], the project participating units and domestic experts in the permafrost field proposed the use of lime squeezed piles to treat the permafrost foundation of the runway extending section in Mohe Airport.

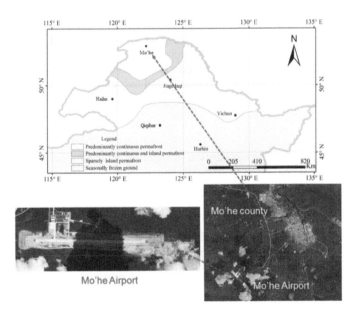

Figure 1. Location of Mohe Airport [Mao et al. 2019].

Based on the test section of lime-squeezed pile foundation treatment in the extending area of Mohe Airport's runway and combined with the empirical theories of domestic linear engineering in permafrost regions, this paper expounds the lime-squeezed pile reinforcement mechanism and proposes the test and inspection plan based on the characteristics of the permafrost engineering in this area. Through the analysis and research of test data, the article evaluates the treatment of permafrost foundations with the method of lime squeeze-expanded pile construction.

2 REINFORCEMENT MECHANISM AND TEST PURPOSE

2.1 *Reinforcement mechanism*

Engineering construction in areas with unstable high-temperature permafrost [GB50324-2014 2014; Liu & Zhang 2012] has destroyed the original cold and heat balance of frozen soil. At the same time, the trend of global temperature warming is bound to cause the permafrost under the foundation to thaw. The frozen soil foundation is completely thawed [Li *et al*. 2010; Short *et al*. 2014] in advance by artificial methods, so the frozen soil foundation is in a thawed state during the construction and the use of the airport runway, so as to meet the technical requirements of the runway for the foundation. 1 kg of fully matured quicklime can release more than 1,000 J of heat, and the volume expansion is about 2 times. The permafrost reinforcement mechanism [Mao *et al*. 2019] of lime squeeze piles derived from the chemical reaction characteristics of the quicklime material can be summarized as:

(1) The mechanical damage during pile formation and the hydrothermal reaction of quicklime (CaO) with the premixed water and free migration in the soil are used to disturb the permafrost so that the thawing permafrost is completely thawed and the water content of the foundation soil is reduced, and the thaw settlement after construction can be eliminated.

$$CaO + H_2O = Ca(OH)_2 + Q(Exothermic) \qquad (1)$$

(2) The reaction volume expansion of lime is utilized to compact the deep foundation soil to improve the compactness of the foundation soil.
(3) The principle of the composite foundation is used to improve the bearing capacity and deformation modulus of the foundation soil and reduce the compression deformation of the soil after thawing.

2.2 *Test purposes*

According to the reinforcement mechanism, through the indoor model test and field test, the following objectives are expected to be achieved:

(1) The ground temperature before and after the test is observed, and the effect of the lime squeeze-expanded pile scheme is analyzed on the ground temperature disturbance of the permafrost foundation.
(2) During the hole formation process, the soil moisture content before treatment is tested, and after the test, holes are drilled to test the soil moisture content, and the changes in soil moisture content are compared and analyzed. After the test, a dynamic penetration test or standard penetration test is carried out to evaluate the bearing capacity of the foundation.
(3) After the experiment, the foundation-bearing capacity of the soil between the piles and the composite foundation is tested, and the degree of improvement of the foundation strength by the construction method compared to the investigation data is analyzed.
(4) After the test, the pile body and soil settlement between the piles are observed to analyze whether the foundation settlement can meet the deformation requirements of the airport runway.

(5) The construction technology is summarized, the design parameters are checked, and the construction parameters are recorded to verify whether the airport can meet the construction requirements of non-stop flights and provide guiding test results for the subsequent design and construction.

3 EXPERIMENTAL RESEARCH WORK

3.1 *Overview of relying on projects in the test section*

According to the survey data, it has been proved that the area of the runway to be extended in Mohe Airport is located in a permafrost development zone. The destruction of the natural environment and the stripping of surface vegetation have caused island-like thawing areas and the lower limit of permafrost in the area, which is a degraded permafrost. During the survey, no groundwater was found to be exposed. The distribution type of permafrost in this area is divided into island-shaped permafrost. It is divided into high-temperature permafrost and extremely unstable permafrost according to ground temperature, and it is mainly divided into less ice-rich permafrost according to ice content. Most of the foundation soil in the frozen area is a weak and non-thaw settlement, and there is a partial thawing settlement area.

Distribution characteristics: From a plane perspective, the thawing permafrost is distributed between 50 and 150 m outside the runway extension; from a vertical perspective, the burying depth of the thawing permafrost is between 8 and 15 m.

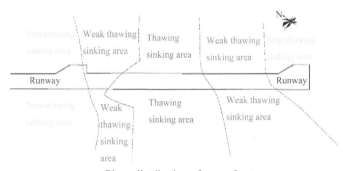

a. Plane distribution of permafrost

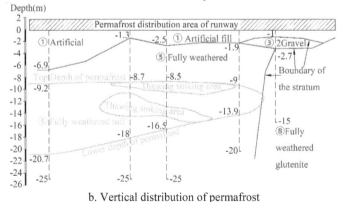

b. Vertical distribution of permafrost

Figure 2. The engineering geological map of the permafrost foundation at Mohe Airport.

Engineering characteristics of permafrost soil foundation: in Table 1, we can the thawing settlement category, thawing settlement grade, and main physical and mechanical properties of the foundation soils that make up each layer of permafrost. The temperature of permafrost in the extended area of the runway is between -0.10 and $0.00°C$, which is extremely unstable permafrost with high temperatures. It is predicted that the frozen soil will thaw by about 85% within 30 to 50 years after the completion of the airport runway extension.

Table 1. The classification of thawing and the main physical and mechanical indexes of permafrost.

Layer number	Strata	Total water content (%)	Thaw coefficient (%)	Thaw and sink grade	Thaw and sink category	Frozen soil type	Penetration level	Characteristic value of bearing capacity after thawing and consolidation (KPa)	Compression modulus (MPa)
⑤	Fully weathered tuff	15.9~32.4	1.43	II	Weak thawing	Frozen soil	Slightly permeable	180	6.4
		16.5~32.5	3.65	III	Thawdown	Rich frozen soil	Slightly permeable	180	6.4
⑥2	Strongly weathered andesite	–	0.00	I	Unsinkable	Less frozen ground	-	450	–
⑦	Fully weathered glutenite	11.7~22.7	0.55	I	Unsinkable	Less frozen ground	Weakly permeable	180	5.5
		10.4~28.9	1.39	II	Weak thawing	Frozen soil	Weakly permeable	180	5.5
⑦1	Fully weathered mudstone	11.2~39.9	0.44	I	Weak thawing	Frozen soil	Slightly permeable	160	5.3
⑧	Fully weathered glutenite	36.3	0.00	I	Unsinkable	Less frozen ground	Weakly permeable	300	–
⑧1	Strongly weathered glutenite	13.0~15.8	0.26	I	Unsinkable	Less frozen ground	Weakly permeable	400	–

The calculation of permafrost foundation thawing deformation using the recommended formula[MH/T 5027-2013, 2013; GB 50021-2001 2009] is shown in Table 2. The maximum allowable post-construction cumulative settlement of the airport runway to the foundation is not more than 30 cm, the differential settlement is not more than 0.15% (the measurement length of 50 m), and the natural foundation doesn't meet the requirements.

$$\delta_0 = \Delta h / h_0 \rightarrow \Delta h = h_0 \times \delta_0 \qquad (2)$$

Where δ_0 refers to the thaw coefficient, Δh refers to the thawing deformation, and h_0 refers to the initial thickness of permafrost.

3.2 *Indoor model test research*

According to the reinforcement mechanism and test purpose, the geological conditions are analyzed and the experience of lime-soil pile foundation treatment in permafrost on the Qinghai-Tibet Railway is used for fuzzy calculation of the heat required for the thawing of

Table 2. Calculation of ground settlement in the runway extension area.

Survey hole	25	27	28	29	31	40	42
Main lithology	Fully weathered tuffaceous sandstone	Fully weathered tuffaceous sandstone	Fully weathered tuffaceous sandstone	Fully weathered tuffaceous sandstone	Fully weathered glutenite	Fully weathered glutenite	Fully weathered glutenite
Thickness of frozen soil h0 (cm)	14	10	8	7	0	3.5	5
Thaw coefficient $\delta 0$	0	0.3~4.1	0.4~2.1	3.6~4.9	0	0~0.5	0
Thawing deformation Δh (cm)	0	28.6	4.6	9.3	0	0.8	0
Differential settlement (%)	–	5.2	9.6	1.9	1.9	0.3	0.15

permafrost foundation. First, the model test of simulating lime pile thawing frozen soil in Figure 3 was carried out indoors [Cheng et al. 2019]. A lime pile with a diameter of 20 cm was selected to measure the heat release and effective components of quicklime. Lime was prepared (the quicklime was mixed with sand, soil, and cement). For pile body backfill material, the content of quicklime is between 30% and 40%, and the mix ratio of the pile body material is adjusted according to the moisture content of the boring soil on site. Measuring elements and measuring lines of three thermistor temperatures are arranged along the pile circumference at a radial interval of 5 cm, from the ground surface in the depth direction to the lower limit of the freezing depth, in order to obtain the data of the influence radius of the lime pile exothermic reaction. The results of the thermistor test model in Figure 4 show that the thawing effect of frozen soil is related to the pile diameter, pile distance, and effective components of the backfill material. The lime content is determined according to the changes in the soil moisture content of the site foundation [Lin & Liu 2003]. According to the test results, the center distances of s = 2.5 d and S = 3 d piles are the best in terms of the amount of lime material using the thawing effect of frozen soil and have the technical conditions for the field test, which can guide the mix ratio of lime material and pile position layout in the field test section.

 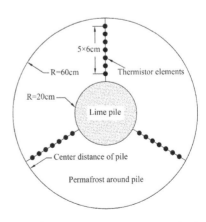

Figure 3. Experimental model arrangement.　　Figure 4. Temperature measurement point layout.

3.3 Field test research

The most unfavorable permafrost distribution area outside the southeastern end of the runway, which has been proven by survey data, is selected to carry out field tests on lime squeeze pile foundation treatment, focusing on the depth range of 8~15 m permafrost distribution. Rotary drilling is used to form holes, and the diameter of the lime squeeze-expanded pile is d = 0.4 m, and two sets of pile center distances s = 2.5 d and s = 3 d are selected, and the lime piles are arranged in a regular triangle. The diameter of the ground temperature test hole, the sampling hole after the test, and the settlement observation hole is d = 0.1 m. The lime material is used as the pile hole filler according to the indoor test to determine the construction mix ratio, and it is backfilled and tamped in layers according to the requirement of a 0.97 compaction coefficient. The detailed layout of various holes is shown in Figure 5.

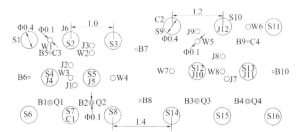

Legend: W-Ground temperature test hole; S-Lime squeeze test pile; C-Bearing capacity test hole; Q-Sampling hole after test; J-Settlement observation hole; B-Standard penetration test hole.

Figure 5. The design scheme of the field test (Unit: m).

In order to evaluate the technical feasibility of the foundation treatment plan, the inspection and testing items carried out before and after the test include: ground temperature test, soil water content and density test, settlement observation, bearing capacity test, pile dynamic penetration test, and standard penetration of soil between piles test. The specific procedures and descriptions are as follows:

(1) The ground temperature test holes W1~W8 are set, and the automatic temperature acquisition system is immediately used to observe the ground temperature changes after burying the thermistor temperature test elements.
(2) Lime squeeze test piles S1~S16 are arranged, and the lime material is backfilled in layers to 4 m below the ground, and the silty clay is backfilled from the surface to 4 m below the ground. Among them, S7 and S9 are also used as the bearing capacity test holes C1 and C2; lime squeeze-expanded test piles S2, S4, S5, S10, S12, and S13 are also used as settlement observation holes J4~J6 and J10~J12. During the hole formation process, starting from 5 m below the ground surface, we take soil samples every 2 m to test the soil moisture content, and the average soil moisture content of the pile body is obtained to guide the design of the pile lime backfill mix ratio.
(3) After the construction of the lime squeeze-expanded test pile, the construction settlement observation hole J1~J12 is completed (4 m depth below the surface). Among them, the lime squeeze expansion test piles S2, S4, S5, S10, S12, and S13 are backfilled to 4 m below the ground surface. A settlement plate is embedded on the top of the lime material as settlement observation holes J4~J6 and J10~J12. Subsidence observation will begin after the end of the process (2), which will be observed once a day in the initial stage and once every 15 days in the later stage.

(4) The lime material of the lime squeeze test pile is digested, and the sampling holes Q1~Q4 are arranged after the test, and the soil samples at the depth of 11~15 m (1 m interval) are taken to test the soil moisture content between the piles. After the sampling is completed, holes Q1~Q4 are also used in the hole position of the standard penetration test.
(5) After the lime material of the test pile is digested, S7 and S9 are used as the composite foundation bearing capacity test piles C1~C2, and the plate load test is used to test the soil bearing capacity between the piles at positions C3~C4.
(6) After the lime material of the test pile is digested, S7 and S9 are used as the composite foundation bearing capacity test piles C1~C2, and the plate load test is used to test the soil bearing capacity between the piles at positions C3~C4.

4 ANALYSIS AND EVALUATION OF TEST DATA

According to the test plan, the continuous test of various indicators was carried out on the foundation treatment test section, and the test data was sorted and analyzed, and then the evaluation of the construction method was as follows:

(1) We use the thermistor temperature test element, and select two groups of test pile group pile centroid positions of the ground temperature test holes W3 and W8 in the 10 m, 12 m, and 14 m permafrost temperature test data within the distribution depth of the permafrost, and draw the difference between the typical boreholes. The depth temperature change curve is shown in Figure 6. Before the lime squeeze test pile is backfilled with lime material (from November 2 to November 8), the original ground temperature is between $-0.07 \sim -0.015°C$, which is a high temperature unstable freezing earth.

Thawing stage: after the lime squeeze test pile, it is backfilled with the lime material for 8 days. Due to the exothermic reaction of the lime material, the ground temperature rises, and the permafrost has become a thawing state.

Heating stage: the heating stage time is 8~14 days after the start of the observation, and the highest temperature reaches 38°C. Due to the different proportions of the lime-soil backfill materials of the two test piles, the ground temperature rises differently. At

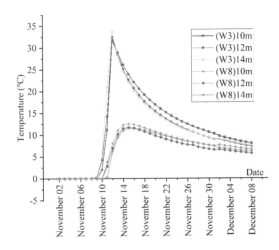

Figure 6. Temperature variation curves of typical boreholes at different depths.

the same time, the natural moisture content of the foundation soil will also affect the changing magnitude of temperature.

Cooling stage: It starts about 15 days after the observation, due to the influence of the ground temperature conduction effect of the surrounding soil in the test area, the thawing soil enters the cooling stage with a maximum drop of 30°C. The later stage of the cooling process is slower, and the heat storage effect makes the ground temperature approach a 5°C steady state.

From the comprehensive temperature test results, it can be seen that under the conditions of pile spacing, pile diameter, and backfill material mix ratio proposed in the field test, the change range of ground temperature is affected by the effective lime content. The more significant, the stable stage of cooling tends to a similar temperature state under the influence of the surrounding soil temperature. The lime extruded and expanded pile achieves the purpose of melting the deeply buried permafrost, and the thawed runway foundation can be treated as the general foundation.

(2) In order to eliminate the interference of the seasonally frozen ground layer on the settlement observation, the foundation deformation monitoring program with a settlement plate buried 4 m below the surface is adopted, and the settlement observation holes J1, J2, J7, and J8 at the center of the pile group are selected for monitoring. The data drawing deformation curve is shown in Figure 7.

The settlement monitoring points are arranged on the top of the pile and the soil between the piles. The observed data shows that the average daily deformation within one month after construction is about 1 mm, the cumulative uplift deformation of the soil between the piles is between 1.8 and 2.8 cm, and the cumulative settlement deformation is within 1 cm, and the deformation is basically in a stable state. The distribution of settlement deformation is irregular, so it can be judged that it is caused by the disturbance of construction machinery. The reason for the uplift deformation is that after the pile hole is backfilled with the lime material, the lime reacts with water to expand in volume, which shows that the construction method has compacted and strengthened the soil between the piles to make the foundation soil compact, and at the same time, the foundation soil is uplifted and deformed. After the permafrost has thawed, the foundation has no thawing settlement deformation. The lime squeeze-expanded pile has achieved the purpose of compacting the foundation, and the settlement deformation of the foundation is within the limit of deformation of the airport runway foundation [Mao et al. 2019].

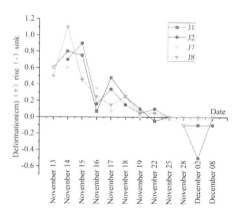

Figure 7. Typical borehole settlement deformation curve.

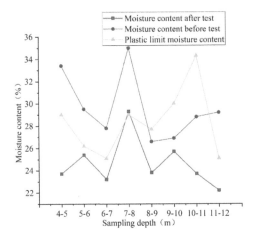

Figure 8. Typical borehole water content contrast curve.

(3) According to the ground temperature test data, it is judged that the lime material of the test pile was digested on November 12, 2017. The position of the standard penetration test hole B9 was selected to sample the soil moisture content between the test piles and changes, and the survey data are shown in Figure 8. The natural moisture content of the foundation soil in the depth range of 4–12m below the surface is between 26.6% and 35%. Because the reaction of quicklime consumes the free water in the soil, the moisture content of the foundation soil is reduced to between 22.2% and 29.3%. The average reduction is 5%.

In accordance with the specification for the grading standard of permafrost thawing [GB 50021-2001 2009], due to the decrease in the moisture content of the foundation soil, the thawing settlement of the foundation soil has transitioned from grade III to grade II weak thawing settlement or from grade I non-thaw settlement to thawing settlement. The reduction of the coefficient calculates the thawing settlement deformation smaller, and the method of lime squeeze-expanded pile foundation treatment has improved the foundation soil.

(4) After the lime material of the pile body is digested, the lime squeeze test piles S7 and S9 are selected to test the characteristic values of the single pile bearing capacity of 35 KN and 40 KN, and the load test bearing capacity test hole positions C3 and C4 are carried at 1 m below the surface of the ground between the piles. The characteristic values of the force are 126 KPa and 108 KPa, which are converted to 180 KPa and 162 KPa at a position 4 m below the ground. Compared with the survey report on natural foundation bearing capacity, the increase rate is about 10%, which meets the strength requirements [Sun et al. 2020] of the airport runway foundation.

(5) After the lime material of the extruded and expanded test pile has been digested, the data of standard penetration test holes B5, B7, and B9 are selected and arranged as shown in Table 3. It can be seen that below the surface depth of 4 m, the standard penetration stroke number after correction is above 17, and the strength of the fully weathered tuff foundation soil is significantly improved compared with the index provided by the survey data. Through the water absorption and exothermic reaction of the lime piles, the permafrost foundation is completely melted, and the moisture content of the soil is reduced. At the same time, under the compaction effect of the lime piles, the compactness of the foundation soil is improved, thereby increasing the foundation compression modulus and reducing the post-construction compression deformation.

Table 3. Statistics of the soil between piles of standard penetration test data.

The standard penetration test hole B5		The standard penetration test hole B7		The standard penetration test hole B9	
Start and end position of the test (m)	Corrected hits	Start and end position of the test (m)	Corrected hits	Start and end position of the test (m)	Corrected hits
4~4.45	17.48	4~4.45	24.84	2~2.45	9.2
6~6.45	34.4	6~6.45	17.2	4~4.45	23
8~8.45	29.97	8~8.45	16.2	6~6.45	21.5
10~10.45	22.68	10~10.45	40.5	8~8.45	17.82
14~14.45	97.09			10~10.45	33.21
				12~12.45	57.75
				14~14.45	109.5

5 NUMERICAL SIMULATION ANALYSIS OF THE GROUND TEMPERATURE FIELD

Through indoor model experiments and field engineering test verification, the reinforcement mechanism of the lime squeeze pile method is verified. After treatment, the bearing capacity and deformation modulus of the permafrost foundation soil can meet the requirements of airport runway construction, and the reduction of water content improves the permafrost thawing sedimentation level. Because ground temperature is one of the decisive factors of the nature of frozen soil engineering, which is affected by engineering activities and the climate environment, the temperature field of the foundation during the construction process and after the completion of the project will show a warming-cooling-steady change. The multi-physics coupling analysis of the software COMSOL multiphysics is used to analyze and deduct numerical simulations of temperature field changes with time.

5.1 Basic equation of the temperature field

The heat in the soil is transferred in the form of heat conduction, heat convection, and heat radiation. Frozen soil is composed of a soil skeleton, ice, unfrozen water, and gas multiphase structure, and the heat transfer process is complicated. The temperature field of frozen soil under the airport runway is mainly the temperature change caused by heat conduction. The numerical simulation of the temperature field of frozen soil mainly analyzes the internal heat conduction of frozen soil and the effect of ice and water phase change.

The heat conduction equation of the plane strain model dominated by phase change factors can be expressed by the following formula:

$$C\frac{\partial T}{\partial t} = \frac{\partial}{\partial x}\left(\lambda\frac{\partial T}{\partial x}\right) + \frac{\partial}{\partial y}\left(\lambda\frac{\partial T}{\partial y}\right) \tag{3}$$

Where C refers to the specific heat capacity of the soil, T refers to the transient temperature of the soil, λ refers to soil thermal conductivity, and t refers to time.

The phase change between water and ice in frozen soil causes changes in soil volume parameters, which in turn causes changes in thermal parameters, such as specific heat capacity and thermal conductivity. The respective temperature field equations are established based on the freezing or thawing state of the soil itself.

The soil in a frozen state:

$$C_f\frac{\partial T_f}{\partial t} = \frac{\partial}{\partial x}\left(\lambda_f\frac{\partial T_f}{\partial x}\right) + \frac{\partial}{\partial y}\left(\lambda_f\frac{\partial T_f}{\partial y}\right) \tag{4}$$

The soil in a molten state:

$$C_u \frac{\partial T_u}{\partial t} = \frac{\partial}{\partial x}\left(\lambda_u \frac{\partial T_u}{\partial x}\right) + \frac{\partial}{\partial y}\left(\lambda_u \frac{\partial T_u}{\partial y}\right) \tag{5}$$

The phase change interface of the two:

$$\lambda_f \frac{\partial T_f}{\partial n} - \lambda_u \frac{\partial T_u}{\partial n} = L\frac{ds(t)}{dt} \tag{6}$$

In Formulas (4) ∼ (6):

Where C_f refers to the coefficient of specific heat capacity, λ_f refers to thermal conductivity, T_f refers to temperature, etc. in frozen soil areas.

Where C_u refers to the specific heat capacity, λ_u refers to thermal conductivity, T_u refers to temperature, and other coefficients of the thawing frozen soil area.

On the phase change interface, it can be seen that the above-mentioned temperature field is difficult to obtain relevant analytical solutions, and a numerical model is established for the finite element analysis and solution.

5.2 Numerical model establishment

For the numerical analysis of the temperature and deformation of the frozen ground, we mainly use the structural mechanic's module and the heat transfer module of the multiphysics coupling analysis software COMSOL multiphysics. The software uses the PDE partial differential equation model to perform multi-physics coupling solutions.

The composite foundation of the lime pile foundation treatment zone is made of nonpermeable, low frost heave, and fully weathered rock and lime materials. The influence of the water migration and seepage field is negligible, and only the phase change caused by the temperature change is simulated. The deformation of the temperature field is caused by the volume change of the soil.

The observation hole of ground temperature change of the lime pile field test section is used as the section for numerical simulation inversion analysis, and the soil layer is dominated by fully weathered tuff.

5.3 Model boundary conditions

(1) Temperature boundary

The temperature boundary of the model directly specifies the temperature value on each boundary surface of the model or expresses it as a temperature function that changes with time. For areas directly receiving solar radiation, such as the road surface and soil surface area, the relationship between surface temperature and time is expressed as a sine-like relationship. After fitting the original observation data, the following formula is obtained:

$$T(t) = T_0 + G(t) + A\sin(wt + \varphi) \tag{7}$$

Where T_0 refers to the temperature of the soil surface, A refers to the soil surface temperature that is half of the annual difference (°C), ω refers to the temperature change cycle, generally, one year, φ refers to the initial phase which determines the starting moment, $G(t)$ refers to the surface temperature as a function of time.

(2) Model initial conditions

Before the foundation treatment, the annual average temperature value of frozen soil temperature is $-0.5°C$, and the temperature value at the bottom of the frozen soil layer model is $0.1°C$.

(3) Determination of calculation parameters

According to the geological survey report, the thermodynamic parameters of the soil and lime pile materials are as follows.

Table 4. Thermodynamic parameter table of soil and lime pile.

Type of soil layer	Density kg/m^3	After thawing thermal coefficient W/(m^3/°C)	After frost heave thermal coefficient W/(m^3/°C)	Before thawing specific heat capacity kJ/(m^3/°C)	After thawing specific heat capacity kJ/(m^3/°C)	Dry density kg/m^3
Lime pile	1.85	1.76	2.36	1708	2334	1.66
Fully weathered tuff	1.84	1.30	1.18	2153.4	2611.1	1.57

(4) Predict simulation results

Through software simulation, the variation pattern of the temperature in the trench area, the soil surface area with time obtained, and the variation of the ground temperature in the 20 years after the construction are predicted. The duration of the melting stage and the heating stage is short, and the heat source introduced from the outside is presented in the foundation soil layer. With the continuous increase of time, due to the influence of the ground temperature conduction effect of the surrounding soil in the test area and the natural climatic conditions, the thawing soil enters the cooling stage. It is estimated that within 20 years after construction, the depth of atmospheric influence will hardly reach the upper limit of the original permafrost layer. It can be inferred that the permafrost has completely disappeared and will no longer cause thawing deformation of the runway. The lime extrusion and expansion method can achieve the expected foundation treatment purpose.

6 CONCLUSIONS

(1) In this paper, relying on the test project of permafrost foundation treatment of Mohe Airport, the lime squeeze pile method is proposed to treat deep-buried, high-temperature, and unstable permafrost foundations. Judging from the field test data, the treatment method has achieved the expected test purpose and has the technical conditions to be applied to the foundation treatment project of the planned extension of the runway. However, from the perspectives of technology, economy, and construction period, the replacement method also has technical applicability. Compared with the lime pile method, the project cost is saved by 10%, and the construction period is shortened by more than 3 months. The available construction time in Mohe is short, and the runway of Mohe Airport is extended. The project adopts the foundation treatment plan of excavating permafrost and backfilling with rubble in 2020.
(2) Since no groundwater was encountered during the survey of the permafrost area at Mohe Airport, the lime squeeze-expanded pile construction method has achieved good test results, and groundwater is often the main factor inducing the deformation of permafrost foundations. In practice, the water of the permafrost foundation should be noted.
(3) Through theoretical analysis and experimental verification, the lime squeeze pile method is feasible to treat the permafrost foundation of the extended section of the Mohe Airport runway. The permafrost is completely thawed, the foundation soil is compacted and

reinforced, and the foundation conditions meet the requirements of the construction runway settlement and uneven settlement, which has a certain promotion significance for airport construction in permafrost areas in the future.

(4) The wide-width endothermic effect [Wang & Chen 2008] and vibration heating effect [Zhang *et al.* 2021] of the runway are strong. Since the cohesive force between the soil skeletons can bear the self-weight stress for a certain period after the water dispersion of the cohesive soil is lost, the thawing and sedimentation deformation of the cohesive permafrost layer has a hysteresis effect. The effect of lime extrusion piles on thawing cohesive soil permafrost foundation needs to be tested in practice. If the foundation soil layer is coarse-grained soil with small cohesion, the thawing deformation can be developed and completed in a short time after the foundation soil water dispersion is lost, and the permafrost foundation treatment method of lime extrusion piles is feasible.

(5) The airport runway is a key security area, and the nature of permafrost engineering varies greatly in different regions. The influence of temperature, water content, and confining pressure on the mechanical parameters of the permafrost [Ling & Zhang 2004; Xiao *et al.* 2015]is considered. At the same time, the influence of the negative accumulated temperature of the permafrost foundation and the degradation of permafrost in Northeast China [Feng & Li 2016] are predicted. The promotion of lime squeeze-expanded piles for permafrost foundation treatment methods still needs to be studied in depth in combination with the actual conditions of permafrost in various regions.

REFERENCES

"Code for Engineering Geological Investigation of Frozen Ground": GB 50324—2014. Planning Press of China, Beijing (2014) (in Chinese).

"Code for Geotechnical Engineering Design of Airport": MH/T 5027—2013. Civil Aviation Press of China, Beijing (2013) (in Chinese).

"Code for Investigation of Geotechnical Engineering": GB 50021—2001. China Architecture & Building Press, Beijing (2009) (in Chinese).

Cheng J., Xiong Z. W., Jin W., Cai H. C., and Meng J. B. "Experimental Study on Treatment of Patchy Permafrost Foundation with Quicklime Piles". *Railway Engineering*, 59 (9): 88–92 (2019) (in Chinese). DOI: 10.3969/j.issn.1003-1995.2019.09.22.

Dou M. J. and Hu C. S. "The Design Principles of the Embankment of Highway in Permafrost Regions and Their Application". *J Glaciol Geocryol*, 23 (4): 402–406 (2001). DOI: 10.3969/j.issn.1000-0240.2001.04.011.

Fang J. H., Li D. Q., Xu A. H., and Dong C. J. "*Measures Application Technology of Special Road Foundation Engineering in Permafrost Regions*". Lanzhou University Press, *Lanzhou* (2016) (in Chinese).

Feng Y. Y. and Li Z. L. "Spatial and Temporal Evolution of the Negative Accumulated Temperature in Northeast China from 1957 through 2007". *J Glaciol Geocryol*, 38 (6): 1529–1537 (2016) (in Chinese). DOI: 10.7522/j.issn.1000-0240.2016.0178.

He R. X., Jin H. J., Chang X. L., Lv L. Z., Yu S. P., Yang S. Z., Wang S. L., and Sun G. Y. "Degradation of Permafrost in the Northern Part of Northeastern China: Present State and Causal Analysis". *J Glaciol Geocryol*, 31 (5): 829–834 (2009) (in Chinese). DOI: 10.1016/S1003-6326(09)60084-4.

Li G. Y., Li N., and Ma W. "Cooling Effects and Mechanisms of Crushed Rock Protective Slopes Combined with Shading Board on the Embankment in Warm Permafrost Regions". *Rock and Soil Mechanics*, 31 (1): 165–173 (2010) (in Chinese). DOI: 10.3969/j.issn.1000-7598.2010.01.029.

Lin T. and Liu Z. D. "Study on Indoor Tests of Fly Ash and Quick Lime Improving Soft Soils". *Rock and Soil Mechanics*, 24 (6): 1049–1052 (2003) (in Chinese). DOI: 10.3969/j.issn.1000-7598.2003.06.039.

Ling F. and Zhang T. J. "A Numerical Model for Surface Energy Balance and Thermal Regime of the Active Layer and Permafrost Containing Unfrozen Water". *Cold Reg Sci Technol*, 38 (1), 0–15 (2004). DOI:10.1016/s0165-232x(03)00057-0

Liu S. W. and Zhang J. M. "Review on Physic-mechanical Properties of Warm Frozen Soil". *J Glaciol Geocryol*, 34 (1): 120–129 (2012) (in Chinese). DOI: ir.casnw.net/handle/362004/8906.

Liu W. B., Yu W. B., Chen L., Yi X., Han F. L., and Hu D. "Techniques of Airport Runway Construction in Permafrost Regions: A Review". *J Glaciol Geocryol*, 37 (6): 1599–1610 (2015) (in Chinese). DOI: 10.7522/j.issn.1000-0240.2015.0177.

Ma W., Cheng G. D., and Wu Q. B. "Thoughts on Solving Frozen Soil Engineering Problems in the Construction of QingHai-Tibet Railroad". *Science and Technology Review*, 23 (1): 23–28 (2005) (in Chinese). DOI: 10.3321/j.issn.1000-7857.2005.01.006.

Mao Y. C., Li Ma W., Mu Y. H., Wang F., Miao J., and Wu D. Z. "Field Observation of Permafrost Degradation Under Mohe Airport, Northeastern China from 2007 to 2016". *Cold Reg Sci Technol*, 161: 43–50 (2019). DOI:10.1016/j.coldregions.2019.03.004.

Mao Y., Li G., Ma W., Mu Y., Wang F., Miao J., and Wu D. "Field Observation of Permafrost Degradation under Mohe Airport", Northeastern China from 2007 to 2016. *Cold Reg Sci Technol*, 161, 43–50 (2019). DOI:10.1016/j.coldregions.2019.03.004.

Oldenborger G. A. and LeBlanc A.M. "Geophysical Characterization of Permafrost Terrain at Iqaluit International Airport". *Nunavut. J Appl Geophys*, 123, 36–49 (2015). DOI: 10.1016/j.jappgeo.2015.09.016.

Short N., LeBlanc A.M., Sladen. W., Oldenborger G., Mathon Dufour V., and Brisco B. "RADARSAT-2 D-InSAR for Ground Displacement in Permafrost Terrain, Validation from Iqaluit Airport", Baffin Island, Canada. *Remote Sens Environ*, 141, 40–51 (2014). DOI:10.1016/j.rse.2013.10.016.

Sun K., Tang L., Zhou A., and Ling X. "An Elastoplastic Damage Constitutive Model for Frozen Soil Based on the Super/sub Loading Yield Surfaces". *Comput Geotech*, 128, 103842 (2020). DOI: 10.1016/j.compgeo.2020.103842.

Wang F. "Research on Treatment Technology for Subgrade Settlement in the Segregated Permafrost Region of Boketu-Yakeshi Expressway". Ph.D. thesis, Chang'an University (2011) (in Chinese). DOI: 10.7666/dy2132836.

Wang L., Marzahn P., Bernier M., and Ludwig R. "Sentinel-1 InSAR Measurements of Deformation Over Discontinuous Permafrost Terrain", Northern Quebec, Canada. *Remote Sens Environ*, 248, 111965 (2020). DOI: 10.1016/j.rse.2020.111965.

Wang S. J. and Chen J. B. "Nonlinear Analysis for Dimensional Effects of Temperature Field of Highway Embankment in Permafrost Regions on Qinghai-Tibet Plateau". *Acta Geomechanics*, 30 (10): 1544–1549 (2008). DOI: 10.3321/j.issn:1000-4548.2008.10.021.

Xiao D. H., Ma W., and Zhao S. P. "Study of the Dynamic Parameters of Frozen Soil: Achievements and Prospects". *J Glaciol Geocryol*, 37 (6): 1611–1626 (2015) (in Chinese). DOI: 10.7522/j.isnn.1000-0240.2015.0178.

Zhang X., Yang Z., Joey, Chen X., Guan J., Pei W., and Luo T. (2021). "Experimental Study of Frozen Soil Effect on Seismic Behavior of Bridge Pile Foundations in Cold Regions". *Structures*, 32, 1752–1762. DOI: 10.1016/j.istruc.2021.03.119.

Water Conservancy and Civil Construction – Oke & Ahmad (Eds)
© 2024 The Author(s), ISBN: 978-1-032-58618-2

Study of tunnel excavation position on the stability of the overlying loess

Peihao Li* & Jinhai Gao
School of Energy Science and Engineering, Henan Polytechnic University, Jiaozuo, Henan, China

ABSTRACT: With the complex topography of the northwest region of China and regional economic development lagging behind that of the eastern region, tunnel construction is increasing as the country gradually promotes construction in the west. At the same time, the impact of disturbance of tunnel excavation on slope stability is a key issue that must be considered for construction safety. Based on the conditions of loess geology in the western region, this paper uses a numerical model to simulate and analyze the effect of tunnel excavation location on the stability of slopes in loess geology based on a tunnel project in the northwest region, providing a qualitative or quantitative analysis of tunnel excavation in this type of geology and making the actual engineering design and construction more economical and reasonable.

1 INTRODUCTION

With the booming economy in China, the country is paying more and more attention to the construction of tunnels in the northwest. At the same time, the construction of tunnels is inseparable from the slope works and it is necessary to choose a reasonable location for the tunnel excavation to ensure the safety of the project. Whether in the construction process or after the completion of the project, the stability of the slope is the key to determining the quality of the project. The main factors affecting the stability of slopes include human activities and natural effects. According to experience and past lessons, some areas with complex terrain have experienced the phenomenon of slope instability accidents, which have caused large economic losses. If the slope stability is not studied during the development process, it will be likely to cause similar engineering problems. Therefore, it is of great theoretical and practical significance to research tunnel slope problems to ensure engineering safety and project quality. For the stability of slopes, scholars at home and abroad have conducted more research and achieved positive results. Qin Lin (Qin 2012) studied and analyzed the influence of tunnel excavation on slope deformation and the deformation law after excavation. Li Guan (Li *et al.* 2022) analyzed the influence of the excavation of foot-of-slope buildings on the stability of slopes of existing highway roadbeds based on on-site monitoring and numerical simulation analysis. Liudunli (Liu 2018) studied the stress and displacement changes of the surrounding rock and slope during tunnel excavation by using model tests and numerical simulations and then analyzed their effects on the stability of the slope. Li Xuan (Li 2006) analyzed the interaction between landslides and tunnels when constructing tunnels at different locations of landslides, and came up with principles for selecting tunnel locations in landslide areas. Liu Fei (Liu *et al.* 2017) established a numerical model of the loess slope during tunnel excavation through FLAC3D software, and

---

*Corresponding Author: lph19971015@163.com

DOI: 10.1201/9781003450832-20

systematically studied the deformation and damage patterns and strain distribution characteristics of the loess slope at different slope angles during tunnel excavation. Jiang Xin (Jiang et al. 2011) showed that a small excavation at the foot of the slope could also lead to large deformation of the slope near the excavation face. Luo Xiaoyi (Luo 2019) studied the stability of ancient landslides during tunnel excavation from theoretical analysis, numerical simulations, and large indoor direct shear tests. Numerical simulations by Chang Yongtao (Chang 2019) showed that the tunnel excavation would extend the sliding range to the top and foot of the slope, analyzed the factors affecting the stability of the tunnel entrance, and summarized the damage pattern and support measures of the slope at the tunnel entrance. Song Jia (Song 2021) explained the importance of the slope of the tunnel entrance in the entire tunnel construction and operation period, analyzed the factors affecting the stability of the tunnel entrance, and summarized the damage pattern and support measures of the slope of the tunnel entrance section. Fang Pan (Fang 2018) analyzed the characteristics of the step method, CD method, CRD method, and double sidewall guide pit method commonly used in large-section tunnel excavation, and suggested how to choose a reasonable excavation method to reduce the disturbance to the slope.

Based on the geological conditions of loess in the western region, this paper establishes a numerical model of the loess slope during tunnel excavation by using FLAC3D software to systematically study the impact of different tunnel excavation locations on the loess slope.

2 FINITE ELEMENT MODELING

2.1 *Slope profile*

The slope is 26.36 m high, with a toe of 30°, and the tunnel is perpendicular to the sliding direction of the slope. The vertical distance between the top of the tunnel and the slope is 20 m, and the maximum span of the tunnel is 10 m. The project only considers self-weight stress, and no other loads are considered.

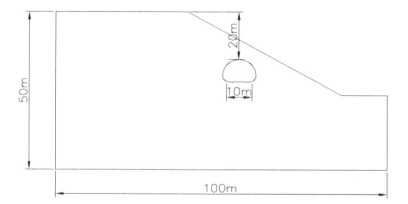

Figure 1. The schematic diagram of the side slope tunnel.

2.2 *Loess mechanical parameters*

The Mohr-Coulomb strength criterion in FLAC3D was used for the numerical model. The physical and mechanical parameters of the loess are density ρ = 1800 kg/m^3, bulk modulus K = 3.3 GPa, shear modulus G = 0.37 GPa, cohesion C = 21 KPa, internal friction angle φ = 24.5°, Poisson's ratio ν = 0.35, and young modulus E = 0.1 GPa.

2.3 Model building

To address the special characteristics of loess, porous and wet sink nature, FLAC3D finite element software was used to build a model for different excavation positions during tunnel excavation and to analyze the horizontal displacement characteristics and vertical position distribution characteristics of the loess slope during tunnel excavation.

The soils are modeled in solid units, with a depth of 30 m, which are considered as Mohr-Coulomb ideal elastic-plastic materials. Horizontal constraints are applied on both sides of the slope, and vertical and horizontal constraints are applied at the bottom. According to the actual situation, excavation locations of vertical distances of 15 m, 20 m, and 25 m from the top of the slope were selected, and the thickness of the overlying loess at the top of each excavation tunnel was ensured to be 10 m, as shown in Figure 2. Monitoring points 1, 2, and 3 were set up at the top, waist, and foot of each model to analyze and compare the effect of three different excavation positions on the horizontal and vertical displacements of the slope under loess slope conditions.

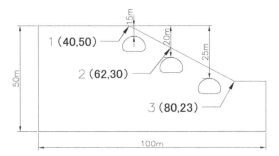

Figure 2. The diagram of the tunnel in different excavation positions.

3 ANALYSIS OF RESULTS

3.1 The vertical distance of 15 m from the top of the slope at the top of the tunnel

Figures 3 and 4 show the equivalent displacement cloud of the top of the cave 15 m from the top of the slope. The maximum horizontal displacement of the slope is about 10 cm, which is located at the top of the tunnel and has a tendency to move to the left; the maximum vertical displacement of the slope is about 40 cm, which is also located at the top of the tunnel and has a tendency to move downwards. It can be seen that after the tunnel excavation, the horizontal displacement of the slope is less disturbed, but the vertical displacement of the slope is more disturbed. There is a risk of vertical instability of the slope, meanwhile, in the actual tunnel project, the tunnel excavation position is lower, and this excavation position is rarely used.

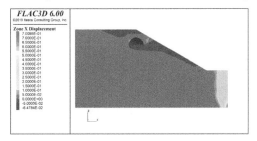

Figure 3. Contour cloud of horizontal displacement at the top of the tunnel at 15 m from the top of the slope.

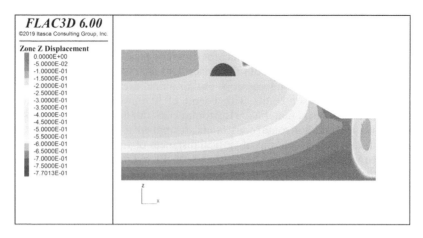

Figure 4. Contour cloud of vertical displacement at the top of the tunnel at 15 m from the top of the slope.

3.2 *The vertical distance of 20 m from the top of the slope at the top of the tunnel*

Figures 5 and 6 show the 20 m displacement equivalents of the top of the tunnel from the top of the slope. The maximum horizontal displacement of the slope is about 12.5 cm, which is located near the top of the tunnel and tends to move to the right; the maximum vertical displacement of the slope is about 15 cm, which is located at the slope directly above the tunnel and tends to move downwards. It can be seen that after the tunnel excavation, the horizontal displacement of the slope is increasing and the vertical displacement of the slope is decreasing.

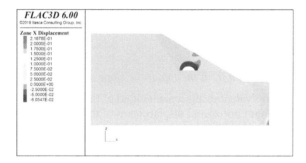

Figure 5. Contour cloud of horizontal displacement at the top of the tunnel at 20 m from the top of the slope.

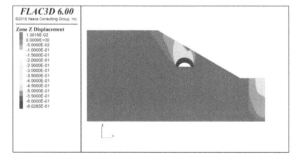

Figure 6. Contour cloud of vertical displacement at the top of the tunnel at 20 m from the top of the slope.

3.3 The vertical distance of 25 m from the top of the slope at the top of the tunnel

Figures 7 and 8 show the 25 m displacement equivalents of the top of the tunnel from the top of the slope. The maximum horizontal displacement of the slope is about 15 cm, which is located on the slope above the tunnel and tends to move to the right; the maximum vertical displacement of the slope is about 10 cm, which is also located on the slope above the tunnel and tends to move downwards. It can be seen that after the tunnel excavation, the horizontal displacement of the slope is increasing but the vertical displacement of the slope is decreasing. It can be seen that the horizontal and vertical displacements of the slope are relatively little disturbed after the tunnel excavation.

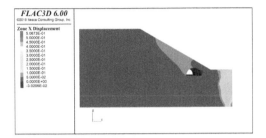

Figure 7. Comparison of horizontal displacements at monitoring points.

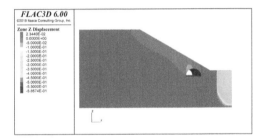

Figures 8. Comparison of vertical displacements at monitoring points.

 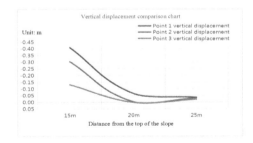

Figure 9. Contour cloud of horizontal displacement at the top of the tunnel at 25 m from the top of the slope.

Figure 10. Contour cloud of vertical displacement at the top of the tunnel at 25 m from the top of the slope.

4 CONCLUSION

The study shows that the tunnel has different disturbance effects on the slope at different excavation positions. The horizontal displacement diagram shows that as the excavation position deepens, the disturbance to the bottom of the slope is smaller, while the disturbance to the middle of the slope gradually increases and the change to the top of the slope is more obvious. The vertical displacement diagram shows that as the excavation position deepens, the vertical displacement of the monitoring points at different locations shows a decreasing trend, and the disturbance to the slope is the smallest at 25 m from the top of the slope.

In summary, as the depth of the tunnel location increases, the horizontal displacement of the monitoring points all increase to varying degrees, but in general does not exceed 15 cm, which is within the controllable disturbance range. Vertical displacement gradually decreases and is minimal at 25 m from the top of the slope, which is negligible.

Therefore, slope reinforcement and protection of loess tunnels are particularly important. Different structural forms such as anti-slip piles, slope protection, and mini-piles can be adopted to increase the shear and tensile strength of the soil and improve the cohesion of the loess to maintain the stability of the slope.

REFERENCES

Chang Yongtao. Stability Analysis of Surrounding Rrock-slope in Large-section Small-clearance Tunnel [D]. *Liaoning University of Engineering and Technology*, 2019. DOI: 10.27210/d.cnki.glnju.2019.000379.

Fang Pan. Numerical Analysis of Slope Disturbance by Excavation Method of Shallow Buried Adjacent Mountain Tunnel [J]. *Shanxi Construction*, 2018, 44 (28): 136–137. DOI:10.13719/j.cnki.cn14-1279/tu.2018.28.072.

Jiang Xin, Wang Jianguo, and Shang Yuequan. Numerical Simulation Analysis of Sex of Slope Stability Affected by Excavation at the Foot of Slope [J]. *Low-Temperature Construction Technology*, 2011, 33 (01): 89–91.

Li G., Lu H., and Qian N. Gui. Influence of Excavation at the Foot of the Slope on the Stability of Existing Road Base Slope [J]. *Guangdong Highway Traffic*, 2022, 48 (04): 28–33.

Li Xuan. *Selection of Tunnel Locations for Landslide Sites [D]*. Southwest Jiaotong University, 2006.

Liu Dunli. The Law of Influence of Tunnel Excavation on Slope Stability [J]. *Railway Construction*, 2018, 58 (07): 68–71.

Liu Fei, Zhang Hongqing, and Zhang Jianying. Study on the Influence of Tunnel Excavation on the Stability of Loess Slope [J]. *Journal of Xinyu College*, 2017, 22 (03): 15–18.

Luo Xiao Yi. Analysis of the Influence of Tunnel Excavation on the Stability of Ancient Landslides [D]. *Hunan University*, 2019. DOI: 10.27135/d.cnki.ghudu.2019.001130.

Qin Lin. Analysis of the Impact of Tunnel Excavation on the Stability of Mountain Slopes and the Study of Reinforcement Measures [D]. Zhongnan University, 2012.

Song Jia. *Study on Slope Stability and Support Optimization of Highway Tunnel Cavity Entrance* [D]. Kunming University of Technology, 2021. DOI: 10.27200/d.cnki.gkmlu.2021.000098.

Water Conservancy and Civil Construction – Oke & Ahmad (Eds)
© 2024 The Author(s), ISBN: 978-1-032-58618-2

Research and engineering application of roadbed settlement prediction method based on measured data

Zhi Liu*
College of Civil Engineering, Tongji University, Shanghai, China

ABSTRACT: Excessive settlement or uneven settlement of highway and high-speed railway roadbeds will lead to deterioration of line operation conditions, reduce passenger comfort and even endanger traffic safety. Hence the control of the post-construction settlement of the roadbed becomes more stringent with the increase in operating speed. Accurate and timely settlement prediction can provide significant guidance for reasonable planning of construction schedule, determination of pavement laying time, and scientific arrangement of height, interval time, and pre-pressing time of roadbed layered filling. In this paper, according to the analyses and summaries of related research results, the prediction of roadbed settlement after load stabilization is carried out by the gray GM (1,1) model and BP neural network based on measured settlement data. The prediction method of roadbed settlement is carried out by modeling with a joint curve fitting-genetic algorithm considering the deformation mechanism of roadbed settlement under graded loading.

1 INTRODUCTION

In the process of highway or high-speed railway construction, if the foundation is not handled properly, it is easy to cause excessive roadbed settlement after construction, which will affect the use quality of the highway and railway. In order to prevent undesirable phenomena such as "bridge jumping", pavement cracking, and pavement depression, the observation, and control of roadbed settlement during the construction has become an important part, in which the use of measured settlement data to predict late settlement is the key technology of settlement control. This paper uses the gray GM (1,1) model and BP neural network to predict the settlement of roadbeds after load stabilization based on the measured settlement data on-site. For the settlement of roadbeds under graded loading with "sudden change", a joint modeling method of curve fitting and genetic algorithm is used, and the advantages and disadvantages of this model are evaluated.

Although the existing settlement prediction methods have been widely used in engineering practices, there is a certain gap between predicted results and measured data. In order to control the embankment construction and reasonably determine the pavement paving time, according to the preliminary settlement observation data, this paper uses dynamic methods to predict the settlement of the roadbed when the load is stable. Considering the deformation mechanism of roadbed settlement under graded loading, a combination of static and dynamic methods is used to predict the settlement of roadbeds under graded loading. Based on the above points, the following works are carried out: (1) The BP neural network program is compiled and the settlement prediction model based on the artificial neural network is established by using the program. The effect of the

---

*Corresponding Author: Zhi_Liu@tongji.edu.cn

DOI: 10.1201/9781003450832-21

number of input vectors, the time interval of monitoring data, and the prediction period on prediction accuracy is discussed through engineering examples. (2) To address the defect of the curve fitting method, a joint BP neural network-gray theory modeling is used to predict the settlement of roadbeds after load stabilization, and the applicability of the model is discussed by comparing it with measured settlement data. (3) The empirical formula method commonly used in engineering practices under graded loading is discussed, and its theoretical basis, the scope of application, and advantages and disadvantages are analyzed. (4) The main reasons for the deviation between the predicted and measured values are analyzed when using the growth curve and exponential curve models to predict the over-step settlement of foundations. The way to improve the accuracy of the growth curve and exponential curve models is studied. The applicability of the two prediction models for predicting settlement under multi-stage loading is evaluated through engineering examples. (5) The influence of the number of measured settlement values at each load level on the accuracy of the model built by the joint curve fitting-genetic algorithm is discussed in the context of the engineering examples.

2 CURRENT STATUS OF RELATED RESEARCH DOMESTICALLY AND ABROAD

Roadbed settlement analysis mainly involves settlement calculation and prediction. The current settlement analysis methods can be summarized into two main categories: theoretical formula calculation method and numerical analysis method. Many theoretical formula methods are based on one-dimensional analysis. As early as 1925, Terzaghi established the basic differential equation for one-dimensional consolidation of saturated soil and obtained the analytical solution under certain initial and boundary conditions. Since the 1960s, Qian Jiahuan had gradually derived the analytical solution for the consolidation of rheological soil.

In the field of the one-dimensional consolidation theory of saturated soft clay soil, Xie Kanghe derived a complete analytical solution of arbitrary layer foundation under variable load, so that the one-dimensional elastic consolidation theory of the layered foundation of saturated soft clay soil has been perfected. In the late 1980s, Zhao Weibing used the generalized Kelvin model to derive a universal analytical solution for viscoelastic soil and proposed a simplified method for calculating the primary and secondary consolidation settlement of soft clay soil. These research results have laid a solid theoretical foundation for the one-dimensional analysis of the settlement of soft clay soil roadbeds.

There are two main methods for calculating settlement versus time. The first category is the finite element method based on consolidation theory, combining with various soil constitutive models, such as the finite element method considering the nonlinear elastic model, the elastic-plastic model, the viscoelastic-plastic model of soil, the finite element method considering structural damage model, and the large deformation consolidation finite element method. However, due to the large number of calculation parameters, which are generally determined by the triaxial test, the finite element method is hard to be applied as the main method of settlement calculation for each section. It is only applicable to calculate the settlement of important projects and key sections. The second type is the prediction method of projecting the settlement amount in relation to time according to the measured data, which is listed as follows:

(1) In-situ test method: settlement can be estimated by measuring the corresponding parameters through in-situ tests. In-situ tests that are usually used to predict roadbed settlement include flat plate load test, cone penetration test, standard penetration test, and pressuremeter test. However, these tests are subject to various factors, such as the accuracy of the instrumentation and the influence of the external environment, which can bring critical errors to the test results and affect the estimated settlement values.

(2) Projection from the initial field measured data: there are many methods of such projection, which can be divided into settlement prediction methods under single-stage loading and graded loading according to different loading methods. By analyzing the measured settlement data in the early stage, it has an important application value in practical projects for the prediction of settlement development law, the projection of settlement amount in the later stage, and the guidance of construction.

The methods of settlement prediction based on partially measured data can be roughly divided into three categories: (1) Curve fitting methods that generally use simple functions to fit the development law of settlement. Some methods have a certain theoretical basis, such as the ASAOKA method, exponential curve method, and growth curve method, while some mainly use mathematical formulas for simple fitting, such as the hyperbolic method. (2) System theory methods, such as the time series method, gray theory method, and neural network method. (3) Numerical analysis method that generally uses the finite element method to predict the late settlement according to the inversion calculation parameters.

Many scholars have used the system theory method to predict post-construction settlement. Lin Chuan used BP neural network method to predict the post-construction settlement of the soft foundation by observing the settlement rate. The result was better than that of the hyperbolic method, but it was only suitable for the soft foundation with a short pre-pressuring period. He Mingman proposed a method to predict roadbed settlement by using the momentum BP algorithm and established a BP neural network model in combination with specific engineering examples. Zhao Junming used the gray theory to predict the settlement of the soft foundation of the highway and established the prediction model GM (1,1) with unequal time spacing. The reliability of the model was checked by the posterior difference method, which made up for the shortage of the equal time spacing GM (1,1) model. Compared with the measured data of the Changzhou section of the riverine expressway, the practicality of the method was verified.

3 MECHANISM OF SETTLEMENT AND DEFORMATION OF SOFT SOIL ROADBED

The settlement during construction can be appropriately compensated by elevation control. For example, the construction settlement of a bridge pier foundation can be compensated by thickening the height of the pier cap, and the construction settlement of the road foundation can be compensated by increasing the earthwork volume. For structural design, the most important concern is the post-construction settlement. Therefore, it is necessary to deeply understand the occurrence and development of the settlement. The development law of settlement with time is an essential part of settlement analysis. The settlement deformation of roadbeds may be caused by many reasons, such as building load, environmental load, and other factors not directly related to the load. The commonly used calculation methods are generally aimed at the settlement caused by the load. Settlement not directly caused by loads needs to be prevented or mitigated by careful site selection, foundation pretreatment, or other structural measures.

4 SETTLEMENT CALCULATION METHOD OF SOFT SOIL FOUNDATION

The calculation methods of foundation settlement are various and can be roughly divided into three categories: (1) the traditional theoretical calculation method; (2) the numerical method based on Biot consolidation theory combined with soil constitutive models; (3) the time data projection method based on empirical formulas fitting from measured settlement data. Each method has its advantages and disadvantages but complements each other. They have become the main methods for settlement calculation at present.

When the compression layer is thick or the soil compressibility is not uniform along the depth, in order to take into account the nonlinearity of the compression stress distribution along the depth and the different soil properties, the entire soil layer should be divided into several soil strata (1 to n) according to the inhomogeneity or customary thickness (generally 0.4 B, where B is the size of the short side of the foundation). The stresses in the soil are calculated according to the elastic theory, and the foundation settlement is obtained by using the principle of superposition of the stratified layers according to the deformation parameters provided by the test. It is convenient to take into account the non-uniformity of the soil layer and the complex factors that exist such as changes in the underground water table.

Skempton and Bjerrum, to consider the three-dimensional deformation characteristics of general foundation, replaced the additional pressure Δp in the one-dimensional compression formula with the pore water pressure increment Δu caused by the instantaneous loading on saturated soil under undrained conditions, and considered the settlement process as the dissipation process of Δu. From this perspective, the stress path of consolidation deformation of saturated cohesive soil under loading was analyzed, and a correction is proposed to the calculation model of one-dimensional consolidation settlement plus initial settlement.

The finite element analysis is theoretically tighter by selecting reasonable intrinsic structure models, which can consider complex boundary conditions, the nonlinearity of soil stress-strain relationship, stress history of soil, coupling effect of stress on water and soil skeleton, anisotropy of soil, the time factor, etc. It can also simulate the field loading step by step, consider the influence of lateral deformation and three-dimensional infiltration on settlement, and figure out the changes in settlement, horizontal displacement, pore pressure, and effective stress at any moment. At present, the principal models used to analyze the stress-strain characteristics of soft clay soil are the elastic model, elastic-plastic model, and viscoelastic-plastic model considering the time effect. For roadbed engineering, the pore pressure, vertical displacement, and horizontal displacement are generally taken as the basic unknown quantities in finite element analysis The Biot consolidation theory is used to establish a set of equations based on the stiffness and permeability of the soil under roadbed loading, to figure out the pore pressure and displacement.

What affects the safe use of the structure is the post-construction settlement, which makes the long-term settlement characteristics of the foundation particularly significant. It is important to have real and credible settlement observation data during construction and to speculate on future settlement values according to development trends. This method is generally based on a specific empirical formula model, which uses the previous settlement observation data combined with specific mathematical methods to optimize the parameters of the model and calculate the settlement of the structure over its service life.

5 ANALYSIS OF FOUNDATION SETTLEMENT PREDICTION METHODS

As long as the calculation parameters are reasonably determined, various theoretical and numerical methods can be used to predict the settlement of the foundation more accurately. The foundation parameter method and the theoretical formula method are commonly used. The foundation parameter method is to establish a forward model based on soil consolidation theory and intrinsic structure relationship, using the measured settlement data and optimization methods to back-analyze the foundation soil parameters. Finally, the established forward model is used to predict the late settlement. The foundation parameter method can consider the effect of variable loads with a clear mechanical meaning. Due to the large computational workload of the finite element method, it is mainly used for the analysis of key sections. The theoretical formula method is based on the one-dimensional consolidation theory of Terzaghi combined with the elasticity theory, which is widely used in the prediction of settlement under one-dimensional loading. However, the one-dimensional consolidation theory of Terzaghi is accurate only in one-dimensional cases. It is not accurate for two-dimensional and three-dimensional problems and thus is not conducive to improving the accuracy of prediction results. This chapter focuses on some methods for predicting late settlement based on the previously measured settlement information.

According to the consolidation theory proposed by Terzaghi, the relationship between pore water pressure and time can be described by exponential expression. For soil with linear elastic property, the consolidation degree can be expressed as follows:

$$U = 1 - \alpha e^{-\beta t} \tag{1}$$

where α and β are consolidation parameters. Then according to the measured settlement data, the average consolidation degree can be expressed as:

$$U = \frac{s_t - s_d}{s_f - s_d} \tag{2}$$

where $_{it}$ is the measured settlement at time t; s_d, and s_f are immediate and final consolidation settlements. In the exponential curve fitting method, three data points on the curve of settlement versus time are selected and the time interval between the points must be the same, that is:

$$\Delta t = t_3 - t_2 = t_2 - t_1 \tag{3}$$

We should substitute the three points into Equation (2), and three equations can be obtained and the expressions of β, s_f, and s_d are shown as follows:

$$\beta = \frac{1}{\Delta t} \ln\left(\frac{s_2 - s_1}{s_3 - s_2}\right) \tag{4}$$

$$s_f = \frac{s_3(s_2 - s_1) - s_2(s_3 - s_1)}{(s_2 - s_1) - (s_3 - s_2)} \tag{5}$$

$$S_d = \frac{S_t - S_f(1 - \alpha e^{-\beta t})}{\alpha e^{-\beta t}} \tag{6}$$

The above equations are the control equations in the exponential curve. The stress-strain curve is nonlinear for soft clay with high compressibility, so the consolidation process of soft clay does not necessarily conform to the exponential curve. When $0.6 < U < 0.9$, the relation between U and time factor T can be expressed by an equilateral hyperbola. The settlement of soft clay at time t can be calculated by the following equation:

$$s = s_0 + \frac{t - t_0}{\alpha_0 + \beta(t - t_0)} \tag{7}$$

The Poisson curve is the logistic curve, which is suitable for incremental or decayed "S" shaped curves. The expression of the commonly used Poisson curve is given as:

$$s_t = \frac{k}{1 + ae^{-bt}} \tag{8}$$

where a, b, and k are parameters that should be determined by measured data. The parameters in the expression of the Poisson curve can be fitted by using the measured s_t curve. The main procedure is to take a time interval, and the total number of time series is 3

n. Then we should divide the time series into three sub-series and each sub-series has several n data points.

The gray prediction method is always referred to as the GM model. The most common GM Model has an isometric time sequence. Its expression is given by the following differential equation:

$$\frac{dx^{(1)}}{dt} + cx^{(1)} = u \tag{9}$$

The logistic model is a kind of growth model that is widely used in the field of ecology and demography. The differential form of the logistic model is as follows:

$$\frac{ds}{dt} = rs\left(1 - \frac{s}{K}\right) \tag{10}$$

The logistic curve contains three parameters that should be determined. If the measured settlement data can be obtained, there will be three methods used to determine these three parameters: three section method, the gray model, and the nonlinear regression method. The main processes of the three methods can be found in other literature. The Gompertz model is a kind of growth curve, and the mathematical expression of which shows as follows:

$$y = me^{-e^{n-lt}} \tag{11}$$

where m, n, and l are constants, and t is the time series. The total settlement of soft clay roadbed contains three components: immediate settlement, primary consolidation settlement, and secondary consolidation settlement. The development of settlement can be divided into the following four stages (as shown in Figure 1):

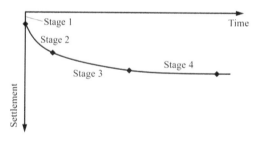

Figure 1. The Sketch diagram of settlement versus time.

Shanghai-Jiaxing-Huzhou (Shen-Jia-Hu) highway connects Zhejiang province with Shanghai, which was built in September 2005, completed in May 2007, and opened to traffic at the end of 2007. Most of the roads are built on the soft subgrade with a thickness of 20–30 m. The underground water table was found about 1 m below the ground surface. The soft soil layer of the subgrade has the characteristics of high water content, high compressibility, low strength, and poor permeability. The basic properties of the soft soil are as follows: natural water content, w_n = 44.3%–51.6%; specific gravity, G_s = 2.73; liquid limit, w_L = 45%; plasticity limit, w_P = 21%; and initial void ratio, e_0 = 1.23–1.34. The settlement date of the highway can be obtained from a general report made by Zhejiang Transportation Planning and Design Institute. Figure 2 presents the settlement results of three monitoring sections (K67 + 625, K67 + 750, and K68 + 410) during construction and operation periods. The predicted settlement is shown in Figure 2 as well. From the figure, it is found that the

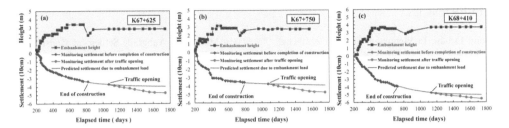

Figure 2. Comparison of measured data and predicted results.

theoretical curve agrees well with the measured settlement curve. Therefore, it is feasible to use the Gompertz model to predict the settlement of soft clay.

Although all of the above methods have been applied to predict the settlement of roadbeds successfully, each method has its application scope. The exponential curve conforms to the one-dimensional consolidation theory proposed by Terzaghi and the parameter β has a definite physical meaning. However, the method of determining the parameters α and β according to three data points is subjective, which will influence the accuracy.

According to Terzaghi's one-dimensional consolidation theory, the hyperbolic method will make the relationship between consolidation degree U and time factor T simpler than the exponential curve. Therefore, the hyperbolic method is more suitable for engineering practices. However, the disadvantage of this method is that the consolidation parameters cannot be reflected. Compared with the consolidation theory of Terzaghi, it is found that the Asaoka method predicts the settlement trend well once the degree of consolidation exceeds 0.6. Furthermore, the final settlement of the soft clay can also be predicted well. The Poisson curve can describe well the S-shaped relationship between settlement and time, but the applicability of the Poisson curve for other settlement versus time curves is questionable. The four methods mentioned above are traditional methods used for predicting the settlement of soft clay, while the gray prediction method, logistic model, and Gompertz model are new methods suitable for special kinds of soft clay or different boundary conditions.

6 CONCLUSION

The problem of roadbed settlement and post-construction settlement prediction has been a hot issue in the field of geotechnical engineering and has attracted much attention with the rapid development of highway and high-speed railway construction. Due to time constraints, this paper has not been able to conduct a more comprehensive study, and the following issues are still to be explored deeply: (1) the loads in engineering projects are applied to the foundation in a graded way. In this paper, the graded load is treated as an instantaneous load at $T/2$ (T is the construction period, i.e., the time when the load increases from zero to constant load q), and then solved by the instantaneous applied load method. The consolidation stress is assumed to be uniformly distributed, and the influence of the distribution form of consolidation stress in the soil consolidation process is not considered. For these reasons, the current research on the consolidation of the foundation during graded load is not perfect. (2) In engineering projects, overload pre-pressure is often used to reduce the amount of post-construction settlement of the roadbed. This paper mainly discusses the problem of prediction of settlement under graded loading, while the problem of prediction of settlement in the case of "unloading and reloading" has not been studied, which needs to be further discussed.

REFERENCES

Dong Lishan, Li Wenqiang, and Liu Zhisheng. "Prediction Method of Aubgrade Settlement Considering Long-term Creep Effect of Loess", *Highway*, 67 (08), 1–7 (2022).

Jing Xudong. "Prediction and Control of Soft Soil Subgrade in High-speed Railway", *Jiangxi Building Materials*, (04), 68–69 (2022).

Li Chunwei. "Application of Chemical Grouting Technology in the Settlement Treatment of Bridge and Culvert Abutment Subgrade", *Sci-Tech & Development of Enterprise*, (07), 93–95 (2020).

Li Yuefeng and Sun Zihao. "Study on Settlement Prediction of the Composite Soft Foundation of Cement Soil mixing Pile", *Journal of Shandong Institute of Commerce and Technology*, 22 (03), 96–98 (2022).

Liu Zhi, *et al.* "Settlement Prediction Method and Engineering Application of Deep Soft Soil Subgrade", *Highway*, 67 (07), 103–108 (2022).

Qiu Hongsheng, Yang Yu, Li Dongjian, and Luo Gang. "Subarade Settlement Prediction Based on Equidistance Processing and IOWHA Operator", *Science Technology and Engineering*, 22 (20), 8884–8892 (2022).

Wang Caijin, *et al.* "Subgrade Settlement Prediction Method for Highway Expansion Projects Based on CPTU test", *Journal of Ground Improvement*, 4 (04), 271–278 (2022).

Wang Fang and Lu Changlong. "Application of Hierarchical Combination Model in Settlement Prediction of Mudstone Subgrade", *Journal of Henan University of Urban Construction*, 31 (03), 45–49 (2022).

Yang Yu, Qiu Hongsheng, Li Dongjian, and Luo Gang. "Application of Combined Model Based on IOWGA Operator in Subgrade Settlement Prediction", *Journal of Wuhan University of Technology*, 44 (04), 36–42 +63 (2022).

Zhou Wei and Wang Qian. "Prediction of Subgrade Settlement Based on GA-BP Neural Network", *Land & Resources Herald*, 19 (03), 76–80 (2022).

Water Conservancy and Civil Construction – Oke & Ahmad (Eds)
© 2024 The Author(s), ISBN: 978-1-032-58618-2

The application of traditional elements in modern architectural design

Rui Wang & Jiang Jin
China Northwest Architectural Design and Research Institute Co., Ltd., JingKai, Xi'an, Shaanxi, China

Yuetao Li
Powerchina Northwest Engineering Corporation Limited, Yanta, Xi'an, Shaanxi, China

Mengyan Tan*
School of Civil Engineering and Architecture, Xi'an University of Technology, Beilin, Xi'an, Shaanxi, China

ABSTRACT: Chinese civilization has given birth to a fine traditional culture. Traditional culture has a great influence on all walks of life. For example, in modern architectural design, the traditional culture and elements penetrate all aspects of modern architectural design, greatly improving the quality of modern architectural design. The text describes the expression principle of traditional elements in modern architectural design. With a comprehensive understanding of the traditional architectural culture, taking traditional architectural elements in Northwest China as an example, we can combine them with the contemporary architectural design to innovate. Finally, it promotes the integration of tradition and modernity.

1 INTRODUCTION

With the increasingly fast pace of China's opening up to the outside world and the accelerated development of global economic integration, the cooperation and exchange between China and other countries in the social, economic, and cultural fields have become increasingly ordinary, resulting in cultural convergence in many cities. To build a modern and international image of cities and win more business opportunities, urban architecture often pursues modernization and symbolism in form. The commercial center of the city is full of tall buildings with the same shape, and takes "more dazzling" and "higher" buildings as the focus of attention, while the traditional local characteristics of the cities disappear in these dazzling buildings. Under the new era and background, the modern building design should actively introduce the new traditional building elements, adopt the new traditional building materials, and apply the rich and colorful traditional building elements directly into the modernization civilization, to add luster to the modernization building design of our country.

2 THE EXPRESSION PRINCIPLE OF TRADITIONAL ELEMENTS IN MODERN ARCHITECTURAL DESIGN

2.1 *Principle of unity with function and form*

Modern architecture should meet its functions. In ancient society, the development of architectural functions was not very mature due to the differentiation of human habitation

---

*Corresponding Author: 2511385868@qq.com

DOI: 10.1201/9781003450832-22

and other activities. The architectural forms in ancient China could be applied to the main functional buildings at that time. However, with the rapid development of society, culture, science, and technology, people's lifestyles and behavior have undergone earth-shaking changes compared with ancient times, and modern architecture bears more extensive, complex, and detailed functional requirements. The first factor that architects should consider when designing buildings is that buildings need to meet people's functional needs in daily life, study, work, and entertainment. Therefore, while expressing traditional elements, it is a major principle that should be adhered to in architectural design to handle the unified and coordinated relationship between architectural forms and architectural functions.

2.2 Principle of element suitability

Traditional elements can't be copied completely. Traditional architectural culture spans thousands of years of history, and sometimes it is too cumbersome and complicated for modern buildings that are practical and simple. Moreover, tradition is a living thing. In the course of history, tradition meets with new or foreign things, and they are absorbed in the collision. The addition of new ingredients will cause the reorganization, renewal, and variation of tradition. Traditional culture has evolved for thousands of years, and some of them are no longer the mainstream. If we copy them without any choice, architecture will be out of step with the development of the whole modern society. To convey the traditional meaning in modern architectural design, we need to refine the traditional architectural culture.

2.3 Modern aesthetic principle

Aesthetic concept, like other social ideologies, is influenced by the development level of social production, and at the same time, it affects the overall ideology of society. Modern aesthetics is developed from tradition, but development is not equal to inheritance. Development is to break the old tradition and create new things. People's aesthetics is also changing, and the new modern aesthetics should gradually replace the old traditional aesthetics. As far as architecture is concerned, traditional aesthetics emphasizes static beauty, stability, balance, massiness, and symmetry. With the development of society, especially the emergence of high technology in post-industrial society, people's ideas have changed dramatically. Regardless of the concept of society and people, it shows a dynamic and open situation. Therefore, the performance of traditional culture must respect the aesthetic standards of modern society and meet the aesthetic requirements of modern society.

2.4 Regional principle

The regional principle is an important principle that the expression of traditional elements in modern architecture should follow. China has a vast territory and vast resources, and there are differences in different degrees from the natural environment to the human environment in various provinces and regions. The architecture is produced according to the topography and climate, materials and technology, and life and customs of that place at that time. The significance of architecture comes from the local context, explains the local context, and is connected with the future of the region, representing the consensus reached by residents in this area in terms of psychological and cultural cognition. This determines that the traditional architecture of China not only follows the common principles and ideas but also has the characteristics of local regional culture with the differences between regions and cultures. Therefore, in modern architectural creation, we should pay attention to respecting the local regional culture, look for elements and inspiration from the local traditional architecture, and never apply the traditional elements which are inconsistent with the local regional style characteristics to architectural creation.

3 EXAMPLES OF THE APPLICATION OF TRADITIONAL ARCHITECTURAL ELEMENTS IN MODERN ARCHITECTURAL DESIGN——THE TRADITIONAL ARCHITECTURAL ELEMENTS IN THE NORTHWEST REGION (TONGZHOU) ARE TAKEN AS AN EXAMPLE

"Tongzhou" is the ancient name of Dali County, Weinan City, Shaanxi Province. At the beginning of the Zhou Dynasty, it was an ancient country. Since the end of the Eastern Han Dynasty, counties, states, governments, and departments have been set up. Qin, Han, Sui, and Tang Dynasties established Chang'an as their capital, and "Tongzhou" became an important town in the east of Kyoto. As the government of the Ming and Qing Dynasties, it governed ten cities in Weinan today. Although the government system was abolished in the Republic of China, as a symbol of culture and tradition, the name "Tongzhou" has been preserved, and it has become the name of Dali County with a long influence at home and abroad. People of Dali often refer to themselves as "a native of Fuli" and "a native of Tongzhou" according to the old tradition.

3.1 Traditional architectural elements in Northwest China (Tongzhou)

3.1.1 Building volume

In the process of investigation, it is found that the traditional architecture of Tongzhou is simple in shape and thick in volume. This feature is evident in the tower as a landmark building. Taking the Golden Dragon Temple Tower in Song Dynasty as an example, the dense eaves towers in Guanzhong East Prefecture (Figure 1) are selected, and the ratio of the bottom diameter of the tower to the tower height is compared as a reference object. It can be known that the Golden Dragon Temple Tower in Tongzhou is more honest and steady in shape than the Qianjin Tower the same time or other close-eave towers at different times (Figure 2), which reflects the simple, introverted, and honest personality of the working people in Tongzhou.

Figure 1. Photos of Guanzhong dense eaves towers.

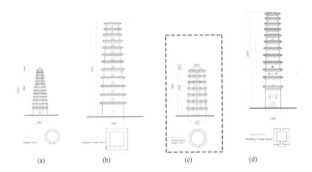

Figure 2. Comparison of dense eave towers in the Guanzhong Dongfu area.

3.1.2 Plane shape

"Room" is the basic unit of building monomer in traditional dwellings. Based on the measurement and statistics of traditional dwellings in Tongzhou, it is known that the width of most main rooms and mansion rooms is 3.2 m. In some buildings, influenced by the traditional concept of respecting the center, the size of the center will be larger, generally between 3.3 and 3.5 m. Through the investigation and analysis of Sanheyuan in Fengtu Yicang, it is known that official architecture also conforms to this composition principle.

3.1.3 Facade form-three stage

Three-stage architecture is a common feature of traditional architecture. By analyzing the roof, body, and platform base of the main building, official building, Daici Cen building, and Zhuwengong temple in Tongzhou folk houses, we can know that Daici Cen building needs a high platform base due to the performance, and other structure is 300~500 mm height. By analyzing the three-stage proportional relationship, it is known that the proportion of roof and house gradually decreases with the improvement of building shape (Figure 3).

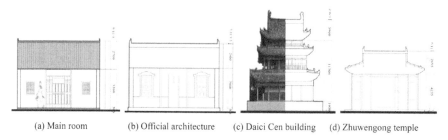

(a) Main room (b) Official architecture (c) Daici Cen building (d) Zhuwengong temple

Figure 3. Three-stage traditional architecture in "Tongzhou". (a) Main room (b) Official architecture (c) Daici Cen building (d) Zhuwengong temple

3.1.4 Roof modeling

The roofs of "Tongzhou" traditional buildings have various forms, such as the hard mountain roof, pavilion roof, and xieshan mountain roof. (Figure 4). The combination of roofs is relatively fixed, so the architectural form is also relatively fixed. Among them, sloping roofs often appear in residential buildings, pavilion roofs in ancient pagodas, and xieshan roofs in sacrificial buildings because of their high level.

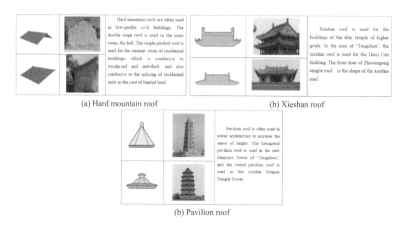

Figure 4. Roof modeling. (a) Hard mountain roof (b) Xieshan roof (c) Pavilion roof

3.2 The application example— the architectural creation of the Dali high-speed railway new district is taken as an example

"Tongzhou" is Dali County, Weinan City, Shaanxi Province, located in Guanzhong Plain. Southwest of the county is 125 kilometers away from Xi'a City, and 59 kilometers away from Weinan City. The Yellow River, Weihe River, and Luohe River converge, and the ecological resources are abundant. The new district is located in the northwest of the present Chengguan Town, and its base is located at Daxi high-speed railway and G108 national road. During this period, many main urban roads are gathered, and the traffic is convenient. The subway station is located in the gateway area of the new high-speed rail area, which provides an important opportunity for "Tongzhou" to create a new city image.

3.2.1 Commercial public building

3.2.1.1 Form and rhythm of architecture

In the traditional architectural form, Fengtu Yicang is similar to the volume of modern commercial buildings. In the project, according to the different levels of abstraction, three ways are adopted to transform and express its architectural form: the first scheme is a direct imitation of its outline (Figure 5); second, based on imitating the outline of traditional architectural elements, the local deformation makes the architectural form more lively and conforms to the temperament of commercial buildings (Figure 5); third, the scheme is based on two points, enclosure, and concise lines. Although it is different from traditional architecture, it inherits the impression of traditional architecture as a whole (Figure 5).

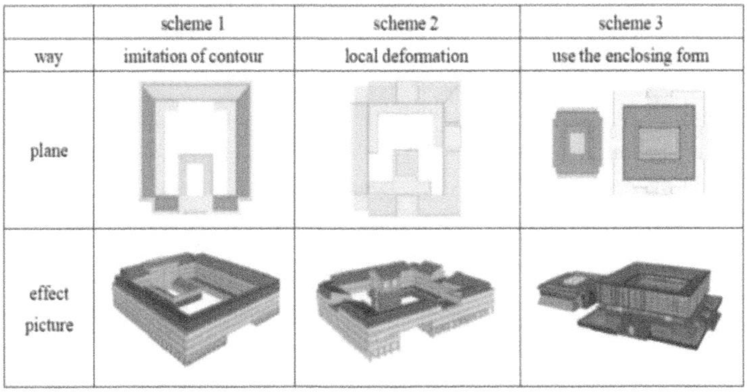

	scheme 1	scheme 2	scheme 3
way	imitation of contour	local deformation	use the enclosing form
plane			
effect picture			

Figure 5. The commercial building scheme diagram.

3.2.1.2 Abstraction and simplification of components

The main entrance and colonnade of Fengtu Yicang are selected as the basis of element evolution. The warehouse body is divided into three parts longitudinally: the roof, the housing body, and the decoration of the basement. In Schemes 1 and 2, the three-stage proportion is deformed and inherited. The facade of the main entrance is divided into two parts, the illuminated wall and the entrance and exit, which simplifies the division form of the illuminated wall and the entrance and exit, and is used in the main entrance of the building in Scheme 1 (Figures 6 and 7).

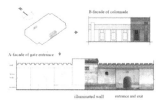
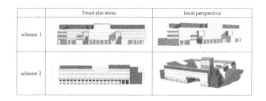

Figure 6. Facade elements of Fengtu Yicang.

Figure 7. Commercial building scheme diagram.

3.2.2 Office buildings

3.2.2.1 Form and rhythm of architecture

The office building is arranged in the form of enclosure and symmetry of the central axis, with the main building on the central axis and the auxiliary functional buildings symmetrically arranged on both sides of the central axis (Figure 8). Moreover, the height of the main building is higher than the auxiliary building, which reflects the architectural thought of "moderation" and "respecting for the middle" in traditional buildings, and also reflects the honest and regular architectural temperament of "Tongzhou" buildings.

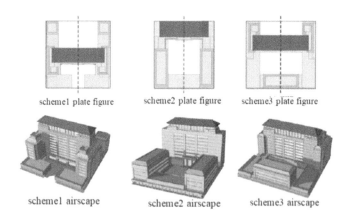

Figure 8. Office building form generation diagram.

3.2.2.2 Abstraction and simplification of components

In the design, the traditional element roof is simplified, that is, the main roof adopts a hipped roof, and the proportional relationship between the length of the roof ridge and the house floor is 1.3/1, which is determined by the deformation of the proportional relationship between the length of the roof ridge and the house floor of Xieshan. The slope is adjusted according to the slope of traditional houses, which is more suitable for modern construction technology as shown in Figure 9. An auxiliary roof is determined by the superposition of square blocks and sloping roofs (Figure 10).

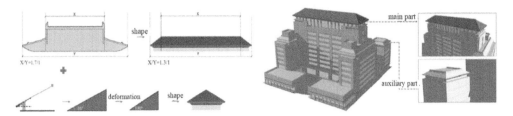

Figure 9. Main building roof modeling.

Figure 10. The commercial building model diagram.

4 CONCLUSION

Under the background of globalization, cultural convergence and the disappearance of architectural features have prompted us to reflect on traditional culture. In China, traditional culture and architecture have gradually become the focus of people's attention. In this context, books and academic papers on how to embody traditional elements in modern architectural design also emerge in endlessly. To a certain extent, the discussion and research of these architectural theories have guided and promoted the development of this kind of architectural practice. However, as the current research is mostly at the level of the whole Chinese nation, it has not been implemented in a specific area or region, and there is still a lack of practical technical methods and methodology for the expression of traditional elements in modern architecture. Therefore, this paper tries to make up for this academic gap and studies the modern expression methods of traditional elements in Northwest China (Tongzhou). This paper puts forward the design principle of the expression of traditional elements in modern architecture and summarizes the typical elements of traditional architecture in Northwest China (Tongzhou). Through the analysis of actual cases, this paper expounds on how traditional architectural elements are expressed in modern architecture. It is hoped that it will give some inspiration to similar architectural designs and creations in the future.

REFERENCES

Cao Z. (2016). *Study on the Design of the Traditional Block of Tongzhou Ancient Town Folk Culture Expo Park*. Xi'a University of Architecture and Technology, Xi'an.

Fang G.G. and Chen Y. (2012). A Preliminary Study on the Characteristics of Traditional Dwellings in Xi'an. *Architectural Culture*, 152–156.

Gao D. (2016). *The Inheritance Model of Local Culture in Dali Urban-Rural Integration Development Zone*. Xi'an University of Architecture and Technology, Xi'an.

Li H. (2019). Analysis of the Application of Traditional Architectural Decoration Elements in Modern Architectural Design. *China Architectural Decoration*, (11): 119.

Liu K., Wang B. X., and Zhou J. L. (2021). Inheritance and Application of Traditional Architectural Elements in Modern Architectural Design. *Housing and Real Estate*, (33): 39–40.

Wu H.P. (2020). Application Strategy Analysis of Traditional Elements in Modern Architectural Decoration Design. *Building Materials and Decoration*, (14): 79+81.

Water Conservancy and Civil Construction – Oke & Ahmad (Eds)
© 2024 The Author(s), ISBN: 978-1-032-58618-2

Structural design of the power distribution building

Zhen Xie*, Xilong Chen & Yan Yan
Economic and Technical Research Institute of State Grid Fujian Electric Power Co., Ltd.,Fuzhou, China

ABSTRACT: As an important part of the substation, the power distribution building mostly adopts concrete structure or brick-concrete structure, while the proportion of steel structure is relatively low. This paper takes a 110 KV substation power distribution building under the complex geological condition as the research object and carries out the targeted design of its superstructure and foundation, which has certain reference significance.

1 INTRODUCTION

As the industry with the largest carbon emission in China, the emission reduction of the power industry is of great significance to the sustainable development strategy of the country (Shu *et al.*, 2021). At present, power distribution building mostly adopts concrete structure or brick-concrete structure. Some areas have seen a shortage of natural aggregate. Therefore, how to gradually reduce the use of natural aggregate has become an important research content in the structural design of the power distribution building.

With the proposed concept of a new power system, the construction of low-carbon substations should be promoted to ensure that the carbon peak is on schedule. Because the steel can be recycled and the construction period of the steel structure is short, the application of steel structure in the structural design of the power distribution building can make a compromise between economic benefit and environmental benefit (Miao *et al.*, 2020).

This paper will take a 110 KV substation power distribution building under the complex geological condition as the research object, and carry out the targeted design of its superstructure and foundation respectively. Considering that the 110 KV substation is a common type, it can provide a certain guiding significance for similar engineering designs.

2 ENGINEERING SITUATIONS

2.1 *Proposed building*

This engineering aims to construct a power distribution building with a total floor area of 1,040 m^2, a design service life of 50 years, and an importance coefficient of 1.0. According to its location, the basic wind pressure is 0.7 kN/m^2 based on the 50-year return period, the ground roughness is Class B, the seismic fortification intensity is 6 degrees, the design basic seismic acceleration is 0.05 g, and the design earthquake group is the third group.

2.2 *Geological condition*

There are many types of soil layers on the site, and the geological condition is complex. During the investigation, the drilling, sampling, and in-situ testing of 17 boreholes were

---

*Corresponding Author: 773719259@qq.com

completed. Two of the boreholes, K1 and K2, were selected, and the equivalent shear wave velocity of the soil layer was obtained according to the thickness of each soil layer and the corresponding shear wave velocity within the depth range of 20 meters. It can be obtained from Table 1 that the V_{se} values are all less than 150 m/s, and the thickness of the site covering layer is in the range of 15 – 80 meters. Based on this, the category of the site is determined to be Class III (GB 50011-2010).

Table 1. Calculation of equivalent shear wave velocity of soil layer.

Borehole	Calculation depth (d_0)	Layer thickness (d_i)	Shear wave velocity (V_{si})	Equivalent shear wave velocity of the soil layer (V_{se})
K1	20 m	2.90 m	200.00 m/s	122.82 m/s ($V_{se} \leq 150$ m/s)
		12.00 m	100.00 m/s	
		5.10 m	180.00 m/s	
K2	20 m	3.70 m	200.00 m/s	111.56 m/s ($V_{se} \leq 150$ m/s)
		15.80 m	100.00 m/s	
		0.50 m	180.00 m/s	

After a thorough investigation, it was discovered that the site is an alluvial plain with no active fault zones or adverse geological effects like landslides or debris flows nearby. Additionally, there are no buried boulders, abandoned wells, or other objects that would be detrimental to the engineering. Because the site is distributed with a weak sludge layer and the local sludge layer is thick, the site stability of this engineering is poor. The site is only appropriate for engineering construction on the assumption of reasonable foundation design. Therefore, the foundation design is also an important content of the structural design for the power distribution building.

3 STRUCTURAL DESIGN

3.1 *Superstructure design*

The finite element model was created by YJK Software, as shown in Figure 1. The two floors above the ground are steel frames, and the roof uses the 130 mm steel-bars truss deck. Both steel columns and steel beams are made of hot-rolled H-shaped steel, and the steel columns use a semi-embedded type column base. The nodes of the secondary beam are hinged connections, and the other nodes are rigid connections. To transfer the load better, the cable tunnel is considered in the underground layer composed of concrete columns when modeling, and the top surface of the cable tunnel uses the 250 mm cast-in-place plate. The beam-column size is shown in Table 2.

Figure 1 Superstructure model.

Table 2. Summary of beam-column size.

Material	Component type	Component size (mm)
Q345	Steel column	HW400x400
	Steel beam	HM488x300, HN400x200, HN350x175
C30	Concrete column	900×900

Live load values are determined according to "Load Code for the Design of Building Structures" (GB 50009-2012) and "Technical Code for the Design of Substation Buildings and Structures" (DL/T 5457-2012), as shown in Table 3. Referring to the "Code for Seismic Design of Buildings" (GB 50011-2010), the seismic calculation of the Category III building with 6-degree fortification and less than 50 m in height is not required, and the seismic rating of the structure is Level 4.

Table 3. Live load values.

Location	Main transformer room	Radiator room	Capacitor room	10KV power distribution room	GIS room	Non-equipment room	Roof
Live load (kN/m^2)	40	20	9	7	10	3	0.5

After calculation, all indicators meet the requirements, and Table 4 lists the significant indicators. From the perspective of the main vibration modes, the first four vibration modes are all translational, the fifth vibration mode is torsional, and the ratio of the first torsional period (0.4558 s) to the first translational period (0.7148 s) is 0.64. Regardless of the X-direction or Y-direction, the ratio of the maximum displacement to the average displacement of the layer is less than 1.2 (GB 50011-2010), indicating that the plane stiffness of the structure is evenly distributed. In addition, the maximum interlayer displacement angle is less than 1/400 (GB 50017-2017), which can prevent excessive horizontal displacement from affecting the normal use of the structure.

Table 4. Summary of calculation indicators.

Indicator		Indicator value
Translation coefficient/corresponding period (s)	First vibration mode	1.00/0.7148
	Second vibration mode	0.76/0.4974
	Third vibration mode	0.96/0.4687
	Fourth vibration mode	1.00/0.4619
	Fifth vibration mode	0.44/0.4558
Maximum interlayer displacement angle	X-direction	1/2548
	Y-direction	1/1389
The ratio of the maximum displacement to the average displacement of the layer	X-direction	1.04
	Y-direction	1.11
1st torsion period /1st translation period		0.64

3.2 *Pile foundation design*

The geotechnical engineering performance of fully weathered tuff lava, sandy strongly weathered rock, and fragmentary strongly weathered rock is better, which can be used as the bearing strata of the proposed power distribution building. The soil parameters are shown in Table 5. Considering that the deformation of the shallow foundation cannot meet the design requirements, and the ground treatment is not economically reasonable, the pile foundation is adopted (Zhao 2020).

In addition, the site has convenient transportation and no underground pipeline. All types of construction machinery can enter the field smoothly, with mechanical construction conditions, so the two types of prestressed pipe pile and punched pile can be used. As the mechanized construction speed of the prestressed pipe pile is fast, and the bearing capacity of a single pile is high (Yan *et al.*, 2022), the prestressed pipe pile with a pile diameter of 500 mm is selected, and the design parameters are shown in Table 6.

Table 5. The soil parameters.

Soil layer	The characteristic value of bearing capacity (KPa)	Natural unit weight (kN/m^3)	Compression modulus (MPa)	Width and depth correction factor	
				η_b	η_d
Miscellaneous fill	90	18.0	3	0.0	1.0
Medium sand	150	18.5	8	3.0	4.4
Silty clay	160	18.9	5.6	0.3	1.6
Sludge	70	15.6	1.95	0.0	1.0
Medium sand	180	18.5	8	3.0	4.4
Silty clay	140	18.0	4.91	0.0	1.0
Fully weathered tuff lava	320	19.5	20	0.5	2.0
Sandy strongly weathered rock	450	20.0	35	1.0	2.5
Fragmentary strongly weathered rock	550	22.0	45	3.0	4.4
Moderately weathered tuff lava	2,000	25.0			

Table 6. Design parameters of prestressed pipe pile.

Bearing strata	Pile side resistance (KPa)	Pile tip resistance (KPa)	Uplift coefficient
Fully weathered tuff lava	100	4,000	0.65
Sandy strongly weathered rock	110	5,500	0.70
Fragmentary strongly weathered rock	130	7,000	0.70

The pile foundation design adopts the combination of pile-raft foundation and pile-supported stand foundation. The pile foundation model is shown in Figure 2. Due to the different depths of the cable tunnel in the GIS room and the interior cable trench, the pile-top elevations are varied, which is a challenge in foundation modeling. It is important to note that due to elements like self-weight consolidation, the settlement of the soil surrounding the pile is larger than that of the pile itself. The soil will exert a downward force on the pile, which is negative frictional resistance (Mao 2022). The effect of negative frictional resistance cannot be ignored during design. To consider the influence of negative friction, the pile length is calculated by Lizheng Software. The pile foundation statistic is shown in Table 7. Due to the complex engineering geological condition, the designed pile length is only for reference, and the final pile length shall be based on the actual effective pile length constructed on site.

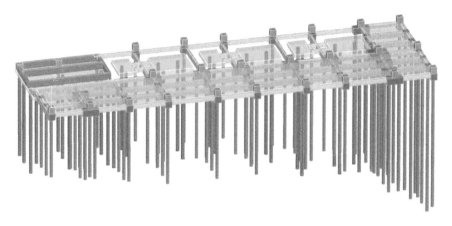

Figure 2. The pile foundation model.

Table 7. Statistical summary of pile foundation.

Pile length (m)	Pile No.	The standard value of the ultimate bearing capacity of a single pile (kN)	The characteristic value of bearing capacity of a single pile (kN)	Bearing strata	Quantity
9~24	ZH1	2,000	1,000	Sandy strongly weathered rock	87
	ZH2	2,000	1,000	Fragmentary strongly weathered rock	2
	ZH3	2,000	1,000	Fragmentary strongly weathered rock	1
	ZH4	2,100	1,050	Sandy strongly weathered rock	5

4 STRUCTURAL DESIGN SUGGESTIONS

In general, the power distribution building realizes the rational design of the steel frame structure and overcomes the complexity of the engineering geological condition through the design of the pile foundation. Given the design process of the steel frame and pile foundation, the specific structural design suggestions are as follows:

(1) If the stress ratio is too low, it is simple to waste steel; if the stress ratio is too high, there is not enough room for structural safety. For the sake of economy and safety, under the

condition of reasonable numerical modeling, complete load arrangement and correct parameter setting, the stress ratio of most steel components should be controlled within $0.75 \sim 0.85$.

(2) The plane size of the power distribution building is long, and the impact of temperature cannot be disregarded. Once the concrete cracks, it will affect the use of the steel bars truss deck. The floor expansion belt can be used to compensate for concrete shrinkage to control the cracking of concrete caused by thermal expansion and contraction (Jiang & Chen 2020), which is also essential to guarantee the long-term performance of the floor system.

(3) If a bad geological phenomenon is discovered after determining the stratum distribution and geotechnical features of the site, the source and distribution range should be examined to further estimate its growth trend and degree of damage. If there is a side slope in the station area, the side slope stability calculation should be carried out, and the corresponding side slope treatment strategy should be suggested.

(4) Because of the "squeezing effect" of the precast piles, which causes the driven piles to be displaced by the extrusion of the post-pile driving, the speed of pile driving should be controlled. In the process of pile driving, the penetration degree or pile pressing force should be the main control, the elevation of the pile end should be the secondary control, and the surrounding environment should be monitored within the control range of pile driving.

(5) The variation of pile length is complicated due to the large difference in soil layer distribution. To facilitate subsequent construction, an appropriate combination of similar pile lengths can be carried out to avoid excessive types of pile lengths. In general, the pile length should not be too long. Otherwise, it is not conducive to the lifting and transportation of the piles.

5 CONCLUSION

Considering economic and environmental benefits comprehensively, steel structure has a broad application prospect. To promote the construction of low-carbon substations, the application of steel structures in the power distribution building is one of the important means. The structural design should first determine the basic design parameters and the geological condition, then pay more attention to the conceptual design. The reasonable design scheme depends on the appropriate conceptual design to a certain extent. Based on the design concept of the total life cycle of the structure, the elastic-plastic deformation of the steel frame, the synergistic effect between the steel frame and the foundation, and the effect of soil between piles can be considered in the future design, to realize the performance-based design of the structure. Therefore, it is necessary to start from the performance design goal to improve the design practice at the present stage: we should choose the appropriate technical scheme for the superstructure and foundation according to the actual engineering condition, further optimize some of the nodes or components so that it not only meets the requirement of safety and economic but also meets the performance requirement.

REFERENCES

DL/T 5457-2012. 2012. *Technical Code for the Design of Substation Buildings and Structures. S.* Beijing: China Planning Press.

GB 50009-2012. 2012. *Load Code for the Design of Building Structures. S.* Beijing: China Architecture & Building Press.

GB 50011-2010. 2016. *Code for Seismic Design of Buildings. S.* Beijing: China Architecture & Building Press.

GB 50017-2017. 2017. *Code for Design of Steel Structure. S.* Beijing: China Architecture & Building Press.

Jiang F. and Chen Y. 2020. Summary of Structural Design for a Power Distribution Building. *J. Low-Temperature Architecture Technology*, 42 (09): 70–71+80.

Mao L.M. 2022. The Calculation and Treatment Measures of Negative Friction Resistance of Pile Foundation Under Large Area Soil Filling. *J. Engineering Technological Research*, 7 (18): 164–166.

Miao Y., Zhou H. B., and Yan L. J. 2020. Total Life Cycle Cost Analysis of Steel Structure Substation and Reinforced Concrete Substation. *J. Mechanical and electrical information*, (32): 95–97.

Shu Y. B., Zhang L. Y., Zhang Y. Z., Wang Y. H., Lu G., Yuan B., and Xia P. 2021. Research on the Path of Carbon Peak and Carbon Neutralization for China's power industry. *J. China Engineering Science*, 23 (06): 1–14.

Yan B., Li Z. M., Cao Y., and Zhao J. 2022. Analysis of Foundation Selection of a Teaching Building. *J. Anhui Architecture*, 29 (11): 157–159.

Zhao X. Y. 2020. Comprehensive Analysis of Pile Foundation Technology. *J. Shanxi Architecture*, 46 (15): 65–67.

Water Conservancy and Civil Construction – Oke & Ahmad (Eds)
© 2024 The Author(s), ISBN: 978-1-032-58618-2

Calculation and analysis of foundation pit excavation near river embankment

Xu Tao & Gong Chunjuan
Nanjing Water Planning and Designing Institute Co., Shenzhen Branch, Ltd. Shenzhen Branch, Shenzhen GHY Environment Water Co., Ltd.

Wang Keliang*
Jiangsu Surveying and Design Institute of Water Resources Co., Ltd., Yangzhou, China

ABSTRACT: Embankment is an important defense line to ensure the safety of buildings and people on the bank. It is of great significance to study the influence of foundation pit excavation near the embankment on the seepage stability of the embankment. Taking an example of foundation pit excavation within the scope of the embankment as the background, this paper analyzes the prototype embankment and compares the seepage and anti-sliding stability of the embankment after taking engineering measures based on finite element calculation results, giving some useful analysis and treatment measures.

1 INTRODUCTION

With the development of urban construction and the increasing shortage of land, engineering construction near the embankment is inevitable, which puts forward new challenges to geotechnical engineering. The construction of a deep foundation pit near the embankment is likely to be threatened by seepage instability. Dike is the last line of defense for flood control. Once it breaks due to instability, it will cause serious consequences (Hu 2013; Zhang 2019). According to the existing research data, there are few studies on seepage and anti-sliding stability analysis of the foundation pit excavated near the embankment (Liu 2013; Li 2011). It is necessary to study anti-sliding stability and anti-seepage measures of the foundation pit excavated near the embankment. (Chen 2012; Liu 2020). Taking an example of foundation pit excavation within the protection range of embankment, this paper analyzes the seepage stability and anti-sliding stability of embankment and puts forward corresponding countermeasures for this example.

2 PROJECT OVERVIEW AND GEOLOGICAL CONDITIONS

The embankment involved in this paper is the Beijing-Hangzhou canal embankment. The urban development and construction in the east occupy management of some floodgates and embankments on both sides, which has a certain impact on safety. The embankment is a grade 2 embankment, with a top elevation of 9.20 m, a top width of 33.0 m, and a slope of 1:3 on the upstream side. The back slope of the original design embankment adopts a gentle slope to connect with the rear urban land. According to the construction drawing design of

---

*Corresponding Author: 33922449@qq.com

DOI: 10.1201/9781003450832-24

the project in Yangzhou, the urban development buildings occupy the embankments on both sides of the floodgate, resulting in the reduction of the width of the embankment top to 24.8 m, and the ground elevation of the water street behind the embankment is 5.3 m, which is inconsistent with the gentle slope of the original design. In addition, according to the geological survey, the silty sandy soil layer and silty sand layer with medium permeability are buried in the soil layer of the site. The buried depth is 1.9-18.9 m, and the smaller the buried depth is, the looser the soil is and the lower the mechanical strength is. Therefore, the excavation of the foundation pit is likely to affect the impermeability stability, and safety of the embankment.

3 SAFETY ANALYSIS

3.1 Seepage analysis of prototype embankment

The maximum water level of the embankment outside the Beijing-Hangzhou canal during the reconstruction period is 8.5 m. The calculation adopts the hydraulic structure finite element analysis system to establish the plane calculation model and semi-automatic element grid division. The embankment section is several quadrilateral elements, and the seepage boundary is determined according to the most unfavorable water level combination of embankment seepage calculation, as shown in Figure 1. The isobaric equipotential diagram, seepage velocity vector diagram, and seepage hydraulic gradient diagram of embankment seepage are obtained through calculation, as shown in Figures 2–4. It can be seen that the seepage overflow point is located in the silty sand layer, and the maximum overflow gradient at the outlet section is 0.25, which meets the requirements of the specification. The maximum seepage gradient of the horizontal section of the embankment body is 0.09, and its seepage stability can not meet the stability requirements. In case of rainfall or other bad weather, the slope of the foundation pit very easy to collapses, affecting the safety of the embankment, flood retaining gate, retaining wall, and other built projects in this section.

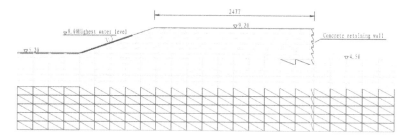

Figure 1. The calculation model of seepage stability in the flood period.

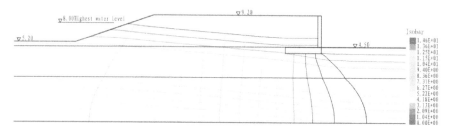

Figure 2. The Isopiestic and equipotential diagram of embankment seepage in the flood period.

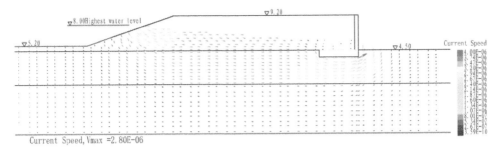

Figure 3. The Vector diagram of seepage velocity of embankment in the flood period.

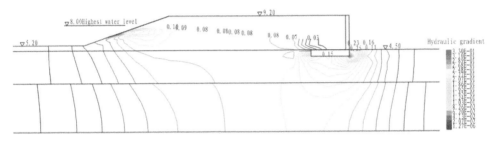

Figure 4. Hydraulic gradient of embankment seepage in flood period.

3.2 *Seepage analysis of dam after engineering treatment*

According to the above calculation, the embankments on both sides of the flood retaining gate can not meet the requirements of seepage stability. It is proposed to use cement soil mixing piles with a diameter of 40 cm within 20 m on both sides of the side pier. The spacing between pile centers is 30 cm, and the thickness of the formed wall shall not be less than 22 cm. P.O 42.5 ordinary portland cement shall be selected as the cement (Wang 2012; Yuan 2018), and the content shall not be less than 18% of the mass of the reinforced soil. The design pile top shall be 8.0 m high, the bottom elevation shall be -5.0 m, the permeability coefficient of the pile body shall not be greater than 2×10^{-6} cm/s, and the strength of the pile body shall not be less than 1.5 MPa. High-pressure grouting shall be adopted between the end of the pile casing and the side pier of the flood retaining gate to ensure the formation of a closed anti-seepage system.

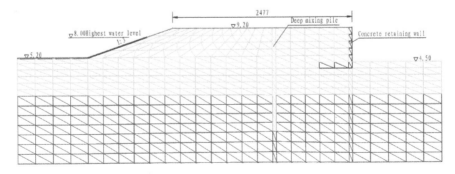

Figure 5. The calculation model of seepage stability in the flood period after treatment.

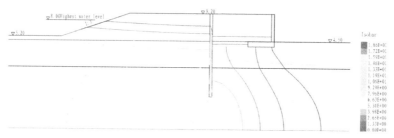

Figure 6. The isopiestic and equipotential diagram of embankment seepage in the flood period after treatment.

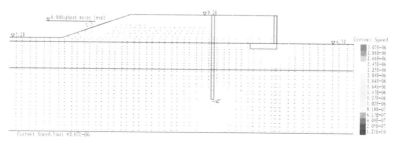

Figure 7. The vector diagram of seepage velocity of embankment in flood period after treatment.

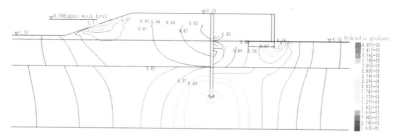

Figure 8. Hydraulic gradient of embankment seepage in flood period after treatment.

3.3 *Dam seepage analysis*

The Swedish arc method is proposed to be used for the anti-sliding stability analysis of the dam (Gan 2022; Lu 2017; Liu 2019). Considering the most unfavorable conditions, the water level is the level in the flood period, and the dam is calculated according to the saturated soil. In the anti-sliding stability analysis of embankment slope, the earth rock dam slope stability analysis system is used for calculation. The calculation and the results are shown in Figure 9. It is obtained that the anti-sliding stability safety factor of the embankment is 1.23, which is greater than 1.15, meeting the stability requirements.

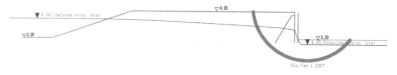

Figure 9. Calculation of embankment anti-sliding stability in flood period.

4 ANALYSIS OF ENGINEERING MEASURES

According to the above analysis, a row of deep mixing piles shall be jacketed within 20 m on both sides of the floodgate. High-pressure jet grouting is adopted between the end of the pile body and the side pier of the flood retaining gate to form a closed anti-seepage system, which plays an obvious role in the anti-seepage of the embankment.

In addition, the foundation pit backfill behind the downstream retaining wall shall be backfilled with gravel, and drainage measures shall be taken, and large-scale mechanical compaction is not allowed, so manual compaction is recommended. At the same time, light good point dewatering is adopted outside the foundation pit to prevent the excessive water level difference between the inside and outside of the foundation pit during rainfall, which will affect the support safety. At the same time, overload is not allowed on the embankment top during long-term rainfall.

5 CONCLUSION

After the stability analysis of the foundation pit slope, it was proved that the excavation slope of the foundation pit can't meet the requirements of seepage stability. The treatment with deep mixing pile and jet grouting is adopted, which can improve the seepage and sliding stability of river embankment.

REFERENCES

Chen Yunbin and Ma Deqiang (2012). Engineering Practices of Preventing Seepage Damage of Excavations in Coastal Thick Sand Layer Region [J]. *Chin. J. Geo. Eng.*, 34 (S1): 461–464.

Gan Lei, Wu Jian, Dai Shouye, et al. (2022). Seepage and Stability Analysis of Embankment with Construction Cracks in Tunnel Crossing Section [J]. *J. Hohai Univ. (Nat. Sci.)*, 50 (1): 85–91.

Hu Fengqiang, Hen Yumin, and U. Liming (2013). Numerical Simulation Analysis of the Embankment Stability of the Yangtze River Near the Foundation Construction [J]. *J. Nanchang Univ. (Nat. Sci.)*, 6:594–600.

Li Shuwei, Xu Jian,Yan Bin (2011). Influence Analysis of Deep Foundation Pit in North Bank of Qiantang River on Flood Embank [J]. *Water Resources and Power*, 29 (2): 93–95.

Liu Ming, Liu Aihua, Zou Jiaqiang, *etc.* (2020). Control Measures of Influence of Foundation Pit Construction on Stability of Soil Embankment in Soft Soil Area [J]. *Science Technology and Engineering*, 20 (9): 3726–3731.

Liu Jian and Wang Dan. Deep Foundation Excavation of the Dyke of the Xiangshuijian Lower Reservoir in Mucky Soil [J]. *Construction Technology*, 42 (19): 40–42.

Liu Yanyan, Li Xiulei, and Feng Mingzhu (2019). Stability Analysis Method for Double Safety Factors of Soil Slope Failure [J]. *Highway Engineering*, 44 (4): 118–123.

Lu Yulin, Bo Jingshan, and Chen Xiaoran (2017). Analysis of a Horizontal Slice Model Using the Swedish Arc Method [J]. *China Earthquake Engineering Journal*, 39 (3): 496–501.

Wang Hongwei, Zhou Jian, and Jia Mincai (2012). Bayesian Networks Analysis on Soil-cement Mixing Wall [J]. *J. TongJi Univ. (Nat. Sci.)*, 40 (1): 27–32.

Yuan Wenjun, Cai Ziqi, Xie Song, Zhang Junrong, etc. (2018). Study on Construction Technology Coordination Parameters of Cement Soil Mixed Piles Based on Strength Tests [J]. *J. Hunan Univ. (Nat. Sci.)*, 45 (S0): 46–51.

Zhang Mingshu, Chen Haiyong, Shen Leilei, et al. (2019). Influences of Underground Comprehensive Pipe Gallery Construction on Seepage Stability of Adjacent Dikes[J]. *J. Civil Engineering and Management*, 36 (3):107–110,118.

Analysis of the influence of different control water levels on the upstream drainage of Hongze lake in flood season

Dawei Zhu*
Jiangsu Surveying and Design Institute of Water Resources Co., Ltd., Yangzhou, China

Xiaolin Zhong
Yangzhou Surveying and Design Institute of Water Resources Co., Ltd., Yangzhou, China

ABSTRACT: The controlled water level of Hongze Lake in flood season is a very important and sensitive indicator, which directly affects the reasonable solution of the contradiction between flood control and benefits. In order to make full use of the flood resources of the Huaihe River, and ensure the economic and social development in northern Jiangsu and the smooth operation of the East Route of the South to North Water Transfer Project, this paper analyzes the influence of water level control in flood season on the upstream drainage of Hongze Lake, providing a basis for water resources, flood control, and drainage regulation of Hongze Lake.

1 INTRODUCTION

Hongze Lake is the largest lake reservoir in the Huaihe River basin and one of the five largest freshwater lakes in China. It is located in the middle west of the northern Jiangsu Plain and at the junction of the middle and lower reaches of the Huaihe River. The control water level of Hongze Lake in flood season is a very important and sensitive indicator, which is directly related to the flood control safety of the Huaihe River basin, the water supply safety of the East Route of the South to North Water Transfer, the water source protection of urban and rural residents living, industrial and agricultural production, and the Grand Canal shipping and other water in northern Jiangsu Province. On the premise of ensuring the safety of flood control and waterlogging drainage, how to maximize the benefits of Hongze Lake is an important issue for flood and drought control dispatching. In order to solve this problem, this paper studies the influence of different control water levels in flood season on upstream drainage.

2 ANALYSIS AND CALCULATION METHOD

The influence of raising the control water level of Hongze Lake during the flood season on the drainage of the upstream area in the Huaihongxinhe area was analyzed according to the river and terrain characteristics. The main analysis methods are as follows:

1) Hydrologic calculation: The design drainage discharge of each area shall be calculated by using the frequency analysis of measured hydrological data or directly using the relevant engineering design hydrological results;
2) According to different starting water levels of Hongze Lake (12.5 m, 12.8 m, 13.0 m, 13.3 m, 13.5 m), the waterlogging drainage water surface profile, the backwater endpoint, and the influence range of each river channel with different frequencies are calculated.

---

*Corresponding Author: 215610600@qq.com

3) According to the hydraulic calculation results, combined with the landform of each area along the line, the layout and application of the drainage system as well as the scope and extent of possible impacts are analyzed.

3 DESIGN DRAINAGE FLOW AND WATER LEVEL

For the influence of Hongze Lake Control Water Level in Flood Season on the Drainage of Huaihongxin River Basin, the influence on the drainage of the trunk river from the Hexiang sluice to the Shuanggou outlet should be mainly analyzed. The upstream drainage flow adopts the hydrological results in the Feasibility Study Report on the Continued Construction of New Huaihong River and the Feasibility Study Report on the Treatment of Baohui River and New Huaihong River Mainstream in Anhui Province. We can see Figure 1 below for the design discharge results of the drainage of the Huaihongxin River basin with a return period of 3-10 years. The outlet water level at the lower reaches of the Huaihongxin River is obtained by using the measured water levels at Linhuaitou and Jiangba stations in the flood season and by establishing the water level correlation. The design drainage flow results can be seen in Table 1.

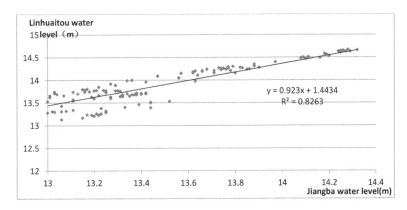

Figure 1. Results of water level correlation between the Jiangba station and Shuanggou station.

Table 1. Results of water level correlation between the Jiangba-Linhuaitou station in flood season (Unit: m).

Station	Water level				
Jiangba	12.50	12.80	13.00	13.30	13.50
Linhuaitou	12.66	12.97	13.18	13.49	13.69

4 ANALYSIS OF THE IMPACT ON THE DRAINAGE OF THE UPSTREAM AREA

4.1 Technical means of analysis

The MIKE 11 modeling system is applied to establish a one-dimensional constant flow drainage mathematical model for the Huaihongxin River. The hydrodynamic module uses the 6-point Abbott Ionescu finite difference scheme to solve the Saint Venant equations, and its numerical calculation adopts the traditional "catching up method".

The basic equations of Mike 11's mathematical model are Saint venant's equations:

$$\frac{\partial A}{\partial t} + \frac{\partial Q}{\partial x} = q$$

$$\frac{\partial Q}{\partial t} + \frac{\partial}{\partial t}\left[\alpha \frac{Q^2}{A}\right] + gA\frac{\partial h}{\partial x} + g\frac{Q|Q|}{C^2 AR} = 0$$

Where Z, Q, A, and R respectively represent the water level, flow, cross-section discharge area, and hydraulic radius at the moment; C is the Chezy coefficient; g is the acceleration of gravity.

The solution of the equations requires the basin description, the measured topography of the river channel and beach area, the hydrologic survey data at the boundary, the hydrologic survey data for calibration verification, the design parameters of water conservancy projects, and the operation rules of dispatching.

4.2 *Analysis of drainage impact*

(1) Analysis and calculation of drainage impact

According to the one-dimensional constant flow calculation water level changes along Huaihongxin River in Hexiang Gate- Shuanggou Section(Table 2), in case of 3-year, 5-year, and 10-year flood drainage, Jiangba controls the water level of 12.5 m, 12.8 m, 13.0 m, 13.3 m, and 13.5 m in flood season. The calculation results show that when the water level of Hongze Lake is adjusted from 12.5 m to 12.8 m, 13.0 m, 13.3 m, and 13.5 m in flood season, the impact on the water level along Huaihongxin River is smaller as it moves upstream.

Table 2. Results of drainage water level along Huaihongxin River with a 10-year return period (Unit: m).

Water level of Jiangba Water level along the line	Ground elevation on both banks	12.50	12.80	13.00	13.30	13.50
Hexiang gate	17	18.82	18.83	18.83	18.83	18.83
Xinhuwa sluice (Xiehekou)	17	18.12	18.12	18.12	18.13	18.13
Jiuwan (Huihekou)	13	17.82	17.82	17.82	17.83	17.83
Shanxi Zhuang	13	16.66	16.66	16.67	16.68	16.70
Xibakou (Tuohu, Tuohe)	13	16.39	16.40	16.41	16.42	16.43
Shizigang (Beidianzi)	15	16.18	16.18	16.20	16.21	16.23
Tianjin lake outlet (provincial boundary)	15	16.02	16.03	16.04	16.06	16.08
Fengshan Town	13.5	15.73	15.74	15.75	15.77	15.79
Shuanggou river diversion inlet	14	15.34	15.35	15.37	15.39	15.41
Linhuaitou	13	12.66	12.97	13.18	13.49	13.69

In the case of 10-year flood drainage, the water level at the inlet of Shuanggou River Approach rises by 0.01~0.07 m; the water level in Fengshan Town rises by 0.01~0.06 m; the water level at the outlet of Tianjin Lake rises by 0.01~0.06 m; the water level in Shizigang is raised by 0.01~0.05 m; the water level at the west dam mouth rises by 0.01~0.04 m; the water level in Shanxi Zhuang rises by 0.00~0.04 m; the water level in Jiuwan Bay rises by 0.00~0.02 m; the water level of Xinhuwa sluice rises by 0.00~0.02 m; the lower water level of Hexiang Gate is raised by 0.00~0.01 m.

4.3 *Countermeasure*

After the control water level of Hongze Lake is raised during the flood season, the adverse impact of the current situation on the drainage of the Huaihongxin River system is mainly in the mainstream of the Baohui River and the main depression of Huaihongxin River.

If 0.05 m is taken as the statistical influence range of the backwater pinch point, the maximum influence range of raising the flood season control water level of Hongze Lake on the Huaihongxin River basin is 3-year flood drainage to Shanxi Zhuang, 5-year flood drainage to Xibakou, and 10-year flood drainage to Tianjin Lake outlet. By comparing the drainage water level of the mainstream of the New Huaihong River with the ground elevation of both banks, it can be seen that when the New Huaihong River basin encounters small and medium floods for drainage, the current ground elevation has also been lower than the designed drainage water level, and the waterlogged water on both banks can no longer be discharged into the mainstream of New Huaihong River by themselves.

According to the implementation of the treatment project for the impact of raising the water level of Hongze Lake and the control project in the planning for the elimination of waterlogging in the key plain depressions of the Huaihe River basin, the Tuo River, Beituo River, Tang River, Shiliang River, and other water system depressions under Zhongshan Xizhuang of the Huaihongxin River system in Anhui Province have been comprehensively controlled, and the current drainage standard has reached a 5-year return period. Therefore, after raising the control water level in the flood season of Hongze Lake, the adverse impact on the drainage of the Huaihongxin River system is mainly to increase the pumping head and operation time of the drainage pump station in the dry river depression below Shanxi Zhuang.

5 CONCLUSION

After the flood season control water level of Hongze Lake is raised, the impact scope is mainly in the depressions along the main stream of New Huaihong River. At present, the depressions on both banks of the New Huaihong River need to be pumped when small and medium floods are drained. Therefore, the adverse impact of raising the flood season control water level of Hongze Lake on its drainage is only to increase the maximum pumping head of the pump stations along the New Huaihong River to 0.13 m.

To make use of flood resources, the current flood season control water level the upstream of Xibakou sluice of Huaihongxin River is 14.67 m, and that at Nushan Lake sluice of the Chi River is 13.50 m. Therefore, raising the flood season control water level of Hongze Lake has a more limited impact on the actual drainage of the Huaihongxin River and Chi River.

REFERENCES

Dittmann R., Froehlich F., Pohl R., *et al.* Optimum Multi-objective Reservoir Operation with Emphasis on Flood Control and Ecology [J]. *Natural Hazards & Earth System sciences*, 2009, 9 (6): 1973–1980.

Hsu N. and Wei C. A Multipurpose Reservoir Real-time Operation Model for Flood Control During Typhoon Invasion [J]. *Journal of Hydrology*, 2007, 336 (3–4): 282–293.

Li J., Zhong P., Yang M., *et al.* Intelligent Identification of Effective Reservoirs Based on the Random Forest Classification Model [J]. *Journal of Hydrology*, 2020, 591: 125324.

Niewiadomska Szynkiewicz E., Malinowski K., and Karbowski A. Predictive Methods for Real-time Control of Flood Operation of a Multi-reservoir System: Methodology and Comparative Study [J]. *Water Resources Research*, 1996, 32 (9): 2885–2895.

Wang F., Valeriano O. C. S., and Sun X. Near Real-time Optimization of Multi-reservoir During Flood Season in the Fengman Basin of China [J]. *Water Resources Management*, 2013, 27 (12): 4315–4335.

Wei C. and Hsu N. Multireservoir Real-time Operations for Flood Control Using Balanced Water Level Index Method [J]. *Journal of Environmental Management*, 2008, 88 (4): 1624–1639.

Windsor J. S. Model for the Optimal Planning of Structural Flood Control Systems [J]. *Water Resources Research*, 1981,17 (2): 289–292.

Water Conservancy and Civil Construction – Oke & Ahmad (Eds)
© 2024 The Author(s), ISBN: 978-1-032-58618-2

Construction technology analysis of 80 meters deep underground diaphragm wall under complex geology

Li Yefu, Wang Rong* & Zhang Shoujia
China Harbour Engineering Co. Ltd., Beijing, China

ABSTRACT: The acceleration of the urbanization process requires the development of construction costs, and the construction technology of the underground diaphragm wall under complex conditions is very important. The purpose of this method statement is to describe the method of constructing the temporary circular diaphragm wall and ensure that the work is carried out in accordance with the relevant drawings, specifications, standards, and safety regulations. And this study statement outlines the proposed work method for the construction of 80 meters circular diaphragm wall. The diaphragm wall is constructed with a sequence of adjacent panels connected by a joint-forming method to create a continuous structural element.

1 INTRODUCTION

To develop a city, it needs to transition to civilization and modernization, and buildings are the main representatives. The quality of buildings that can become city representatives meets the requirements and standards. Therefore, the foundation of the underground diaphragm wall is more stable and reliable. Therefore, the development and growth of the diaphragm wall are mainly based on drilling wells and oil drilling, using mud and underwater pouring of concrete. The construction operation characteristics of the wall and the construction of the underground diaphragm wall are suitable for basements, underground shopping malls, parking lots, underground laboratories, and underground oil depots, as well as the maintenance of deep foundations of high-rise buildings, with certain anti-seepage, retaining soil to prevent soft soil. Therefore, the acceleration of the urbanization process requires the development of construction costs, and the construction technology of the underground diaphragm wall under complex conditions is very important.

This study statement outlines the proposed work method for the construction of 80 meters circular diaphragm wall. The diaphragm wall is constructed with a sequence of adjacent panels connected by a joint-forming method to create a continuous structural element. Diaphragm wall panels follow a construction process in three phases: excavation, installation of reinforcement cages, and concreting. The excavation of the diaphragm wall panel is carried out from ground level where a guide wall has been cast previously to ensure the positioning reference. During excavation, the sides of the trench are supported by bentonite slurry. Each panel is adequately excavated according to the design drawings by Cable operated Clamshell, Hydraulic Clamshell, Heavy Chisel, and Trench Cutter suspended from crawler base machines. Should obstructions be encountered during excavation, the grabbing method with the clamshell is employed to retrieve the obstacles. Depending on the nature and size of the obstruction, the clamshell would be assisted by Heavy Chisel to weaken and

---

*Corresponding Author: wangrong@chec.bj.cn

break the obstruction. For completion of the excavation, bentonite slurry contaminated by the excavation will be substituted with appropriate bentonite slurry or recycled through a desanding unit to meet the requirements prior to concreting. In any case, the contaminated bentonite slurry is recycled and treated before being stored for further reuse in excavation. In the case of diaphragm wall construction, prior to reinforcement cage installation, the panel is prepared according to the joint forming method to comply with the panel sequence. Then, by means of a lifting appliance, the rebar cage is lowered into the bentonite slurry-filled trench at the required depth. The cage might be constructed in a singular cage or in multiple cage sections connected to form a continuous structural element. The connection is typically realized using U-bolts and Coupler (if necessary). At the final stage, each cage section is installed, and the cage is supported by suspension beams resting on the top of the guide wall. Tremie pipes are then installed to the base of the panel and sound concrete is cast from the panel toe up to the required cut-off level. The tremie pipe level will follow the rising concrete level, ensuring a continuous embedment of the tremie pipe into the concrete to prevent contamination with the bentonite slurry. During the concreting, the displaced bentonite slurry is drawn off the panel and recycled for reuse.

2 SITE PREPARATION

As part of the preparation for diaphragm wall construction, the following preliminary activities must be prepared, as shown in Figures 1 and 2.

1. Bentonite plant assembly and commissioning;
2. Establishment of rebar fabrication and storage area;
3. Working platform design and approval;
4. Construction of working platform;
5. Guide wall construction;
6. Trench stability calculation and approval.

Furthermore, site preparation shall cater for area demarcation of working machines, pedestrian walkways, vehicle routes, and excavation area barricading to prevent falling into the excavated panel.

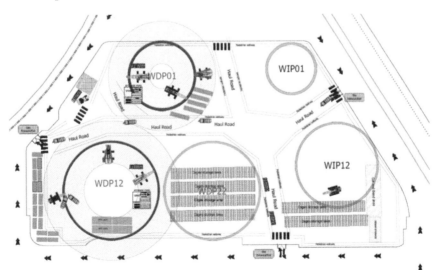

Figure 1. Sample of site operation layout inclusive of the pedestrian walkway.

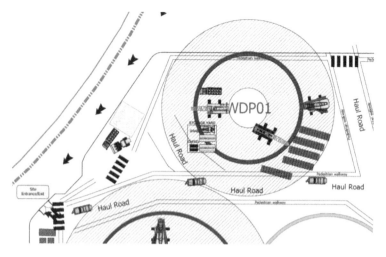

Figure 2. Sample of site operation layout inclusive of the pedestrian walkway.

3 WORKING PLATFORM

Before work, the working platform is to be designed and constructed to accommodate the excavation equipment load and bearing pressures, as defined by the equipment operation manuals. These loads are factored to allow for site conditions, as shown in Figure 3.

The platform bearing capacity must be verified through testing to confirm that it meets these requirements. The defined load cases are used for normal operation and salvage cases. If the existing ground does not meet the requirements, it will depend on available fill materials, and the required platform capacity can normally be achieved with compacted fill. If the underlying material is unsuitable, a concrete slab will be a possible solution. Since the salvage load case is hard to achieve for such a large working area, in the rare event that this additional capacity is required, engineered support mats can be mobilized to the site and then removed again once the machine is back to normal operating.

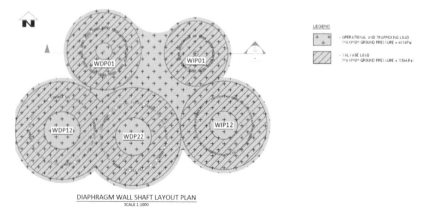

Figure 3. Diaphragm wall shaft layout plain.

4 CONSTRUCTION SEQUENCE

4.1 *Guide wall construction*

The guide walls are temporary works that consist of a pair of parallel concrete walls located on both sides of the proposed diaphragm wall.

4.1.1 *Guide wall purposes*
The purposes of the guide walls are to:

1. ensure the positioning and the construction tolerances of the trench excavation;
2. prevent the platform material and other objects to collapse into the trench;
3. ensure slurry Bentonite reservoir of the trench;
4. support the reinforcement cages, tremie pipes, and other equipment suspended in the trench.

4.1.2 *Guide wall construction process*
The construction process of the guide wall is exposed below:

1. The surveyor will set out the alignment of the diaphragm wall;
2. All excavation areas will be fenced off;
3. The line of the diaphragm wall is excavated;
4. Any required ground-bearing capacity checks are to be defined by the guide wall designer. Verification can be by plate test or other, as specified;
5. A concrete blinding layer will be placed;
6. The alignment of the diaphragm wall on the concrete blinding will be set out;
7. Reinforcement and formwork will be placed;
8. Concrete is discharged directly from the concrete truck or using a backhoe bucket;
9. Once the concrete has sufficiently set (normally the following day) and the formwork is removed;
10. The guide wall is then braced by using timber in necessary and the voids behind and between are backfilled up to the ground level.

4.1.3 *Guide wall typical details*
Typical details of the guide wall are shown as shown in Figure 4.

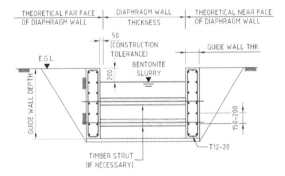

Figure 4. Typical guide wall details.

The design and calculations of the guide walls shall be submitted with endorsement separately on final confirmation of the Working Platform Design.

4.2 Bentonite slurry

The slurry is based on a mixture of bulk bentonite powder and fresh water. The main purpose of the bentonite slurry is to act as a support fluid, supporting the walls of the trench during the panel construction until the completion of concrete placement.

4.2.1 Bentonite slurry purposes
The essential functions of the bentonite slurry are summarized below:

1. to ensure the trench stability;
2. to reduce the soil permeability;
3. to suspend and transport excavated ground particles. The contents below detail the technical specifications related to the bentonite slurry and its management during all stages of production.

4.2.2 Bulk bentonite powder and storage
The proposed bulk bentonite powder type is Bentonil GTC 4 or Bentosund 120E 45, or a similar bentonite grade. The bulk Bentonite will be stored in individual bags of 25 kg stacked on pallets, weighing 1,050 kg at its maximum capacity of 42 bags or in individual bags of 1.1 tons. According to the quantity of bentonite slurry to be produced, the bulk storage silo or bentonite pool of about 50 m^3 or 350 m^3 capacity might be employed respectively.

The bentonite slurry will be prepared in high turbulence mixers under the ratio of approximately 20 kg to 40 kg of bulk bentonite (per m^3 of water). Water to be used for bentonite slurry shall be close to neutral in pH and shall not be contaminated by salts. Then the freshly mixed bentonite slurry is pumped into storages, such as silos, pools, or tanks. Silos and pools will be installed and anchored onto a slab.

Dust nuisance related to bulk bentonite is to be controlled by the following procedures:

1. Individual bulk bentonite bags are protected with multiple paper bag wrappers and one shrink-fit watertight polythene cover. Additionally, the pallets are fully wrapped with polythene sheets;
2. The empty bags will be placed directly into a skip;
3. The bulk bentonite pallet storage and the mixing area will be enclosed and ventilation with a dust control system shall be installed.
4. All workers appointed for mixing the bentonite slurry must wear suitable personal protective equipment during operation. (i.e. dust mask, goggles, gloves).

A laboratory equipped with the appropriate apparatus to control the bentonite slurry properties will be provided on-site. Those apparatuses have to be maintained and checked before use by the Bentonite Supervisor. The frequency of testing, typical testing requirements, and the normal range of acceptable results are detailed in Table 1.

5 CONSTRUCTION SEQUENCE OF DIAPHRAGM WALL

5.1 Stop-ends joint former method

The stop-end joint former method ensures structural continuity by interlocking the panels together due to the stop-end positioned before casting. 3 types of panels are related to the construction sequence of the diaphragm wall: primary, successive, and closing panels.

1. Primary panels: These panels are constructed at first in the sequence. The excavation is carried out directly into the ground with no adjacent panel. Before the cage installation and concreting, the primary panels will be equipped with 2 Stop-Ends.

Table 1. Typical testing requirements and the normal range of acceptable results.

Property to be measured	Test method and apparatus	Fresh mixture (1)	During excavation	Sample from the panel before concreting (1)	Before re-use (1)
Density (g/ml)	Mud balance	< 1.10	1.08 –1.25	< 1.15	< 1.25
Viscosity (seconds)	Marsh cone	32 to 60	32 to 50	32 to 60	
Fluid loss (ml): CC for 30 min. test	Baroid filter press	< 30	< 50	N/A	< 50
Filter cake thickness (mm)		< 3	< 6	N/A	< 6
Sand content (%)	Wet sieve analysis	N/A	N/A	< 4%	N/A
PH	Electric PH meter	7 to 11	7 to 12	N/A	7 to 12
Frequency		Minimum once daily (sample from silos of fresh bentonite)	Minimum once daily (sample from excavating panel)	Once immediately before concreting	Once daily (sample from silos of used bentonite)

2. Successive panels: These panels are constructed successively to a primary or a successive panel. The excavation is carried out between an adjacent panel and the ground. Before the cage installation and concreting, the Stop-End previously installed in the adjacent panel has to be removed. Additionally, one Stop-End former has to be equipped on the other end of the panel.
3. Closing panel: The closing panels are constructed between two previously constructed panels. The previously constructed panels may be primaries or successive panels depending on the layout. The stop-ends are removed prior to cage installation and concreting.
4. The stop-end toe will be a short distance above the rock head level or a maximum of 35.5 m. The Stop-End is supported on the top of the guide wall when equipped in a primary or a successive panel before concreting. The dimensions of stop-end in use are presented below:

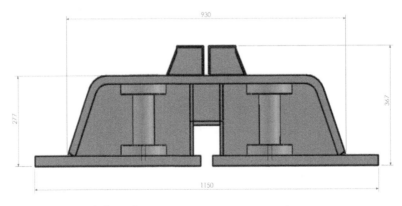

Figure 5. Typical stop-end dimensions.

The removal of the Stop-End (Figure 6) Joint Former for the successive or closing panels is executed laterally. First, the stop end is cleaned by using the purpose of built tools lowered into the exaction by an excavation crane. Then the stop end is peeled away carefully, again using the excavation crane and slewing slowly towards the center of the excavation. The self-weight of the stop end causes it to gradually debond from the previously cast pane. Once the free stop end is lifted out and relocated to another panel or storage as necessary.

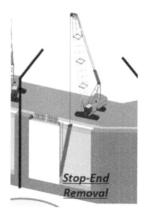

Figure 6. Typical testing requirements and the normal range of acceptable results.

5.2 *Milled joint method*

The milled joint method ensures the structural continuity of the diaphragm wall constructed to shape and the prevention of water leakage between the panels thanks to an interface overcut by means of a Trench Cutter, as shown in Figure 7.

1. Primary panels: These panels are constructed at first in the sequence. The excavation is carried out directly into the ground with no adjacent panel. Prior to the cage installation and concreting, the primary panels will be equipped with 2 Stop-Ends for the first 24 m. The stop-ends ensure smooth gliding of the trench cutter during closure panel construction.
2. Closing panels: These panels are constructed last in the sequence by the Milled joint method. The excavation of the closing panel will be carried out between the stop-ends on the adjacent primary panels. Then the stop-ends are removed, allowing the excavation to continue by a trench cutter and creating an overcut of concrete in the 2 adjacent primary panels previously cast.

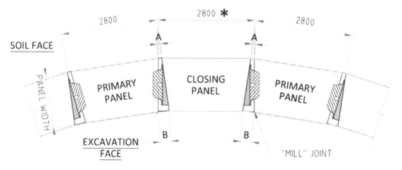

Figure 7. Milled joint details.

5.2.1 *Excavating tools*

The excavating tools are guided by the guide wall to enter the bite to be excavated. The cable-operated or hydraulic clamshells are suspended from and operated by a Heavy Duty Crawler Crane type Liebherr 855 or similar. The excavation process with clamshell relies on repetitive cycles of soil grabbing to remove from the panel. Suitable round or box chisels will be employed to overcome rock or hard layers of excavation and to form the rock socket when required. The excavation process with a chisel consists of cyclic impacts to fracture hard layers encountered. The excavation process with the trench cutter relies on a cutting of the ground by means of rotating wheels equipped with cutting teeth. The excavated cuttings of soil and rock are transported with the bentonite slurry suctioned from the bottom of the trench by means of a mud pump. Out of the panel, the bentonite slurry is transported to the cutter desanding unit to be separated from the excavated cuttings.

Figure 8. Cable-operated clamshell.   Figure 9. Trench cutter.

5.2.2 *Removal of excavated spoil*

The spoil excavated from the trench will be deposited directly into dump trucks or tanks. The spoil will then be transported to a dumping area.

Figure 10. Disposal of excavated spoil with a dump truck.

6 CONCLUSIONS

1. Verticality checks are performed during excavation to ensure that the completed panel is constructed to the required tolerance. The Diaphragm wall panel shall be vertical within a tolerance of 1:200. Verticality will be monitored and maintained within the specification tolerance. Any excessive deviation will be rectified back to within tolerance. In some instances, this may lead to local over break which can be removed if required during excavation.
2. Mixing bentonite with the added spoil by the clamshell is used in the open trench. The backfilling and mixing allow the fine materials to fill the small voids of the soil and it is the common method to cope with the typical bentonite slurry loss.
3. When it is determined that the bentonite slurry loss has stopped, the excavation can restart immediately, with ongoing monitoring of the bentonite level. If minor loss continues the process of backfilling, refilling can be repeated to re-plug new voids, until resolved. Backfilling of the open trench with soil or cement to a level situated above the "voids" level together with continuous discharge of bentonite slurry until the loss of bentonite slurry is stopped. NCR is to be raised to the main contractor. Possible causes are to be identified, and remediation measures are to be proposed and agreed upon before the recommencement of excavation.

REFERENCES

Shuai Huang, Mingming Huang, and Qingjie Qi. A Simplified Calculation Method of Ice–Structure–water Dynamic Interaction Under Earthquake Action. *Extreme Mechanics Letters*, 2021, 43:1011782021.

Shuai Huang, Mingming Huang, and Yuejun Lyu. A Novel Approach for Sand Liquefaction Prediction Via Local Mean-based Pseudo Nearest Neighbor Algorithm and its Engineering Application. *Advanced Engineering Informatics*, 2019, 41: 1–12.

Shuai Huang, Mingming Huang, and Yuejun Lyu. Seismic Performance Analysis of a Wind Turbine with Monopile Foundation Affected by Sea Ice Based on the Simple Numerical Method. *Engineering Applications of Computational Fluid Mechanics*, 2021, 15 (1): 1113–1133.

Shuai Huang, Mingming Huang, Yuejun Lyu, and Liwei Xiu. Effect of Sea Ice on the Seismic Collapse-resistance Performance of Wind Turbine Tower Based on a Simplified Calculation Model. *Engineering Structures*, 2021, 27: 111426.

Shuai Huang, Yuejun Lyu., *et al.* Seismic Performance Assessment of Unsaturated Soil Slope in Different Groundwater Levels. *Landslides*, 2021, 18: 2813–2833.

Tan Y. and Li M. Measured the Performance of a 26 m Deep Top-down Excavation in Downtown Shanghai. *Canadian Geotechnical Journal*, 2011, 48(5): 704–719.

Whittle A. J., Hashash Y. Ma., Whitman R. V. Analysis of Deep Excavation in Boston. *Journal of geotechnical engineering*, 1993, 119 (1): 69–90.

Xnathkaos P. P. Excavation Support Methodist [M]. *Gortmd Control and Improvement*, John Wiley and Sons Pub, 1994.

Water Conservancy and Civil Construction – Oke & Ahmad (Eds)
© 2024 The Author(s), ISBN: 978-1-032-58618-2

Foundation trench excavation and basin dredging technology in port waterway construction

Xingquan Jiang*
Guangzhou Salvage Bureau of the Ministry of Transport, Guangzhou, Guangdong, China

ABSTRACT: With the development of the economy, the construction of ports and waterways is very important to the international trade of our country. This paper puts forward the technical research of foundation trench excavation and basin dredging in the construction of ports and waterways. Taking the actual project as an example, the foundation trench was excavated by the dredger in the form of layers, sections, and strips. The harbor basin was dredged by the millisecond blasting technology. The multi-beam was used to measure the back silting after the construction of the foundation trench. The maximum back silting thickness at the bottom of the foundation trench was 1.29 m, which was within the controllable range, indicating that the construction quality was good, and it had an important contribution value to promoting the construction quality of the port and waterway.

1 INTRODUCTION

With the rapid development of the social economy, maritime transportation has been paid more and more attention by international trade. Generally speaking, transportation projects crossing river basins and overseas regions have extremely strict requirements on construction design and technology. Countries around the world have adopted immersed tunnels to build ports and waterways, and there are many successful examples of immersed tunnels. The immersed tunnel construction technology in China started late. Compared with some European and American countries, it is still not yet mature. The immersed tunnel engineering of ports and waterways involves many disciplines. The construction process is complex and arduous, so it is necessary to continuously improve construction technology in immersed tunnel engineering. In the construction of port and waterway engineering, foundation trench excavation and harbor basin dredging is the premise for the smooth development of the follow-up project. Only by doing these two constructions well can the landfill construction of immersed tube tunnel be done well, to ensure the construction quality of the port and waterway. Based on this, this paper makes an in-depth study of the foundation trench excavation and basin dredging technology in the construction of ports and waterways, hoping to contribute to the expansion of China's port scale and the development of water transport.

2 PROJECT OVERVIEW

The coastline of a port waterway project is 500 m long, and the total water area is 386,500 m². The actual dredging scope of the project mainly includes foundation trench excavation and basin dredging, and the total dredging volume is about 680,000 m³. In the construction of the port and

---

*Corresponding Author: 1871513861@qq.com

DOI: 10.1201/9781003450832-27

waterway, the construction is carried out in the way of one-time excavation to the design bottom elevation in combination with the soil characteristics of the construction site and the excavation depth of the foundation trench. Due to the amount of dredging in the foundation trench, the silt deposition in the water area is serious, so the excavation of the foundation trench attaches great importance to urgency and flexibility (Mo 2021). The bearing stratum of the foundation trench of this port channel project is mostly a weathered rock layer. If rocks are encountered during the excavation of the foundation trench, it is necessary to carry out the dredging process of the harbor basin through reef blasting to ensure safe and smooth construction.

3 EXCAVATION OF FOUNDATION TRENCH

The excavation depth of the foundation trench in the construction of the port channel is 30 cm (Li et al. 2021). We can refer to the construction design scheme. The cutter suction dredger cannot be directly used to excavate to the design elevation. In the process of foundation trench excavation, the dredger is first used for foundation trench dredging. Then, the dredged sludge is transported to the fixed discharge area by the mud barge, and finally, the cutter suction dredger is used to backfill the foundation trench to the design height (Feng & Zhao 2020). The specific construction process of foundation trench excavation is as follows: due to the special geographical location of the port waterway project, to avoid the influence of foundation trench siltation, it is necessary to set the cutter suction dredger in front of the wharf. Then, we can use the dredger to carry out dredging construction in the order of the natural section of the dredging area and excavate in layers, sections, and strips. The construction personnel drives the dredger to the designated position of the foundation trench excavation at a stable speed (Gu & Liang 2020). After controlling the direction of the dredger, they start to prepare the rake, put the rake arm of the dredge into the water, connect the elbow of the rake arm with the sludge suction inlet accurately, and then control the low-concentration discharge valve of the dredge pump and the dredge boat, so that the clean water in the compartment can be discharged into the foundation trench. After all the clean water is discharged, the construction personnel operates the dredger and controls the drag head to enter the foundation trench sludge. Under the condition that the dredge pump is at normal speed, the foundation trench dredging construction is officially started. According to the actual situation of the port channel construction, Tongxu is divided into six sections for construction, and the length of each section is controlled at 150 m. In combination with the design width parameters of the foundation trench, each Tongxu section is subdivided into six sections for construction. The width of each strip shall be controlled within 25 ~ 30 m concerning the type of dredger. There shall be an overlapping area between adjacent strips because the thickness of silt in the foundation trench excavation area of the port channel is not evenly distributed. To ensure the smooth excavation of the foundation trench, the excavation of the foundation trench shall be carried out through layered control, as shown in Figure 1.

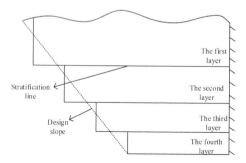

Figure 1. The Schematic diagram of layered dredging construction.

As shown in Figure 1, it is necessary to control the thickness of each mud layer between 3.0 and 4.0 m, and the mud layer with the largest thickness is the first construction sequence. In the process of dredging, to control the quality of the side slope, the excavation of the side slope shall be completed by the step excavation method according to the design gradient. The principle of excess at the bottom and under balance at the top shall be referred to, to ensure that the side slope of the foundation trench meets the requirements of port and waterway construction. At the same time, by controlling the amount of dredging, it is ensured that no shallow points are left after the excavation of the foundation trench. In the process of dredging with a dredger, the construction personnel needs to adjust the depth and speed of the dredger according to the parameters, such as sludge concentration, flow rate, and pressure. At the same time, the construction personnel driving the dredger need to observe the instrument panel in real-time. When the indicated concentration rises, the construction personnel should close the low-concentration discharge valve in time, and then open the loading valve to start loading (Xu *et al.* 2020). When the dredger reaches the end of the foundation trench, we should stop the dredger and retract the rake arm. At this time, when the sludge concentration decreases, we shall close the loading valve again, open the discharge valve to discharge clean water, and discard the excavated sludge to a fixed position before backfilling until the depth of the foundation trench reaches the design elevation. Due to the under-excavation and over-excavation in the foundation trench excavation, the errors between the dredger type, the tidal data of the port, and the design parameters, it is difficult to improve the accuracy of the longitudinal slope of the foundation trench excavation. Therefore, in the process of foundation trench excavation, it is necessary to combine the monitoring and measurement work. Every time a layer of mud is excavated, the sonar range finder is used to measure the underwater topography to ensure the quality of the foundation trench excavation.

4 DREDGING CONSTRUCTION OF HARBOR BASIN

Due to the unsatisfactory excavation effect of the hard rock stratum in the foundation trench excavation construction, before the construction of the immersed tube tunnel of the port channel, to ensure that the buried depth of the pipeline meets the expectation, it is necessary to assist the dredging construction of the harbor basin (Liu & Zeng 2022). Considering the construction cost and period, this paper uses the millisecond blasting technology to carry out the dredging construction of the harbor basin (Chen *et al.* 2021). First, we can refer to the "Blasting Safety Regulations" to determine the distance and dosage of each hole.

$$d = 0.8\frac{z}{33} \times \sqrt{\frac{\rho k}{qms}} \times m \tag{1}$$

$$R = rdlh \tag{2}$$

Where d represents the distance parameter between each gun hole in harbor dredging construction; z represents the diameter parameter of the gun hole; ρ denotes the density parameter of explosive used in blasting; k represents the power coefficient value of explosive used in blasting; q represents the tilt coefficient value of each hole; m is the density coefficient value of the hole layout; s represents the rock constant value to be dredged; R represents the explosive quantity parameter used in each hole of harbor dredging construction; r represents the unit explosive consumption parameter in harbor dredging construction; l indicates the row spacing parameter of the hole layout; h represents the rock layer thickness parameter to be dredged. According to Equations (1) and (2), the blasting design scheme is strictly controlled to avoid engineering accidents. At the same time, the broken pipe is arranged in the form top-down (Guo & Lu 2022), and the pipe orifice is sealed with the sand column to prevent the occurrence of serial explosion between each hole, thus affecting the effect of the blasting reef. When the blasting platform is arranged underwater, a hole network with a length of 3 m and a width of 2.8 m is adopted to arrange the holes. The diameter of the drilling hole is controlled to be 100 mm and the depth to be 1.8 m. Then the blasting is carried out at different times and regions two times every day. According to the stress wave interference hypothesis, to enhance the blasting effect, the time difference between the two blasting intervals is as follows:

$$\Delta T = \frac{\sqrt{d^2 + 4D^2}}{V} \qquad (3)$$

In the formula, D the minimum resistance line parameter of explosive explosion; V is the velocity parameter of blasting stress wave propagation. For each blasting, it is necessary to divide the dredging area of the harbor basin into two areas with an interval of more than 30 m to reduce the blasting vibration (Wei et al. 2021). Because of the complex construction environment of the port and waterway, to reduce the harm of reef blasting to the surrounding buildings, it is necessary to arrange a bubble curtain on the reef-blasting ship. The high-pressure air wall generated by the bubble curtain is used to buffer the explosion stress wave. Before the dredging of the harbor basin, the test blasting operation shall be carried out, and the vibration monitoring points shall be arranged in the surrounding buildings. The relevant parameters of underwater reef blasting construction shall be adjusted according to the vibration detection data of the test blasting, and then the reef blasting construction shall be completed according to the design parameters (Zhong & Wen 2020). After the construction is completed, the explosives shall be cleaned up.

5 INSPECT THE CONSTRUCTION QUALITY

Due to the long construction period of foundation trench excavation and harbor basin dredging in harbor channel construction, back silting is inevitable in the construction process. If the back silting intensity is large, it will have an impact on the subsequent harbor channel construction. Therefore, this section will judge the construction quality by detecting the back silting intensity. To detect the back silting after the excavation of the foundation trench, four monitoring sections are set in the foundation trench of the port channel, and then the multi-beam measurement is used to obtain the distribution of back silting. The multi-beam measurement is to carry out plane positioning in the excavation area of the foundation trench. The back silting thickness data is obtained after measurement, as shown in Figure 2.

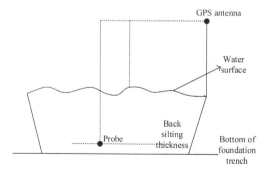

Figure 2. Multi-beam measurement for back silting thickness.

The thickness of the back sitting at the bottom of the foundation trench within two months of each monitoring section by multi-beam measurement is shown in Figure 3.

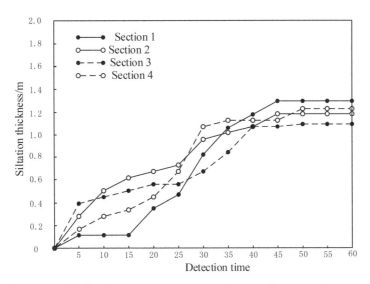

Figure 3. The curve of siltation intensity at the bottom of the foundation trench.

In the process of the foundation trench excavation, considering the factors such as manual operation and storm tide, the back sitting at the bottom of the foundation trench will be affected. In the process of foundation trench excavation, it is only necessary to slow down the mud layer slope or set up a back silting trough nearby, which can effectively reduce the back silting intensity. It can be seen from Figure 3 that the back sitting at the bottom of the foundation trench in this construction is mainly divided into two stages. 45 days before the construction, amounts of back silting occurs at the bottom of the foundation trench with the passage of the construction period, in which the monitoring Section 1 has the largest amount of back silting, with a thickness of 1.29 m. After 45 days of construction, the back siltation at each section tends to be stable, with no significant increase in thickness. The siltation after the excavation of the foundation trench is still within the controllable range. It is only necessary to clean up the silt in time to avoid a large amount of silt deposition caused by too long a time, which will not affect the subsequent processes. Therefore, it can be seen that the construction quality is good and meets the construction requirements of the port and waterway.

6 CONCLUSION

At present, the immersed tube tunnel has the advantage of superior waterproof performance and low project cost and has gradually become the first choice for port and waterway construction in the world. In this paper, a port channel project is taken as the research object, and the key construction technologies of foundation trench excavation and basin dredging in the construction process are studied in depth, which provides a reference for the subsequent immersed tunnel construction projects. In the excavation of the foundation trench, the initial stability of the foundation trench is guaranteed using layered, sectional, and strip dredging. In the dredging of the harbor basin, the safety of dredging is improved using millisecond blasting technology. In the construction of the port channel, the above technologies were adopted and the construction was completed. Because the excavation of the foundation trench and the dredging of the harbor basin is a complex project, it is inevitable to appear the phenomenon of siltation in the process. Through the detection of the siltation intensity at the bottom of the foundation trench, this paper proves that the construction quality is good, which can be used as a reference for the construction of immersed tube tunnels in the construction of ports and waterways in China.

REFERENCES

Chen Jian, Wang Yue and Xie Jiankai, 2021. Application of Millisecond Priming Technology in Reef Blasting Construction of Meizhou Bay Channel Third Phase (Stage II). *Port & Waterway Engineering*, **9** 201–205.

Feng Yutao and Zhao Jinfeng, 2020. Treatment Technology of Siltation Layer and Large Boulder in Steel Cofferdam Foundation Trench under Deep-water Conditions. *Highway*, **65** 104–108.

Gu Renguo and Liang Jianxun, 2020. Finite Element Analysis and Model Test of Bridge Displacement Caused by the Excavation of Immersed Pipe Foundation Trench. *Journal of China & Foreign Highway*, **40** 157–162.

Guo Qingchao and Lu Qin, 2022. Study on Scale and Effect of River Dredging Aiming at Reducing Flood Level. *Journal of Hydraulic Engineering*, **53** 496–503.

Li Mengguo, Han Zhiyuan, Xu Ting, et al., 2021. Study on Sediment Problems of Harbors and Channels in Lingdingyang Bay. *Port & Waterway Engineering*, **9** 1–8.

Liu Birong and Zeng Chengjie, 2022. Countermeasures Against Siltation of No.1 Harbor Basin of Tongzhou Bay Port Area. *Port & Waterway Engineering*, **2** 76–82.

Mo Wenhe, 2021. Optimization of the Design Scheme and Construction Technology of the Zawr Comprehensive Port Development Project. *Port & Waterway Engineering*, **8** 220–224.

Wei Hanhui, Zhai Yugang, and Li Hanting, 2021. The Dredging Method of a Large Trailing Suction Hopper Dredger in Shallow Water. *Journal of Hydraulic Engineering*, **2** 195–198.

Xu Yanliang, Li Zhen, and Zhou Chun'er, 2020. Application of Tree Root Pile in Support for Excavation of Foundation Trench of Gravity Wharf. *Port & Waterway Engineering*, **11** 82–87.

Zhong Weilin and Wen Chunpeng, 2020. Feasibility Study on Navigable Depth in Harbor Basin and Inner Channel of Shenhua Huanghua Port. *Journal of Hydraulic Engineering*, **8** 42–47.

Water Conservancy and Civil Construction – Oke & Ahmad (Eds)
© 2024 The Author(s), ISBN: 978-1-032-58618-2

Study on the interaction and construction time sequence between interchange construction and subway construction

Lei Chen* & Shihu Qi
Nanjing Urban Construction Management Group Co. Ltd, Nanjing, China

ABSTRACT: Using numerical simulation, taking a practical project in Nanjing as the background, the finite element analysis model of bridge pile foundation, a new tunnel, and foundation soil is established by Midas/GTS software, and the influence of shield tunnel construction on existing bridge pile foundation in a stratum is simulated. The changing trend of the pile foundation and surrounding soil when the shield passes through the existing bridge pile foundation is obtained from the simulation results, providing an important reference for monitoring scheme and construction scheme design, as well as an effective reference for similar projects.

1 INTRODUCTION

Traffic problems have become increasingly prominent since the scale of cities has been expanding. Modern cities are gathering places where human beings live, educate, work, treat, and entertain. But the rapid urbanization process makes the urban transportation system face severe situations and challenges (Guo 2019; Guo *et al.* 2011). Underground tunnel traffic has become a kind of urban transportation tool to meet the requirements of modern city operation because of its advantages of large volume, high speed, safety and reliability, punctual operation, and no occupation of ground roads. The main methods of urban subway construction include the shield method, cut and cover method, mine method, shallow underground excavation method, immersed tube method, mixed method, and so on. Shield tunneling is a reasonable construction method when urban subway tunnels pass through buildings and bridges. (He *et al.* 2020)

With the research and solution of problems such as long distances, large diameters, largely buried depths, and complex sections in the aspect of shield construction technology, the shield method has been paid more and more attention and favored (Ding *et al.* 2014; Wang *et al.* 2020). Using the shield method to build a large number of urban subways is bound to be restricted by the existing building environmental conditions, resulting in many new subway lines that can only extend along the overpass line, interweaving through the foundation piles or pile groups of overpasses. (Xu *et al.* 2015)

Tunnel construction will break the original soil balance, and have a lot of adverse effects on the buildings on the ground, bringing great difficulties to design and construction in engineering technology (He *et al.* 2021; Li *et al.* 2021). Some scholars have carried out related research on the influence of subway tunnel construction on bridge pile foundations. (Nematollahi & Dias 2019; Yang *et al.* 2007; Wang *et al.* 2018) However, it is often limited to the settlement change and stress and strain change of pile foundation when studying the

---

*Corresponding Author: chl-nj@163.com

DOI: 10.1201/9781003450832-28

influence of shield tunnel construction on adjacent pile foundation, and does not effectively combine with the movement law of stratum to find the common ground.

Therefore, the evaluation method of bearing capacity and settlement of existing bridge foundations affected by disturbance of newly-built adjacent subway tunnels is deeply studied. It is of great engineering and academic significance to classify and evaluate the influence law in depth to provide a scientific basis for the safety and reinforcement of bridges adjacent to tunnels and minimize tunnel excavation's influence on surrounding pile foundations. This paper mainly adopts the method of numerical simulation to study the influence of subway tunnel construction on pile foundations.

2 MODELING METHOD

2.1 Selection of soil layer parameters

According to the geological report given by the construction unit, there are 7 layers of soil from top to bottom at SW0 (Figure 1) where the subway passes through the interchange, namely 1 layer of miscellaneous fill, 1 layer of plain fill, 3 layers of silty clay, 1 layer of strongly weathered argillaceous siltstone, and 1 layer of moderately weathered siltstone. The total thickness is about 45 m, and the thickness of each layer is 3.2 m, 1.8 m, 4.1 m, 13.4 m, 1.5 m, 4.5 m, and 15.5 m respectively.

Figure 1. Model diagram of intersection area between interchange and subway.

2.2 Shield excavation process and pile foundation

2.2.1 Simulation of the shield excavation process

The shield tunnel is excavated step by step with a full section, and the buried depth at the bottom of the tunnel is 24 m. The diameter of the shield is excavated with a circular section with D = 6.3 m in a typical stratum. The tunnel runs through the silty clay layer. The tunnel is divided into 33 slab units on average, each with a length of 1.5 m. Segment lining shall be installed in tunnels according to the network division. The segment is simulated by plate-shell element, with a thickness of 0.3 m, and C30 concrete is used. During construction, the grouting pressure is 150 KPa and the excavation pressure is 120 KPa. The segment length is the same as that of each tunnel network.

2.2.2 *Simulation of pile foundation parameters*

A pile element is applied to simulate the existing pile foundation. Pile elements can well simulate the deformation characteristics of pile foundations and more truly reflect the interaction between pile and soil. The action of pile soil can exert pile contact around the pile foundation and couple with the surrounding soil. Shear stiffness modulus and normal stiffness modulus are set to reflect the deformation between the pile and soil in contact.

The diameter of the pile is 1.5 m, and C30 concrete is used for pouring. The load on the pile top adopts a concentrated force load. The finite element software **MIDAS GTS** is used to simulate this section. The length of the model is 50 m, the width is 150 m, and the depth is 75 m. It is divided into 8 layers of soil. There are 50945 grids and 288971 nodes in the partitioned model.

2.3 *Model diagram*

The method of size seeding is adopted for tunnel and pile foundation in the process of network division, and the network of the solid tunnel, segment, and pile foundation is divided densely. The soil body adopts the function of automatic network partition in the software, and its network partition is slightly sparse compared with the grid of tunnel and pile foundation. The final model is shown in Figure 2. This paper simulates the whole process of excavation, drilling, and grouting in subway construction, which is divided into 38 steps, including initial stress balance (1 construction step), pile, cap, pier construction (3 construction steps), left side tunnel excavation (17 construction steps) and right-side tunnel excavation (17 construction steps).

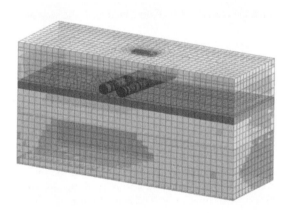

Figure 2. Simulated construction model.

2.4 *Deformation control standard of adjacent pile foundation*

To classify and discriminate the settlement, it is necessary to establish the settlement control standard of adjacent pile foundations in subway tunnel construction. For the pile foundation whose settlement exceeds the control standard, corresponding reinforcement measures are adopted to control the influence caused by subway tunnel construction within the allowable range. At present, the allowable ground deformation caused by subway tunnel construction is only provided empirically, and the allowable displacement value is usually used to control the settlement value of adjacent pile foundations when formulating the control standard. However, it involves many factors, which not only meet the requirements of bearing capacity but also meet the allowable settlement value of pile foundation superstructure.

Because the engineering background of this paper is the hard clay stratum, and the stratum is relatively weak, referring to the settlement control standard of pile foundation near Japan (Table 1), taking 10 mm as the settlement limit value, taking 5 mm of the minimum value of settlement range as an alarm value, and taking 3 mm, 60% of the warning value of 5 mm, as the early warning value. Similarly, the limit value of horizontal displacement is 8 mm, the warning value is 5 mm, and the corresponding warning value is 3 mm.

Table 1. Control standard of pile foundation deformation in this paper.

Standard	Early warning value	Alarm value	Limit value
Settlement value (mm)	6	12	25
Horizontal displacement (mm)	3	5	10

3 ANALYSIS OF STRATUM DEFORMATION LAW

3.1 *Lateral surface subsidence*

In the process of tunnel construction, the settlement of the stratum above the tunnel starts from the axis above the tunnel center and gradually decreases to both sides. After exceeding a certain range, it has little or no influence on the settlement of the stratum. The Peck formula is used to compare and verify the maximum settlement value at the same time.

The calculation method of stratum loss per unit length of shield tunnel is as follows: $Vi = \pi R^2 \eta$, where η is the stratum loss rate, which is usually 0.5%–2.0%. The friction angle of the soil layer where the tunnel is located is 15°. Combined with the stratum selected in this paper, $\eta = 1\%$.

$$V_i = \pi R^2 \eta = \pi \times 3^2 \times 1\% = 0.283 \tag{1}$$

$$i = \frac{z}{\tan(45° - \varphi/2)\sqrt{2\pi}} = 10.91 \tag{2}$$

$$S_{V, max} = \frac{V_i}{i\sqrt{2\pi}} \approx \frac{V_i}{2.5i} = \frac{0.283}{2.5 \times 10.91} = 0.01038\text{m} \tag{3}$$

After a tunnel excavation calculated by the Peck formula, the maximum settlement value of the ground surface is 0.01038 m, which is about 10 mm, and the maximum settlement value calculated by numerical simulation is 9.08 mm (Figure 3). The two results are similar, so it is considered that the longitudinal settlement of the stratum conforms to the curve of the Peck formula.

From the data of numerical simulation, we can also see that the settled law of the whole is that the settlement is the largest at the central axis of the tunnel, and it is symmetrically distributed along both sides of the tunnel axis. When the horizontal clear distance between the points in the soil and the central axis of the tunnel is defined as Δx, $\Delta x \geq 3D$ (D is the diameter of the tunnel), the curve change begins to ease, and the influence on the settlement of the soil is already very small, which can be regarded as a slight influence area.

3.2 *Longitudinal surface displacement*

The longitudinal displacement of the ground surface is the change of surface settlement along the tunnel axis during the shield moving forward. As can be seen from the previous analysis, the longitudinal displacement of the ground surface during shield excavation is

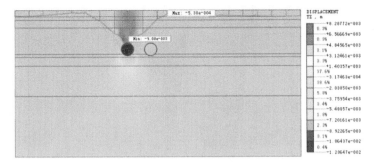

Figure 3. Surface displacement after a tunnel runs through.

divided into five parts. (1) Initial settlement. (2) Deformation when the excavation face arrives. (3) Ground deformation when shield passes through. (4) Settlement of the gap at the shield tail. (5) Late consolidation and settlement of strata. Figure 4 is a schematic diagram of the influence of shield excavation on stratum longitudinal settlement and segment settlement. It can be seen from the figure that when the left tunnel was initially excavated, it had no influence on the pile foundation, and it was 25 m away from the pile foundation at this time.

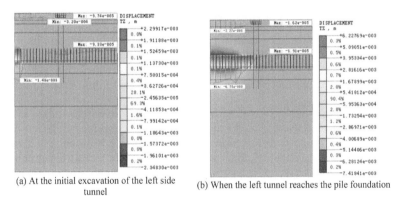

(a) At the initial excavation of the left side tunnel

(b) When the left tunnel reaches the pile foundation

Figure 4. Influence of shield excavation on longitudinal settlement of stratum and segment settlement.

When the shield reaches the pile foundation, it can be found that there is a sudden change of settlement at the junction of the shield shell and segment though the maximum settlement occurs at the leftmost side of the model. The maximum settlement increases from 6.75 mm to 6.87 mm when the shield tunneling passes through the pile foundation. When the first tunnel leaves the pile foundation for 6 segments, the deformation of the whole soil tends to be stable, and the land settlement is about 10 mm after the construction of the left tunnel is completed.

4 ANALYSIS OF STRESS AND DEFORMATION LAW OF PILE FOUNDATION

Figure 5 is a bottom view of the positional relationship between the shield segment and the pile foundation. It can be seen from the figure that the left tunnel passes through the left pile side, while the right tunnel passes directly under the viaduct. The disturbance to the pile foundation of these two tunnels is not the same when passing through SW0.

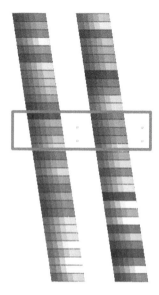

Figure 5. Bottom view of position relationship between shield segment and pile foundation.

The maximum axial force of the pile foundation is 32.9 kN and the minimum value is 76.25 kN when the left tunnel passes through the pile foundation. The maximum and minimum values here reflect the axial force's direction, and the minimum value's absolute size is still very large. When the left shield passes through the pile foundation, the maximum axial force increases by 5.7%, and the minimum axial force decreases by 5.2%.

When the right shield arrives, the grouting pressure and jacking force of the shield have a certain pulling effect on the soil. The maximum axial force continues to decrease, while the minimum axial force (in the same direction as the jacking force of the shield) increases a lot, from 76 kN to 152 kN. Therefore, the influence of the right tunnel construction on the existing pile foundation is much greater than that of the left tunnel.

Figure 6 shows the change of bending moment of the pile foundation when the shield arrives and leaves. It can be seen from the figure that the right tunnel passes through the pile foundation and has a much greater influence on the pile foundation than the left tunnel. The maximum bending moment before the arrival of the left shield is below 0.2 kN·m, while the bending moment when the right tunnel crosses directly becomes 2.8 kN·m, which is more than 14 times.

Figures 7 and 8 show the deformation patterns and vertical displacements of four SW0 piles after the completion of tunnel construction. The simulation results show that the horizontal displacement of the pile foundation after the tunnel passes through the pile foundation exceeds the limit value. Although the left tunnel does not exceed the limit when passing through the pile foundation, the right tunnel will affect it when passing through the pile foundation. Although it can be seen from the Mises stress of SW0 four piles after the completion of tunnel construction in Figure 9 that the maximum stress is 140 kPa, that is, 0.14 MPa, which does not exceed the strength of concrete, it is still suggested that the subway construction unit must reduce the jacking pressure and control the grouting pressure and grouting amount to ensure the safety of the pile foundation.

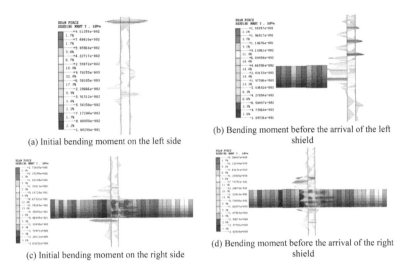

(a) Initial bending moment on the left side

(b) Bending moment before the arrival of the left shield

(c) Initial bending moment on the right side

(d) Bending moment before the arrival of the right shield

Figure 6. Bending moment of pile foundation when shield arrives and leaves.

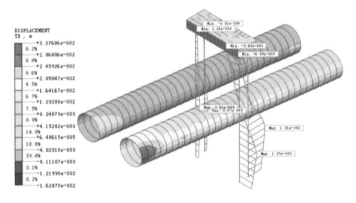

Figure 7. Deformation patterns of SW0 four piles after tunnel construction.

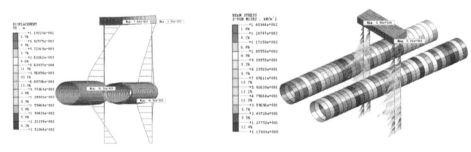

Figure 8. Vertical displacement of SW0 four piles after tunnel construction.

Figure 9. Mises's stress of SW0 four. piles after tunnel construction.

5 CONCLUSIONS

This paper simulates the influence of shield tunneling in stratum on the pile foundation of existing bridges, and the main conclusions are as follows:

1. The state of the two soil layers is mainly a plastic failure state and an unloading / reloading state after the first tunnel is penetrated.
2. The longitudinal displacement of the stratum is the largest at the central axis of tunnel excavation. The stratum begins to settle when the shield moves to the pile foundation. When the first shield leaves the 6-ring segment of the pile foundation and the second tunnel leaves the 8-segment distance of the pile foundation, the stratum tends to be stable.
3. The second tunnel crossing the pile foundation has a greater influence on the stratum than the first tunnel.
4. The horizontal displacement of the pile foundation after the tunnel passes through the pile foundation exceeds the limit value. Although the left tunnel does not exceed the limit when passing through the pile foundation, the right tunnel will affect it when passing through the pile foundation.
5. When passing through the pile foundation in the process of jacking, the subway construction unit must reduce the jacking pressure and control the grouting pressure and grouting amount to ensure the safety of the pile foundation.

REFERENCES

Ding L, Zhang L, Wu X, et al. Safety Management in Tunnel Construction: A Case Study of Wuhan Metro Construction in China[J]. *Safety Science*, 2014, 62: 8–15.

Guo H. A Review of Metro Tunnel Construction Methods[C]//*IOP Conference Series: Earth and Environmental Science*. IOP Publishing, 2019, 218(1): 012110.

Guo Ji-fu, Liu Ying, Yu Liu. Traffic Congestion in a Large Metropolitan Area in China [J]. *Urban Transport of China*, 2011, 3:8–14. (In Chinese)

He S, Lai J, Li Y, et al. Pile Group Response Induced by Adjacent Shield Tunneling in Clay: Scale Model Test and Numerical Simulation[J]. *Tunnelling and Underground Space Technology*, 2021: 104039.

He S, Lai J, Wang L, et al. A Literature Review on Properties and Applications of Grouts for Shield Tunnel[J]. *Construction and Building Materials*, 2020, 239: 117782.

Li Z, Chen Z, Wang L, et al. Numerical Simulation and Analysis of the Pile Underpinning Technology Used in Shield Tunnel Crossings on Bridge Pile Foundations[J]. *Underground Space*, 2021, 6(4): 396–408.

Nematollahi M, Dias D. Three-dimensional Numerical Simulation of Pile-twin Tunnels Interaction–Case of the Shiraz Subway Line[J]. *Tunnelling and Underground Space Technology*, 2019, 86: 75–88.

Wang L, Li C, Qiu J, et al. Treatment and Effect of Loess Metro Tunnel under Surrounding Pressure and Water Immersion Environment[J]. *Geofluids*, 2020, 2020.

Wang Z, Zhang K, Wei G, et al. Field Measurement Analysis of the Influence of Double Shield Tunnel Construction on the Reinforced Bridge[J]. *Tunnelling and Underground Space Technology*, 2018, 81: 252–264.

Xu Q, Zhu H, Ma X, et al. A Case History of Shield Tunnel Crossing through Group Pile Foundation of a Road Bridge with Pile Underpinning Technologies in Shanghai[J]. *Tunnelling and Underground Space Technology*, 2015, 45: 20–33.

Yang C, Huang M, Liu M. Three-dimensional Numerical Analysis of Effect Tunnel Construction on Adjacent Pile Foundation[J]. *Chinese Journal of Rock Mechanics and Engineering*, 2007, 26: 2601–2607.

Research on key technologies of underbalanced drilling

Jianqiu Feng*

PetroChina Coalbed Methane Company Limited, Lvliang Shanxi, China

Xiangbing Pei, Tan Zhang, Ke Jiang & Xiaonan Wang

PetroChina Coalbed Methane Company Limited, Lvliang Shanxi, China

ABSTRACT: Process underbalanced drilling refers to keeping the bottom hole underbalanced during the whole drilling, testing and completion operations, and tripping-in process, so as to achieve the purpose of discovering and protecting reservoirs. However, if it is used improperly in practical application, it will cause huge losses to surface equipment and downhole equipment, and more seriously, it will cause fatal injuries to drilling workers. In this paper, a low-density drilling fluid system was developed, and the formate solid-free drilling fluid system, oil-in-water drilling fluid system, and hollow glass bead composite drilling fluid system were tested, respectively. Combined with the actual needs of the field, the cement slurry system was optimized pertinently. At the same time, this paper discusses the key technologies of underbalanced drilling and analyzes the risk factors in the process of underbalanced drilling, which is beneficial to further popularize and apply the wide application of underbalanced drilling technology.

1 INTRODUCTION

Underbalanced drilling technology began at the end of the 1980s, which catered to the current form of oil and gas exploration and development in the world, and then quickly became popular all over the world. According to the forecast of the US Department of Energy, the application of underbalanced drilling technology will continue to grow steadily in the next 10 years, mainly because the technical characteristics of underbalanced drilling have solved some difficult problems in oil and gas exploration and development at present [1,2]. From the analysis of the application practice of underbalanced drilling technology at home and abroad, it is found that underbalanced drilling technology has many advantages, such as protecting formation, preventing lost circulation, reducing cost, and improving drilling speed, which all show that Qianping drilling has obvious advantages compared with conventional drilling. Facts have proved that underbalanced drilling has low economic cost and good social benefits and has important significance and broad application prospects [3].

China's underbalanced drilling technology has been tested in some areas in the 1960s, such as Da'anzhai in the Moxi structure of Sichuan Oilfield. After that, in the 1990s, underbalanced drilling technology accelerated its development, such as Jiefang 128 well in Tarim Oilfield and the Lungu series underbalanced wells, which made the application of underbalanced drilling gradually popularized. In the 21st century, the application of underbalanced drilling is gradually moving toward a more mature stage. Only at the end of 2009, Shengli Oilfield successfully completed the construction operations of well DP14, well DP19,

---

*Corresponding Author: pxb_cbm@petrochina.com.cn

DOI: 10.1201/9781003450832-29

well S-92, well Caogu 100-19, etc. Among them, the penetration rate of nitrogen drilling is increased by 4~5 times, and the drilling cycle is shortened by more than 30 days. Practice shows that underbalanced drilling technology has been successfully applied in complex geological conditions and achieved good results, which successfully accelerates the popularization and application of underbalanced drilling technology [4].

Although underbalanced drilling technology has been widely used, due to its inherent technical defects, external unpredictable geological conditions, and general management environment with low safety awareness, there are many risks and potential safety hazards in the application process of balanced drilling technology [5]. At present, the knowledge system and theory of underbalanced drilling are not perfect, and there are only a few underbalanced research institutes in China. Most of the drilling crew members mainly drill conventional drilling, and there are few professional underbalanced drilling crews [6]. This paper analyzes the underbalanced drilling technology, finds out the hidden danger factors in the blowout accident caused by underbalanced drilling, and puts forward corresponding methods to minimize the hidden danger factors in the process of underbalanced drilling and effectively guarantee the popularization and application of underbalanced drilling technology.

2 UNDERBALANCED DRILLING FLUID SYSTEM

(1) Formate solid-free drilling fluid

Solid-free drilling fluid, also known as clay-free drilling fluid, is the third generation of drilling fluid developed on the basis of low-solid drilling fluid after clear water and mud. The original composition of the system does not contain clay, and its main components are organic polymers and salts, which can eliminate the formation damage caused by artificial clay minerals. Formate drilling fluid is a new type of solid-free drilling fluid, which has obvious advantages in environmental protection, reservoir protection, formation rheology inhibition, high-temperature resistance, and pollution resistance. Compared with common halogen salt solid-free drilling and completion fluid, formate drilling fluid has superior characteristics.

Formate drilling fluid is a kind of bentonite-free, low-solid organic salt drilling fluid system, which is mainly composed of viscosifier CXV, fluid loss reducer CXF, reservoir protector CXG, auxiliary inhibitor, and formate. According to the needs, polyalcohol, defoamer, water-based lubricant, and other treatment agents can be added. The basic formula is: 1.5%-1.0% CXV + 3-5% CXF + 1-2% auxiliary inhibitor + 5% CXG + formate.

The inhibition of the system is evaluated by a core recovery experiment. The experimental conditions are as follows: adding 50g core into the drilling fluid, rolling at 120°C for 16h, and recovering it with a sieve with a hole diameter of 0.45 mm after leaving the tank; the formate drilling fluid used in the experiment is prepared indoors, and its density is 1.10 g/cm (the following experiments are the same). The recovery of formate drilling fluid is 40.5 g, and the recovery rate is 81%, which can well inhibit the hydration expansion and dispersion of shale.

The experimental core is made of second-class bentonite through a sieve with a pore diameter of 0.154 mm, which is dried at 105°C for 2 hours and then pressed according to the standard. The instrument used is NP-01 shale expansion tester, and the expansion amount is measured for 8 hours. It can be seen from Table 1 that the expansion of formate drilling fluid is small, which indicates that formate drilling fluid has good inhibition.

Table 1. Expansion experiment of drilling fluid.

Project	Expansion quantity	Core expansion reduction rate, %
Distilled water	6.56	/
Formate drilling fluid	2.13	62

Adding soda ash and anhydrous CaCl into formate drilling fluid respectively, the anti-pollution ability experiment of formate drilling fluid is carried out, and the experimental results are shown in Table 2. It can be found that formate drilling fluid has the ability to resist calcium invasion completely, and when soda ash is used to treat calcium invasion, the drilling fluid performance will not be affected.

Table 2.   Experiment on salt pollution resistance of formate drilling fluid.

Formula	φ 600mm	φ 300mm	FL, ml
Original drilling fluid	60	43	5.5
Original drilling fluid + 1.12% soda ash	61	43	5.3
Original drilling fluid + 2.75% CaCl$_2$	61	44	5.4
Original drilling fluid + 2.75% CaCl + 1.12% soda ash	62	44.5	5.3

The core is used to evaluate the permeability recovery value. The instruments are a static core flow tester and a dynamic water loss meter. The core permeability before and after pollution is measured by a core flow tester, and the core is polluted by a dynamic water loss meter. Therefore, the permeability recovery value of formate drilling fluid is over 80%, which has a good effect on protecting oil and gas reservoirs.

(2) Oil-in-water drilling fluid system

With the deepening of oil and gas exploration and development, underbalanced drilling technology in Mesozoic and Paleozoic low-pressure formations has been applied extensively. Oil-in-water drilling fluid is suitable for underbalanced drilling in this kind of formation because of its stable density, convenient construction, high safety, and moderate cost, and has become one of the key technologies of underbalanced drilling in deep wells. Oil-in-water drilling fluid is a solid-free oil-in-water emulsion that disperses a certain amount of oil in freshwater or brine with different salinity and forms a water-continuous phase and oil-dispersed phase. Besides water and oil, its components include water phase viscosifier, fluid loss reducer, and emulsifier. Its density can be adjusted by changing the oil-water ratio and adding different types and different mass fractions of soluble inorganic salts, and the lowest density can reach 0.9 g/cm.

There is no essential difference between low-toxic or non-toxic oil-in-water drilling fluid and diesel-based drilling fluid. The main difference is that the base oil used in preparing drilling fluid is different, and the base oil and other auxiliary agents are required to be low-toxic or non-toxic. Low-toxic or non-toxic oil-based drilling fluids use refined oil (called white oil). The most significant difference between diesel oil and white oil is the content of aromatic hydrocarbons. Diesel oil has high aromatic hydrocarbon content and high biological toxicity, while white oil has low aromatic hydrocarbon content and low toxicity or non-toxicity. The selection of base oil not only considers its toxicity to meet the requirements of environmental protection departments but also considers its flash point and ignition point to meet the requirements of well construction safety.

Emulsifier selection is first based on the HLB value of the emulsifier needed to form oil-in-water drilling fluid and then based on the reduction of oil-water interfacial tension. After these selections, the oil and water were emulsified by the selected emulsifiers, and then the evaluation method of emulsification stability and the distribution of emulsion droplet size were used to evaluate.

The above methods were used to optimize the emulsion of octyl phenol polyoxyethylene ether OP-10, fatty alcohol polyoxyethylene ether AEO3, polyoxyethylene sorbitan fatty acid ester T-60 and compound FPR emulsion (0.4 AEO3 + 0.3 OP-10 + 0.3 T-60), and the FPR emulsion was optimized. FPR emulsion has high surface activity, low surface tension, low critical micelle concentration, large micelle aggregation number, good emulsifying ability, low foaming performance and low toxicity, so it is suitable for preparing oil-in-water drilling fluid systems. Add 1% evaluation agent to 400ml of clear water, and measure the readings of 600r/min and 300r/min with a six-speed viscometer. The experimental results are shown in Table 3.

Table 3. Viscosity increasing effect of four kinds of CMC.

Evaluation agent	φ 600mm	φ 300mm
1 # CMC	14	7
2 # CMC	6	2
3 # CMC	7	3

GJL-2 is a kind of cellulose fluid loss reducer, which has a good fluid loss reduction effect. Moreover, the fluid loss reducer has a certain viscosity-increasing effect, which can increase the external viscosity of oil-in-water drilling fluid and enhance the emulsion stability of oil-in-water drilling fluid.

LV-CMC is sodium carboxymethyl cellulose, and the viscosity of 2% aqueous solution is less than 50mPa · s at 25°C. It is mainly used as a fluid loss reducer for weighting drilling fluid, so as not to cause excessive viscosity.

XC is a kind of biopolymer, which is a light yellow powder. After stirring, it can be completely dissolved in water to form a viscous solution. It can effectively improve the viscosity and shear force of drilling fluid. The characteristics of drilling fluid treated with XC are that it has outstanding rheological properties and shear viscosity reduction characteristics, increases the dynamic-plastic ratio, is beneficial to rock carrying, and improves mechanical efficiency.

The temperature resistance of XC is 100–120°C.

KW-868 is a kind of high-temperature resistant starch, which can effectively reduce the filtration of drilling fluid. The temperature resistance is higher than 120°C, and it has a certain viscosity-increasing effect after adding an oil-in-water system.

The optimum formula of oil-in-water drilling fluid system is 200g water + 0.1% 2 # CMC + 0.15% XC + 1% KW-868 + 2% GJL-2 + 200g 5 # white oil + 7.5% emulsifier.

(3) Hollow glass bead composite drilling fluid system

The density of drilling fluid is usually less than 1.0 g/cm when drilling in low-pressure and depleted oil and gas reservoirs or underbalanced drilling. At present, most drilling fluids with a density between 0.8 g/cm and 1.0 g/cm are oil-based drilling fluids, while almost all drilling fluids with a density below 0.8 g/cm contain nitrogen, air, or natural gas, which often leads to following problems when in use: 1) it increases the cost of drilling equipment; 2) it is difficult or even impossible to apply conventional MWD; 3) it shows serious corrosion of drilling tools; 4) it is easy to cause underground fire or explosion; 5) the hydraulic calculation is difficult, which will affect the displacement and rock carrying effect, and will affect the hydraulic performance of downhole power drilling tools to a certain extent; 6) Generally, the friction coefficient is higher than that of conventional water-based or oil-based drilling fluids. Therefore, it is necessary to develop low-density drilling fluids without gas to meet the requirements of low-pressure reservoir drilling and underbalanced drilling.

Back in the early 1960s, the Soviet Union tried to use hollow glass beads (HGS) as density lighter in the drilling fluid to prevent leakage. After the 1990s, the US Department of Energy and some foreign oil companies began to use HGS and other density-reducing materials to prepare low-density drilling fluids and low-density cement slurries and made field applications. In recent years, glass bead low-density water-based or oil-based drilling fluids have been successfully applied in many low-pressure oil and gas wells and underbalanced wells in the United States and Russia. The application results show that the hollow glass microsphere drilling fluid has the characteristics of low density, good dispersion, good lubrication drag reduction and sand cleaning effect in horizontal section, easy performance control, high penetration rate, and safe preparation. The adjustment method is similar to that of conventional drilling fluid, and it is easy to maintain and handle on-site.

HGS are monocyte base lime boron silicates, which are white powder in appearance, similar to calcium soda salts of silicate, insoluble in water and oil, incompressible, and low alkaline. At present, it is the lightest lightning agent and wellbore lubricant with the highest strength and lightest density in the world. HGS are chemically inert, resistant to high temperature and high pressure, insoluble in solutions except for hydrofluoric acid, pH value is 9 $\sim$ 10, volatility is not more than 0.5% of the total mass, and the bulk density/true particle density of flexible packaging begins to be 55 $\sim$ 70% at 600°C. Field application results show that it can effectively reduce the density of cement slurry and prevent formation leakage.

Through laboratory research, it is necessary to meet the requirements of underbalanced drilling. The density of hollow glass microspheres should be controlled at 0.3–0.4 g/cm as the lightening agent of base paste with a density of 1.0–1.1 g/cm. For different products, the lower the product density, the less the amount of drilling fluid used when reducing to the same density, but the lower its compressive strength. Considering all kinds of factors, we finally adopt HGS5000, a high-strength product made in America. Its average compressive strength is 35MPa and its average density is 0.38 g/cm. After testing, the particle size distribution ranges from 0.5 to 125 μ m, and 90% of the glass beads have a particle size of 8 to 85 μ m. The rupture compressive strength tests of HGS5000 at room temperature and 120°C at 30MPa were carried out. The compressibility coefficients were 7.05 $\times$ 10MPa and 5.88 $\times$ 10MPa, respectively, which indicated that the hollow glass microspheres were incompressible. At present, among the hollow glass microsphere series products, HGS18000 has the strongest compressive strength, and its compressive strength reaches 128MPa, so it can be used in deep well drilling and even ultra-deep well drilling.

Because of the large number of HGS and the relatively high cost of slurry preparation, whether it can be recycled and its wear resistance has become an important factor restricting the popularization and application of glass bead drilling fluid. Therefore, it is necessary to ensure that the beads are not damaged as much as possible under the circulating pressure of the drilling pump and solid control system. After completion, the cost of hollow glass bead drilling fluid can be roughly equal to that of conventional water-based drilling fluid only when the recovery rate of the bead is 50 $\sim$ 100%.

Because drilling fluid belongs to non-Newtonian fluid, the settling velocity of the solid phase is mainly affected by fluid rheology. Therefore, the commonly used separation and recovery method for water-based drilling fluid with HGS is to dilute it with water to transform it into a Newtonian fluid. The water-based drilling fluid containing 3% bentonite, 35% HGS, and 2% sand was diluted with water (the dilution ratio is 20 $\sim$ 100%) in a laboratory, and the recovery rate of HGS was measured. The experimental results show that the recovery of HGS is only 40% after 15 minutes of resting when the dilution ratio is 20%; when the dilution ratio is 100%, the recovery rate of HGS can reach 70% after 15 min. Therefore, the dilution ratio directly affects the recovery effect of HGS.

3   KEY TECHNOLOGY OF DRILLING TECHNOLOGY

1. Shaft wall collapse

The wellbore pressure of underbalanced drilling is much smaller than that of conventional drilling fluid. Low wellbore pressure may cause mechanical instability of the wellbore, especially in soft formation drilling. In addition, if the pore pressure of shale is abnormal, ion hydration and drilling fluid systems are not suitable for formation characteristics, and it is easy to collapse early.

2. Underground fire and explosion

   Gas phase underbalanced drilling is easy to produce a downhole explosion and ignition. Air and nature contain a lot of combustible gas, especially when the downhole pressure is too high, it is easy to produce an explosion.

3. Corrosion technology of drilling tools

   When air drilling is used in underbalanced operations, it is easy to be corroded by drilling tools due to the formation of water, including oxygen corrosion, electrochemical corrosion, and bacterial corrosion. If the gas produced in the formation contains HS, the corrosion problem will be more serious.

4. Blowout

   When the pressure balance between the formation and wellbore circulation system is destroyed in underbalanced drilling, it will easily lead to a blowout. Blowout will not only lead to casualties and equipment losses but also make precious oil and gas suffer losses.

5. Sticking drill

   In underbalanced drilling, the oil, gas, water, etc. in the formation will flow into the wellbore from the formation and come into contact with the circulating drilling fluid, which may result in a viscosity increase of drilling fluid. Second, if emulsified drilling fluid is used, the crude oil from the reservoir may produce stable flowers with high viscosity after invading the wellbore. The viscosity increase of drilling fluid and the emulsion produced will lead to the accumulation of drilling cuttings and sticking.

4 DRILLING WELL-CONTROL TECHNOLOGY

1. Well-control technology

   Usually, when using underbalanced drilling technology, when drilling oil and gas reservoirs, we can judge whether drilling tools reach reservoirs by virtue of their "venting" status, which is a kind of early prompt; to start the calculation according to this time, it is necessary to calculate the time when the gas escapes from the wellbore, which is the time to deal with the emergency procedures in advance. In order to achieve the purpose of "drilling while blowout" without runaway blowout, it is necessary to obtain the accurate relationship between bottom hole pressure difference and formation fluid velocity reaching the wellbore, which needs to be calculated by gas equation and reservoir seepage theory; in order to adjust the velocity of formation fluid, the bottom hole negative pressure can be controlled and adjusted by relevant special equipment, such as liquid-gas separator. Because of the high risk of underbalanced drilling operations, it involves many factors such as safety, environment, and engineering. Therefore, it is very important to control the bottom hole negative pressure in order to prevent the blowout from getting out of control. Therefore, corresponding well-control measures must be taken.

2. Well-killing technology in underbalanced drilling

   In the early stage of determining the underbalanced drilling fluid system, the related well-killing problems have been considered. Because of the complexity of the well control technology, the well-killing fluid system needed in actual operation has also been studied and demonstrated, and the well-killing fluid system matched with the underbalanced drilling fluid system has been developed.

   Underbalanced well killing is special; it is fundamentally different from conventional drilling and well killing. During conventional drilling, after killing the well, killing fluid can be recycled as conventional drilling fluid. However, the well-killing fluid after underbalanced well killing can not be used continuously in drilling operations, so it must be replaced with the low-density drilling fluid to continue drilling. Therefore, the problem to be solved is how to switch between them quickly, which needs to be realized by a corresponding system.

5 CONCLUSION

1. The high-strength hollow glass microsphere drilling fluid is incompressible as a continuous medium, which avoids many shortcomings of gas drilling fluid in use. Its density can be adjusted in the range of 0.6–1.0 g/cm, and it is suitable for low-pressure reservoir drilling and underbalanced drilling.
2. Generally, the comprehensive drilling cost of glass bead drilling fluid is lower than that of air or compressed ammonia drilling. If no leakage occurs, the recovery rate of glass beads is 70 ~ 95%. With the popularization and recycling of glass beads, the cost of slurry preparation is expected to be greatly reduced.

ACKNOWLEDGEMENTS

This work was not supported by any funds. The authors would like to show sincere thanks to those techniques who have contributed to this research.

REFERENCES

[1] Yilmaz. O, Okka. M.A, Effect of Single and Multi-channel Electrodes Application on EDM Fast Hole Drilling Performance, 51st ed., vol. 1–4. International Journal of Advanced Manufacturing Technology, 2010, pp. 185–194.

[2] Pedersen. T, Godhavn. J.M, Schubert. J., *Supervisory Control for Underbalanced Drilling Operations*, 48th ed., vol. 6. IFAC PapersOnLine, 2015, pp,120–127.

[3] Jiang. J, Rabbi. F, Wang. Q, Underbalanced Drilling Technology with Air Injection Connector, 4th ed., vol. 3. *Journal of Petroleum Exploration and Production Technology*, 2013, pp. 275–280.

[4] Haba. B, Morishige. Y, Novel Drilling Technique in Polyimide using Visible Laser, 66th ed., vol. 26. *Applied Physics Letters*, 1995, pp. 3591–3593.

[5] Liu. Y.G., Effect of Annular Cuttings on Negative Pressure in Under-balanced Drilling Process, 97th ed., vol. 5–8. *Acta Petrolei Sinica*, 2005, pp. 96–99.

[6] Wu. P, Meng. Y, Li. G, Chen. Y, Wei. N., Development and Application of a Monitoring-while-drilling System for Underbalanced Drilling, 31st ed., vol. 5. *Natural Gas Industry*, 2011, pp. 77–79.

Study on optimal and fast drilling technology of extended reach wells

Hongwei Liu*, Peng Wei & Mengyi Liu
No.4 Drilling Engineering Company, BHDC, Cangzhou, China

ABSTRACT: After years of technical research and development, the drilling technology of extended-reach directional wells in China has also made great progress and has preliminarily possessed the technical ability to carry out directional wells, highly deviated wells, and horizontal wells with horizontal displacement above 3000 m. To further improve the drilling efficiency, shorten the drilling cycle, and provide technical support for the high-quality, efficient, and rapid development of horizontal wells. In this paper, the reservoir formation characteristics in the study area are analyzed, and the key technologies of extended-reach drilling are discussed. Using a cationic polymer drilling fluid system and combining it with the actual situation of field drilling, the field drilling fluid system formula is formed. This system has the advantages of low filtration, good inhibition, strong ability to control formation slurry production, no spalling phenomenon on wellbore wall, good reservoir protection effect, stable performance, and easy maintenance. The suspension sand carrying capacity is strong, and the tripping and tripping are smooth, which meets the requirements of wellbore purification in extended-reach wells. At the same time, the technical measures of field wellbore purification are formulated, which makes the field wellbore purification easy to operate and implement, and the cuttings of wells return normally without the phenomenon of drilling tool blocking, which proves that the technical measures of well bore purification can effectively inhibit the generation of cutting bed, ensure the safety of drilling process, and thus improve the penetration rate.

1 PREFACE

With the development of science and technology, and the continuous improvement of drilling and completion technology for highly deviated wells and horizontal wells, a number of advanced tools and instruments have been developed, making them an important technology with revolutionary significance in exploration and development. Displacement well drilling technology comprehensively embodies the frontier technology of international drilling technology at present. In recent years, the drilling technology of extended reach wells has been further developed, which is mainly reflected in the application of Log While Drilling (LWD), Steerable Rotary Drilling System(SRD), Pressure While Drilling(PWD), and other high-tech technologies in extended reach wells; multi-objective extended reach wells appeared. Drilling of super-large reach wells, etc. With the development of extended reach well technology, the emphasis of extended reach well research gradually focuses on the following aspects: trajectory design, directional control, hydraulics and wellbore purification, and reducing friction and resistance [1].

---

*Corresponding Author: liuhw@cnpc.com.cn

DOI: 10.1201/9781003450832-30

After years of technical research and development, the drilling technology of extended-reach directional wells in China has also made great progress and has preliminarily possessed the technical ability to carry out directional wells, highly deviated wells and horizontal wells with horizontal displacement above 3000 m. In recent years, domestic technicians have researched and developed special tools for extended-reach wells and made great progress in the development of antifriction and torsion reduction, trajectory control, and casing running tools. The drilling technology of extended-reach wells implemented in Nanhai Oilfield and Bohai Oilfield in China is relatively leading. The reason why the excellent and fast drilling technology of horizontal wells can be widely used in China is mainly due to its high input-output ratio. Compared with the traditional vertical well production method, it has the advantages of delaying bottom water coning, increasing the oil drainage area of oil and gas wells, improving oil and gas well production and recovery ratio, increasing recoverable reserves, etc [2]. Especially, it can realize efficient development of difficult-to-produce reserves.

With the development of domestic horizontal well technology, it is very important to develop various types of reservoirs efficiently and quickly by using horizontal well technology, and some supporting technologies are gradually applied, mainly including optimal design of horizontal well bore structure; optimizing high-performance bit and roller bit; PDC bit + screw compound drilling; horizontal well trajectory control technology. For extended reach wells, efficient sand carrying, reducing the formation of cutting beds, reducing friction and torque, and lubricity of drilling fluid are the problems that must be solved in drilling fluid technology of this kind of wells [3]. According to the characteristics of the oilfield area and in order to meet the requirements of the environmental protection index, it is the main content of drilling fluid system optimization research to optimize the appropriate drilling fluid system and study the drilling fluid formula according to the formation and field conditions. In the construction of extended-reach wells, there are some complicated situations caused by the fact that cuttings can not be carried out from the wellbore in time. If cuttings are not cleaned in time, a stable cutting bed will be formed, resulting in repeated grinding of cuttings, resulting in a series of problems, such as increasing the solid content of drilling fluid and causing drilling sticking accidents [4]. Combined with the site conditions of the selected block, suitable, excellent, and fast drilling technical measures are formulated, and a set of feasible drilling technical measures are formed through research and field application.

2 STRATIGRAPHIC CHARACTERISTICS OF THE STUDY AREA

The reservoir in the study area has the characteristics of low permeability, multi-layer, heterogeneity, and fracture development. The distribution of oil, gas, and water is controlled by stratum structure and stratum lithology, which is complex. Sandstone thickness is small, and a single well generally hits 3–5 layers of oil-bearing horizon. Complex reservoir conditions cause the following difficulties in the operation of extended reach wells: the thickness of sandstone in oil layers is thin, the average thickness is about 1.2 m, and the thickness less than 1.0 m accounts for more than 30%. The vertical depth of actual wellbore trajectory is often limited within 0.5 m, so it is difficult to control wellbore trajectory; the target horizon contains multi-layer formation, and the target points can reach more than a dozen, with large sag and more landing times; the well section with large deviation angle is relatively long, and the curvature of wellbore trajectory often changes, so it is difficult to transfer WOB downhole, which increases the risk factor.

Stratigraphic lithology is described as follows:

Quaternary system: The uppermost part is gray-black humus soil, the lower part is gray-yellow silty clay and silty flowing sand layer, and the bottom part is a variegated gravel layer.

Mingshui Formation of Upper Cretaceous: The second member of the Ming Dynasty is composed of purplish red, green-gray mudstone and variegated glutenite with different thicknesses, while the first member of the Ming Dynasty is composed of gray mudstone and variegated glutenite with different thicknesses.

Sifangtai Formation of Upper Cretaceous: Purple-red mudstone and gray siltstone are interbedded with unequal thickness, and they are in unconformity contact with underlying strata.

Nenjiang Formation of Lower Cretaceous: The upper part is interbedded with green-gray, gray mudstone, gray argillaceous siltstone, and siltstone of varying thicknesses; the middle part is interbedded with gray, dark gray mudstone and gray siltstone of varying thicknesses; and the lower part is a large section of black-gray, gray-black mudstone, and ostracod-bearing mudstone with black-brown oil shale.

Yaojia Formation: The middle and upper parts are dark gray and black gray mudstone with thin layers of gray argillaceous siltstone, while the lower parts are interbedded with green-gray mudstone and gray siltstone of varying thicknesses. It is in false integration contact with the underlying strata.

Qingshankou Formation (not drilled): A large section of dark gray and black-gray mudstone.

3 ANALYSIS OF DRILLING TECHNOLOGY

(1) Drilling fluid technology

As the blood of drilling engineering, the importance of drilling fluid can be imagined. For extended-reach wells, many key factors are closely related to the drilling fluid system, which is an important content of extended-reach well drilling technology. At present, the research on drilling fluid technology for extended reach wells mainly focuses on three aspects: reducing friction and torque, reducing the formation of cutting bed, and maintaining wellbore stability. In the drilling process of extended reach wells, torque increase and sticking account for a large proportion, which can be caused by cutting bed, wellbore instability, keyway, and excessive drilling pressure difference. Therefore, it is necessary to consider the problem comprehensively to reduce friction and torque.

At present, the drilling fluids used in extended-reach wells are mainly oil-based drilling fluids, water-based drilling fluids, and new drilling fluids. Different types of drilling fluids should be selected according to the field conditions in the specific drilling process. In order to effectively reduce friction and torque, oil-based drilling fluids with excellent lubricating performance are almost used in ultra-long distance extended reach wells. At present, oil-based drilling fluids are used in most extended-reach wells in the world. However, due to the improvement of environmental protection requirements and the high cost of oil-based drilling fluids in recent years, the use of oil-based drilling fluids in the past should be minimized. If we want to meet the requirements of environmental protection, reduce the drilling cost and make the drilling site easy to operate, water-based drilling fluid should be considered as the drilling fluid for extended-reach wells under the conditions permitted by geological conditions. In recent years, water-based drilling fluids mainly include polyalcohol drilling fluids, KCl/polymer drilling fluids, silicon-based anti-sloughing drilling fluids, positive gel drilling fluids, and other systems.

During drilling, the drilling fluid has friction between the drill string and wellbore. If you want to reduce friction and torque, you must avoid these frictions as much as possible. The application of lubricant should be able to effectively avoid the above friction, so as to reduce friction and torque. For oil-based drilling fluid, the lubricant has little effect on its performance, mainly because the oil-water ratio can influence its performance. Oil-based drilling fluid with a high oil-water ratio can reduce the friction between interfaces by nearly 50%, but lubricant has little effect on it. Therefore, increasing the oil-water ratio can obviously improve the lubricity of oil-based drilling fluid. Inert solid lubricants such as graphite and plastic pellets are mixed into drilling fluid, which can effectively change the friction state between drill string, casing and wellbore, and greatly improve the lubricating performance of drilling fluid. Its principle is to effectively reduce the friction coefficient between them. At present, developing more efficient lubricants has become one of the main methods to improve the lubricating performance of water-based drilling fluids in the world.

The rheological performance of drilling fluid is an important factor affecting effective rock carrying in extended-reach wells, and its key is to maintain wellbore cleanliness. The friction coefficient of drilling fluid mainly includes solid content, viscosity, and flow rate of drilling fluid, Therefore, it is necessary to reduce the clay content in drilling fluid as much as possible, make full use of solid control equipment, remove harmful solid phase in drilling fluid, optimize the type of fluid loss reducer, adjust the wall-building property of water loss and the particle size distribution of solid phase particles, improve the quality of mud cake, and reduce the thickness of mud cake. Reasonable use of a flow pattern regulator can reduce viscosity, so as to ensure the normal lubrication performance of drilling fluid.

(2) Cleaning of cuttings

The cleaning effect of cuttings directly affects the torque and friction in the drilling process. For the extended reach, high deviation, and horizontal well section, the cuttings can be carried out of the well efficiently and timely, which reduces the generation of cutting bed and makes the wellbore fully purified, which determines whether the extended reach well can be successfully completed. Horizontal pipeline equipment simulation test shows that oil-based drilling fluid and water-based drilling fluid have the same cuttings carrying capacity when their rheological properties are similar, so their rheological properties are an important factor of drilling fluid's wellbore cleaning capacity when other conditions are the same, and have nothing to do with the basic composition of drilling fluid.

When the annulus return velocity is constant, the drilling fluid flow pattern and optimization of drilling fluid rheological parameters become the key to wellbore purification. Due to the limitation of field equipment and the large depth and deviation angle of extended reach wells, it is still difficult to apply turbulence which is easy to carry rock from the perspective of an annular drilling fluid flow pattern. When the flow pattern of drilling fluid is laminar flow, it is necessary to adjust its rheological properties, reduce the occurrence of flat laminar flow and make peak laminar flow dominant, so as to improve the rock carrying capacity of drilling fluid. When the dynamic plastic ratio is high, the pump pressure will increase obviously because of the increase of dynamic shear force; however, when the dynamic-plastic ratio is too low, the peak laminar flow will appear again. Therefore, under appropriate conditions, a high dynamic-plastic ratio will make the drilling fluid system form the required flat laminar flow, which is beneficial to efficient borehole purification. At the same time, whether the flow pattern of drilling fluid is turbulent or laminar, increasing the viscosity of the drilling fluid system will be beneficial to carry cuttings. The vertical component of annulus return velocity decreases with the increase of inclination angle in the large deviated and horizontal interval, while the vertical component is zero in the horizontal interval, and the cuttings carrying capacity is greatly reduced, which leads to the bottom subsidence of cuttings well and cutting bed,; therefore, it is necessary to optimize the appropriate minimum annulus return velocity as much as possible.

Adjusting the rheological property of drilling fluid can obviously improve the rock-carrying capacity of drilling fluid, but under the existing conditions, it is technically impossible to prevent cuttings from sinking in drilling fluid. Therefore, other methods must be developed to clean the wellbore efficiently. At present, a non-traditional drilling fluid-weak gel drilling fluid has been studied at home and abroad, which can form colloids at lower temperatures in a short time. It also has a high dynamic-plastic ratio, low shear viscosity, good thixotropy, and suspension ability. Although foam drilling fluid is used in horizontal wells extended reach wells and other special technology wells there is little research on foam fluid-carrying rock which needs further experiment and practice.

(3) Wellbore stability

As for wellbore stability, there are obvious differences between extended reach wells and vertical wells, which are not only affected by the stress state of surrounding rock and rock strength of the formation but also directly affected by many factors such as in-situ stress orientation and wellbore trajectory. If drilling fluid is mainly considered, because shale is easy to collapse and sandstone is stable, drilling fluid should mainly stabilize the wellbore of shale section to ensure downhole safety. Because some water phase of drilling fluid will enter the formation, the drilling fluid column and formation pressure will be rearranged, and the wellbore instability will occur. In actual construction, the wellbore instability of extended reach wells is usually in the upper wellbore of the deviated section, and the deviation angle greatly affects the formation fracture pressure gradient, and the larger the deviation angle, the smaller the fracture pressure gradient. Generally speaking, the inclination angle has a great influence on wellbore stability, and the larger the inclination angle, the smaller the safe density window of drilling fluid. To prevent wellbore instability in extended reach wells, the following aspects should be taken into account: optimizing wellbore trajectory, optimizing drilling fluid system suitable for field conditions, appropriate density value, and appropriate drilling fluid filtration. From the field, if the wellbore instability occurs, it will inevitably cause the friction between drilling tools and the wellbore to increase and even cause drilling sticking. Furthermore, when the wellbore is unstable, it may lead to formation leakage at the same time, which will affect safety. Once the wellbore collapse occurs, the consequences are more serious and the accident treatment is more complicated than conventional wells. Therefore, it is a comprehensive project to ensure the stability of the wellbore in order to make the drilling work smoothly.

(4) Wellbore cleaning technology

During drilling, there may be a sedimentary moving layer, stable sedimentary layer, pseudo-homogeneous flow layer, and heterogeneous suspension flow layer in annulus separation at the same time, and the rheological property and flow pattern of drilling fluid are very complex, and the drilling fluid suspension is uneven. Cuttings in the lower borehole wall of the section with a large deviation angle will sink when the pump is stopped, which will lead to the sand bridge sticking. It is easy to form a cutting bed in deviated well section, and the greater the deviation angle, the greater its thickness. With the increase in inclination angle, the drilling tool will be close to the lower borehole wall, and the annular gap at the lower part of the drilling tool will become smaller, which makes it more difficult to remove the cutting bed and easily causes complex situations. Therefore, it is necessary to study the wellbore purification technology.

Compared with vertical wells, the downhole conditions of extended-reach wells are more complicated. Besides the stable deviation angle and the well trajectory conforming to the design, gravity, friction torque, and cutting bed will cause a series of other problems. In the process of extended reach wells, the cuttings will migrate along the axial direction of the wellbore and deposit in the lower part of the well when the deviation angle exceeds 30 degrees due to the axial component of gravity. If the cutting bed is to be eliminated, it is required that the drilling fluid has good suspension ability and strong rock-carrying ability, so it is difficult to form a cutting bed in a short time; on the contrary, it is easy to produce cutting bed in a short time. Practice shows that when the deviation angle is between 30 and 60, it is easy to produce a stable cutting bed. When the sediment is too thick, the debris in the annulus slides down as a whole, resulting in sand settling and drilling sticking. Wellbore purification is an important influencing factor in extended-reach well drilling. Deep well depth, large horizontal displacement, and large deviation angle make wellbore purification more important in extended-reach wells. Unclean wellbore cleaning will lead to the generation of cutting beds. Cutting bed is a common factor leading to underground accidents and complexity.

4 APPLICATION OF DRILLING TECHNOLOGY

(1) Drilling fluid formula

Through the experiments of inhibition, anti-pollution ability, and lubricity, the formulation of the system was determined as follows: bentonite (5%) + soda ash (0.3%) + HX-D (0.5%) + NW-3 (0.5%) + NPAN (1.3%) + YD-II (1.1%) + SMP (0.5%) + FT-80 (0.5%) + SF-260 (1.2%) + KOH (0.05%) + DRH3D (5%–7%) + residual oil (10%–15%) + HY205 (1.2%) + graphite (1.2%) + XC (according to the condition of sand return).

The newly developed drilling fluid is used in W2 and W3 wells. Compared with the W1 well using the original drilling fluid, the average penetration rate of the first opening is increased by 16.2 m/h, the average penetration rate of the second opening is increased by 1.6 m/h, and the average penetration rate of the whole well is increased by 21.6 m/h, as shown in Table 1.

Table 1. Comparison of penetration rate.

Well number	Average penetration rate, m/h		
	As soon as it opens	Second opening	Triple opening
W1	41.9	14.5	18.9
W2	24.6	13.2	15.3
W3	26.8	12.6	14.9

(2) Wellbore cleaning of extended-reach wells

According to the study block, the drilling displacement is increased to improve the circulation speed of drilling fluid and accelerate the discharge of cuttings in the annulus. Based on following the design displacement and according to the drilled data, the hydraulic parameters are properly adjusted, mainly increasing the displacement, and certain results are achieved. See Table 2 for hydraulic parameters. For the study block, the main indexes of the deflection section are to ensure the balance of drilling fluid density, low fluid loss, strong lubricity, and strong inhibition. While the drilling fluid with high viscosity is used for a short time in the highly deviated section, when the deviation angle is greater than 50, the drilling fluid with high viscosity will weaken the wellbore purification effect, and the viscosity needs to be adjusted in time. Therefore, when adjusting the rheological properties of drilling fluid, it is necessary to comprehensively consider various parameters such as rock carrying, viscosity, and thixotropy, so as to avoid adjusting a single parameter. Therefore, under the premise of the reasonable control of drilling fluid rheological properties, strictly monitor the rheological properties of drilling fluid, and make timely adjustments according to the needs of the site. When the drilling tool penetration rate is high, it can agitate the cuttings, makes the cuttings re-enter the drilling fluid, destroys the cutting bed, and at the same time, facilitates the cuttings carrying out. However, high speed will harm drilling rigs and directional equipment, so the speed must be combined with the existing site conditions. At the same time, during the tripping of extended-reach wells, all the tripping holes are used. In the process of reverse stroke, the speed must be controlled well, and the optimal speed must be determined by combining other parameters.

Table 2. Hydraulic parameter table.

Parameter	Vertical interval	Oblique section
Well depth, m	*140 ~ 200*	*180 ~ 1900*
Discharge capacity, L/s	*35*	*35 ~ 40*
Pump pressure, MPa	*2 ~ 3*	*5-15*
Circumferential return speed, m/s	*1.3*	*1.3*

5 CONCLUSION

1. In order to ensure the well trajectory changes evenly and smoothly, the drilling process must be strictly monitored and controlled, and sharp bends must be eliminated as much as possible. Extended-reach wells have higher deviations and more complex trajectories than conventional wells. Controlling the actual trajectory in time during drilling can optimize the trajectory and ensure the smoothness of tripping and running casing.
2. The research results can be easily implemented in the field by cooperating with the specific technical measures of excellent and fast drilling and the installation of excellent and fast drilling equipment. After strictly observing the technical measures, the complicated accident did not occur, while the standardized installation of well control equipment can prevent the timely adoption of pressure control after a well control failure, thus creating conditions for speeding up. The drilling practice of several extended-reach wells proves that these measures are effective, reduce the complicated situations easily occurring in the drilling of extended-reach wells, and have an obvious speed increase effect.

REFERENCES

[1] Yilmaz. O, Okka. M.A, Effect of Single and Multi-channel Electrodes Application on EDM Fast Hole Drilling Performance, 51st ed., vol. 1–4. *International Journal of Advanced Manufacturing Technology*, 2010, pp. 185–194.

[2] Haba. B, Morishige. Y, Novel Drilling Technique in Polyimide Using Visible Laser, 66th ed., vol. 26. *Applied Physics Letters*, 1995, pp. 3591–3593.

[3] Xia. W, Li. Z, Chen. M, Zhang. Y, Zhao. W, Study on Electrode Vibration in the Touch-down Stage of Fast Electrical Discharge Machining Drilling, 109th ed., vol. 5–8. *International Journal of Advanced Manufacturing Technology*, 2020, pp. 1–11.

[4] Wang. Z, Hao. T, Yong. L, Li. C, Dielectric Flushing Optimization of Fast Hole EDM Drilling Based on Debris Status Analysis, 97th ed., vol. 5–8. *International Journal of Advanced Manufacturing Technology*, 2018, pp. 1–9.

Water Conservancy and Civil Construction – Oke & Ahmad (Eds)
© 2024 The Author(s), ISBN: 978-1-032-58618-2

Stability analysis of the foundation pit based on the three-dimensional limit equilibrium method

Haitao Wei
AVIC Institute of Geotechnical Engineering Co., Ltd., Beijing, China

Ning Jia*
CABR Foundation Engineering Co., Ltd., Beijing, China

Bin Jia
Resource & Environment Engineering co., Ltd., Beijing, China

ABSTRACT: Foundation pit engineering is a kind of high-risk and high-difficulty engineering, and it is influenced by many other factors. The global slide is one of the major failures of the foundation pit. Currently, slice-based two-dimensional limit equilibrium models are widely used in evaluating the stability of foundation pits. However, two-dimensional models have limitations: (1) the shear resistance along the two sides of slip mass, end effects, and lateral non-homogeneity are neglected. (2) the spatial distribution of the slip body cannot be considered.

The above-mentioned drawback can be solved by a 3D model. The stability of the foundation pit depends on the complex spatial distribution of terrain, stratum, and groundwater table. This paper studies the three-dimensional stability analysis of the foundation pit. The 3D safety factor and critical slip surface can be obtained. The reliability of the proposed system is verified by one case.

1 INTRODUCTION

1.1 Global stability of foundation pit

Foundation pit engineering is a system engineering with high risk, high difficulty, and many influencing factors. Global stability of the foundation pit refers to the ability of the soil around the foundation pit or the soil to maintain stability together with the envelope system. The purpose of checking the global stability of the foundation pit support system is to prevent the global sliding instability and failure of the foundation pit support structure and surrounding soil. It is a checking content that needs to be often considered in the design of the foundation pit support (Liu 2010; Tang 2006).

The limit equilibrium method is widely used in global stability analysis, and the stability index is a well-known safety factor. The safety factor of slope stability F can be defined as follows:

$$F = \frac{\text{Shear strength of soil}}{\text{Shear strength required when the slope reaches the limit equilibrium}} \tag{1}$$

The slice method based on the limit equilibrium theory is commonly used in the slope stability analysis. The analysis of slope stability by the slice method is statically indeterminate, so various assumptions are generally made for the interforce between slices in the application. For example, some assumptions must give the lateral force and the shape of the

---

*Corresponding Authors: whtsay@163.com, ninjanin@yeah.net and 173048522@qq.com

DOI: 10.1201/9781003450832-31

sliding surface. Despite the above limitations, the limit equilibrium method is still the most widely used analysis method because of its simplicity, fast calculation, and easy implementation. Most limit equilibrium methods are based on one-dimensional, two-dimensional, and three-dimensional models. The difference between the three models lies in the introduction of different assumptions and research objects.

The one-dimensional model (infinite slope model, Figure 1) assumes that the slope extends infinitely in all directions, and the sliding occurs along the plane parallel to the slope. This model is suitable for shallow failure. The one-dimensional model assumes that the slope extends infinitely in all directions, and the sliding occurs along the plane parallel to the slope (Tang 2006), which is suitable for shallow failure.

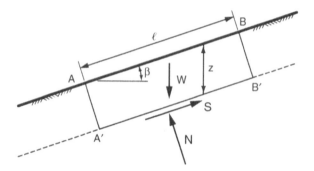

Figure 1. Schematic diagram of the one-dimensional model.

The two-dimensional model is called the slice method (Figure 2). It studies the stability of the main section along the sliding direction. The soil above the sliding surface is divided into many vertical bars. According to different assumptions, the sliding surface can be a circular arc or any shape. The two-dimensional slice method is the most widely used the global stability analysis method in foundation pit engineering. However, in the two-dimensional model, because the sliding body is simplified as an infinite extension in the direction perpendicular to the main section, it can not reflect the spatial distribution of geological structures such as terrain, stratum, and groundwater. Moreover, because the two-dimensional model ignores the shear capacity and end effect between strips, the safety factor of the two-dimensional model is conservative.

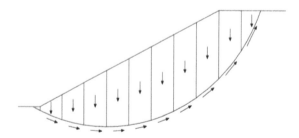

Figure 2. Schematic diagram of the two-dimensional model.

The above shortcomings can be solved by the three-dimensional limit equilibrium method. Three-dimensional models rose in the 1960s (Hovland 1977; Hungr 1987, 1990; Lam 1993; Leshchisky 1992), most of which are based on cylindrical elements (Figure 3). These three-dimensional models solve the problems of one-dimensional and two-dimensional models. Their stability depends on the complex spatial distribution of terrain, stratum, geotechnical parameters, and groundwater. This spatial distribution information is processed in the three-dimensional stability analysis program.

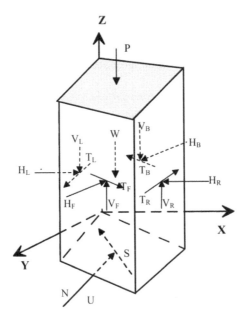

Figure 3. Cylindrical element of 3D model.

2 THREE-DIMENSIONAL SLOPE STABILITY ANALYSIS MODEL

2.1 *Algorithm of the 3D model*

In the three-dimensional model, a sloped body is usually divided into many column elements. Figure 3 shows the stress of a typical column element. Most three-dimensional algorithms satisfy force balance or moment balance, or both. Since the number of equations is less than the number of unknowns, some assumptions are usually introduced. This study applies the following three-dimensional model based on column elements to analyze the three-dimensional stability of the slope. The external forces considered include distributed load, water pressure, anchor force, and seismic load.

2.2 *Extension of the 3D Janbu model*

This model is based on the work of Hungr (1990), which is a simplified Janbu model without considering the three-dimensional expansion of the correction coefficient. This model assumes that there is no shear stress between columns. The advantage of the Janbu model is that it is applicable to almost all slope bodies.

$$N = \frac{P + W + SF_{3D}^{-1} u_w \tan\phi' \sin\theta_{Avr} - SF_{3D}^{-1} c'A \sin\theta_{Avr}}{\cos\theta + SF_{3D}^{-1} \tan\phi' \sin\theta_{Avr}} \quad (2)$$

$$SF_{3D} = \frac{\sum_j \sum_i [c'A + (N - u_w A)\tan\phi']\cos\theta_{Avr} + E}{\sum_j \sum_i [N \sin\theta \cos(Asp - AvrAsp) + kW]} \quad (3)$$

where N = the normal stress; P = the vertical force acting on each column; W = the weight of one column; SF3D = the 3D slope safety factor; ϕ' = the effective friction angle; θAvr = the apparent dip in the main inclination direction of the slip surface; c' = the effective cohesion; θ = the dip (the normal angle of the slip surface); A = the area of the slip surface; j = the row numbers of the grid in the range of slope failure; i = the column numbers of the grid in the range of slope failure; uw = the pore-water pressure acting on the slip surface of each column; E = is the result for all the horizontal components of the applied point loads; Asp = the dip orientation of the grid column slip surface; AvrAsp = the average dip orientation of the slip surface; and k = the horizontal earthquake acceleration factor.

2.3 Search for the most unfavorable sliding surface

In the limit equilibrium method, it is necessary to first generate a sliding surface and calculate the safety factor of the sliding surface. In this study, Monte Carlo random simulation method is used to search the three-dimensional most unfavorable sliding surface. It is assumed that the initial sliding surface is the lower half of the ellipsoid. Each randomly generated sliding surface changes according to different soil strengths and weak surface conditions. In this way, the most unfavorable sliding surface is obtained, so the minimum safety factor corresponding to the most dangerous sliding surface is also obtained.

The most unfavorable sliding surface is realized by trial search and calculation of safety factor. Here, five parameters controlling the shape and size of the most dangerous sliding surface are used as random variables of Monte Carlo simulation: three-axis variables "a, b, c", ellipsoid center c, and ellipsoid inclination θ. These five parameters are uniformly distributed. If the position of the random sliding surface generated based on the bottom of the ellipsoid is lower than the internal edge interface of the weak surface or hard rock, the edge interface of the weak surface or hard rock is considered to be the sliding surface. The sliding surface shown in Figure 4 is part of the bottom surface of the lower half of the ellipsoid.

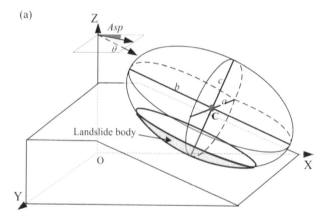

Figure 4. Random search of the sliding surface by Monte Carlo method.

2.4 Technological process

In order to search for the most unfavorable sliding surface, random variables are generated by Monte Carlo simulation, and the corresponding three-dimensional safety factor for each sliding surface can be calculated by a three-dimensional model. In each search, the six parameter values are randomly obtained and a sliding surface is formed. The three-dimensional safety factor corresponding to the sliding surface can be obtained by bringing the sliding surface into the three-dimensional models. Through a sufficient number of simulations, the sliding surface corresponding to the minimum safety factor is the most unfavorable sliding surface of the sliding body.

3 ALGORITHM CASE

A foundation pit support project is located in Changping District, Beijing. The depth of the pit is about 25 m. The stratum soil in the influence range of the foundation pit is described as follows:

The surface layer is a layer of pebble fill, gravel fill ① with a general thickness of 0.70–3.20 M and a layer of silty clay plain fill and clayey silt plain fill①1.

Below the artificial accumulation layer is the Quaternary sedimentary layer: Layer ② is slightly dense to medium dense pebbles and round gravel; Layer ②1 is medium dense fine sand and silty sand; Layer ③ is wet slightly wet medium dense clayey silt and silty clay; Layer ③1 is hard clayey silt sandy silt; Layer ③2 is medium dense pebble and round gravel; Layer ③3 is medium dense fine sand; Layer ④ is medium dense pebble and round gravel; Layer ④1 is hard clayey silt and silty clay; Layers ④2 are dense fine sand; Layer ⑤ is dense pebble.

Table 1. Mechanical parameters of each soil layer.

Layer	Soli	Bulk density (kN/m³)	c'(kPa)	$\varphi'(°)$
①	Filling	20	0	15
①₁	Filling	20	10	10
②	Pebbles	20	0	36
②₁	Fine sand	20	0	28
③	Silty clay	19.6	17	20
③₁	Silty clay	20.7	14	24
③₂	Pebbles	20	0	37
③₃	Fine sand	20	0	30
④	Pebbles	20	0	40
④₁	Silty clay	20.4	20	24
④₂	Fine sand	20	0	32
⑤	Pebbles	20	0	42

The buried depth of groundwater is 11 m below the surface. Layer ③,④,⑤ are aquifers.

On the premise of ensuring the safety of foundation pit excavation, comprehensively analyzing the economy and technology and considering the needs of structure and building use, the foundation pit adopts the support form of pile anchor and three-stage slope. The plan of the foundation pit is shown in Figure 5, and the three-dimensional model of the foundation pit can be shown in Figure 6.

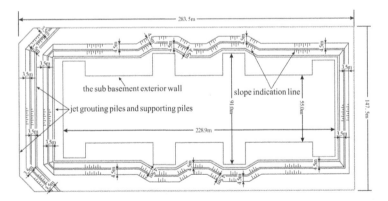

Figure 5. Analysis area.

Using the three three-dimensional stability models and the search method for dangerous sliding surfaces introduced in this paper, the safety factor (Table 2) and the most unfavorable sliding arc under different working conditions can be obtained (Figure 7).

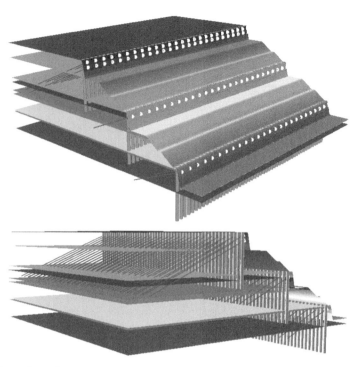

Figure 6. Three-dimensional view of the foundation pit support structure.

Table 2. Safety factors under different working conditions.

Safety factor	Janbu Model normal	Considering seismic
The groundwater drops to 1 m below the base	1.83	1.56
The final water level is 6 m above the base	1.71	1.47

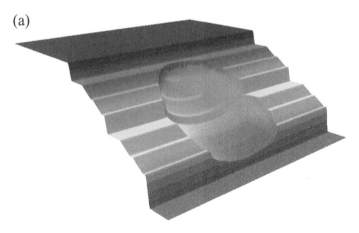

Figure 7. (a) The most unfavorable sliding surface searched in the global working condition; (b) The most unfavorable sliding surface searched in the substep working condition.

Figure 7. (Continued)

4 CONCLUSION

Due to the complex geological and hydrological conditions, the traditional two-dimensional stability analysis method can not accurately evaluate the global stability of the foundation pit. In this study, the three-dimensional stability model is used to analyze the global stability of the foundation pit. The applied system can comprehensively consider the spatial distribution of various geological structures to evaluate the three-dimensional global stability of the foundation pit. The influence of the earthquake effect, various supporting structures (supporting pile, anchor bolt, soil nail), and external load (uniformly distributed force and concentrated force) on the stability of the foundation pit is considered. The reliability of the system is verified by an engineering example.

REFERENCES

Hungr O. An Extension of Bishop's Simplified Method of Slope Stability Analysis to Three Dimensions. *Geotechnique*, 1987, 37(1): 113–117.

Hungr O. Momentum-transfer and Friction in the Debris of Rock Avalanches-discussion, *Canadian Geotechnical Journal* 1990, 27:697–701.

Hovland H J. Three-dimensional Slope Stability Analysis Method. Journal of the Geotechnical Engineering, *Division Proceedings of the American Society of Civil Engineers* 1977, 103(GT9): 971–986.

Liu Guobin, Wang Weidong.(2010). *Excavation Engineering Handbook*. M. China Architecture & Building Press.

Lam L, Fredland D G. A General Limit Equilibrium Model for Three-dimensional Slope Stability Analysis. *Canadian Geotechnical Journal* 1993, 30: 905–919.

Leshchisky D, Huang Ching-Chang. Generalized Three-dimensional Slope Stability Analysis, *Journal of Geotechnical Engineering* 1992, 118(11): 1748–1763.

Tang Hua. *Information Visualization and Intelligent Analysis and Management System To Safety Monitoring In Underground Powerhouse. D. Wuhan Institute of Geotechnical Mechanics*, Chinese Academy of Sciences 2006.

Taylor D. W. *Fundamentals of Soil Mechanics*, Wiley, Hoboken, NJ. 1948.

Water Conservancy and Civil Construction – Oke & Ahmad (Eds)
© 2024 The Author(s), ISBN: 978-1-032-58618-2

A mechanical constitutive damage model for non-penetrating crack rock-like specimen

Cheng Pu*
Corresponding Author PowerChina Northwest Engineering Corporation Limited, Xi'an, Shaanxi, China

Yuetao Li & Jiaxin Wen
PowerChina Northwest Engineering Corporation Limited, Xi'an, Shaanxi, China

ABSTRACT: In this study, rock-like materials are used to simulate the rock mass, sheet metal is used to simulate the non-penetrating crack, and indoor uniaxial compression tests are used to explore the mechanical properties of non-penetrating crack rock. Meanwhile, a constitutive relationship based on damage mechanics is proposed. The experimental results show that the strength of non-penetrating crack specimens is generally influenced by the angle of the crack, and the constitutive relationship proposed in this study can be well adapted to the experimental results.

1 INTRODUCTION

The damage mechanics was first applied to metal materials and first introduced into the geotechnical in 1976 (Dougill 1976), which opened a new field of rock damage mechanics research. Applying the concept of damage, Dragon (1979) put forward an elastic constitutive relationship to reflect the strain-softening behavior of rock and established a continuum damage mechanical model. A comprehensive study on the constitutive equation of rock-like materials was made based on the thermodynamics theories (Krajcinovic 1981) and the microcrack damage theories (Costin 1985; Ortiz 1985). By introducing parameters into the damage mechanics area, the influence of stress transfer was considered (Swoboda 2008) and the rock mass damage model was modified (Kawamoto 1998; Yuan 2013). The damage models of rock mass were established to simulate the triaxial mechanical behavior of granite rock (Golshani 2006; Zhu 2008) and the uniaxial mechanical behavior of saturated rocks under the condition of dry-wet cycles (Shao 1999). Most researchers focused on the area of the anisotropic rock mass. An anisotropic damage constitutive model was constructed based on the directional tensor invariants (Shao 1999). Using anisotropy indices $k\beta$ and $\alpha\beta$, Saroglou (2008) and Shi (2016) modified the H-B criterion for describing the triaxial strength anisotropy of layered rock mass, respectively. Regarding sandstone and Slate as particle aggregate, the anisotropy strength characteristics of layered rock mass were described by combining the local strength criterion and the micromechanical formula of contact deformation (Pouragha 2020). Using the uniaxial compress tests, triaxial compress tests, and Brazilian splitting tests, the elastic parameters and strength characteristics of Slate in different bedding directions were evaluated by Gholami (2006). For isotropic rock mass, the size effect of slate was studied by Li (2021), and the elastic-plastic constitutive model was constructed by Wang (2018) based on the D-P criterion.

---

*Corresponding Author: 1909488416@qq.com

252

DOI: 10.1201/9781003450832-32

In this paper, considering the actual environment of rock engineering in the Three Gorges Reservoir Region, rock-like materials are used to simulate rock mass, and embedded sheet metal is used to simulate the non-penetrating crack rock mass. The influence of crack angle on the strength of the specimen is analyzed. Finally, a constitutive relationship is studied based on the damage mechanics.

2 EXPERIMENT

2.1 *Specimen creation*

Cement mortar was chosen to make rock-like specimens due to its stability and ease of production. To create a material similar to the rock, several comparison tests were repeated to identify a ratio of composite 32.5 R cement, standard sand produced by Xi'an Hesha Company, and distilled water was determined to be 1:2.5:0.5. Two cracks were prefabricated in the center of the mold. A metal plate with a thickness of 0.5 mm and a width of 20 mm was inserted into the prefabricated groove position and the mold. The materials were mixed well according to the above proportions, and the mixed materials were poured into the mold. At the initial setting of the cement mortar specimen, the metal sheet was removed and then non-penetrating cracks with a certain crack angle and length were formed on the specimen. 24 hours later, the mold was removed, and then the sample was cured for 28 ds under a normal environment.

The samples were processed into cylinders with a height of 100 mm and a diameter of 50 mm. The crack angles were 0°, 30°, 45°, 75°, and 90°, respectively. Figure 1 shows the rock-like samples, in which Figure 1 (a) is a schematic diagram of the cracks distribution and Figure 1 (b) shows the prepared samples.

(a) Schematic diagram of crack

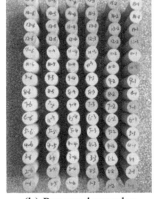
(b) Prepared samples

Figure 1. Rock-like samples. (a) Schematic diagram of crack (b) Prepared samples.

Table 1. Physics-mechanical parameters of a rock-like material.

	Density/g.cm^{-3}	Peak strength/MPa	Elastic modulus/GPa	Poisson ratio
Specimen	2.25	14.9	3.04	0.29
Rock	2.20	16.0	3.51	0.27

2.2 *Testing machine*

The indoor test was performed with a WDT-1500 large multifunctional testing machine, as shown in Figure 2 (a). The displacement loading was adopted with a speed of 0.5 mm/min. Figure 2 (b) shows the installation of the specimens. The specimens and pads were wrapped with heat-shrinkable tubes and the axial or radial strain of the specimen was measured by axial and radial extensometer, respectively.

(a) WDT-1500 testing machine   (b) Extensometer

Figure 2. Testing machine. *(a) WDT-1500 testing machine (b) Extensometer*

3 STRESS-STRAIN RELATIONSHIP

The stress-strain curves of specimens with different crack angles are shown in Figure 3. Based on the uniaxial compression tests, the stress-strain curves of the cracked specimens all went through an initial compaction phase, a linear elastic stage, a yield stage, and a failure stage. In the compaction stage, the original pores and fissures of the cracked specimens were compacted. The stress of the samples slowly increased with increasing strain. Its stress-strain showed a trend of the "upper concave" type. With the increasing stress, samples entered into the elastic deformation phase. The deformation of the specimen was mainly shown as recoverable elastic deformation. The stress-strain of the samples conformed to the generalized Hook law and the slope of the stress-strain curve was the average elastic modulus. In the

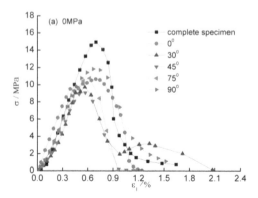

Figure 3. Stress-strain curves of rock-like specimens.

yield phase, the deformation of samples transformed from elastic deformation to elasto-plastic deformation. The stress-strain curve also transformed from linear growth to non-linear growth. When the tensile stress at the crack tip caused by the compressive stress exceeded the crack stress of the sample, cracks near the tip started to develop and gradually penetrated the sample, eventually leading to the destruction of the sample. In the after-peak deformation phase, the deformation of samples was mainly based on the relative slip between blocks. Meanwhile, the samples were damaged but did not completely lose their bearing capacity. Under uniaxial compression, the specimens showed some brittleness in deformation after the peak of deformation.

The peak intensity of the cracked specimens is generally not as strong compared with the complete specimens. This indicates that the crack has reduced the peak strength of the samples, and with the increasing crack angle, the after-peak strength first decreases and then increases. When the crack angle is 45°, the peak strength of the non-penetrating crack specimens is the lowest. The residual strength of the crack specimens is nearly 0 MPa under uniaxial compression. The stress rapidly falls after it reaches peak strength. The residual strength of the complete specimen and the cracked specimen with a crack angle of 90° is relatively high, while the after-peak strength of the cracked specimen with a 45° or 75° crack angle is relatively low.

4 CONSTITUTIVE RELATION

4.1 *Constitutive relation using damage variable*

Assuming the micro-element strength of the rock mass is subjected to Weibull distribution, its probability density function is as Equation (1).

$$P(F) = \frac{m}{F_0} \left(\frac{F}{F_0} \right)^{m-1} \exp\left[-\left(\frac{F}{F_0} \right)^m \right] \tag{1}$$

Where $P(F)$ is the Distribution function of micro-element strength, F is the distribution variables of random distribution, and m and F_0 is the distribution parameter, respectively.

Similarly, the damage variables of the complete rock mass and the statistical constitutive model of its damage are shown as Equations (2) and (3).

$$D = 1 - \exp\left[-\left(\frac{F}{F_0} \right)^m \right] \tag{2}$$

$$\sigma_1 = E\varepsilon_1 \exp\left[-\left(\frac{F}{F_0} \right)^m \right] + \mu(\sigma_2 + \sigma_3) \tag{3}$$

In this study, $\sigma_2 = \sigma_3 = 0$, and Equation (4) can be obtained.

$$\frac{\sigma_1}{E\varepsilon_1} = \exp\left[-\left(\frac{\varepsilon_1}{\varepsilon_0} \right)^m \right] \tag{4}$$

Taking the logarithm of both sides, we can obtain Equation (5).

$$\ln\left[-\ln\frac{\sigma_1}{E\varepsilon_1} \right] = m \ln \varepsilon_1 - m \ln \varepsilon_0 \tag{5}$$

Where σ_1 is the principal stress of rock mass, ε_1 is the principal strain of rock mass, and m and ε_0 is the distribution parameter, respectively.

Equation (5) is a linear relation, where m is the slope of the line, and $-mln\varepsilon_0$ is the intercept. It is assumed that the natural rock mass is a material and contains lots of initial micro-fractures. The crack is large compared with the original micro-fractures in the rock mass. The elastic modulus of the cracked rock mass with macro and micro damage, the elastic modulus of the rock mass with only microscopic damage, the elastic modulus of the rock mass with only macroscopic damage, and the elastic modulus of the rock mass without damage are marked as E_{12}, E_1, E_2, and E_0, and ε_{12}, ε_1, ε_2, and ε_0 are for their strains, respectively. According to the Lemaitre strain equivalence hypothesis, their relationship is shown in Figure 4, which can be written as Equations (6) and (7).

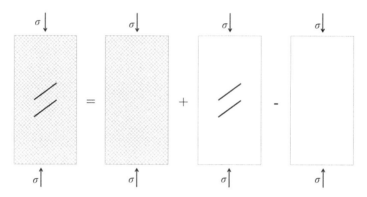

Figure 4. Strain equivalent diagram.

$$\varepsilon_{12} = \varepsilon_1 + \varepsilon_2 - \varepsilon_0 \tag{6}$$

$$\frac{\sigma}{E_{12}} = \frac{\sigma}{E_1} + \frac{\sigma}{E_2} - \frac{\sigma}{E_0} \tag{7}$$

This leads to:

$$\frac{1}{E_{12}} = \frac{1}{E_1} + \frac{1}{E_2} - \frac{1}{E_0} \tag{8}$$

The damage variables in the direction of the loading action of macroscopic damage and microscopic damage are D_1 and D_2 and the coupling damage variably is D_{12}. From the Lemaitre hypothesis, we can get the following equations.

$$E_{12} = E_0(1 - D_{12}) \tag{9}$$

$$E_1 = E_0(1 - D_1) \tag{10}$$

$$E_2 = E_0(1 - D_2) \tag{11}$$

Applying Equations. (9) to (11) to Equation (8), we can get the following equations.

$$\frac{1}{1 - D_{12}} = \frac{1}{1 - D_1} + \frac{1}{1 - D_2} - 1 \tag{12}$$

Where:

$$D_{12} = 1 - \frac{(1-D_1)(1-D_2)}{1-D_1 D_2} \quad (13)$$

D_2 can be calculated as Equation (2) and D_1 can be calculated as Equation (14) (Li, 2001).

$$D_1 = \begin{cases} 0 & \tan\alpha < \tan\phi \\ 1 - \dfrac{1}{1+\dfrac{12\rho_v m_0 a^2 B}{k^2}\tan\dfrac{\pi k}{2}\sec\left(\dfrac{\pi a}{w}\right)} & \tan\alpha \geq \tan\phi \end{cases} \quad (14)$$

Where $m_0 = \cos^2\alpha(\sin\alpha - \cos\alpha \tan\phi)$ and $k = 2a/b$ \quad (15)

From Equations (14) to (16), α is the angle of the crack, φ is the internal friction angle of the crack, $2a$ is the length of the crack, b is the space between two cracks, B is the depth of the crack, A is the surface area of the crack, $A=2Ba$, ρ_v is the Surface density, and w is the breadth of the section.

We substitute the data in Figure 3 to Equation (5) and we can obtain the value of m and ε_0. According to Equation (14), the results of m, ε_0, and D_1 are shown in Table 2.

Table 2. Calculated results.

	Complete	0	30	45	75	90
m	5.87	6.12	3.25	7.15	1.55	2.43
ε_0 / %	0.28	0.22	0.11	0.07	0.11	0.13
D_1	–	0	0.272	0.261	0.061	0

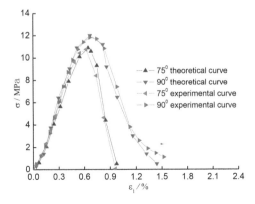

Figure 5. Comparison between the theoretical and experimental curves of specimens.

We substitute the results in Table 3 into Equation (13), the constitutive relation of the crack specimen can be expressed as Equation (16).

$$\sigma_1 = E_\theta \varepsilon_1 D_{12} \tag{16}$$

Taking the specimens with 75° and 90° cracking as an example, Figure 5 shows a comparison between the theoretical and experimental curves of the specimens. It can be concluded that the damaged constitutive relation is in good agreement with the experimental results.

5 CONCLUSION

The stress-strain curve of specimens all went through an initial pressure-dense phase, a linear elastic stage, a yield stage, and a failure stage. A constitutive relationship based on damage mechanics was proposed, and the calculation results can be well adapted to the experimental results.

REFERENCES

Costin, L. S. (1985). Damage Mechanics in the Post-failure Regime. *Mech. Mater.* 4, 149–160.

Dougill, J. W., Lau, J. C., and Burt, N. J. (1976). Toward a Theoretical Model for Progressive Failure and Softening in Rock, Concrete and Similar *Materials. Mech. in Engng., ASCE-EMD*, 335–355.

Gholami, R., Rasouli, V. (2014). Mechanical and Elastic Properties of Transversely Isotropic slate [J]. *Rock Mech. Rock Eng.* 47(5), 1763–1773.

Golshani, A., Okui, Y., Oda, M. (2006). A Micromechanical Model for Brittle Failure of Rock and its Relation to Crack Growth Observed in Triaxial Compression Tests of Granite. *Mech. Mater.* 38(4), 287–303.

Kawamoto, T., Ichikawa, Y., Kyoya, T. (1998). Deformation and Fracturing Behavior of Discontinuous Rock Mass and Damage Mechanics Theory. *Int. J. Numer. Anal. Met.* 12(1): 1–30.

Krajcinovic, D. (1981). Continuum Damage Theory of Brittle Materials. *Appl. Mech. Rev.* 48(4), 809–815.

Lemaitre, J. (1985). A Continuous Damage Mechanics Model for Ductile Fracture. *J. Eng. Mater. Tech.*, 107, 83–89.

Li, K., Yin, Z. Y., Han, D. (2021). Size Effect and Anisotropy in a Transversely Isotropic Rock Under Compressive Conditions. *Rock Mech. Rock Eng.* 54(9), 4639–4662.

Li, N., Chen, W., Zhang, P. (2001). The Mechanical Properties and a Fatigue-damage Model for Jointed Rock Masses Subjected to Dynamic Cyclical Loading. *Int. J. Rock Mech. Min.* 38(7), 1071–1079.

Ortiz, M. (1985). A Constitutive Theory for the Inelastic Behavior of Concrete. *Mech. Mater.* 4(1), 67–93.

Pouragha, M., Eghbalian, M., Wan, R. (2020). Micromechanical Correlation Between Elasticity and Strength Characteristics of the Anisotropic Rock. *Int. J. Rock Mech. Min.*, 125: 104–114.

Saroglou, H., Tsiambaos, G. (2008). A Modified Hoek-Brown Failure Criterion for Anisotropic Intact rock. *Int. J. Rock Mech. Min.*, 45(2), 223–234.

Shao, J. F., Hoxha, D., Bart, M. (1999). Modeling of Induced Anisotropic Damage in Granites. *Int. J. Rock Mech. Min.*, 36(8), 1001–1012.

Shao, J. F., Lu, Y. F., Lydzha, D. (2004). Damage Modeling of Saturated Rocks in Drained and Undrained Conditions. *J. Eng. Mech.* 130 (6), 733–740

Shi, X., Yang, X., Meng, Y., (2016). Modified Hoek-Brown Failure Criterion for Anisotropic Rocks. *Environ. Earth Sci.*, 75(11), 1–11.

Swoboda, G., Yang, Q. (1999). An Energy-based Damage Model of Geomaterials-I, *Formulation and Numerical Results. Int. J. Solids.* 36, 1719–1734.

Wang, Z., Zong, Z., Qiao, L. (2018). Elastoplastic Model for Transversely Isotropic Rocks. *Int. J. Geomech.* 18(2), 04017149.

Yazdani, S., Schreyer, H. L. (1988). An Anisotropic Damage Model with Dilatation for Concrete. *Mech. Mater.* 7(3), 231–244.

Smart city and environmental planning and management

Water Conservancy and Civil Construction – Oke & Ahmad (Eds)

Contributing factors to urban transport carbon dioxide emissions and reduction measures

Yue Hao, Yugang Liu & Shuai Zheng

School of Transportation and Logistics, Southwest Jiaotong University, Chengdu, Sichuan Province, China
National Engineering Laboratory of Integrated Transportation Big Data Application Technology, Southwest Jiaotong University, Chengdu, Sichuan Province, China
National United Engineering Laboratory of Integrated and Intelligent Transportation, Southwest Jiaotong University, Chengdu, Sichuan Province, China

Linqiang Fan*

Institute of Transportation Development Strategy and Planning of Sichuan Province, Chengdu, Sichuan Province, China

ABSTRACT: In response to the climate crisis, China has vowed to achieve carbon neutrality by 2060. Shockingly, the transportation industry is expected to be the largest emitter of greenhouse gases by 2050. Hence, identifying the main influencing factors of urban traffic carbon dioxide emissions and formulating efficient urban traffic carbon dioxide emission reduction measures are crucial to urban green development. Urban carbon dioxide emission prediction and emission reduction analysis involve many urban characteristics, such as industrial structure, spatial and temporal characteristics, population size, economic level, urbanization level, and energy intensity. It should be noted that urban carbon dioxide is greatly influenced by region, and there is a lack of a universal method to analyze the carbon dioxide of various urban transportation modes. To solve this problem, this paper considers the differences between different cities, innovatively screens out different significant influencing factors according to the characteristics of cities, and establishes a regression analysis model in line with the actual situation to get targeted emission reduction programs. It is concluded that the impact of single measures is limited and that a suitable combination of measures is needed to achieve efficient emission reductions. Energy management and control measures are the core elements of various emission reduction measures. The most fundamental way to reduce emissions is to improve the cleanliness of energy sources and transform the energy mix into a cleaner one.

1 INTRODUCTION

In response to the climate crisis, more than 195 countries worldwide have signed the Paris Agreement to enhance their capacity to address climate change. 2020 China proposes to increase its autonomous national contribution, adopt stronger policies and measures, strive to peak CO_2 emissions by 2030, and work towards carbon neutrality by 2060 (Xinhua News Agency 2020). With the rapid growth of cities, the rapid growth of the transport sector has led to significant CO_2 emissions, and the transport sector is both a key economic sector and a significant source of CO_2 emissions. According to the International Energy Agency (IEA),

---

*Corresponding Author: 252804276@qq.com

DOI: 10.1201/9781003450832-33

about 72% of the transport sector's greenhouse gas emissions come from road transport (Fan Y. *et al.* 2018). Developing countries in Asia are the main source of greenhouse gas emissions (Asian Development Bank 2019).

In this context, reducing transport CO_2 emissions in cities is one of the important ways to achieve the goal of carbon peaking and carbon neutrality. To address the issue of urban transport CO_2 emissions, scholars at home and abroad have researched various aspects such as its sources, influencing factors, hazards, emissions projection, and emission reduction measures. The characteristics of industrial structure (Du *et al.* 2017), spatio-temporal characteristics (Song & Wu 2020), population size, economic level, urbanization level (Yang *et al.* 2019), and energy intensity (Guo *et al.* 2014) are all used to predict the carbon dioxide of urban transportation. The social dynamics model and statistical model have been widely used in urban carbon dioxide emission prediction and emission reduction measures analysis.

Through different studies on the current status of national CO_2 emissions, it is found that developed countries have achieved good results in their efforts to reduce CO_2 emission levels through the use of green energy, different forms of transport, and improved and efficient modes of transportation. Most developing countries, on the other hand, whose main development goal is to increase economic opportunities, still rely heavily on energy-intensive modes of transport, putting enormous pressure on their transport CO_2 emission reduction targets (Wang *et al.* 2020; Wang *et al.* 2020). Studies on the current status of CO_2 emissions in different regions and provinces in China have shown that improving energy efficiency was a key link to decoupling China's economy from carbon emissions (Du *et al.* 2017). Still, significant regional and inter-provincial differences exist in the characteristics of CO_2 emissions and their drivers in China's transport sector (Yan 2015). The proposed CO_2 reduction measures should be tailored to local conditions and cannot be generalized.

City-level climate action is necessary to mitigate global warming, and low-carbon development is a win-win solution for both the economy and the climate. Many countries around the world are currently making policy interventions in transport CO_2 reduction, with various transport policies helping to save energy and reduce pollution emissions (Zhou *et al.* 2020), with significant synergies between policies as the type and number of policy combinations increase (Yang *et al.* 2020). In addition, urban characteristics such as population, economy and road network size, and transport characteristics such as commuting mode, promotion rate of new energy transport equipment, and share of public transport trips also have important effects on transport CO_2 emissions (Axsen *et al.* 2020; Bigazzi 2020; Du *et al.* 2017; Wang *et al.* 2020; Yan 2015). However, the influence of various factors varies greatly in different regions or cities, which does not have a universal law.

In summary, the results obtained from different levels of research vary, and urban transport CO_2 emissions are strongly influenced by regional factors. At present, there is a lack of a universal method for the targeted analysis of carbon dioxide from transport in different cities. Therefore, the study takes cities as the object of research, specifically analyses different factors influencing urban transport CO_2 emissions in cities, and by predicting urban transport CO_2 emissions under different development scenarios, conducts analyses to obtain the most effective measures or combinations of standards. The study considers the differences between different cities, innovatively screens out different significant influencing factors according to the characteristics of cities, and establishes a regression analysis model in line with the actual situation, to get a tailored emission reduction program.

The article is structured as follows: Section 2 details the calculation of urban transport CO_2 emissions for two cities, Beijing and Shanghai. Section 3 details an analytical model of the factors influencing the urban transport CO_2 emissions in the two case cities. In Section 4, future urban transport CO_2 emissions are projected under different emission reduction scenarios. Finally, the conclusions of the paper are given in Section 5.

2 CALCULATION OF CO$_2$ EMISSIONS FROM URBAN TRANSPORT

Beijing and Shanghai are selected for the study. As two mega first-tier cities in China, Beijing and Shanghai, are located in the northern inland city and the southern coastal city, respectively; both are characterized by rapid development, well-developed transportation industries, serious urban traffic congestion problems, and an urgent need to improve the atmospheric environment. As two representative cities in China, studying their transport CO$_2$ emissions can provide a reference for many cities.

The top-down approach, recommended in the "IPCC 2006 Guidelines for National Greenhouse Gas Inventories," is used to calculate the transport carbon emissions of a case city from the consumption of various types of energy at the energy terminals due to its advantage of better data availability and more accurate calculation results. The top-down method is shown in Equation (1).

$$Emission = \sum_j F_i \times EF_i \tag{1}$$

where i is the variety of energy consumed by urban transport in a year; F_i is the consumption of energy i by urban transport, using the calorific value of energy as the consumption indicator; EF_i is the CO$_2$ emission factor of energy i; and *Emission* is the annual CO$_2$ emission of urban transport.

We collect the consumption of the main energy species in the transport sector in the two cities from the Beijing Statistical Yearbook and the Shanghai Statistical Yearbook. We convert the consumption of various energy species into standard coal consumption. Traditional coal, also known as coal equivalent, is a calorific value standard. There is currently no uniform international standard for calculations related to standard coal; the study standardized the units based on the coefficients converted to traditional coal from the China Energy Statistical Yearbook.

China specifies a calorific value of 7000 kcal per kg of standard coal, converted from the calorific value as follows:

$$7000 kcal/kg = 7000 \times 4.1868 kJ = 29307.6 kJ/kg \tag{2}$$

The calorific value consumption of each energy source is calculated from the conversion to standard coal consumption. Then, the carbon dioxide emissions of each category of energy source and the total annual carbon dioxide emissions are calculated according to Equation (1).

CO$_2$ emissions from urban transport in Beijing increased from 22,964,500 tonnes in 2010 to 31,540,400 tonnes in 2018, an increase of 8,850,900 tonnes in nine years or 37.37% more than in 2010. From 2010 to 2018, the energy types with the highest CO$_2$ emissions in Beijing were paraffin, diesel, and electricity, from highest to lowest, with paraffin producing much higher CO$_2$ emissions from urban transport than the other types of energy produced during use. Figure 1 depicts the historical carbon dioxide emissions from Beijing urban transportation as well as the carbon dioxide emissions from Beijing urban transportation by energy.

Shanghai's urban transport CO$_2$ emissions increased from 43,312,000 tonnes in 2010 to 53,784,600 tonnes in 2018, indicating an increase of 10,472,500 tonnes in nine years and an increase of 24.18% from 2010. From 2010 to 2017, the energy types with the highest CO$_2$ emissions in Shanghai were, from highest to lowest, fuel oil, paraffin, and diesel. This changed in 2018, with the energy types with the highest CO$_2$ emissions, from highest to lowest, being paraffin, fuel oil, and diesel. Figure 2 depicts the historical carbon dioxide emissions from Shanghai urban transportation and the carbon dioxide emissions from Shanghai urban transportation by energy.

Comparing the results of the calculations, the main energy categories of the transport sector in Beijing were more abundant than those in Shanghai, and the types of energy increased in 2016. Urban transport CO$_2$ emissions in Beijing were lower than in Shanghai in all years from 2010 to

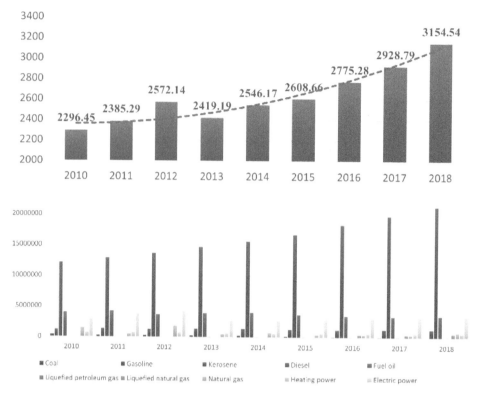

Figure 1. The historical emissions of carbon dioxide from urban transport in Beijing and emissions of carbon dioxide from urban transport in Beijing by energy (million tonnes).

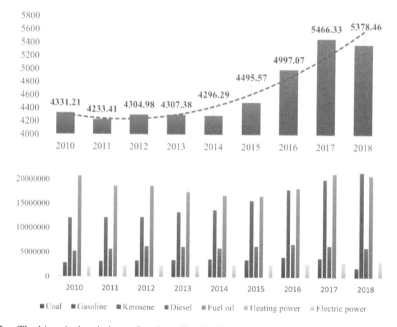

Figure 2. The historical emissions of carbon dioxide from urban transport in Shanghai and emissions of carbon dioxide from urban transport in Shanghai by energy (million tonnes).

2018, and the difference was large. The difference in transport CO_2 emissions between the two cities reached a high of 25, 375, 300 tonnes in 2017 and a low of 17, 328, 400 tonnes in 2012. The growth rate of urban transport CO_2 emissions in Shanghai was lower than that of Beijing. Due to the larger base of urban transport CO_2 emissions in Shanghai, the overall growth in urban transport CO_2 emissions was still slightly higher than that of Beijing.

3 STUDY ON THE FACTORS INFLUENCING CO_2 EMISSIONS FROM URBAN TRANSPORT

3.1 *Selection of indicators*

According to existing studies, many factors influence the CO_2 emissions of urban transport. The study divides the influencing factors into five categories: urban development, urban structure, public transport, energy structure, and transportation, and collects 16 influencing factors as indicators according to the main emission reduction measures studied.

Table 1. Indicators of factors affecting carbon dioxide emissions.

Categories	Influencing Factors
Urban Development	GDP, GDP per capita, Resident population, General public budget.
Urban Structure	The number of urban centers, Average commuting time, and average commuting distance.
Public Transport	Private car ownership, Taxi ownership, Bus operating vehicles, and urban rail operating vehicles.
Energy Structure	Clean energy consumption.
Transportation	Passenger traffic, Freight traffic, Passenger turnover, Freight turnover.

3.2 *Influencing factor analysis model*

The dependent variable of the thesis is the urban transport CO_2 emissions of Beijing and Shanghai for the calendar years 2010–2018 calculated in Section 2, and the independent variables are a series of indicators selected in Section 3.1. The study chooses multiple regression analysis to build a mathematical model between various variables.

As so many influencing factors are to be explored in the thesis, it uses a stepwise regression analysis model to analyze the influencing factors of urban traffic carbon dioxide emissions and obtain the most significant influencing factors. The commonly used stepwise regressions are the forward, backward, and stepwise methods. The stepwise method combines the advantages of the forward and backward methods, where each new independent variable is re-computed after being introduced forward to check whether it is retained. This method is used as a basis for alternately introducing or removing each independent variable until no new variables can be raised or removed, which is a more accurate method.

The stepwise regression model using the stepwise method is shown below:

For p, the independent variables of the regression $X_1, X_2, \ldots, X_P$ with the dependent variable Y build a one-dimensional regression model.

$$Y = \beta_0 + \beta_i X_i + \varepsilon, i = 1, 2 \ldots, p \tag{3}$$

The value of the variable X_i in the corresponding regression test coefficient F test statistic is calculated and denoted as $F_1^{(1)}, \ldots, F_p^{(1)}$, and the maximum value $F_{i1}^{(1)}$ is taken.

$$F_{i1}^{(1)} = \max\left\{F_1^{(1)}, \ldots, F_p^{(1)}\right\} \tag{4}$$

The given significance level serves as α and the corresponding critical value is denoted as $F^{(1)}$, if

$$F_{i1}{}^{(1)} \geq F^{(1)} \tag{5}$$

then X_{i1} introduces a stepwise regression model, and the set of indicators for the variables that have been chosen to be included is I_1.

A subset of the independent variables $\{X_{i1}, X_1\}, \ldots, \{X_{i1}, X_{i1-1}\}, \{X_{i1}, X_{i1+1}\}, \ldots, \{X_{i1}, X_p\}$ is created with the same dependent variable Y build a binary regression model to establish that there are $p-1$ subsets of variables. The regression coefficients of the variables F are calculated to test the value of the statistic; the stepwise method involves re-testing the independent variables that have been substituted, noting $F_k{}^{(2)}$ ($k = 1, 2, \ldots, p$) and choosing the value of the dependent variable other than $F_{i1}{}^{(1)}$ the maximum value other than $F_{i2}{}^{(2)}$.

$$F_{i2}^{(2)} = \max\left\{F_1^{(2)}, \ldots, F_{i1-1}^{(2)}, F_{i1+1}^{(2)}, \ldots, F_p^{(2)}\right\} \tag{6}$$

The given significance level serves as α and the corresponding critical value is denoted as $F^{(2)}$, if

$$F_{i1}^{(2)} \geq F^{(2)}, F_{i2}^{(2)} \geq F^{(2)}, \tag{7}$$

then the variable X_{i1} is kept and the variable X_{i2} is introduced, and vice versa; the variables are removed.

A new subset of variables I_2 is obtained. Then, the researchers repeat the above steps, alternating between introducing and removing independent variables, until there are no unknown variables to introduce or remove.

3.3 *Calculation results*

Calculations were carried out using stepwise regression to obtain a standardized impact factor analysis model for Beijing.

$$y = 0.806x_1 + 0.093x_2 + 0.131x_3 + 0.253x_4 + 6189355.528 \tag{8}$$

where y is the urban transport CO_2 emissions, x_1 is the passenger turnover (million kilometers), x_2 is the volume of freight transported (million tonnes), x_3 is the share of clean energy (%), and x_4 is the GDP (billion yuan).

According to the model, all four influencing factors and CO_2 emissions are positively correlated, indicating that as passenger turnovers, freight volume, clean energy share, and GDP increase, urban transport CO_2 emissions increase. Passenger traffic turnover has the largest effect on CO_2 emissions, followed by GDP, and then the share of clean energy and freight volume has the smallest effect factor. The results show that the economic activity effect is the main driver of the transport sector's increase in carbon dioxide emissions. The clean energy used in urban transport in Beijing generally shows a trend of first decreasing and then increasing, which is instead positively proportional to the increase in CO_2 emissions from urban transport, indicating that the current rate of increase in the share of clean energy in Beijing is not large enough, but instead is like the rate of growth in CO_2 from urban transport, to have a significant effect on emission reduction.

Calculations are carried out using stepwise regression to obtain a standardized model for analyzing influencing factors in Shanghai.

$$y = -0.438x_1 + 0.736x_2 + 0.123x_3 - 0.082x_4 + 39344292.533 \tag{9}$$

where y is the urban transport CO_2 emissions, x_1 is the average commuting time (min), x_2 is the number of passenger trips (million), x_3 is the number of buses operating (vehicles), and x_4 is the volume of freight transported (million tonnes).

According to the model, passenger volume, bus-operating vehicles, and CO_2 emissions are positively correlated, indicating that as passenger volume and bus-using vehicles increase, so do urban transport CO_2 emissions, whereas average commuting time, freight volume, and CO_2 emissions are negatively correlated, indicating that as average commuting time and freight volume increase, so do urban transport CO_2 emissions. Again, this proves that the economic activity effect is the main driver of total CO_2 emissions from the transport sector. The increase in freight volume instead contributes to reducing CO_2 emissions from urban transport, mainly because the structure of freight transport in Shanghai has changed significantly from 2010 to 2018. Although the increase in freight volume increases CO_2 emissions, a reduction in energy consumption can be achieved by adjusting the freight structure, and the reduction in energy intensity partially offsets the increase in CO_2 emissions from the transport sector. The growth in the number of vehicles operating on buses and railways shows that Shanghai has been developing and promoting public transport. However, the current development of public transportation has not been effective enough to attract more private trips to public transportation. The data shows that although Shanghai's average commute time and average commute distance have been among the highest in the country over the years, the work-life balance has improved. However, as the population grows, so do commuter trips. As commuting times and distances decrease, trips become more frequent, resulting in a reduction in average travel times but still an increase in CO_2 emissions.

A comparison of the impact factor analysis models for Beijing and Shanghai shows that the impact of transport economic activity on urban transport CO_2 emissions is significant in both cases. Still, there are regional differences in the effects of the same indicator on urban transport CO_2 emissions.

4 STUDY ON MEASURES TO REDUCE CARBON DIOXIDE EMISSIONS FROM TRANSPORTATION

Based on Sections 2 and 3, urban transport CO_2 emissions are still growing at a high rate in Beijing and Shanghai. This section projects future urban transport CO_2 emissions under different scenarios and explores which measures or a combination of measures will effectively reduce CO_2 emissions from urban transport.

4.1 Business as usual

Assuming business-as-usual (BAU) development in both cities with no new development policy interventions, projections will be made using the model obtained from the stepwise regression analysis in Section 3. The results of predicting CO_2 emissions from urban transportation in Beijing and Shanghai from 2020 to 2060 are shown in Figure 4. Using the stepwise regression model, it is assumed that the influencing factors always changed in accordance with the trend from 2010 to 2018. In that case, the transport CO_2 emissions of both cities will increase to about three times the 2010 emissions by 2060, as well as urban transport CO_2 emissions in Beijing will be as high as 78,579,200 tonnes, 342.2% higher than in 2010. Shanghai's urban transport CO_2 emissions in 2060 will be 118,930 million tonnes, 274.5% of the 2010 level.

4.2 Emission projections under the effect of a single measure

A wide range of policies can directly or potentially reduce urban transport CO_2 emissions. Axsen *et al.* (Axsen *et al.* 2020; Bigazzi 2020) considered an approach used by policymakers in California and elsewhere that classified greenhouse gases generated in transport into three categories of mitigation measures: switching to low-carbon fuels, improving vehicle efficiency, and reducing vehicle trips. Based on the results of the transport CO_2 emissions calculations in Section 2 and the analysis of influencing factors in Section 3, and concerning the above road transport policy classification, the study projects the following urban transport CO_2 emissions under the effect of mitigation measures such as those in Figure 3.

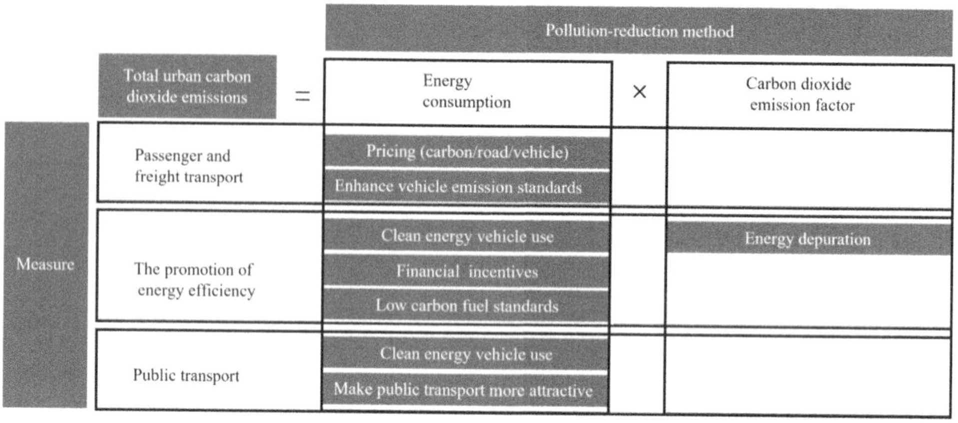

Total urban carbon dioxide emissions	=	Pollution-reduction method		
		Energy consumption	×	Carbon dioxide emission factor
Measure	Passenger and freight transport	Pricing (carbon/road/vehicle)		
		Enhance vehicle emission standards		
	The promotion of energy efficiency	Clean energy vehicle use		Energy depuration
		Financial incentives		
		Low carbon fuel standards		
	Public transport	Clean energy vehicle use		
		Make public transport more attractive		

Figure 3. Classification of carbon emission reduction measures.

The effects of the emission reduction measures in Figure 3, except for enhancing the attractiveness of public transport, are quantified based on available statistics, relevant research findings, and imperative policy requirements.

Table 2. Effectiveness of emission reduction measures.

Measures	Abbreviations	Effectiveness
Pricing	P	Setting a carbon price of US$40–100/t for 2020–2060 and maintaining a uniform growth rate
Enhance Vehicle Emission Standards	EVES	Requiring a 5% reduction in CO_2 emissions per year from 2020 to 2060 for all new vehicles
Clean Energy Vehicle Use	CEVU	Requiring 60% of clean energy vehicle sales by 2060, with even growth between 2020 and 2060
Financial incentives	FI	The shift from non-clean energy vehicles to clean energy vehicles is required to reach 10% of current non-clean energy vehicles and private transport to public transport to reach 5% of current private transport trips by 2060, with the shift maintained at an even rate between 2020 and 2060.
Low-carbon Fuel Standard	LCFS	Requiring a 20% reduction in the lifecycle carbon intensity of transport fuels for cars sold in cities by 2060, with a constant reduction rate between 2020 and 2060
Energy Depuration	ED	Requiring electricity to become clear at the primary energy end and for all other energy sources to have a 5% reduction in CO_2 emissions by 2060, with an even shift between 2020 and 2060

Based on the effectiveness of economic incentives, it can be seen that economic incentives are complementary measures to promote the use of clean energy vehicles and enhance the attractiveness of public transport implementation. They are used in conjunction with command-based policies to accelerate the rate of CO_2 reduction in urban transport and are not discussed separately.

As shown in Table 3, the three measures, P, CEVU, and LCFS, have the same reduction rate in Beijing and Shanghai, while there are urban differences in the effects of EVES, LCFS, and ED.

According to the data, the most effective emission reduction measure by 2060 is using clean energy vehicles, with a transport CO_2 reduction rate of 32.0% in both Beijing and Shanghai. The next most effective emission reduction measure is using low-carbon fuel standards, with Beijing and Shanghai achieving a 20.0% reduction in CO_2 emissions from transport. The least effective abatement measure is upgrading vehicle emission standards, with a reduced rate of only 0.88% and 1.70% in Beijing and Shanghai, respectively.

Table 3. Carbon emission reduction rates under single measures.

	Measures	2030	2035	2040	2045	2050	2055	2060
Beijing	P	4.33%	5.24%	5.95%	6.54%	7.05%	7.49%	7.88%
	EVES	0.64%	0.68%	0.72%	0.76%	0.80%	0.84%	0.88%
	CEVU	11.8%	15.1%	18.5%	21.9%	25.3%	28.6%	32.0%
	FI	5.5%	8.0%	10.5%	13.0%	15.5%	18.0%	20.0%
	LCFS	5.7%	7.6%	9.6%	11.5%	13.5%	15.4%	17.4%
	ED	4.33%	5.24%	5.95%	6.54%	7.05%	7.49%	7.88%
Shanghai	P	4.33%	5.24%	5.95%	6.54%	7.05%	7.49%	7.88%
	EVES	0.98%	1.10%	1.22%	1.34%	1.46%	1.58%	1.70%
	CEVU	11.8%	15.1%	18.5%	21.9%	25.3%	28.6%	32.0%
	FI	5.5%	8.0%	10.5%	13.0%	15.5%	18.0%	20.0%
	LCFS	3.4%	4.6%	5.8%	7.0%	8.2%	9.4%	10.6%
	ED	4.33%	5.24%	5.95%	6.54%	7.05%	7.49%	7.88%

Based on the above-predicted results and the comparison of carbon emission reduction rates under the effect of single measures, it can be seen that the effect of single measures on the reduction of carbon dioxide emissions from urban transport is limited and insufficient to achieve the goal of carbon neutrality.

4.3 *Emission projections under the effect of combined measures*

To enhance the effect of CO_2 reduction in urban transport, the six single measures in Table 2 are combined to obtain the effect of the combined measures, as shown in Table 4.

Table 4. Emission projections under the effect of combined measures.

Combined Measures	Serial Number	Effectiveness
P + CEVU	①	Carbon pricing of $40–100/t between 2020 and 2060, with 60% of clean energy vehicle sales by 2060.
CEVU + ED	②	60% of clean energy vehicles sold by 2060 will be electric. Electricity becomes clean on the primary energy side, and CO_2 emission factors for all other energy sources are reduced by 5%.
P + LCFS + EVES	③	There will be a carbon pricing of US$40–100/t between 2020 and 2060, a 5% reduction in annual CO_2 emissions from all new vehicles, and a 20% reduction in the lifecycle carbon intensity of transport fuels for cars sold in cities by 2060.

(continued)

Table 4. Continued

Combined Measures	Serial Number	Effectiveness
P + CEVU + ED	④	There will be a carbon pricing of US$40–100/tonne from 2020 to 2060, 60% sales of clean energy vehicles by 2060, clean electricity at the primary energy end, and a 5% reduction in CO_2 emission factors for all other energy sources.
P + LCFS + EVES + CEVU	⑤	There will be a carbon price of $40–100/tonne from 2020 to 2060, a 5% reduction in annual CO_2 emissions from all new vehicles, a 20% reduction in the lifecycle carbon intensity of transport fuels for cars sold in cities by 2060, and 60% sales of clean energy vehicles.
P + LCFS + EVES + CEVU + ED	⑥	There will be a carbon price of US$40–100/tonne from 2020 to 2060, a 5% reduction in annual CO_2 emissions from all new vehicles, a 20% reduction in the life cycle carbon intensity of transport fuels for cars sold in cities by 2060, 60% sales of clean energy vehicles, clean electricity on the primary energy side, and a 5% reduction in CO_2 emission factors from all other energy sources.
P + LCFS + EVES + CEVU + ED + FI	⑦	Carbon pricing of US$40–100/tonne from 2020 to 2060; a 5% reduction in annual CO_2 emissions from all new vehicles; an economic incentive to reduce the lifecycle carbon intensity of transport fuels for cars sold in cities by 30% by 2060; 75% sales of clean energy vehicles; electricity becoming clean on the primary energy side; and a CO_2 emission factor for all other energy sources of 5%.

The effects are incorporated into the regression model-based urban transport CO_2 emission projection model in 4.1 to obtain the projected transport CO_2 emissions in Beijing and Shanghai for 2020–2060 under the effect of the seven combined measures in Table 5, and compare with the projected transport CO_2 emissions under the BUA development scenario, as shown in Figure 4.

Table 5. Carbon emission reduction rates under combined measures.

	Measures	2030	2035	2040	2045	2050	2055	2060
Beijing	①	15.6%	19.6%	23.4%	27.0%	30.5%	34.0%	37.4%
	②	16.7%	21.6%	26.3%	30.9%	35.3%	39.6%	43.8%
	③	10.2%	13.4%	16.4%	19.3%	22.1%	24.8%	27.0%
	④	20.3%	25.7%	30.7%	35.4%	39.9%	44.2%	48.3%
	⑤	20.7%	26.5%	31.9%	37.0%	41.8%	46.3%	50.3%
	⑥	25.2%	32.1%	38.4%	44.2%	49.6%	54.6%	59.0%
	⑦	30.2%	38.0%	45.1%	51.5%	57.4%	62.8%	67.7%
Shanghai	①	15.6%	19.6%	23.4%	27.0%	30.5%	34.0%	37.4%
	②	14.8%	19.1%	23.3%	27.4%	31.4%	35.3%	39.2%
	③	10.5%	13.8%	16.9%	19.8%	22.6%	25.3%	27.6%
	④	18.5%	23.3%	27.8%	32.1%	36.2%	40.2%	44.0%
	⑤	21.0%	26.8%	32.2%	37.3%	42.1%	46.7%	50.7%
	⑥	23.7%	30.2%	36.2%	41.7%	46.9%	51.7%	56.0%
	⑦	28.8%	36.3%	43.1%	49.3%	55.1%	60.4%	65.3%

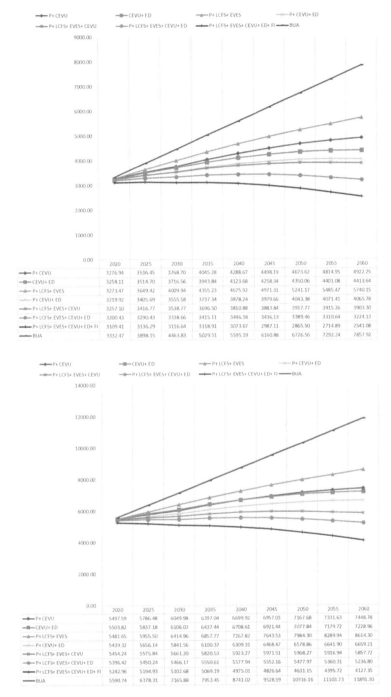

Figure 4. Projected transport CO_2 emissions in Beijing (the picture on the left) and Shanghai (the picture on the right) under a combination of measures.

Comparing the effects of the above emission reduction measures, pricing, and new industry standards are not enough to reduce emissions. Cleaner energy vehicle use, pricing, incentives, enhanced vehicle emission standards, and low-carbon fuel standards are all policies that serve this goal.

The quantitative impact indicators used in the study are conservative, and cities could set higher standards for emissions reductions to achieve peak and neutral carbon targets more quickly, considering their actual circumstances.

In addition to the combination of measures such as P + LCFS + EVES + CEVU + ED + FI, achieving carbon neutrality in urban transportation necessitates the concerted efforts of various measures such as strengthening the construction of green transport infrastructure, improving the attractiveness of public transportation, encouraging walking and cycling, strengthening the promotion of low-carbon green transport concepts, and increasing city greening coverage. Cities should establish their characteristics.

5 CONCLUSIONS

The study obtains the current status of transport carbon dioxide emissions, the main influencing factors, and effective emission reduction measures or combinations of measures in Beijing and Shanghai through the calculation of urban transport carbon dioxide emissions, the study of the influencing factors of urban transport carbon dioxide emissions, and the prediction of urban transport carbon dioxide emissions under different development scenarios. The main results of our work are as follows:

(1) The CO_2 emissions and emission characteristics of urban transport in Beijing and Shanghai are calculated and analyzed. The standardized impact factor analysis model for Beijing and the standardized impact factor analysis model for Shanghai are obtained through stepwise regression. If business continues in both cities, transport CO_2 emissions in both cities will increase to more than 10 times the 2010 emissions by 2060.
(2) After forecasting urban transport CO_2 emissions under five single measures (pricing, upgrading vehicle emission standards, clean energy vehicle use, low carbon fuel standards, and clean energy treatment), it is concluded that the most effective emission reduction measure by 2060 is clean energy vehicle use, with a 32.0% reduction in transport CO_2 emissions in both Beijing and Shanghai. The next most effective measure is low carbon fuel standards, with a reduced rate of 20.0% in both Beijing and Shanghai.
(3) The projection of urban transport CO_2 emissions under the effect of seven combination measures concludes that the effect of the combination of measures is significantly better than that of single measures. The most effective combination of measures is P + LCFS + EVES + CEVU + ED + FI, which reduces CO_2 emissions by 67.7% and 65.3% in Beijing and Shanghai, respectively, by 2060, and saves 53,168,400 t and 77,639,500 t compared to the BUA scenario. A reasonable combination of abatement measures is needed to achieve a reasonable and efficient carbon reduction. The combination of P + LCFS + EVES + CEVU + ED + FI can be used as a template for targeted improvements. Many abatement measures take a long time to take effect, and governments need to plan and consider the long term when enacting measures and be mindful of the timeliness of the measures.
(4) The research innovatively screens out different significant influencing factors based on city characteristics, in accordance with the actual situation of the regression analysis model established, in order to put forward more effective and more targeted measures to reduce traffic carbon dioxide emissions, reduce the city's transportation carbon dioxide emissions, and finally realize the city's green development. This method has good generalization.
(5) The study does not consider the cost of achieving the abatement measures, and subsequent studies could consider the impact of the cost of abatement measures to analyze the most effective abatement measures or combinations of measures for the same budget.

ACKNOWLEDGMENTS

The authors of this paper are supported by the National Natural Science Foundation of China under Grant No. 51774241 and No. 71704145, and Sichuan Youth Science and Technology Innovation Research Team Project under Grant No. 2020JDTD0027, and the Institute of Transportation Development Strategy and Planning of Sichuan Province Basic Research Operations Fund Project No. 2022JBKY03.

REFERENCES

Longxu Yan A., "Quantifying and Analyzing Traffic Emission Reductions from Ridesharing: A Case Study of Shanghai," *Transportation Research Part D: Transport and Environment* 89 (2015).

Asian Development Bank, "*Addressing Climate Change in Transport*," 2019, <https://www.adb.org/sectors/transport/key-priorities/climate-change/>.

Axsen J., Pltz P., Wolinetz M., "Crafting Strong, Integrated Policy Mixes for Deep CO_2 Mitigation in Road Transport," *Nature Climate Change*, 10: 1– 10 (2020).

Bigazzi A., "Marginal Emission Factors for Public Transit: Effects of Urban Scale and Density," *Transportation Research Part D Transport and Environment*, 88: 102585 (2020).

Du K., Xie C., Ouyang X., "A Comparison of Carbon Dioxide (CO_2) Emission Trends Among Provinces in China," *Renewable and Sustainable Energy Reviews*, 73: 19– 25 (2017).

Guo B., Geng Y., Franke B., "Uncovering China's Transport CO_2 Emission Patterns at the Regional Level," *Energy Policy*, 74: 134– 146 (2014).

Litman T, "Comprehensive Evaluation of Energy Conservation and Emission Reduction Policies," *Transportation Research Part A*, 47: 153–166 (2013).

Song M., Wu J., "Spatiotemporal Regularity and Spillover Effects of Carbon Emission Intensity in China's Bohai Economic Rim," *Science of The Total Environment*, 740: 140184 (2020).

Van Fan Y., Perry S., Kleme., Jií Jaromír., "A Review on air Emissions Assessment: Transportation," *Journal of Cleaner Production*, 194: 673– 684 (2018).

Wang C., Wood J., Wang Y., "CO_2 Emission in Transportation Sector Across 51 Countries Along the Belt and Road from 2000 to 2014," *Journal of Cleaner Production*, 266: 122000 (2020).

Wang C., Zhao Y., Wang Y., "Transportation CO2 Emission Decoupling: An Assessment of the Eurasian Logistics Corridor," *Transportation Research Part D Transport and Environment*, 86: 102486 (2020).

Wang L., Yu Y., Huang K., "The Inharmonious Mechanism of CO_2, NOx, SO_2, and PM 2.5 Electric Vehicle Emission Reductions in Northern China," *Journal of Environmental Management*, 274: 111236 (2020).

Xinhua News Agency, "*Speech at the General Debate of the 75th United Nations General Assembly*," 22 September 2020, <https://www.xuexi.cn/lgpage/detail/index.html?id=17443756313685536978&item_id=17443756313685536978>.

Yang L., Wang Y., Han S., "Urban Transport Carbon Dioxide (CO_2) Emissions by Commuters in Rapidly Developing Cities: The Comparative Study of Beijing and Xi'an in China," *Transportation Research Part D Transport and Environment*, 68: 65– 83 (2019).

Yang L., Wang Y., Lian Y., "Factors and Scenario Analysis of Transport Carbon Dioxide Emissions in Rapidly-developing Cities," *Transportation Research Part D Transport and Environment*, 80 (2020).

Zhou X., Xu Z., Xi Y., "Energy Conservation and Emission Reduction (ECER): System Construction and Policy Combination Simulation," *Journal of Cleaner Production*, 267: 121969 (2020).

Design and application of a new gas-bearing gas film in the closed coal yard

Xiaolong Sheng & Honghui Wang
Shaanxi Yanchang Petroleum Fuxian Power Generation Co., LTD., Yan'an, Shaanxi Province, China

Ming Shang
ZC Space (Shenzhen) Intelligent Technology Co., Ltd, Shenzhen, Guangdong Province, China

Weisheng Li
Shaanxi Yanchang Petroleum Fuxian Power Generation Co., LTD., Yan'an, Shaanxi Province, China

Sensen Wang
ZC Space (Shenzhen) Intelligent Technology Co., Ltd, Shenzhen, Guangdong Province, China

ABSTRACT: Because of its fast construction speed, low cost, high investment performance, good environmental performance, and other characteristics in recent years rapidly applied in the domestic open yard environmental protection closed and rapid development, it has become the first choice of many production enterprises. Due to the lack of relevant national standards and specifications, air film is also facing many problems in the promotion and application of gas film, especially in the design and application of firefighting facilities are the focus of users.

1 OVERVIEW

1.1 Basic information

Shaanxi Yanchang Petroleum Fuxian power Plant 21000MW project is the gas film coal shed project. The gas film coal shed size is 191 m (long), 182 m (wide), and 64 m (high), with a construction area of 34762 m². After sealing, the coal storage capacity is about 200,000

Figure 1. Aerial view of the air-dome of Fuxian power plant.

tons. The fixing power plant gas film coal shed project is the first project in Shaanxi province to use a gas-bearing membrane structure to close the coal storage yard.

1.2 *Process description*

In this project, two turbine operating platforms spang the length of the coal shed of the gas film. The width of the turbine foundation platform is 11 m, and two pile pickers are installed on each platform to meet the storage, transfer, and adjustment of coal in the gas film.

2 FIRE CONTROL DESIGN DIFFICULTIES

(1) Because of the gas membrane coal shed use function needs and special form of space structure, the fire partition is difficult to use the firewall and other physical fire partitions to separate two across the coal shed length direction of the turbine rail set causing certain obstacles to evacuation. The evacuation distance of the coal shed internal local area is long, and the gas membrane coal shed is difficult to completely follow the current fire code for fire control design difficulties. It needs its special fire design evaluation study.
(2) The gas film membrane material is of grade B1. Still, the lower part is maintained, the external wall is reinforced concrete, the combustion performance is grade A, and the overall refractory grade of the air film cannot be determined.
(3) There needs to be a clear definition of the structural properties of air-film buildings, and there are no relevant national specifications and standards. Most adopt the specifications and standards of industrial storage and application fields to carry out relevant designs and applications.

3 FIRE RISK ANALYSIS

(1) Spontaneous combustion of coal may cause fire accidents. When the coal in the coal pile contacts the air, the oxidation reaction will occur to release heat, which increases the temperature of the coal pile, accelerates the oxidation reaction speed of the coal, and makes the temperature of the coal pile higher and higher. When the temperature exceeds the spontaneous combustion point of the coal, spontaneous combustion occurs.
(2) Fire spread when coal burning is one of the potential dangers. In the gas film coal shed, it is impossible to set the corresponding physical fire partition for each coal pile, and there is no physical fire separation between the coal pile. If the fire cannot be controlled timely and effectively controlled, the fire may lead to the burning of the adjacent coal pile, and the fire situation occurs.
(3) The fire source left by the site staff may cause fire accidents. Fireworks should be prohibited in the coal shed areas. However, when some staff members violate the discipline, it may cause cigarette butts or other fire sources in the coal shed combustible combustion, causing fire accidents.
(4) Electrical equipment in the coal shed may cause electric shock accidents or fire accidents. Electrical equipment short circuits, overload, poor contact, produce electric sparks, and so on may cause casualties or fire accidents. If the gas concentration in the coal shed exceeds the limit and other unsafe factors, even explosion accidents will occur.
(5) Casualties may be caused due to the effect of smoke. ① When the oxygen content in the flue gas is lower than the normal value, human activity is weakened, and mental confusion and even faint suffocation; When the content of various toxic gas exceeds the minimum concentration of normal physiology, it will cause poisoning death. ② The terror of the smoke will cause psychological panic, especially when the boom, flames, and smoke out of the doors and windows, fire, and smoke billowing. It will make people have great fear, which will cause chaos during evacuation.

(6) High-temperature effects caused by large-area coal combustion during a fire may lead to casualties. Burning emits a lot of heat, and high temperatures can cause skin and respiratory tract burns and even death.

4 GAS FILM COAL SHED SPECIAL FIRE PROTECTION DESIGN RESEARCH SCHEME

The amount of coal stored in the gas film coal shed is large, and the area is large. If a fire occurs in a certain coal pile, it is likely to spread to other coal piles, leading to a fire in the whole coal storage shed and causing significant economic losses. If the fire burns through the air film building, resulting in the collapse of the air film, it affects the evacuation and even causes casualties.

(1) Isolation of the coal piles in the coal storage shed and certain fire isolation belts shall be set between the coal piles and between the coal piles and the side wall to prevent the spread of fire.
(2) Based on the research results of the combustion performance of the membrane material, we study the structural stability of the gas film coal storage shed and determine whether it affects the evacuation of personnel.
(3) Safe evacuation facilities are based on FDS software and Pathfinder software designers.

Table 1. Research scheme for special fire control design of gas film coal shed.

Research contents	Technical proposal	Appraisal procedure
Structural stability	Experiments and theoretical analysis	We analyze the stability of the gas film coal storage shed structure during the fire, and determine whether it will affect personnel evacuation
Fire compartment	We use fire dividers to isolate the coal pile to prevent the spread of the fire.	The combination of theory and FDS software to study whether the fire isolation zone can prevent the spread of fire
Safe evacuation	The required evacuation time is simulated by theoretical calculation and Pathfinder evacuation software.	Fire simulation and calculation will be conducted by comparing the required evacuation time with the available evacuation time simulated by FDS to determine the feasibility of safe personnel evacuation.

5 SELECTION AND APPLICATION OF FIRE PROTECTION DESIGN

5.1 Fire division

The construction area of the gas film coal shed is 34762 m^2. According to the code "Fire Protection Standard for Thermal Power Plant and Substation Design" GB50229-2019, the area of each fire partition should not be greater than 12000 m^2. When the fire partition area is greater than 12000 m^2, the width channel is not less than 10 m, the firewall is more than 3 m above the pile coal surface, and the coal yard should be divided into three fire partitions. Fire partition parameters are shown in Tables 1–2. In this project, the basic platform of a bucket turbine in a gas membrane coal shed is used as a fire isolation channel. 4 meters of channels are reserved on both sides of the platform, which can be used by fire trucks. A 19 m fire isolation belt is set up between the coal storage pile to meet the requirement that the fire of a coal pile will not ignite the adjacent coal pile within a certain period.

Table 2. List of fire protection zones.

Order number	Fire compartment	Area
1	First fire partition	5483.83 m²
2	Second fire partition	9675.58 m²
3	Third fire partition	5483.83 m²

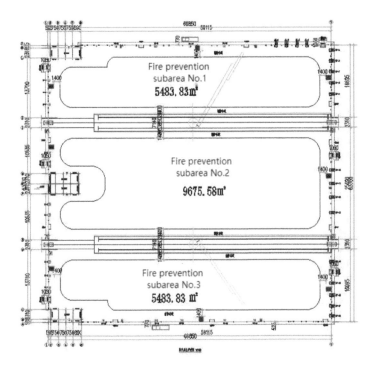

Figure 2. Layout of fire prevention subarea.

5.2 *Fire emergency lighting and evacuation indication system*

(1) Evacuation lighting: Fire emergency lighting shall be provided in crowded places such as evacuation channels. Evacuation indication signs and safety exit sign lights shall be installed at evacuation walkways and safety exits. The emergency lighting and evacuation indicator lamp with battery shall be at least 30 minutes.
(2) Lighting control: The emergency lighting lamps are forced to be lit in case of fire.
(3) Based on the FDS simulation of the flue gas flow of the gas membrane coal storage shed and the surface temperature field of the membrane structure, the available safe evacuation time and the membrane structure instability time are obtained through analysis and research, and the Pathfinder evacuation simulation software is used to simulate the necessary time for personnel evacuation. An emergency evacuation door is set up in this project, and the distance between the two emergency doors is at most 80 m. The size of the emergency evacuation door is 15002100, which can accommodate three groups of people passing through at the same time.
(4) Due to the large building volume of the coal shed, it is challenging for distant people to detect when a fire occurs. Therefore, an automatic fire alarm system is set up in this project,

and a sound and light alarm with a large sound level is set up in the coal shed (the sound pressure level should be at least 15 dB higher than the environmental background noise). When a fire occurs, the automatic alarm system can inform the evacuation of personnel the first time.

5.3 *Automatic fire alarm and fire control linkage control system*

(1) An automatic fire alarm system is set up inside the coal yard. The main control room is equipped with a single area alarm controller connected to the whole factory fire control room through the bus. The fire control room can display all alarm and linkage control status signals.

(2) The bus short circuit isolator is set on the system bus. The bus short circuit isolator protects the fire detector, manual fire alarm button and module, and other firefighting equipment. When the bus crosses the fire zone, the bus short circuit isolator is set at the crossing place.

(3) When there is a hidden situation in the coal yard, we remind the personnel in the coal yard to evacuate in time through the fire broadcast or the hand report button set inside the coal yard. Moreover, the sensing monitoring system in the coal yard is linked with the intelligent control system, which can quickly improve the gas film's internal environment, such as smoke exhaust and gas exchange.

5.4 *Fire protection system and water consumption*

The gas film coal shed of this project is constructed according to the industrial plant, and according to Article 3.0.1 of the Design Fire Protection Standard for Thermal Power Plant and Substation GB50229-2019, the fire risk of indoor coal storage yard is classified as Class C and the fire resistance grade 2. This project selects fire water consumption according to a class C workshop design in an industrial building.

(1) Indoor fire hydrant system

According to Article 7.4.6 of Technical Specification for Fire Water Supply and Fire Hydrant System, the layout of the indoor fire hydrant meets the requirements of 2 filling water columns in the same plane and any part at the same time. The water consumption of the indoor fire hydrant is 40 L/s, arranged in the retaining wall around the gas film coal shed and on both sides of the bucket turbine platform.

(2) Indoor, fire-fighting gun system

According to the fixed fire extinguishing water cannon set in the indoor coal storage yard and Article 7.5.6 of the Fire Protection Standard for Thermal Power Plant and Substation Design GB 50229-2019, the water column of at least one water cannon shall reach any point in the coal yard. The flow rate of each water gun should be at least 20 L/s. The fire rate selected for this project is 40 L/s, the rated pressure is 0.8 MPa, and the designed flow is 40 L/s. It is set in the retaining wall around the gas film coal shed and on both sides of the bucket turbine platform.

(1) Fire extinguisher configuration

Fire extinguishers shall be equipped in the coal shed. According to the Code for The Design of Building Fire Extinguishers (GB50140-2005), a certain number of fire extinguishers shall be arranged along the gas film retaining wall on both sides of the coal conveying belt and the stacker. When there is a local fire, fire extinguishing can be carried out in time for the first time.

5.5 *Ventilation system and combustible gas and coal dust control*

The gas membrane coal yard adopts the mode of mechanical air supply and overpressure air exhaust. It is divided into a forced ventilation state, production state, non-production state, and empty warehouse state. We open no less than 50% fans and guarantee no less than 0.5 ventilation changes per hour. It is automatically realized by controlling the number of fans and changing the operating frequency of dust concentration in the air film to ensure the minimum positive pressure of the safety of the gas film structure and limit the minimum number of open fans. When the dust suppression device fails, the dust concentration in the membrane is too large, and the gas concentration exceeds the standard. The ventilation system is manually turned into the accident state. The air supply fan is all turned on to ensure 0.8~1 ventilation per hour.

The combustible gas inside the gas film coal storage yard is mainly the gas produced by coal accumulation and the coal powder sink generated by the stacking operation, which will produce an explosion risk in the face of an open fire. Therefore, the control mode of combustible gas and coal powder sink should be fully considered in the design.

Gas control: Exhaust valves are set on both sides of the gas film retaining wall, multiple ventilation ports are set on top of the gas film, and the gas gathered at the top of the coal shed is discharged outdoors. At the same time, a gas probe is set inside the coal shed. Once there is gas accumulation, and the concentration exceeds the limit value, the ventilation system automatically runs into the accident state through the automatic control system. All the fans are opened to eliminate gas quickly.

Dust control: A spray dust removal system is set in the air film, which meets the occupational health requirements after dust removal. At the same time, a dust sensor is set inside the gas film. When the coal dust concentration reaches a certain index, the automatic control system will link the fresh air exchange system to increase the air supply and exhaust volume and quickly eliminate the dust to the external collector.

6 CONCLUSION

Gas-type gas film in industrial environmental protection closed application field is used more and more widely, in the field of fire design and application research also gradually improved. With the continuous development of the industry and enterprises and institutions of technical exchanges and cooperation, gas-type gas membranes in fire design and application will gradually lead to standardized, compiled, and published in the field of fire national standard design specification, guiding the development of gas membrane industry specification. Based on the existing fire control design, to further strengthen the management of the gas film inside, for example, add bucket turbine rail belt temperature monitoring and early warning system, strict control between the coal heap and coal pile, and membrane structure between the fire barrier, controlling electrical equipment compliance use, strengthening the maintenance and reliability of water fire extinguishing system and strengthening fire management are the focus of gas membrane in the use process.

REFERENCES

Code for Design of Automatic Fire Alarm System (GB 50116-2013)
Code for Design of Building Fire Extinguisher Allocation (GB50140-2005)
Code for Design of Fixed Fire Fighting Systems (GB 50338-2003)
Code for Fire Protection in Building Design (GB 50016-2014)
de for Design of Automatic Fire Alarm System (GB 50116-2013)
Fire Protection Standard for Thermal Power Plant and Substation (GB 50229-2019)
Project of Shaanxi Yanchang Petroleum Fuxian Power Plant (Miners elimination (Z) -2021002)
Research Report on Special Fire Control Design of Gas Film Coal Shack of 21000MW
Technical Specification for Fire Protection Water Supply and Fire Hydrant System (GB 50974-2014)
Technical Specification for Membrane Structure (CECS 158-2015)

Water Conservancy and Civil Construction – Oke & Ahmad (Eds)
© 2024 The Author(s), ISBN: 978-1-032-58618-2

Design and construction of urban ventilation corridors based on wind circulation in the Liling ceramic valley area

Shan Guan & Ana Hao*
Department of Architecture and Urban Planning, Shenyang Jianzhu University, Shenyang, Liaoning, China

Rujiao Cao
Shenyang Planning and Design Institute Co., Ltd, Shenyang, Liaoning, China

ABSTRACT: The design and optimization of the urban ventilation corridor in the Ceramic Valley of Liling City, Hunan Province, created good circulation conditions for the introduction and dredging of natural wind sources passing through the city via the overall delineation of the urban space, which was mainly based on the macro-spatial scale. The design comprehensively used large-scale spatial elements such as mountains, forests, green spaces, buildings, urban road networks, and urban water systems to create urban ventilation conditions that are suitable for the ventilation of the ceramic valley. In addition, it rationally applied the local circulation theory to construct an effective network structure for unobstructed ventilation inside and outside the city. According to the direction of the air duct in the urban design of the Liling Ceramic Valley area and the physical characteristics of the underlying surface of the Ceramic Valley, we analyzed the characteristics of the urban wind environment in the Ceramic Valley and divided the urban ventilation corridors. In terms of the urban wind environment, an urban natural ventilation system was established in which the ceramic valley interaction space, compensation space, and air guiding channels were unified with each other.

1 BACKGROUND

China's urbanization process has accelerated since 1987, and the urban space has undergone earth-shaking changes. At the same time, the rapid advancement of urbanization also seriously affects the urban environment and spatial quality. The overall urban temperature continues to increase, the wind speed in urban areas generally decreases, and air pollution and smog occur frequently. Urban environmental problems directly affect public health (Leng 2009). Given the above problems, it is necessary to improve urban climate conditions through urban design based on the application of urban climate information in the process of urban planning and design. However, the use of climatology in current urban design is very limited and frequently overlooked. Therefore, given today's huge pressure on the ecological environment caused by urbanization, it is necessary to appropriately change the traditional urban planning and urban design methodologies. Establishing an analytical model of urban spatial ecology to facilitate design is an important way to make the city embark on green and sustainable development (Zhu 1980).

The urban heat island effect and air pollution are two major issues in today's urban climate environment (Chen *et al.* 2014). At this stage, most of the solutions to these two types

---

*Corresponding Author: 530644749@qq.com

of urban health problems were proposed via reducing carbon emissions, such as by reducing coal use during winter heating, closing factories and enterprises with severe pollution, and restricting the frequent travel of motor vehicles. To a certain extent, urban environmental problems can be alleviated, but some urban functions need to be sacrificed. At the same time, due to the urban heat island effect and the complexity and variability of air pollution sources, daily monitoring requires a lot of financial and material resources. Therefore, to further alleviate such issues, improving the urban wind environment is the most effective countermeasure to deal with the urban heat island problem and reduce air pollution.

The continuous maturity of science and technology promotes the continuous development of urban climate environment research, which has become an important means to study urban climate environments. The urban wind environment research techniques and methods can be roughly divided into three stages. The earliest is field measurement, but the research has great limitations. In the intermediate stage, it evolved into the physical simulation, and the research scope is greatly improved compared with field measurement, but the requirements for experimental equipment are high, so the test accuracy is still not enough. The current research on the urban wind environment relies on powerful computer technology, especially computer simulation.

Computational hydromechanics is an independent discipline based on classical hydromechanics and numerical calculation methods (Wei 2010). It quantitatively describes the numerical solution of the flow field in time and space to analyze various flow properties of the fluid using computer numerical calculation and image display. For one thing, the rapid advancement of computer simulation technology is based on the development of computer technology. For another thing, compared with wind tunnel experiments, there is no need to consider space constraints, and there is no limitation of experimental sites. Moreover, various real-time data in the simulation process can be obtained in a relatively short period, and parameters and models can be modified through feedback results to achieve efficient space optimization design. The output of visual charts is also more convenient for displaying the simulation results than the results of wind tunnel tests. At the moment, computer simulation technology is widely used in the study of urban space design with the goal of adapting to the climate environment (Bai 2009).

2 THE CONSTRUCTION METHOD OF URBAN-SCALE VENTILATION CORRIDOR

The German scholar Kress divided the "urban ventilation corridor system" into three components, namely: climate ecological interaction space ("interaction space" in short), climate ecological compensation space ("compensation space" in short), and air guide channels.

Interaction space mainly refers to the urban built-up area or the area to be built that needs to improve the wind environment, especially the areas with severe urban heat island effects or needing to reduce pollution.

Compensation space refers to the source areas of the local circulation system that produce clean or cold air, regional or local climate resource areas, and climate-sensitive areas. The mainland types covered are farmland, cultivated land, grassland, hillside woodland, country green space, and large park green space.

The air guide channel refers to the connecting channel that guides the air from the compensation space to the interaction space. Usually, the space is relatively open with low surface roughness, and there are no tall buildings or trees. The air guide channels are generally linear or have a small arc. The direction mainly conforms to or forms a small angle with the prevailing wind direction or the local circulation wind direction. Its role is to facilitate air transport and diffusion. Especially for weak wind conditions, it plays an important role in strengthening the airflow between the city and its surroundings and within the city, as well as mitigating the climate problem in the interaction space.

Based on previous research, this paper proposed to construct urban-scale ventilation corridors by combining qualitative and quantitative analyses from the perspective of urban ventilation corridor planning. The details are as follows:

(1) The specific content of urban air duct analysis and construction was determined. In addition, the potential interaction spaces, compensation spaces, and air guide channels were analyzed and positioned based on the urban spatial scale and an urban ventilation corridor system was constructed considering the comprehensive characteristics of the dominant wind direction in the city.

(2) The quantitative research method was determined for the construction content of each urban air duct. To determine the interaction and compensation space, the urban surface temperature inversion was used to circumscribe the urban high- and low-temperature areas and combine the urban landscape characteristics, a green space system, and urban construction status. The important factors affecting urban ventilation were comprehensively analyzed and selected and each impact factor was transformed into a single quantifiable one. An evaluation model was constructed, and every single factor was overlapped into a comprehensive impact factor map by using the spatial overlay analysis method of ArcGIS to analyze and judge the potential air guide channels in the city. The wind rose diagram of each site was drawn by considering the distributed automatic weather station data, the dominant wind direction of each point is obtained, and the dominant wind direction characteristics of the city were comprehensively analyzed and determined.

(3) The construction and guidance method of the air duct were proposed. Based on the potential urban air ducts that have been constructed, some control measures were proposed for the potential air ducts from the perspectives of the overall urban layout, the urban road system, urban open space, and the construction of generalized air ducts to further strengthen the ventilation effect of the potential air ducts.

3 ANALYSIS ON THE POTENTIAL OF VENTILATION AND THE SELECTION OF VENTILATION CORRIDORS IN THE URBAN DESIGN OF LILING CERAMIC VALLEY

3.1 *Overview of design goals*

China Ceramic Valley is located in the northern core of Liling Economic Development Zone, Hunan Province. It is located in the "Zhuli Transition Zone" and "City-Garden Junction," which belong to the core of the "two-type society" reform experiment in Changzhutan. The total planned land is 15.4 square kilometers. Based on the urban positioning of the upper-level planning for the north area of Liling City, this plan proposes to build the north area of the city into a city with mountains and water as the base, emphasizing the interaction and integration of industrial-city space and natural landscapes to create an ecologically low-carbon, industrial-city integrated, and charming north area of Liling City. Based on this, the urban design concept of Liling Ceramic Valley is "mountainous." (Figure 1)

3.2 *Determination of the urban design interaction space and compensation space of Liling ceramic valley*

Through the inversion analysis of the surface temperature, the interaction space and compensation space of the ceramic valley were determined. Based on the atmospheric correction method for multi-phase Landsat 8 remote sensing image data, Landsat 8 TIRS inverted the surface temperature in the ceramic valley area (Figure 2). The data came from the Geospatial Data Cloud Platform of the Chinese Academy of Sciences (website: http://www.gscloud.cn).

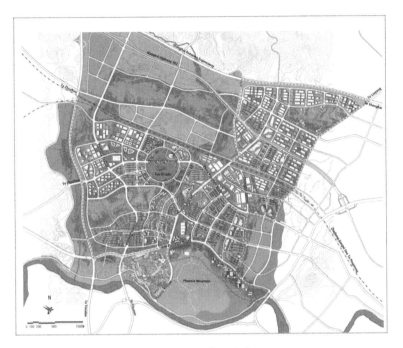

Figure 1. The general plan of the ceramic valley urban design.

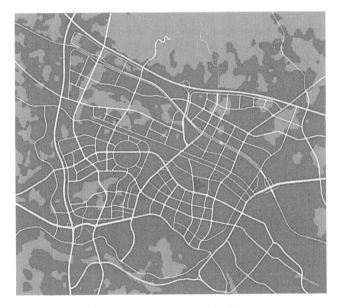

Figure 2. Inversion map of surface temperature in ceramic valley.

According to the surface temperature inversion map, it can be seen that the surface temperature of the core area at the current intersection of Fenghuang Avenue and Guoci Road in Ceramic Valley is significantly higher than that of the surrounding suburbs, showing the characteristics of a high core area and a low peripheral temperature as a whole (Figure 3).

Since the Ceramic Valley was located in the Economic and Technological Development Zone of Liling City, which was classified as a place with hot summers and cold winters according to the national climate division, the increase in heat load in summer would reduce the comfort of residents in the environment. In the subsequent design, we should strive to improve the ventilation capacity of the ceramic valley in summer at the urban level (Chen et al. 2014).

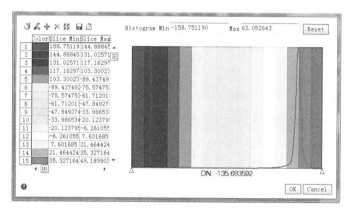

Figure 3. Classification curve of surface temperature in ceramic valley.

According to the inversion of the surface temperature in the ceramic valley area and the analysis of the high-temperature area, combined with the subsequent urban design plan of the ceramic valley, the interaction space of the ceramic valley was decided (Figure 4), which was marked as T1 to T9. T1 was the headquarters' business core area. T2 was the porcelain art and culture experience area. T3 was the industrial park. T4 was the residential and

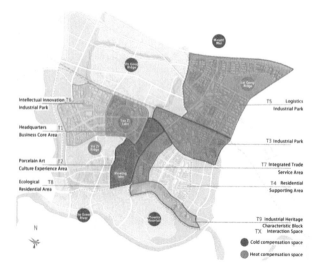

Figure 4. Schematic diagram of the interaction space and the compensation space of the ceramic valley.

supporting area. T5 was the logistics industrial park. T6 was the intellectual innovation industrial park. T7 was the integrated trade service area. T8 was the ecological residential area, and T9 was a characteristic block of industrial heritage.

According to the inversion of the surface temperature in the ceramic valley area and the analysis of the low-temperature area above, combined with the spatial distribution of farmland forests, mountain forests, water areas, green spaces in urban parks, and protective green spaces planned in the subsequent urban design, the compensation space for the ceramic valley was decided. From the perspective of classification, compensation space could be divided into two categories: cold compensation space in the cold air generation area outside the city and heat compensation space in the city.

3.3 *Determination of ceramic valley air guide channel*

Comprehensively considering the air guidance needs within the research scope, we conducted a comprehensive evaluation of the wind circulation potential of the ceramic valley area and determined the assignment and weight of wind circulation impact factors through analyses such as building density, building height, urban green space, urban water bodies, and the urban road network, and established a comprehensive evaluation index system (Table 1).

Table 1. Comprehensive evaluation checklist of impact factors of ventilation potential in the ceramic valley city.

Impact Factor	Evaluation Standard	Hierarchical Assignment	Weight
Average Building Density	BD > 40%	1	0.5128
	30% < BD ≤ 40%	2	
	20% < BD ≤ 30%	3	
	BD ≤ 20%	4	
Average Building Height	H > 80 m	1	0.2615
	40 m < H ≤ 80 m	2	
	20 m < H ≤ 40 m	3	
	H ≤ 20 m	4	
Green Space Rate	G ≤ 30%	1	0.1290
	30% < G ≤ 40%	2	
	40 < G ≤ 50%	3	
	G > 50%	4	
Urban Body of Water	H ≤ 20 m	1	0.0634
	20 m < H ≤ 30 m	2	
	30 m < H ≤ 40 m	3	
	40 m < H ≤ 50 m	4	
	H > 50 m	5	
Urban Road Network	H ≤ 25 m	1	0.0333
	25 m < H ≤ 35 m	2	
	35 m < H ≤ 4 2 m	3	
	42 m < H ≤ 50 m	4	
	H > 50 m	5	

In order to better describe and analyze the impact factor evaluation results of each block, this paper divided the target area into 16 control units, namely: A, B, C, D, E, F, G, H, J, K, L, M, N, P, Q, and X. Specifically, the distribution of building density within each control unit needs to be expressed graphically through the single impact factor of urban ventilation. According to the statistical data of the building density and height of each land unit, the hierarchical assignment of impact factors and the urban green space, water space, and road space of the Ceramic Valley were used to generate the single impact factor analysis diagram of the building density, building height, green space rate, and urban water bodies and urban roads, which was converted into the grid graphic language as well (Figure 5).

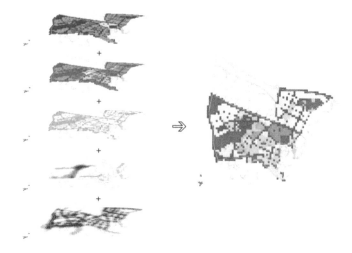

Figure 5. Impact factor overlay analysis chart.

The image map of each single impact factor was superimposed and analyzed according to the weight value above, and the comprehensive impact factor map of the urban ventilation potential was generated. The darker the color is, the greater the potential is for wind circulation in this place, and the lighter the color is, the less conducive it was to the circulation of urban wind. It can be seen from the distribution characteristics of wind circulation potential that the distribution of wind circulation potential in the Ceramic Valley was uneven, and each area had both strong and weak potentials.

3.4 *Construction of the main potential air ducts in the urban design of ceramic valley*

Combined with the analysis of the interaction and compensation space of the ceramic valley and the ventilation potential of the air guidance channel, and taking into account the difference in the dominant wind direction between summer and winter, five main ventilation

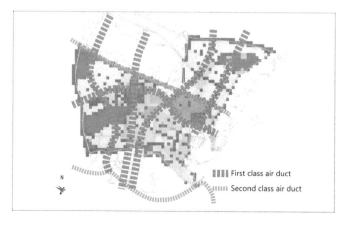

Figure 6. Sketch of urban air duct construction in ceramics valley.

corridors were selected. In addition, five potential primary air ducts (one horizontal and four verticals) were constructed, and two lateral potential secondary air ducts were proposed (Figure 6).

The following are the five potential primary air ducts:

(1) Lujiang River, Phoenix Mountain Park, Porcelain Art Culture Park, Shiziling, Taozi Lake Park, Shigongling Ventilation Corridor

Shigongling is close to the Ceramic Valley planning area, with a forest coverage rate of 90%. It is one of the most important suburban mountain woodlands in the ceramic valley, and it is also an important cold air generation area in the city. The cold air generated at night can pass through Taozi Lake Park and have a good gas exchange with the hot air in the city. Under the action of the south wind, the cold air formed by the Lujiang River passes through Fenghuang Mountain Park, connecting a series of park green spaces such as Porcelain Art Culture Park, Shiziling, and Taozihu Park from south to north, and finally meeting at Shigongling, forming an important ventilation corridor that plays an irreplaceable role in alleviating the heat island effect in the inner core of the city.

(2) Lujiang, Phoenix Mountain, Phoenix Avenue, Wei Mountain

Phoenix Mountain is a large cold air generation area and fresh air delivery source in the southern part of the Ceramic Valley. With the help of the wider red line width of Phoenix Avenue and the greenery on both sides of the road, the cold air and fresh air from the suburbs are transported to the central and northwest parts of the city. This area can effectively alleviate the heat island effect in the central, western, and north-western parts of the Ceramic Valley, especially the high-temperature interaction spaces such as the densely populated southern area.

(3) Lujiang, Phoenix Mountain Park, Shiziling, Panlong Avenue

Relying on Panlong Avenue, it connects Shiziling and Phoenix Mountain Park and extends to the Lujiang River, forming an urban air duct running through the west of the ceramic valley from north to south, which can effectively take away most of the urban heat of Zhichuang Industrial Park. What's more, in the summer, it provides a large amount of cool and fresh air for the ecological residential area in the southwest area of the Ceramic Valley.

(4) Ceramic Valley City East Entrance Square, Changge Road, Leigongling, Wei Mountain

The East Entrance Plaza of Ceramic Valley City is an important source of fresh air transportation in the eastern area. With the help of Changge Road and the surrounding strip-shaped waterfront green space, and with Leigongling, the city's cold source, it can effectively take away most of the city heat of the logistics industrial park.

(5) Dongcheng Avenue, Taozi Lake Park, Guoci Road

As the only road-type urban air duct that runs horizontally from east to west with a wider road red line as the basic condition, it consists of green belts on both sides of the road, Taozi Lake Park, and partially open urban parks. The large area of agricultural green space and woodland on the side of the building continuously transports fresh and cold air into the urban area, which has the effect of alleviating the urban heat island effect and reducing haze pollution.

The following are the two potential secondary air ducts:

(1) Lujiang

The Lujiang River is a first-class tributary of the Xiangjiang River. It has a wide river channel and sufficient water flow. It is a natural channel-type ventilation corridor. It takes away some of the waste gas, heat, and polluted air in the city through the fluidity of the water body and natural wind, which regulates the urban ecological environment. The green belts or roads along the Lujiang River can guide the cool natural wind of the Lujiang River into the urban core area and alleviate the urban heat island effect.

(2) Railway (Zhejiang-Jiangxi Line)

The west section of the Ceramic Valley section of the Zhejiang-Jiangxi railway line is the northern boundary of the city, and the eastern section passes through it. In the winter, the wider protective green belt can act as a good barrier, reducing the invasion of the cold winter wind on the interior of the ceramic valley city. In summer, fresh air can be transported to the ceramic industrial area with the help of urban air ducts, which can alleviate the heat released inside the industrial area and reduce the impact of air pollution such as smog.

4   CONCLUSION

This paper first proposed a new urban air duct construction method. Based on the basic principles of urban climatology, a comprehensive study of the theory of wind circulation at home and abroad, and the physical characteristics of the underlying surface of the ceramic valley urban design city, appropriate impact factors were selected for the potential analysis of wind circulation. On the basis of the superposition of the weights of each factor, the influence map of the wind circulation potential of the ceramic valley was obtained. Through the reservation of compensation space and the analysis of the urban wind direction of the ceramic valley, the air guide channels and the potential ventilation corridors of the city were figured out, so as to achieve the purpose of promoting urban ventilation, preventing the stagnation of polluted air in the city, and relieving the heat island effect in the city.

REFERENCES

Blocken B., Stathopoulos T., Beeck J.P.A.J. van, Hou Enzhe 2016 (05) Pedestrian-level Wind Conditions Around Buildings: Review of Wind-tunnel and CFD Techniques and Their Accuracy for Wind Comfort Assessment [J]. *Building energy efficiency*, China

Chun Bai 2009 *Urban Climate Design: A Way to Achieve Climate Rationality in Urban Spatial Form* [M]. China Architecture and Building Press, China

Feng Ding, Hanqiu Xu 2008, 1 (24): 91–96 Comparison and Analysis of Three Land Surface Temperature Retrieval Algorithms Based on Landsat TM [J]. *Journal of Fujian Normal University (Natural Science Edition)*, China

Guoan Tang, Xin Yang 2012: 217–222 *ArcGIS Spatial Analysis Experiment Course* [M]. Beijing: Science Press, China

Hong Chen, Xuefan Zhou, Fei Dai, *et al.* 2014 (7): 24–30 Study on the Planning of Urban Ventilators in Response to Urban Heat Island Effect and Air Pollution [J]. *Modern City Research*, China

Hong Leng 2009 *Study on the Livability of Urban Environment in the Cold Land* [M]. China Architecture and Building Press, China

Kun Li, Zhuang Yu 2006 (06) Analysis of Urban Ventilation Ducts Based on Climate Regulation [J]. *Journal of Natural Resources*, China

Ruizhao Zhu 1980 (04) Wind and Urban Planning [J]. *Meteorological Science and Technology*, China

Tingting Wei 2010 *Research on Urban Climate Simulation and Climate Adaptability Planning Strategy Based on CFD Technology* [D]. Master's Thesis of Central South University, China

Water Conservancy and Civil Construction – Oke & Ahmad (Eds)
© 2024 The Author(s), ISBN: 978-1-032-58618-2

Study on the accumulation and spatial distribution characteristics of sediment in a reservoir

Hongbo Tao*, Shuai Song, Yun Long & Yue Long
Guizhou Polytechnic of Construction, Guiyang, China

ABSTRACT: There are two main sources of pollution in water bodies: endogenous and exogenous. When the exogenous pollution is effectively controlled, the sediment will gradually be released upward after accumulating to a certain extent, aggravating the water pollution. The essence of sediment pollution is endogenous pollution, and the refractory organic matter and heavy metals in the sediment will be released to the water body under certain conditions, damaging the ecological environment. The reservoir is one of the main sources of tap water supply in the city, but there are more than 200 small and medium-sized coal mines in the catchment area of the reservoir, resulting in high content of total phosphorus, total nitrogen, iron, fierce, sulfate and other heavy metal elements in the water body and sediment. For the treatment of sediment pollutants in reservoirs, it is necessary to fully grasp the spatial distribution characteristics of sediment pollutants according to data analysis. The scope and degree of pollution are determined and the scientific basis for seeking the best plan for sediment treatment is provided.

1 INTRODUCTION

As a large amount of industrial and domestic sewage is discharged into reservoirs and other water bodies, pollutants are gradually deposited in the sediment (Washington 1994; Sondergard 2004), forming an endogenous load of the lake (Bing 2016; Telfeyan *et al.*, 2017). Exogenous pollution refers to the input of external pollutants, and the awareness of exogenous pollution is enhanced according to the path of entry, and exogenous pollution is usually better controlled (Lee *et al.*, 2016); On the contrary, endogenous pollution is a serious challenge to control water pollution. When the self-purification capacity of the water body is exceeded, the pollutants in the water body will be deposited in the bottom mud, the liquid nutrients in the sediment at the bottom of the lake will be released into the overlying water, and the nutrients will be suspended under the action of power to cause endogenous pollution (Zhou *et al.*, 2014). As one of the main sources of water supply, the pollution problem of the reservoir has received great attention from the government and widespread attention from environmental activists (Johnson *et al.*, 2016). The sediment-water interface of lakes is the main channel for the diffusion of lake sediment pollutants to the overlying water body, so studying the concentration and spatial distribution characteristics of pollutants in sediments will have a crucial impact on the release of endogenous pollutants in lakes, and this study aims to provide scientific guidance for the control of endogenous pollution in lakes.

2 SAMPLE COLLECTION AND ANALYSIS METHOD

2.1 *Distribution*

In this project, the measurement points (average density is about 100~150 / km2) are arranged according to the grid method in the area with thick silt at the bottom of the reservoir, and the

---

*Corresponding Author: 862103453@163.com

DOI: 10.1201/9781003450832-36

distribution of the sediment thickness in the total water area is shown in Figure 2 during the investigation process. Figure 1 is based on the actual situation of the water, and the distribution map of 264 measurement points is deployed in the pre-selection area through GPS. The collected sediment column core is 0~5cm, 5~10cm, 10~20cm, 20~30cm, and 30~40cm according to the actual silt thickness of the sediment. Aliquot and bring the obtained samples back to the laboratory. The contents of total phosphorus (TP), total nitrogen (TN), total sulfur (TS), iron (Fe), and manganese (Mn) were determined, and the pollutant concentration data of each measurement point in the pre-selected area were summarized to draw the spatial distribution map of sediment pollutants at different levels in the dredging preselection area of Reservoir. The horizontal and vertical distribution characteristics of TP, TN, FE, MN, and TS contents of the sediment of the Reservoir were revealed.

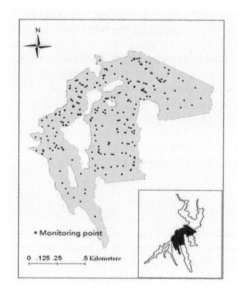 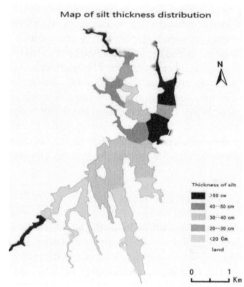

Figure 1. Distribution map of monitoring sites in sediment control area.

Figure 2. Map of silt thickness distribution in all water areas.

2.2 Sample analysis method

Sediment is mainly silt. For the sediment samples collected (thickness of silt 0-30cm), the sediment samples filtered out of the pore water were freeze-dried by vacuum freeze-dryer (Techconp FD-3-85-MP), ground to less than 120 mesh with an agate mortar, and placed in a sealed bag for analysis and reserve. All experiments in this project were completed at the State Key Laboratory of Environmental Geochemistry, Institute of Geochemistry, and Chinese Academy of Sciences.

2.2.1 Determination of total phosphorus (TP) content of sediment

By weighing 0.5 g sample, ashing at 500 °C for 2h, then extracting 25 mL 1mol/L HCl for 16h, extracting the solution through 0.45μm acetate filter, and then using the national standard molybdenum antimony anti-spectrophotometry to determine the content of TP, the minimum detection limit of this method is 0.01 mg/L.

2.2.2 Determination of total nitrogen (TN) and total sulfur (TS) content of sediment

Approximately 5 mg of the sample was weighed and the sediment TN and TS content was analyzed by an organic element analyzer (varia MACRO cube). The combustion temperature of the instrument is 975 °C, the analysis range of nitrogen and sulfur is 0.001-6 mg, and the analysis error is less than 5%.

2.2.3 *Determination of sedimentary iron (Fe) and manganese (Mn) content*

Weighing about 50mg of sample, placing it in a Teflon tube, adding 0.8mLHF+1 mL HNO_3 in a steel tank and seal, heating (180-190 °C) in an oven for 24-30h, cooling the Teflon tube and placing it on a hot plate to evaporate (140°C), then adding a little HNO3 (<1ml), evaporating (dry through) add 2mLHNO$_3$ + 3mL of deionized water, heating (140°C) in a sealed oven for 4-5h, cooling and setting the volume to 100mL, the sample after the fixed volume is analyzed by plasma emission spectrometer (ICP-OES).

3 THE ACCUMULATION AND SPATIAL DISTRIBUTION OF EACH POLLUTANT

3.1 *Spatial distribution characteristics of total phosphorus in sediment*

The average content of total phosphorus in 0~30cm of sediment in the reservoir was 1.07%, and the distribution of high-value areas was similar to that of 0-5cm. The area with total phosphorus content exceeding 2% is located in the upper reaches of the Northeast River Basin, and the total phosphorus content of sediment in the middle reaches of the basin is between 1.5~2%. In the estuary of the northwest basin, the total phosphorus content of the central sediment in the southern lake area is between 1~1.5%, and the total phosphorus content of the sediment in other lake areas is 1% lower.

3.2 *Spatial distribution characteristics of total nitrogen in sediment*

The average content of total nitrogen in the 0~30cm sediment of the reservoir was 0.35%, and its distribution of high-value areas was relatively scattered. Areas with a total nitrogen content of more than 0.4% are mainly located in the upper reaches of the southeast basin, estuaries, and upper and middle reaches of the northeast basin. The total nitrogen content in the lake area near the management office, the northern basin, the lower reaches of the northeast river basin, and the lake area on the south side of the management office are between 0.3~0.4%. In other lake areas, sediment's total nitrogen content is less than 0.3%.

In all sediment samples collected from this reservoir, the lowest value of total nitrogen was 0.19% and the highest value was 0.69%.

3.3 *Spatial distribution characteristics of sediment iron content*

The average iron content in the 0~30cm sediment of the reservoir was 7.14%, and its high-value area distribution was relatively concentrated. The areas with iron content exceeding 7.5% were located in the northwest basin, southwest basin, southern basin, and the middle of the reservoir lake, of which the iron content in the upper reaches of the southwest basin exceeded 10.0%. The iron content of sediment in the middle of the estuarine lake area in the northeast basin and southeast basin is between 5.0~7.5%.

Among all the sediment samples collected in this experiment, the lowest value of iron content was 2.15% and the highest value was 19.75%.

3.4 *Spatial distribution characteristics of sediment manganese content*

The average content of manganese in the 0~30 cm sediment of the reservoir was 0.43%, and the distribution of high-value areas showed the law of "high in the middle and low in the four weeks". The area with more than 0.75% of manganese content is located in the main part of the lake area, of which the manganese content in the estuary of the northwest basin and the middle of the lake area exceeds 1.0%. The manganese content in the southwest, southeast, and northeast basins was less than 0.50%.

In all sediment samples collected from this reservoir, the manganese content was as low as 0.07% and as high as 3.24%.

3.5 *Spatial distribution characteristics of total sulfur content in sediment*

The average total sulfur content in the 0~30 cm sediment of the reservoir was 0.26%, and the distribution of high-value areas was relatively concentrated. The areas with a total sulfur content of more than 0.25% were mainly located in the upper reaches of the southwest basin, the estuaries of the northeast basin, the estuaries of the northwest basin, and the central lake area of the lake body, among which the sulfur content in the upper reaches of the southwest basin and the estuaries of the northwest basin exceeded 0.50%. In most other lake areas, the sulfur content of sediment is less than 0.25%.

In all sediment samples collected from the reservoir, the lowest value of total sulfur content was 0.03% and the highest value was 1.50%. Strong biogeochemical cycling at the sediment-water interface may lead to the release and diffusion of trace (heavy) metals and sulfates, and the remigration of pollutants may induce secondary pollution, resulting in the deterioration of water quality in drinking water source areas.

3.6 *Content of phosphorus, nitrogen, iron, manganese, and sulfur in sediment*

The contents of phosphorus, nitrogen, iron, manganese, and sulfur in the reservoir's 0-30cm sediment are shown in Table 1.

Table 1. Element content was detected in 0-30cm sediment of the reservoir.

Phosphorus content	Nitrogen content	Iron content	Manganese content	Sulfur content
1.07%	0.35%	7.14%	0.43%	0.26%

4 CONCLUSIONS

The depth of sediment silt in the reservoir is 8~76cm, and the average silt thickness is about 39cm. According to the weighted sum of manganese, iron, total sulfur content, and dry weight of each sediment sample, the total amount of manganese, iron, and total sulfur of the sampled product was obtained, and the manganese, iron, and sulfur accumulation per unit volume of silt in Reservoir was obtained (Tables 2 and 3). Combined with the lake area and the average thickness of silt at the lowest level of the Reservoir, the total iron accumulation of the reservoir is about 48,000 tons, the total amount of manganese storage is about 2,000 tons, and the sulfur storage is about 1,700 tons.

Table 2. Estimation results of major pollutants in Reservoir stocks 1.

Average thickness of silt	Silt volume	Phosphorus content per unit volume	Nitrogen content per unit volume	Phosphorus stocks	Nitrogen stock
39 cm	1.3×106 m3	0.53 mg/cm3	1.77 mg/cm3	700 t)	2350 t)

Table 3. Estimation results of major pollutants in Reservoir stocks 2.

Average thickness of silt	Silt volume	Unit volume Iron content	Unit volume Manganese content	Unit volume Sulfur content	Iron content	Manganese content	Sulfur content
39 cm	1.3×10^6 cm	36.1 cm^3	1.58 g/ cm^3	1.33 mg/ cm^3	4.8×10^4 t	2000 t	1700t

The pollutant load of reservoir sediment is relatively serious. The average concentrations of total phosphorus (TP), total nitrogen (TN), total sulfur (TS), iron (Fe), and manganese (Mn) in the whole sediment column core ranged from 651~1085 mg/kg, 0.282%~0.537%, 0.861%~1.718%, 3.05%~11.02%, and 0.24~1.59%, respectively, and the average values were 895 mg/kg, 0.391%, 1.222%, 5.91%, and 0.67%, respectively, of which TP, TN, and TS contents showed the distribution trend of "high on the surface and low in the bottom", while the iron (Fe) and manganese (Mn) contents showed the distribution characteristics of "low on the surface and high in the bottom". The average contents of total phosphorus (TP), total nitrogen (TN), total sulfur (TS), iron (Fe), and manganese (Mn) in the surface 0-5 cm sediment were 986 mg/kg, 0.557%, 4.68%, and 0.49%, respectively, and the content ranges were 717~1794 mg/kg, 0.228%~0.848%, 1.93%~7.78%, and 0.05~2.62%, respectively.

The average concentrations of total nitrogen (TN), iron (Fe), and manganese (Mn) in the surface 2cm sediment were as high as 47.1 mg/L, 1.118 mg/L, and 2.825 mg/L, respectively, especially in the surface 0~10cm sediment. The high iron and manganese content is coupled with the large water depth of the reservoir (10~20m), and the water body of the bottom stagnant zone are in a state of hypoxia and anaerobic for a long time, which is easy to induce the diffusion of nitrogen and phosphorus pollutants through the sediment-water interface (endogenous release), which is manifested as a high release risk, and it is urgent to treat the reservoir sediment.

REFERENCES

Committee on Contaminated Marine Sediments, National Research Council (1997). Contaminated Sediments in Ports and Waterways: Cleanup Strategies and Technologies. Washington D. C.: National Academy Press.

Dexter K S, Ward N L (2004). Mobility of Heavy Metals Within Freshwater Sediments Affected by Motorway Stormwater, *Science of the Total Environment*,334-335;271–227.

Eastern Tibetan Plateau as Reconstructed from Alpine Lake Sediments Over the Last Century (2016). *Chemosphere*, 148: 211–219.

Johnson JE, Webb SM, Ma Chi, et al (2016) Manganese Mineralogy and Diagenesis in the Sedimentary Rock Record. *Geochemical et Cacochymical Acta*,73: 210–231.

Lee S, Oh J, Kim D, et al. (2016) A Sensitive Electrochemical Sensor Using an Iron Oxide /Graphene Composite for the Simultaneous Detection of Heavy Metal Ions. *Talanta*,160: 528–536.

L Y Yang, L Zhang, G W Zhu (2006). Mechanism and Control of Lake Eutrophication[J]. *Chinese Science Bulletin*, 10(3): 82–87.

Roberts D A (2012). Causes and Ecological Effects of Resuspended Contaminated Sediments (RCS) in Marine Environments, *Environment International*, 40:230–243.

Sondergard L M, Jensen R J (2004). Internal Phosphorus Loading in Shallow Danish lakes[J]. *hydrobiology*, 15(6): 145–152.

Telfeyan K, Breaux A, Kim J, et al (2017) Arsenic, Vanadium, Iron, and Manganese Biogeochemistry in a Deltaic Wetland, Southern Louisiana, USA. *Marine Chemistry*, 192: 32–48.

Zhou G, Sun B, Zeng D, et al. (2014) Vertical Distribution of Trace Elements in the Sediment Cores from Major Rivers in East China and its Implication on the Geochemical Background and Anthropogenic Effects. *Journal of Geochemical Exploration*, 139: 53–67.

Effects of water and nitrogen coupling on potato yield, water and nitrogen utilization and soil environment

Chao Liang

College of Agronomy and Agricultural Engineering, Liaocheng University, Liaocheng, China
College of Water Conservancy and Hydropower Engineering, Gansu Agricultural University, Lanzhou, China

Shouchao Yu

College of Agronomy and Agricultural Engineering, Liaocheng University, Liaocheng, China

Hengjia Zhang*

College of Agronomy and Agricultural Engineering, Liaocheng University, Liaocheng, China
College of Water Conservancy and Hydropower Engineering, Gansu Agricultural University, Lanzhou, China

ABSTRACT: To understand the research progress of potato water and nitrogen coupling effect and further improve potato yield and quality, this paper first introduces the coupling effect of water and nitrogen. On this basis, the effects of water and nitrogen coupling on potato yield, quality, water, fertilizer use efficiency, and soil environment were discussed. The results showed that there was a threshold range of water and nitrogen coupling in potatoes. Reasonable water and nitrogen coupling management could effectively improve water and nitrogen efficiency, improve soil environment, and ultimately increase yield and improve quality. Finally, the existing problems of water and nitrogen coupling were analyzed, and future research work has been anticipated.

1 INTRODUCTION

Potato (Solanum tuberosum L.), is rich in nutritional value, and waxy taste, with wheat, corn, and rice as the world's four major food crops (Tang *et al.* 2021). China is a large potato-producing country, and its production area and total output rank first in the world (Tang *et al.* 2021). According to the report, the amount of fertilizer applied in China is about $393.2kg/hm^2$, which is much higher than the international safety threshold of $225kg/hm^2$ (Yang *et al.* 2020), which is 3.1 times and 2.5 times that of the United States and the European Union, respectively (Chen *et al.* 2021). With the excessive or unreasonable application of nitrogen fertilizer, soil nitrogen loss, water pollution, groundwater decline, environmental damage, and other issues have become increasingly prominent, resulting in crop yield and quality decreasing year by year. Moisture not only helps soil nutrient activation, and improves nitrogen availability, but also may increase nitrogen loss or leaching losses, causing environmental hazards (Wang *et al.* 2010). The response of crop nitrogen application and its nitrogen use efficiency depends not only on nitrogen application rate, nitrogen application time, method, etc. but also on irrigation amount, irrigation time, method, etc. Reasonable management of water and fertilizer can make crops fully absorb

---

*Corresponding Author: 596088683@qq.com

water and fertilizer, and then reach the effect of adjusting water with fertilizer, promoting fertilizer with water (Du *et al.* 2017), giving full play to the synergistic effect of water and nitrogen, solving the problem of the low utilization rate of irrigation water resources in China, and alleviating the safety of soil ecological environment caused by excessive fertilization. In this paper, the effects of water and nitrogen interaction on potato yield, quality, water, nitrogen use efficiency, and soil environment were reviewed, to provide a basis for formulating reasonable irrigation and nitrogen application technical measures suitable for potato growth and yield formation.

2 EFFECT OF WATER-NITROGEN COUPLING ON POTATO

2.1 *Effects of water and nitrogen coupling on water and nitrogen use efficiency of potato*

China's current irrigation water use efficiency and crop water production efficiencies are only 45% and 0.8kg/m^3, far below the level of 70% and 2 kg/m^3 in developed countries (Wang 2021). Improving the utilization efficiency of agricultural water resources is an important goal of developing modern efficient water-saving agriculture. Jia Shuai *et al.* (2021) found that the effect of coupling of nitrogen, phosphorus, and potassium on the water use efficiency of potatoes was nitrogen application > phosphorus application > potassium application. When the irrigation amount was 1500 m^3/hm^2 and the nitrogen application amount was 180 kg/hm^2, the water use efficiency was up to 11.06 kg/m^3.Tang *et al.* (2021) showed that the coupling of water and nitrogen could effectively improve the water and nitrogen production efficiency of potatoes, but it was greatly affected by climatic factors. The water use efficiency in humid years increased by 7.72%~9.60% compared with drought years. Nitrogen use efficiency is an important index to evaluate the economic benefit of crops. Sun Fubin *et al.* (2022) found that the irrigation water use efficiency of potatoes increased at first and then decreased in middle and low water treatment, and the changing trend was the opposite under high water treatment. The changing trend of fertilizer partial productivity is the same as that of irrigation water use efficiency. Under certain nitrogen fertilizers, the changing trend of irrigation water use efficiency is inversely proportional to irrigation amount. Li *et al.* (2016) reported that under the same water level, the nitrogen use efficiency of potatoes with high nitrogen was lower than that with medium nitrogen. Under different water and nitrogen levels, which were 52.5 g/g and 143.9 g/g, respectively, it can be seen that reasonable water and nitrogen supply can promote each other, and there is also a threshold reaction between water and nitrogen. Exploring a reasonable water and nitrogen coupling threshold level can effectively reduce resource waste and improve agricultural production efficiency.

2.2 *Effects of water and nitrogen coupling on the soil environment of potato*

The study found that unreasonable water and fertilizer application will have adverse effects on the soil environment, water environment, and atmospheric environment. At present, the recovery and utilization of nitrogen fertilizer in China are relatively low, about 30 % ~ 40 %. The loss and residue in soil are large, and the loss rate and residue rate are 20 % ~ 60 % and 25 % ~ 35 % respectively. Excessive or insufficient irrigation will result in nitrogen loss and reduced utilization. When excessive nitrogen fertilizer is applied, the remaining nitrogen fertilizer will exist in the soil in the form of nitrate, which seriously endangers food health. When sufficient water is encountered, excessive nitrate in the soil will migrate downward with water and lead to eutrophication of the water environment, causing water environment pollution. Zhang *et al.* (2018) found that the apparent loss, loss rate, apparent surplus, and surplus rate of soil increased significantly when the nitrogen application rate of potato exceeded 90 kg/ha under rain-fed conditions, indicating that the applied nitrogen fertilizer was leached to deeper soil or volatilized to the environment in addition to being absorbed by plants and accumulated in 0-100 cm soil. Zhou Nana *et al.* (2005) showed that under different irrigation conditions, soil nitrogen leaching increased with the increase in nitrogen application rate. Multiple low-water irrigations are the best irrigation method for potato planting. It not only increases the nitrogen uptake of potatoes but also reduces the nitrate nitrogen content in deep soil and has little environmental pollution. Therefore, to determine the reasonable amount of nitrogen and irrigation, in addition to considering crop yield and nitrogen use efficiency, the impact of soil nitrogen residue and loss on the environment should also be considered, which meets the nitrogen residue standard (nitrate nitrogen residue in 0-90 cm soil is less than 45.0 kg/ha or inorganic nitrogen residue is less than 50 kg/ha) (VandermPloeg 1997).

2.3 *Effect of water and nitrogen coupling on potato yield*

Fertilizers and water are not only the key factors determining crop yield but also the technical measures that can be controlled artificially. The effect of water and fertilizer coupling on crop yield is different at different water and fertilizer supply levels (Chen 2011). Song Na *et al.* (2013) found that the yield of potatoes was significantly affected by nitrogen application, and there was no significant difference in irrigation. Increasing the soil moisture ratio or increasing nitrogen fertilizer can increase the tuber quality per plant of potato, thereby increasing potato yield. Under the same nitrogen fertilizer conditions, the yield of soil moisture ratio 70% treatment was higher than that of soil moisture ratio 40% treatment, while the water use efficiency was lower than that of soil moisture ratio 40% treatment. Under the same irrigation level, the yield increased first and then decreased with the increase of the N application rate. When the N application rate increased to $135\sim180$ kg/hm^2, the yield reached the maximum. Zhou Nana *et al.* (2005) showed that excessive nitrogen application reduced potato yield when the irrigation amount was the same, and excessive nitrogen application inhibited the formation of potato stem tubers, thereby reducing potato weight per plant and potato commodity rate.

2.4 *Effect of water and nitrogen coupling on potato quality*

Water and fertilizer are the main factors directly affecting crop photosynthesis and biomass accumulation, thus affecting crop yield and quality. Increasing nitrogen application within a certain range can effectively improve crop quality, but beyond a certain range, increasing nitrogen application will lead to reduced crop quality, which is similar to the effect of nitrogen fertilizer on yield (Chen 2011). Gao *et al.* (2017) found that the amount of nitrogen and the amount of nitrogen applied must increase the amount of irrigation, and the changing trend of potato starch content is consistent, both in the form of a secondary parabola. When the amount of irrigation and nitrogen application were variables, there was no significant difference among the treatments. When the irrigation amount was 1710kg/hm^2 and the top-dressing amount of nitrogen was 180 kg/hm^2, the potato starch content reached the maximum value of 14.01%. Hu *et al.* (2022) showed that different water and nitrogen coupling had significant effects on potato starch, reducing sugar and vitamin C content. When the irrigation amount was constant, the vitamin C content of potato tubers increased with the increase of nitrogen application. Zang Wenjing *et al.* (2018) used sprinkler irrigation technology to conduct different levels of water and nitrogen coupling experiments on potatoes. The study found that there was no significant difference in the content of starch and vitamin C in the same potato stems among different water and nitrogen combination treatments. Excessive application of nitrogen fertilizer will also reduce the content of crude protein in potato stems.

3 CONCLUSIONS AND PROSPECTS

The process of water and nitrogen coupling and efficient utilization of potatoes is very complicated and vulnerable to environmental impact. Making full use of the positive synergistic effect of water and nitrogen coupling is an important way to improve potato yield and water and nitrogen use efficiency. Combined with the research results of many scholars at home and abroad, the potato water and nitrogen coupling can be further studied in the following aspects: (1) The biological process of water-nitrogen interaction for high yield and high efficiency of potato was studied, and the mechanism of water-nitrogen interaction was revealed from the aspects of root signal transduction, endogenous hormone, nitrogen metabolism system, and gene expression. (2) The environmental effects of water and nitrogen coupling in potatoes from soil physical and chemical properties, nitrogen migration and transformation, and greenhouse gas emissions are clarified.

ACKNOWLEDGMENTS

This work was financially supported by the Industrial Support Plan Project of the Gansu Provincial Department of Education (No. 2022CYZC-51), the National Natural Science Foundation of China (No. 52269008), and the Key Research and Planning Projects of Gansu Province (No. 18YF1NA073).

REFERENCES

Chen R.Y.: *Effects of Water and Nitrogen Interaction on Yield and Nitrogen Absorption and Utilization Characteristics of Potato*, D. Inner Mongolia Agricultural University. 2011.

Chen R.Y., Meng M.L., Liang H.Q., Zhang J., Wang Y.H., Wang Z.X., Effects of Different Treatments of Irrigation and Fertilization on the Yield and Nitrogen Utilization Characteristic of Potato, *J. Chinese Agricultural Science Bulletin* 2012, 28(03):196–201.

Chen Y., Hu S., Z. Guo, T.H. Cui, L.P. Zhang, C.R. Lu, Y.Q. Yu, Z.B. Luo, H. Fu, Y. Jin, Effect of Balanced Nutrient Fertilizer: A Case Study in Pinggu District, Beijing, China, *J. Science of The Total Environment*. 2021, 754, 142069.

Du J.J., An Y.J., Huang B.Y., Li Y.S., Wang X. A.: Research Progress on Water and Fertilizer Regulation Technology and Functional Fertilizers, *J. Journal of Plant Nutrition and Fertilizers*. 2017, 23, 1631–1641.

Gao F.: *Effects of Water and Nitrogen Coupling on Soil Water and Nitrogen Transport, Yield and Quality of Potato Under Mulched Drip Irrigation*, D. Inner Mongolia Agricultural University. 2017.

Hu P.C., Yi J., Wei X.D.: Effects of Different Water-nitrogen Treatments on Potato Quality and Soil Urease Activity, *J. Jiangsu Agricultural Sciences*. 2022, 50, 87–92.

Jia S., Guo Y.J., Gao H.X., Meng M.G.: Study on the Coupling Effect of Water, Nitrogen, Phosphorus and Potassium on Potato Yield and Efficiency in Arid Areas, *J. Water Saving Irrigation*. 2021, 04, 23–27.

Li W.T., Wang S.W., Deng X.P., Li H.B.: Effects of Different Water and Nitrogen Levels on Yield and Water and Nitrogen Use Efficiency of Potato, *J. Agricultural Research in the Arid Areas*. 2016, 34, 191–196.

Song N., Wang F.X., Yang C.F., Yang K.J.: Coupling Effects of Water and Nitrogen on Yield, Quality and Water Use of Potato with Drip Irrigation under Plastic Film Mulch, *J. Transactions of the Chinese Society of Agricultural Engineering*. 2013, 29, 98–105.

Sun F.B., Yi J., Wei X.D., Ma Z.H., Yang Y.P., Shun S.: Comprehensive Evaluation of Effects of Water-nitrogen Topdressing Ratio on Potato Soil Enzyme Activity, Yield and Irrigation Water and Fertilizer Use Efficiency, *J. Soil and Fertilizer Sciences in China*, 2022, 1–15.

Tang J.S., Xiao P.D., Wang J., Wang R.D., Bo H.Z., Guo F.H., Liu J.F.: Optimizing Irrigation and Nitrogen Management for Potato Production Under Multi-objective Production Conditions, *J. Transactions of the Chinese Society of Agricultural Engineering*. 2021, 37(20).

VandermPloeg R.R., Ringe H., Machulla G., Hermsmeyer D.: Postwar Nitrogen use Efficiency in West Germany Agriculture and Groundwater Quality, *J. Journal of Environmental Quality*. 1997, 26(6), 1203–1212.

Wang X.B., Dai K., Zhao Q.S., Wu X.P., Zhang D.C., Feng Z.Z., Jia S.L., Yang Y.M., Cai D.X.: Relationship between Water and Nitrogen in Farmland and Its Collaborative Management, *J. Acta Ecological Sinica*. 2010, 30, 7001–7015.

Wang X.M.: *Effects of Nitrogen Reduction and Water Saving on Yield and Water and Nitrogen Use Efficiency of Summer Maize in Guanzhong Plain*, D. Northwest A&F University. 2021.

Yang Y., He Y., Li Z.: Social Capital and the Use of Organic Fertilizer: An Empirical Analysis of Hubei Province in China, *J. Environmental Science and Pollution Research*. 2020. 27(13): 15211–15222.

Zang, W.J., Li J.J., Pei S.S., Li Y.J., Yan H.J.: Effects of Different Water-nitrogen Combinations on Potato Water Consumption, Yield and Quality under Sprinkler Irrigation, *J. Journal of Drainage and Irrigation Machinery Engineering*. 2018, 36(08):773–778.

Zhang J.: *Study on Water - Nitrogen Coupling Effect and Rational Utilization Mechanism of Potato in North Yinshan Mountain*, D. China Agricultural University. 2018.

Zhou N.N., Wang G.: Study on Yield and Varieties of the NO3–N Contain with Different Water and N Levels in Potato, *J. Journal of Hainan Tropical Ocean University*. 2005, 05, 54–56.

Research on the evaluation index system for Yangtze River waterway maintenance effectiveness under the new situation

Liu Lei, Duan Guo-sheng* & Mu Bao-yin
The Traffic Department of Transportation Water Science Research Institute, Beijing, China

ABSTRACT: Safe and smooth navigation is the basis for the high-quality development of Yangtze River shipping, and maintenance is an important way to ensure the smooth flow of the Yangtze River waterway. So far, there is no effective evaluation method for the Yangtze River waterway maintenance that would be detrimental to promoting the high-quality development of waterway maintenance. Based on the current situation and characteristics of the Yangtze River waterway maintenance, a systematical evaluation index system for Yangtze River waterway maintenance effectiveness based on the statistical and mathematical analysis method was established, which provides the theoretical reference and technical support for waterway maintenance effectiveness evaluating and policy formulating.

1 INTRODUCTION

To speed up the construction of China's strength in transportation and build a modern and high-quality national comprehensive three-dimensional transportation network, the State Council of the Central Committee of the Communist Party of China has issued *Outlines for the Building a Transport Power* and *Guidelines for the National Integrated Multidimensional Transparent Network*, the Ministry of Transport has formulated and released *Opinions of the Ministry of Transport on Promoting the High-Quality Development of Yangtze River Shipping and the 14th Five-Year Plan for the Development of the Yangtze River Navigation System*. The Communist Party of China's 20th National Congress further emphasized the goal of building transportation power. In recent years, the development trend of large-scale ships on the Yangtze River has become obvious, and its shipping demand has shown diversified characteristics, thus, besides the smooth and efficient development, the safety and green development of the Yangtze River waterway has also become increasingly high.

The Yangtze River waterway starts from Shuifu in Yunnan Province to the estuary of the Yangtze River, with a total length of 2843 km. It is divided with the characteristics of the different river sections of the waterway: the section above Yichang is the upstream channel, the section from Yichang to Wuhan is the middle channel, the section from Wuhan to Liuhekou is the downstream channel, and the section from Liuhekou to Yangtze River Estuary lightship is the Yangtze River Estuary channel. There are three navigable structures in the main channel of the Yangtze River, namely the Three Gorges ship lock, Gezhouba ship lock, and Three Gorges ship lifter. The freight volume on the Yangtze River has increased from 1.75 billion tons in 2012 to 3.5 billion tons in 2021, which has consistently displayed the highest levels in the world's inland shipping for more than 10 years. Besides, the Cargo turnover on the Yangtze River accounts for 60% of the total social volume along

---

*Corresponding Author: 892389667@qq.com

DOI: 10.1201/9781003450832-38

the river and the GDP of the Yangtze River Economic Zone accounts for more than 46% of the national economy (Liu 2022).

After systematic analysis, the main features of the Yangtze River waterway maintenance are as follows:

(1) Heavy workload. The Yangtze River waterway has a long mileage, and every year more than 6,800 navigation marks need to be added, about 2.2 million of them need to be maintained, 40,000 conversion square kilometers channels need to be surveyed, and 15,000 conversion square kilometers regulating structures in the channel need to be surveyed.
(2) Difficulties. The water level of the Yangtze River waterway varies greatly, the shoals evolve drastically, and the locations of navigation marks change frequently and are susceptible to natural conditions such as floods, gale, bank failure, and other human factors such as ship collisions.
(3) Great difference. The upstream is a three-stage channel with complex conditions; the midstream is a sandy riverbed with many shallow bends; the downstream has long mileage and complex channels, which needs maintenance and dredging throughout the year.

2 PRINCIPLES AND IDEAS FOR THE CONSTRUCTION OF THE EVALUATION INDEX SYSTEM

An index system is an organism of all indexes that indicate the overall characteristics and nature of the object to be evaluated (Yue 2014). At present, some evaluation index systems have been developed, such as the evaluation index system for environment management of ecological city (Wang 2012), the comprehensive development evaluation index system for rural revitalization (Zhang 2022), the modernization evaluation index system for waterway maintenance and management (Zhang 2022), and the evaluation index system for waterway service level (Zhang 2006). Its principles include the principles of comprehensiveness, operability, representativeness, scientificity, the combination of comprehensiveness and focus, comparability and feasibility, etc. The general idea of establishment is to use mathematical methods to construct an evaluation index system based on fully studying the evaluation contained in this field, combined with the principle of index system construction.

In terms of the evaluation index system for waterway maintenance and management, China is still in its starting stage and there is no recognized evaluation index system yet, but the above-developed evaluation index systems provide a good reference for the development of this present study.

2.1 *Principles of construction*

To accurately and objectively represent the maintenance effect and scientifically guide the maintenance work in Yangtze River waterways, these listing principles for the index system construction should be followed.

(1) Scientificity. The evaluation content of the maintenance effect should be in line with the national industry policies, regulations, and technical standards, and the selection of evaluation indexes and the establishment of the evaluation index system should use scientific and reasonable methods;
(2) Objective. The evaluation indexes should not only contain the essential elements for evaluating but also have a certain orientation;
(3) Representativeness. The selected evaluation indexes should represent the main elements or their key aspects of Yangtze River waterways maintenance effectiveness;
(4) Operability. The accessibility of the indexes should be considered when selecting them, and then the established index system can be implemented concretely.

2.2 Construction ideas

Firstly, insisting on the combination of problem-oriented and goal-oriented, the content and scope of the waterway maintenance effect evaluation are determined from the demand side according to the content and characteristics of the Yangtze River waterway maintenance work. Secondly, hierarchical analysis (Zhao 2022) and the Delphi method (Zhang 2018) were used to hierarchize the evaluation contents of waterway maintenance effectiveness, and divide the evaluation contents into the project layer and index layer. Then, based on the evaluation contents of the waterway maintenance effectiveness, by the use of a combined method of theoretical analysis, frequency analysis, and expert advice (Wang 2012), the primary and secondary indexes in the index layer are determined according the principles of the index system construction mentioned above. At last, a comprehensive evaluation work for the Yangtze River waterways maintenance will be undertaken by the use of a combined qualitative and quantitative approach (Cui *et al.*, 2022).

3 INDEX SYSTEM FOR EVALUATING THE YANGTZE RIVER WATERWAY MAINTENANCE EFFECTIVENESS

Based on fully sorting out the content of the Yangtze River waterway maintenance, and considering the positive effects and benefits generated by waterway maintenance, the evaluation of the Yangtze River waterway maintenance effectiveness is divided into two aspects: evaluation of the effectiveness of waterway maintenance and evaluation of the benefits of waterway maintenance.

3.1 Evaluation of the waterway maintenance effectiveness

The indexes for evaluating the waterway maintenance effectiveness are mainly used to evaluate the positive effects brought about by waterway maintenance, and they mainly focus on four dimensions: smooth, safe, efficient, and green. There are 4 primary indexes and 17 secondary indexes in each dimension, as shown in Table 1.

Table 1. Index system of evaluation of waterway maintenance effectiveness.

Project	Level 1 indexes	Secondary indexes	Guidelines
Evaluation of the effectiveness of waterway maintenance	Unobstructed	Guarantee rate of navigation scale (%)	98
		Normal rate of buoy maintenance (%)	99.9
		Normal rate of navigation signal command (%)	100
		Analytical coverage of waterway maintenance tracking observation (%)	37.8
		Annual guarantee rate of maintenance of water depth (%)	≥ 98
		Annual navigation time guarantee rate of ship lock (%)	94
	Security	Accident rate of waterway maintenance liability (%)	0
		Average recovery time of navigational marks malfunction (hour)	Upstream ≤ 3; Midstream ≤ 3; Downstream ≤ 6
		Emergency response time for buoy vessels (Minutes)	60
		General natural disasters and accidents caused by the damage to the waterway, blockage repair rescue time (hour)	≤ 48

(*continued*)

Table 1. Continued

Project	Level 1 indexes	Secondary indexes	Guidelines
	Efficient	Digital channel coverage (%)	100
		Maintenance scale information release frequency	1-2 times per week
		Mechanization rate of navigation marks maintenance operations (%)	50
		Coverage rate of automatic monitoring terminals for multiple elements of hydrological sediment in key waterways (%)	30
	Green	Utilization rate of soil dredged from the Yangtze River waterway (%)	1.8
		Percentage of clean energy and new energy vessels among newly built vessels (%)	10
		Greening rate of shore protection project (%)	100

A qualitative evaluation method is used in this evaluation work to compare the corresponding indexes with the guidelines, then the evaluation results are classified into excellent, good, qualified, and disqualified according to the number of achieved secondary indexes. The specific rules are shown in Table 2.

Table 2. Grading scale.

Grade	Percentage of compliance	Secondary index attainment (n)
Excellent	$\geq 90\%$	$n \geq 15$
Good	$80\% \leq n < 90\%$	$13 \leq n < 15$
Qualified	$70\% \leq n < 80\%$	$11 \leq n < 13$
Disqualified	$n < 70\%$	$n < 11$

3.2 Evaluation of the benefits of waterway maintenance

The evaluation indexes are usually used to evaluate the actual benefits brought by waterway maintenance, which focus on two dimensions including economic benefits and service users' satisfaction. Each dimension consists of several primary and secondary indexes, as shown in Figure 1.

The evaluation of economic benefits is carried out by using quantitative methods, while the satisfaction of service users is carried out by using questionnaires, as follows:

(1) Direct economic benefits

The amount of GDP growth can be driven by Yangtze River management and maintenance,

$$\Delta Y = K \times \Delta I \tag{1}$$

where ΔY is the change in revenue; ΔI is the change in maintenance investment; K is the investment multiplier, where:

$$K = 1/(1 - b) \tag{2}$$

302

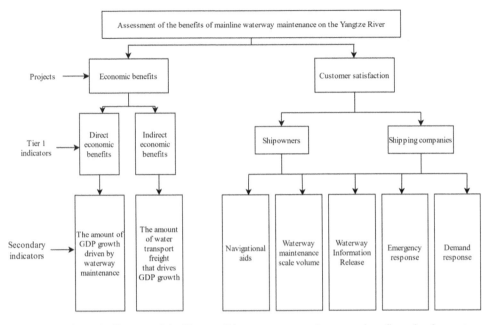

Figure 1. Schematic diagram of the Yangtze River waterway maintenance benefit evaluation system.

where b is the marginal propensity to consume, according to relevant calculations, and the population's marginal propensity to consume in China is about 0.25.

(2) Indirect economic benefit

The increase in the volume of freight transported along the Yangtze River drives the increase in GDP.

$$\Delta Y = \Delta S \times c \times d \times t \qquad (3)$$

where ΔS is the change of cargo volume on the Yangtze River; c is the proportion of the added value of freight volume caused by the maintenance project in the added value of freight volume of the Yangtze River estuary; d is the coefficient relationship between the water freight volume and the port throughput, namely, the port throughput driven by the water freight volume of 1 ton; t is the GDP coefficient driven by the increase of port throughput of 10,000 tons. According to relevant studies, t is often set as 0.03568.

(3) Customer satisfaction

Customer satisfaction was investigated and evaluated in 5 aspects: navigation marks, waterway maintenance scale, waterway information issue, emergency treatment, and demand response. The full score of each aspect was set to 20, and the satisfaction level was evaluated according to the total score, as shown in Table 3.

Table 3. Customer satisfaction level table.

Level	Score (m)
Excellent	$m \geq 90$
Good	$80 \leq m < 90$
Qualified	$70 \leq m < 80$
Disqualified	$m < 70$

In trial operations, we administer questionnaires to more than 100 shipping enterprises or shipping companies and more than 1,000 captains or sailors along the Yangtze River, and then the conclusion of the satisfaction score of the service customers in the Yangtze River waterway is 91 in 2021, which is obtained by the comprehensive evaluation.

Combining the content of waterway maintenance, this work establishes a set of relatively complete index systems for evaluating the waterway maintenance effectiveness, which fills the gap in the evaluation system of waterway maintenance effectiveness in China. In addition, the satisfaction index of service objects for the maintenance of the Yangtze River waterway is tested based on the existing data, as the trial effect shows, the system proposed in this work was more realistic, through which the scientificity and rationality of the research results are demonstrated in part. At the same time, further studies will also pay more attention to the remaining indexes not verified in this survey, thus providing more scientific guidance for the Yangtze River waterway maintenance.

4 CONCLUSIONS

In this paper, based on the actual maintenance content and characteristics of the Yangtze River waterway, and the positive effect and the practical benefits brought by the waterway maintenance, the evaluation index system of the Yangtze River waterway maintenance effectiveness was theoretically established. The evaluation index system not only includes the effect evaluation of the smooth, safe, efficient, and green aspects of the waterway but also pays attention to the economic benefits brought by waterway maintenance and the satisfaction of service customers. For different evaluation contents and indexes, this paper also puts forward specific evaluation methods and criteria and further carries out systematic evaluation in combination with the actual situation, then continuously optimizes and improves the content and methods of indexes in combination with the evaluation situation, to provide theoretical reference for the evaluation and decision-making of the Yangtze River waterway maintenance effectiveness.

REFERENCES

Cui, A., Liu, O., Huang J.E., et al. 2022. Qualitative Evaluation Method of Ship Cabin Equipment Maintainability Based on Virtual Simulation. *Ship Electronic Engineering*, 42(5): 129–133.

Liu, L. & Duan, G. S. 2022. *Evaluation of the Effectiveness of the Yangtze River Waterway Maintenance and Countermeasures for High-quality Development of Waterway Maintenance and Management in the 14th Five-Year Plan period*. Beijing: Institute of Water Transport Science, Ministry of Transport.

Wang, X. Q. 2012. Research and Evaluation of Eco-environmental Management Index system - An Example from Dadukou. *Journal of Chongqing Normal University (Natural Science)* 8–10.

Wang, X. Q. 2012. *The Research and Assessment of City Environmental Management Index System*. Chongqing University.

Yue, C. S. 2014. Construction of Highway Maintenance Performance Evaluation Index System. *SME Management and Technology* (3): 24–25.

Zhang, D. M. 2018. A Study of Delphi Application Based on Two Cases. *Information Studies: Theory & Application* 41(3): 73–77.

Zhang, J.F. & Zhou, Y.P. 2022. Modernization of Waterway Maintenance Management and its Evaluation Index System. *Shipping Management* 44(2): 17–19.

Zhang, J.H & Yue, Q.H. 2006. Establish a Waterway Service Level Evaluation Index System. *Shipping Management* 28(8): 19–21.

Zhang, X.Y. & Xu, S.W. 2022. Study on the Construction of an Evaluation Index System for Agricultural and Rural Modernization in Chin. *Research of Agricultural Modernization* 43(5): 1–9.

Zhao, X. & Sun, C.H. 2022. Evaluation of Urban Ecological Environment Quality Based on Analytic Hierarchy Process. *China Resources Comprehensive Utilization* 40(5): 163–166.

Water Conservancy and Civil Construction – Oke & Ahmad (Eds)
© 2024 The Author(s), ISBN: 978-1-032-58618-2

Basic prototype and evolutionary logic of clan shrines in northern Guangdong

Ziru Ren
South China University of Technology, Guangzhou, China
South China University of Technology Architectural Design and Research Institute Co., China

Zhaohui Tang
South China University of Technology, Guangzhou, China
South China University of Technology Architectural Design and Research Institute Co., China
State Key Laboratory of Subtropical Building Science, South China University of Technology, China

Yao Wang*
Guangdong Poly Real Estate Development Co., Guangzhou, China

ABSTRACT: Shaoguan and Qingyuan in northern Guangdong are called the northern part of Guangdong, and many clan shrines with different styles exist in this region. Based on the architectural shapes and plan features of the typical clan shrines of different folk systems along the ancient route in northern Guangdong, the reference prototypes of clan shrine architecture in northern Guangdong are selected from the typical residential buildings of the surrounding folk systems. The evolutionary logic of clan shrine architecture is sorted out from the unfinished forms found in different times and spaces, and the sources of the prototypes of clan shrine architecture in northern Guangdong are summarized in terms of plan evolution, style, and unique symbols, and six evolutionary features are summarized to further clarify the evolutionary logic of clan shrine architecture in northern Guangdong and analyze the reasons for its creation.

1 INTRODUCTION

North Guangdong refers to the northern part of Guangdong Province, the administrative division includes five prefecture-level cities, Shaoguan, Qingyuan, Heyuan, Meizhou, and Yunfu, and the academic division discusses two prefecture-level cities, Shaoguan and Qingyuan. In the doctoral dissertation of Yan Zeng of the South China University of Technology, the traditional dwellings of Shaoguan Qingyuan were classified as the original settlement dwelling culture area (Zeng 2016). After an in-depth investigation, it was found that the portage route used the northern Guangdong area as a carrier to communicate the culture of Guangfu, Hakka and Xiangchu, and Hakka in Ganan, and a variety of strong folk cultures intermingled and spread in the northern Guangdong area (Figure 1), making the local architecture present some typical folk characteristics at different levels, which is the result of multiple folk influences superimposed on the primitive folk.

Ancestral halls are an important cultural feature of each clan to commemorate their ancestors' understanding of their origins to pass on family ancestral traditions. The function of ancestral halls was clearly defined in the "Sixteen Articles of the Oracle" (Anonymous 1985) issued by the imperial court in the ninth year of Kangxi. The ancestral architecture was an

---

*Corresponding Author: 595302951@qq.com

DOI: 10.1201/9781003450832-39

Figure 1. Distribution of surrounding folk in northern Guangdong.

important means for the ruling class to rule the people from the top down, and an important vehicle for villagers to pass on their cultural achievements from the bottom up. The research on ritual architecture in the Guangdong region is mainly concentrated on Guangfu and Hakka basins, and the ritual architecture in the northern Guangdong region also mostly starts with technical forms, such as Dr. Bi Xiaofang in his paper (Bi 2016), which is one of the important bases of this paper as he has made a technical compendium of the architecture including the clan shrines. In this paper, based on the study of the ancient portage route in South Guangdong (Lu & Cai 2018), we conduct long-term research and sort out the shape and layout of the ancestral shrines along the ancient route located in the northern part of Guangdong, and compare and analyze the cultural characteristics they show. Why are there architectural ancestral shrines with the characteristics of Hunan and Jiangxi in Guangdong? Some of the buildings are shaped in the same way as Guangdong architecture. Why do ancestral halls with strong elements of Guangfu culture appear in northern Guangdong, a frontier region far from Guangfu culture, while there are few examples in the neighboring Qingyuan and Shaoguan regions? Why do the buildings on the old waterway of the Chajeong River, which connects Hunan and Guangdong, have the characteristics of Xiang-Gan ancestral halls? The following article tries to answer these questions through research examples and historical analysis. Reference prototypes of clan architecture are selected from typical surrounding folk systems, and the logic of the evolution of clan architecture is sorted out from the evolutionary unfinished forms found in different times and spaces.

2 PLANAR EVOLUTION

2.1 *Unification of ancestral houses into independent ancestral houses*

In 2008, Mr. Lu Yuanding described a cross-tang house as "a house with two or three rooms, the front one being a foyer and the back one being a hall for ancestor worship. (Lu 2008) In his master's thesis in 2017, Liu Danfeng systematically explained the way of combining tang houses and cross houses and the evolution of the form of cross-tang houses with other types of Hakka dwellings (Liu 2017), and among the Hakka dwellings mentioned in the paper, except for Kang houses, which do not have a tang house, the vast majority of existing wai-long houses, cross-tang houses, and wai-you are set up with a tang space.

When the Hakka ancestors arrived at a place to settle, they would first build a single room to live in. Because of the fear of living in the same room with their ancestors, they built a special table to place the ancestral tablets. After that, the tradition of building a hall or ancestral hall to place the ancestor's spirit before building their own house next to the ancestral hall was formed. As time goes by, people tend to look for a better living environment to live in. When the living space was left unmaintained, the house gradually fell into disrepair. The ancestral hall, located in the center of the cross-tongue house, became the blood tie that connected the clan, and people always returned to the village to gather in the ancestral hall during the Qingming festival. Over time, the ancestral hall space has become an independent symbol.

Many independent ancestral halls belonging to other folk groups were built in northern Guangdong, and the Hakka people inevitably borrowed elements from other cultures when building ancestral halls. During the field research, we found that there are four types of ancestral halls along the old Xiang-Yue Road in Yile and Chengkou, from the unification of houses and sacrifices to independent ancestral halls. The Huang Ancestral Hall in Damenlou, Yingde, is a complete horizontal hall with the middle ritual building still in use and maintained by relevant personnel. The surrounding dwellings are intact, but no one lives there anymore, and the whole building is well preserved. In the central part of Qingyuan and Shaoguan, many cross-tongue buildings begin to disintegrate from the middle to the two sides, such as the Lan's Ancestral Hall in Yingde Gaoqiao Village, Qingyuan, where most of the residents have moved out and the original cross-tongue layout can be seen vaguely. However, no one has repaired the houses, and the area around the ancestral hall is full of wasteland grass. In the case of the Ye Ancestral Hall in Qujiang District, Shaoguan, only the front row of houses in the village is inhabited, and the back part is completely abandoned. However, the author believes that this is the midway point of the evolution of the cross-tongue house to the independent clan ancestral hall. (Figure 2)

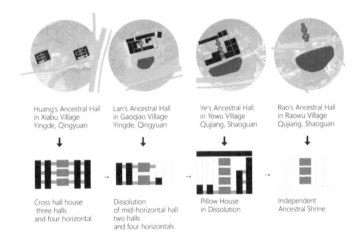

Figure 2. Distribution of surrounding folk in northern Guangdong (Photo credit: Self-drawn by the author).

Shaoguan Qujiang District Rao Ancestral Hall is a new independent ancestral hall. The site around the ancestral hall is open and there are no other buildings adjacent to it, but the builder deliberately kept the wall next to the patio of the ancestral hall open and with a convex boundary during construction. To ensure the integrity of the ritual space, the concave and convex parts were used for storing miscellaneous goods or ritual supplies. Therefore, the independent ancestral hall of the Hakka family is judged to be a ritual space detached from the cross-tongue house, which is an important trace of cultural integration.

2.2 *Independent ancestral halls in the Guangdong-Hunan-Jiangxi area*

Among the ancient roads in northern Guangdong, WuJing and Xi Jing are the oldest roads, and Lianzhou and Nanxiong became the earliest landing points for the migrants in northern

Guangdong. The layout of the ancestors of the Tang clan from Lishui Village, Lianzhou City (Figure 3), and the northernmost point of the Xingzi ancient road, was very similar to that of the Tunisian Langgong Ancestral Hall in Liukeng Village, Dong'an County, Jiangxi Province (Figure 4) (Li 2006), with an enclosed courtyard in front of the house and openings on the sides. The interior of the building is raised by steps. Even the position of the offset columns of the two ancestral halls is the same, and this style of the ancestral hall does not appear in other areas of Lianzhou, while this arrangement is very common in Jiangxi, so it is speculated that the layout of this ancestral hall comes from Jiangxi. Most of the immigrants from Lianzhou and Nanxiong came from Jiangxi, while Lechang, Renhua, and parts of Lianzhou bordered the southern part of Hunan and also had residents who migrated from the Hunan region. The Jinma family and the Tang family were moved from Lanshan, Yongzhou, and Hunan Province to the front of Baihe Mountain.

Figure 3. Plan of Tang Ancestral Hall in Lou village, Lishui village, Lianzhou city, Guangdong Province (Photo credit: Self-drawn by the author).

Figure 4. Plan of Tun Tian Lang Ancestral Hall in Liukeng village, Dong'an county, Jiangxi (Photo credit: Self-drawn by the author).

3 STYLE MODELING

3.1 *Nanxiong Zhuji Lane replicates the classic Qing Dynasty pattern of Guang San road*

The traditional Guang-style ancestral hall shape often adopts a three-way layout, with a three-room open-headed door in the middle, and a three-entry ancestral hall on the left and right sides, separated from the middle ancestral hall by Qing Yun Lane. Qing Yun Lane should belong to the same function as the side porch of the traditional ancestral hall (Lai 2010). The Yang Clan Ancestral Hall in Beishan Village, Zhuhai (Figure 5) is a traditional Guang-style ancestral hall. The Li Clan Ancestral Hall in Zhugui Lane, Nanxiong, and Shaoguan (Figure 6) has no Qingyun Lane, but the overall head gate is all divided into three ways, with a three-entry ancestral hall in the middle. Several ancestral halls stand next to each other in Zhugui Lane, and the layout and shape are off the traditional Guang-style architecture, while no similar examples have been found in other parts of Nanxiong.

Figure 5. Yang Clan Ancestral Hall in Beishan village, Zhuhai (Photo credit: Author's photo).

Figure 6. Ancestral Hall of the Li Clan in Zhuguixiang, Nanxiong, and Shaoguan (Photo credit: Author's photo).

In addition to Nanxiong, only areas such as Qingxin County and Fogang in Yangshan, which are adjacent to Guangzhou, have Guangfu-style ancestral halls, such as Huang's Ancestral Hall in Dongkeng Wai, Lianyao Village, Shuitou Town, Fogang County, which also belongs to the classic Kwong San Road layout, but the building is not on the ancient road along the northern part of Guangdong. It is not surprising that the Guangfu style appears due to its proximity to Guangzhou. It can be seen that outside the ancient road along northern Guangdong, the Guang style appears in Fogang, Yangshan due to its proximity to Guangzhou. Because of its special location, Zhuguixiang became the first stop for immigrants on their way to migrate and an important place for many people to go south to escape the war, and many people settled nearby because of the comfortable living environment, and over time Zhuguixiang became a symbol of their hometown. The emergence of Guangfu factors within the region with cultural symbols of attraction, not the ancient migration brought about by the phenomenon of the intermingling of people.

3.2 *Lechang Renhua and Hunan Rucheng ancestral hall head gates similar in shape*

Both the Tan Ancestral Hall of Wu Wang in the former village of Lechang (Figure 7) and the Shi Ke Ancestral Hall of En Village appear to have a similar architectural style to the head gate of the Yuan Ancestral Hall of Rucheng in Chenzhou, Hunan Province (Figure 8), a shape that is not common in Guangdong. The arch is often located under the eaves on the column, making the projection of the eaves more exaggerated. During the evolution of history, the arch was gradually transformed from a purely structural element to a decorative element, and later became a symbol of architectural rank in ritual architecture, signifying power, and status. Hunan-Guangdong and other places far from the center of power, less influenced by the traditional imperial power, the arch appeared to cross the level of the head door building. Both of the preceding used such modeling practices and the surrounding buildings were not affected by this practice, presumably because craftsmen from Hunan were invited to intervene in the modeling during construction, which was one of the advantages brought by convenient transportation.

Figure 7. Ancestral Hall of the Tan Clan in Wu Wang village, Shaoguan, Guangdong (Photo credit: Author's photo).

Figure 8. The Yuan Family Ancestral Hall in Rucheng, Chenzhou, Hunan (Photo credit: Author's photo).

4 UNIQUE SYMBOL

4.1 *Clan function extends*

During the Ming Dynasty, Kunqu was popular, and the Xiang-Egan region often invited Kunqu troupes to perform in their theatres during festivals to gather together in front of their ancestors. The emergence of opera led to the development of architecture, making the stage an essential part of the local clan shrines. In the Xiang-Engan region, a two-story theater stage was built at the foyer, and the first patio of the ancestral hall was accessed through the bottom of the stage. However, this practice is rare in Guangdong, and the number of theatres in clan buildings is even smaller.

The ancestral hall in Dazitang Village, Lianzhou, is three rooms wide, with a courtyard enclosed by a wall in front of the head gate. Behind the courtyard, there is a very rare theater stage in Guangdong (Figure 9), which has a single eave hibiscus roof and a simple form. It is divided into two parts: the front stage is hollowed out on three sides for performances, and the backstage is used for dressing the actors. The stage faces the gods in a row, and when the opera is sung, it has the double effect of bringing the clansmen together and enjoying with the ancestors. Outside the Tang Ancestral Hall in Mabai Village, there is a single-roofed hillside theater (Figure 10), which is constructed of wood. The building is a combination of a raised beam and a pierced bucket. The date of construction is not known, but the original construction site is not good for feng shui, so it was moved to this place. This theater stage is located outside the Tang family ancestral hall, and there is a similar example of this practice in front of the Tianhou Palace in Huizhou, Guangdong.

Figure 9. The built-in theater stage of the Ancestral Hall of Dazitang village (Image source: The ancient ancestral hall in Qingyuan).

Figure 10. The independent theater stage of the Tang Ancestral Hall in Ma Dai village (Photo credit: Self-drawn by the author).

In addition to the function of the theater, the ancestral hall also served as a place to eat and gather and to hold wedding ceremonies. This practical function adds life to the ancestral hall, enabling the clan to be more cohesive and reinforcing the function of the times.

4.2 *Fusion of Western symbols*

During the Republic of China period, many expatriates working abroad, developed after returning to their hometowns foreign concepts brought a new Western architectural element. Watchtower, located in Kaiping, Jiangmen, Guangdong Province, is the expatriates brought Western style set. Although there are not many expatriates in northern Guangdong, but also Western symbols of architecture. These products of the cultural interchange of immigrants in a particular period are an important witness to the cultural interchange between East and West. Western-style architectural symbols, mainly arch coupons, appear both indoors and outdoors in northern Guangdong, ranging from full Roman-style round arches to simple flat arches, all with a central symmetrical composition. The span of the middle arch voucher is mostly different from the two sides, while the combination of arch vouchers at the two ends is relatively free, and the span and form remain consistent, echoing the balance in classical Chinese architecture.

The ancestral hall in Nianfeng Village, Taiping Town, Qingxin (Figure 11) has a three-way ancestral hall, separated by Qingyun Lane, with a wide style ancestral hall in the middle, and the left and right sides of the lined ancestral hall arch coupons with full arch coupons and thick columns, similar to the Roman Tashkent column style. The Zhajiao Liu Gong Ancestral Hall (Figure 12) in Shaba Village, Xiniu Town, Yingde, was built in the Republican period, using a combination of Chinese and Western construction styles, with a double layer of arch coupons on the pagoda-style façade, columns running through the second floor, and no special decoration on the column header.

Figure 11. Phuong Thuan Chen Gong Ancestral Hall right road lining Ancestral Hall (Photo credit: Author's photo).

Figure 12. Zhajiao Liu Gong Ancestral Archway (Image source: The ancient ancestral hall in Qingyuan).

In addition to the outdoor head door with Western symbols, some of the ancestral halls have interior decorations combining Chinese and Western. The three arched roofs of the Su-style ancestral hall in Xiyao Village, Shantang Town, Qingcheng District (Figure 13) are decorated with green glazed tiles on the ridge of the flying hood. The courtyard of the Deng Hengfu Ancestral Hall mentioned earlier (Figure 14) is flanked by continuous Western-style arcades, and such ancestral halls are found sporadically in northern Guangdong and are not much associated with ancient roads.

Figure 13. Su clan ancestral hall three-arch roof-shaped hanging down flying cover (Image source: The ancient ancestral hall in Qingyuan).

Figure 14. Corridor of Deng Hengfu Ancestral Hall (Photo credit: Author's photo).

5 CONCLUSION

The northern part of Guangdong has been influenced by many immigrants brought by the ancient roads, and clan architecture has evolved in the spatial presentation and expression of multiple cultures.

(1) The evolution of planes is divided into Hakka-style evolution and Xiang-Gan-style evolution. The Hakka style has undergone many changes during the migration process,

and the original cross-tongue spiritual space was separated to form an independent ancestral hall, which was built in modern times and is not a traditional Chinese building, but is also an important example of cultural evolution and integration. The plan of the Xiang-Gan style is mainly found in Shaoguan's Wulun ancient road and Xijing ancient road, and the plan of these clan shrines is the same as the plan of the buildings in the place of migration, without intermediate evolution.

(2) The evolution of the style is mainly spread in the villages and counties bordering the two folk systems connected by the ancient roads. Many Ganzhou immigrants brought the original layout of their hometowns with them and tended to hire craftsmen from their original places of residence to build ancestral halls. Lechang, along the ancient road of Renhua, was infected by Xiangchu culture.

(3) The unique symbolic penetration phenomenon is very common along the ancient post road in northern Guangdong, with the Xiang-Gan style gradually weakening from north to south, and the central part of the ancient road is dominated by the Hakka style, with a mixed architectural style incorporating other cultural elements. Typical Guangfu-style ancestral halls appear in the southern part of the ancient road near the Guangzhou area, especially in Fogang County of Xingyang Mountain, but such ancestral halls are not along the ancient road, so it is speculated that the ancient road has less influence on the migration of Guangfu ancestral halls.

REFERENCES

Anonymous. *The Actual Records of the Saintly Ancestors of the Qing dynasty* (vol. 34) [M]. Beijing, China Book Bureau (1985).

Bi Xiaofang. *Study on the Building Techniques of Ming and Qing Dynasty Wooden Buildings in Northern Guangdong* [D]. The South China University of Technology (2016).

Lai Ying. *A Study on Ancestral Hall Architecture of Guangfu Folk System in the Pearl River Delta* [D]. South China University of Technology (2010).

Li Qiuxiang. *Ancestral Shrines (Native Treasures Series)* [M]. Beijing, Sanlian Publishing House (2006).

Liu Danfeng. *Study on the Architecture of Tang Cross House in Gaosi Village, Jiaoling, Meizhou* [D]. South China University of Technology (2017).

Lu Q, Cai Yijun. Ancient Post Roads and Humanistic Features of Traditional Villages in Southern Guangdong[J]. *China Famous Cities*, 18(04):88–96. (2018).

Lu Yuanding. Characteristics of Hakka Dwellings in Meizhou and Their Inheritance and Development[J]. *Southern Architecture*, 08(02):33–39(2008).

Zeng Yan. *Cultural Geography of Guangdong Traditional Settlements and Their Dwelling Types* [D]. South China University of Technology (2016).

Water Conservancy and Civil Construction – Oke & Ahmad (Eds)
© 2024 The Author(s), ISBN: 978-1-032-58618-2

Location planning method of flood control and drainage pumping station based on improved genetic algorithm

Wei Zhang & Yan Liang*
HOHai University Design and Research Institute Co., Ltd, Jiangsu, Nanjing, China

Zhen Liu
Science and Technology Information Branch of the Eastern Route of South-to-North Water Transfer Jiangsu Water Source Co., Ltd, Jiangsu, Nanjing, China

ABSTRACT: The conventional method of pumping station location planning is mainly based on the drainage model, which does not combine with the surrounding environment to select the location of the pumping station, resulting in a poor drainage effect. Therefore, an improved genetic algorithm is designed for the location planning of flood control and drainage pumping stations. It determines the elevation of the drainage outlet of the pumping station and adjusts the lowest drainage level of the reservoir of the pumping station. To improve the drainage effect, an improved genetic algorithm is taken to construct the model of site selection and planning of the pumping station. The effective site selection and planning of the pumping station are carried out according to the surrounding environmental characteristics. The example analysis verifies that the method has a better drainage effect on the pumping station and better effect on site selection planning, and has higher promotion value.

1 INTRODUCTION

Cities are vulnerable to rainstorms and heavy rainfall, resulting in greater waterlogging problems and affecting the travel environment of residents. Pumping station construction is an effective means of urban drainage, and the effect of its site selection and planning directly determines the effect of urban drainage. Aiming at the problem of urban waterlogging, researchers have designed a variety of methods for the location planning of the pumping stations. Among them, the location planning method of the pumping station based on the reservoir connection project and the location planning method of the Foyukou pumping station is widely adopted. The method of pumping station location planning based on the reservoir connection project is mainly to choose the appropriate pumping station location according to the actual situation of the project, and connect the pumping station with the reservoir to ensure the water diversion and drainage effect of the pumping station (Ding 2021).

The site selection planning method for the Foyukou Pumping Station is mainly based on the wetland type of Foyukou and the location of the pumping station which is more suitable for the regional characteristics. The pumping station has good water quality and low investment and can adapt to the drainage environment of the region (Fu 2022). Although the above two methods have considered the impact of the actual terrain on the pumping station, they have planned a more suitable site for the pumping station (Guo 2020). However, the above two methods are more inclined to the planning effect of low cost and high efficiency, which reduces some drainage processes and leads to poor drainage effect of pumping stations. In general, the

---

*Corresponding Author: 1030201031@hhu.edu.cn

DOI: 10.1201/9781003450832-40

accumulated water can be drained in five days at most in the environment of a planned rainstorm and accumulated water. Because part of the drainage process is missing, it takes seven days or more to drain, which seriously affects the use effect of the pumping station (Li 2020). The improved genetic algorithm is a direct use of the solution form to express the planning path, which can make the planning path more in line with the actual needs (Liu 2022). Therefore, from the perspective of an improved genetic algorithm, this paper designs a method for the location planning of flood control and drainage pumping stations.

2 THE IMPROVED GENETIC PROGRAMMING METHOD FOR THE DESIGN OF FLOOD CONTROL AND DRAINAGE PUMPING STATION LOCATION

2.1 Determine the elevation of the pump station's outlet

Urban construction is inseparable from building construction and public facilities construction. Public facilities mainly include urban greening, municipal drainage, road construction, etc. As the key construction to alleviate the urban waterlogging problem, the drainage effect of municipal drainage is very important (Luo 2021). The change in the drainage outlet of the pumping station is related to the pumping station's drainage volume. In this paper, under the comprehensive consideration of the site selection and planning of the pump station, the highest water level that does not affect the water storage of the pump station is taken as the lowest water level. The rainstorm condition is increased to the maximum, thus determining the elevation of the drainage outlet (Lv 2022). The drainage process of the pump station discharges the accumulated water through the pipe network composed of multi-stage pipe sections, which is regarded as a series structure in this paper. The drainage process is displayed in Figure 1.

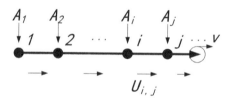

Figure 1. Drainage flow chart.

As shown in Figure 1, A_1 is the first pipe segment; A_2 is the second pipe segment; A_i is the *ith* pipe segment; A_j is the *jth* pipe segment. i and j are constants. v is the drainage pipe segment, and u_{ij} is the confluence of all pipe segments. The drainage process is from A_1 to A_v. After the surface ponding enters A_1, the surface water flows in line and waits. All drainage outlets provide drainage services. The surface ponding is drained in the order of A_1 to A_v to ensure the drainage effect of the ground (Wang 2022). Therefore, in this paper, v is set as the elevation of the drainage outlet planned by the pumping station, which follows the drainage rules of A_1-A_v, and the surface water is completely drained.

2.2 Construction of the pumping station location planning model based on an improved genetic algorithm

In the process of pumping station location planning, the concept of an improved genetic algorithm is introduced in this paper. With the progress of the pumping station location planning process, the greater the crossover probability of the location planning process, the faster the speed of generating new genetic individuals, and the more paths for pumping station site planning (Xiong 2020). The generation of multiple genetic location planning paths makes the planning paths cross. Whether the factor can enter the next generation is judged according to the fitness value of the cross factor, the larger the fitness value is, the more suitable the area of the pumping station location planning is for the local drainage demand. In this paper, the improved genetic algorithm is applied to improve the operators in the early stage of the location population's evolution. The individual fitness value of the location factor is calculated according to the fitness function, and the sorted location individuals are divided into four parts and eliminated in the form of 1/4 (Yan 2020). After several iterations of location planning, the fitness values of location factors are close. Before the self-adaptation is poor, this paper designs the evaluation index of the location factors' difference degree. The index formula is as follows:

$$k = \frac{\overline{H_i}}{H_{imax}} \tag{1}$$

In Formula (1), k is the evaluation index of the different degrees of location factors; $\overline{H_i}$ is the average fitness value of the better individual in the ith location factor; H_{imax} is the maximum fitness value in the *ith* location factor. At the initial stage of the location planning population's improvement, there is a big difference between H_i and H_{imax}. With the evolution of the population, the value of k becomes larger and larger. At this time, the dynamic adjustment formula of the location planning population is:

$$\begin{cases} K_{A_1}(i) = K_{A_1} - \dfrac{K_{A_1} - K_{A_2}}{2} * k \\ K_{A_2}(i) = K_{A_2} \end{cases} \tag{2}$$

In Formula (2), $K_{A_1}(i)$ and $K_{A_2}(i)$ are the upper and lower limits of the crossover probability; K_{A_1} and K_{A_2} are the initial value range of the crossover probability. According to the values of $K_{A_1}(i)$, $K_{A_2}(i)$, K_{A_1}, and K_{A_2}, the location planning model of the pumping station is constructed. The model expression is as follows:

$$\begin{cases} \min[K_{A_1}, K_{A_2}] = \max\{\max(H_i)\} \\ \min[K_{A_1}(i), K_{A_2}(i)] = k \sum_{k \in K} a_k u_k \end{cases} \tag{3}$$

$$\delta_m = kf(k)A_i \tag{4}$$

In Formulas (3-4), $\min[K_{A_1}, K_{A_2}]$ is the minimum value range of crossover probability; H_i is the fitness value of the ith location factor; $\min[K_{A_1}(i), K_{A_2}(i)]$ is the minimum upper and lower limits of K_{A_1} and K_{A_2}; a_k is the rainstorm intensity; u_k is the drainage intensity; δ_m is the location planning model expression. Formula (3) is the objective function of the model, which is adopted together with δ_m to plan the site selection of the pumping station. Under the pumping station planning of the location planning model, the actual location conditions of the pumping station, surface conditions, and other environmental factors can be taken into account. The location planning of the pumping station can be more in line with the drainage demand to maximize the flood control and drainage effect of the pumping stage area.

3 THE CASE STUDY

3.1 *The pump station overview*

To verify whether the pumping station location planning method designed in this paper has practical value, this paper takes the X lake pumping station as an example to analyze the above method. The X Lake Pumping Station, built in 1987, is the only pumping station in the X Lake water system. The pumping station is constructed according to the standard that

250mm with large rainstorms or floods can be discharged within 5 days. The water is directly discharged into Lake X. The current scale of the pump station is 60m³/s, with 10 pumps, and the initial drainage level is 20.12m. Under the rainstorm condition, the pump is built around the catchment area, direct drainage area, and lakeside. There are two pumping stations in X Lake. M pumping station is close to X Lake and the catchment area of X Lake; N pumping station is close to X Lake and the direct drainage area of Dagang of X Lake. The distance between the direct drainage area of X Lake Zhagang and the M and N pumping stations is similar. The site planning of the pump station is shown in Figure 2 below.

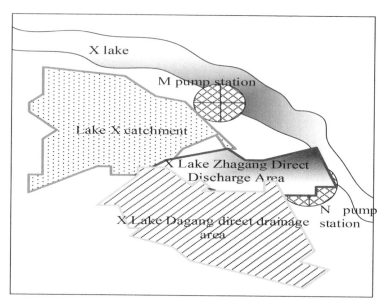

Figure 2. Schematic diagram of the pumping station site planning.

As displayed in Figure 2, the M and N pumping stations planned in this paper are adjacent to the catchment area of X Lake, the direct drainage area of the gate port of X Lake Zhagang, and the direct drainage area of the big port of X Lake Dagang. The water discharged from the pumping station is directly flowed into X Lake to make the pumping and drainage flow of the pumping station optimal. Due to the frequent occurrence of rainstorms in X Lake, the flood control and drainage effect of the pumping station is very important. Using the planning method designed in this paper, the overall design of the pumping station can meet the standards and the development needs of the region after the pumping station is located in the catchment area of X Lake, the gate port direct drainage area of the X Lake Zhagang, and the port direct drainage area of the X Lake Dagang.

3.2 *The application results*

Under the above conditions, this paper chooses the catchment area of X Lake, the direct discharge area of the gate port of X Lake Zhagang, and the direct discharge area of the big port of X Lake Dagang for analysis. Each area is connected to the pump station. Two outlet nodes A and B are selected, and the outlet flow is different. After a rainstorm, the water depth of overland flow around X Lake was analyzed. The lower the depth of overland flow, the better the effect of flood control and drainage, and the better the effect of site selection and planning of pumping stations. In the case of other conditions that are the same, using the improved genetic algorithm based on the pumping station location planning method designed in this paper, the water depth of overland flow around the X Lake is analyzed. The application results are displayed in Table 1 below.

Table 1. Application results.

Positions	Exit node	Outlet flow (m³/s)	Depth of surface overflowing water around Lake X after rainstorm/mm	After using the improved genetic algorithm based on the pumping station location planning method designed in this paper, the surface overflow water depth around the X Lake/mm
Lake X	A	44.32	35.24	12.23
catchment	B	46.58	38.56	15.45
X lake Zhagang	A	57.63	22.19	8.24
direct discharge area	B	61.36	18.62	6.58
X lake Dagang	A	125.44	11.79	3.12
direct drainage area	B	132.87	10.02	2.28
Others	A	55.56	15.92	5.55
	B	58.32	17.63	6.56

As displayed in Table 1, the outlet flow of nodes A and B in the catchment area of X Lake, the direct drainage area of the gate port of the X Lake Zhagang, and the direct drainage area of the port of the X Lake Dagang are different. The depth of the overland flow fluctuates in the range of 10.02mm ~ 38.56mm after the rainstorm. That is to say, when a rainstorm or flood disaster comes, there is more water on the road surface. Under the condition of the same outlet flow, the effect of flood control and drainage on the road surface is not good, which can easily cause greater road disasters and affect people's normal life. After using the method based on the improved genetic algorithm designed in this paper, the overflow water depth around the X Lake has been reduced, fluctuating in the range of 2.28mm ~ 15.45mm. That is to say, when a rainstorm or flood disaster comes, the location of the pumping station planned in this paper can quickly carry out flood control and drainage, reduce the amount of road water, and enable people to live a normal life, which is in line with the purpose of this study.

4 CONCLUSION

In recent years, the framework of urban modernization has gradually become prominent, and people's lives have been improved, which is conducive to the further development of the city. However, in the process of urban development, there are still unavoidable details, and the recent frequent waterlogging problems have seriously affected people's travel life. Heavy rain, heavy rainfall, continuous rainfall, and other conditions sometimes make urban traffic face paralysis. Then, some people question the ability of urban drainage and flood control. Therefore, in this paper, the improved genetic algorithm is applied to design the location planning method of flood control and drainage pumping station. Under the design conditions of the intake elevation and location planning model, the urban drainage problem is perfected, and the water depth of the ground area is reduced to the minimum. Through the above research, the purpose of the paper is to reduce the area of water and provide protection for urban travel.

REFERENCES

Daoxi L., Zhihao S. (2020). Research on the Optimal Design of Pump Station Pressurized Irrigation Pipe Network based on Excel Planning Solution. *China Rural Water Resources and Hydropower*, 2020 (01): 36–38.

Hongyong L., Yichao Z., Yang L., et al. (2022). Research on Key Issues of Energy Saving and Consumption Reducing Operation of Long-distance Water Transfer Project — Take Shandong Section of the East Line of South to North Water Transfer Project Phase I as an example. *Research on Water Resources Development*, 22 (07): 24–31.

Jun X., Yao Z., Bangshuo W., et al. (2020). Selection and Implementation of an Integrated Pump Station in the Solution of Integrated Drainage System in Old Urban Areas. *Construction Technology*, 49 (18): 58–62.

Li W., Changxin W., Rui G., et al. (2022). Comparison and Selection of Overall Design Schemes of a Super Large Pump Station on the Deep Muddy Foundation in Navigable Rivers – Taking Xihe Pump Station in Modaomen Waterway of Xijiang River in the Pearl River basin as an Example. *China Water Resources*, 2022 (06): 68–70.

Ping Y., Dan Z., Ke X., et al. (2020). Research on Evaluation of Water Transfer Lines based on Utilization of Surplus Scenery Resources. *Journal of China Academy of Water Resources and Hydropower Research*, 18 (06): 494–501.

Ting L., Chaohui F., Honghong C. (2021). Emergency Joint Dispatching of Sewage Plants Cluster - Zhuhai Xiangzhou District Sewage System Planning Case. *China Water Supply and Drainage*, 37 (06): 17–23.

Xiaowei L., Minghu H., Xiaohui L., et al. (2022). Prediction Model of Water Level in Front of Pump Station based on GRA-NARX Neural Network. *South to North Water Transfer and Water Conservancy Science and Technology*, 20 (04): 773–781.

Yandong F., Junqiang C., (2022). Zhongdong. Optimization of Pumping Station Scheme for Wohushan Jinxiu Chuan Reservoir Connection Project. *Shandong Water Conservancy*, 2022 (04): 26–28.

Yanhui D., Yueyun J. (2021). Study on the Scheme of Foyukou Pump Station in Yanqing Competition Area of the Beijing Winter Olympics. *People's Yellow River*, 43 (S2), 136–137.

Yongling G., Haichen Z., Xinglin Z. (2020). An Optimization Model for Daily Economic Operation of Single-stage Pump Station based on Dynamic Programming. *China Rural Water Resources and Hydropower*, 2020 (01): 192–196.

Water Conservancy and Civil Construction – Oke & Ahmad (Eds)
© 2024 The Author(s), ISBN: 978-1-032-58618-2

The application of BIM+AR smart construction technology in municipal engineering

Liguo Liu*
Shandong Normal University, Jinan, Shandong, China

ABSTRACT: Today, with the rapid development of science and technology, BIM and AR technologies are more mature and perfect, and are widely used in modern remarkable fields to provide important help for the construction of municipal engineering. This paper analyzes the specific application of BIM and BIM+AR technology in municipal engineering through a brief introduction of BIM and AR technology and then takes a certain project as the research object to lay a solid foundation for the better construction of modern municipal engineering. Based on the traditional BIM technology, this study adds AR technology to further enhance the level of wisdom of modern building construction, which is conducive to the better development of the modern architecture field.

1 INTRODUCTION

Municipal engineering is an important part of the development of modern society, and the effective completion of municipal engineering construction activities can promote the development of modern society in a more favorable direction. At the same time, the scale of modern municipal engineering is more massive, and the quality requirements of engineering construction are higher, which greatly increases the construction difficulty of municipal engineering. In this background, BUM+AR intelligent construction technology is gradually applied in municipal engineering construction, which greatly improves the construction quality and efficiency of municipal engineering.

2 THEORY

2.1 BIM technology

BIM technology is one of the more common technical means in the modern engineering field, and the whole process is building information model technology, which refers to the use of special computer software to build a corresponding engineering model based on various parameters of building indexes, through which the whole building is visually displayed, to provide support for the construction of the building. BIM technology can close the connection between all the participants through the drawn-out 3D model, and provide support for the coordination of all the participants (Cao et al. 2022). It is because BIM has many characteristics mentioned above that it is widely used in the field of modern engineering construction.

2.2 AR technology

Within the modern engineering construction field, AR technology is gradually applied to further improve the quality and efficiency of engineering construction. AR technology, fully

---

*Corresponding Author: tg667788@xzcstudio.com

DOI: 10.1201/9781003450832-41

known as augmented reality technology, first appeared in the early 1990s and was developed based on network technology. AR technology refers to the construction of a realistic visual, auditory, force, touch, motion, and many other senses as one through professional computer software. The virtual environment is used as the basis, and the corresponding sensing equipment is used to immerse the user within this scene to obtain a more accurate feeling through contact with the virtual environment. In essence, AR belongs to a new human-computer interaction technology, and through the application of this technology, a simulated environment close to the real scene can be simulated. When AR technology is specifically applied, it mainly contains three major elements, which are CRV (combining virtual and reality), IRT (instant interaction), and R-3D (3D positioning), and only if these three elements are satisfied, VR technology will be able to play the three key technologies for AR application system development are 3D registration technology, virtual reality fusion technology, and real-time interaction technology as shown in Table 1.

Table 1. Three key technical features of AR technology.

Features	Content
Three-dimensional registration	Computer-generated virtual objects correspond to the real environment and accurate positioning
Virtual reality fusion	Augmented reality scenes by overlaying virtual and real scenes
Real-time interaction	User interaction with the augmented reality scene

2.3 *Advantages of BIM+AR technology*

In the process of modern municipal engineering construction, BIM+AR technology is of great significance, specifically, it is mainly reflected in the following aspects: firstly, through the application of BIM+AR technology, a more realistic and accurate three-dimensional model of the project can be constructed, through which the specific conditions of the project are displayed visually, and the owner, designer, supervisor, and constructor can find the defects in the scheme in time through the observation of the model and optimize the design scheme, thus improving the quality of construction. Through the model, the owner, designer, supervisor, and constructor can find the defects in the scheme and optimize the design scheme to get the best design scheme, which can provide good guidance for the subsequent construction activities and prevent the problem of engineering changes, to improve the quality of engineering construction. Secondly, the number of various materials can be effectively controlled through the analysis of the 3D model during the construction cycle to ensure that the quality and quantity of materials meet the needs of engineering construction based on reducing the use of adoption, reducing the cost investment in materials (Cao et al. 2022) bim+vr.

3 PROJECT OVERVIEW

In this study, a municipal project is selected as the object of study, which is an overpass construction project in our city with a total length of about 8.0 km. The project consists of four parts, which are road construction, drainage channel diversion project, road, and drainage channel under the high-speed railway protection project, and railroad equipment diversion project caused by the underpass bridge. The scale of the project is relatively large, the internal structure is more complex, and the volume of work is higher, not only prone to linear quality problems but also may cause the problem of cost waste.

4 APPLICATION OF BIM TECHNOLOGY

4.1 Reading link

When a certain municipal project is applied to BIM technology, it can be used to read the drawings. According to the different contents of a project drawing, it can be divided into two parts, one part is the aqueduct section and the other part is the high-speed section, which, within the high-speed section, can be further divided into the east-facing west section and the west-facing east section. In the process of designing the new section, the original road should be re-shifted for the specific conditions of the construction site, thus creating more viaducts and re-shifted roads on the high-speed section (Gao et al. 2020).

4.2 Modeling link

BIM technology contains many functions, and three-dimensional model construction is one of the most important functions, mainly using professional BIM modeling software to complete. Through the software in the family building function, the corresponding "family" builds out the mold, after that, in turn, each mold is transferred to Rcvit for the drawings in the elevation, location, and other information, these molds together, to get a complete three-dimensional engineering model and to achieve the purpose of assembly-type construction, first making the model components in assembling of the components. For the first two problems, the direction and slope of the mold can be adjusted in the family first according to the demand of 3D modeling, and then transferred to Revit; for the last problem, one model is transferred first, and then based on this, other molds are imported in turn and put together with the previous model to get the final 3D model. Specifically, the modeling process is shown in Figure 1 (Shang & Liu 2020).

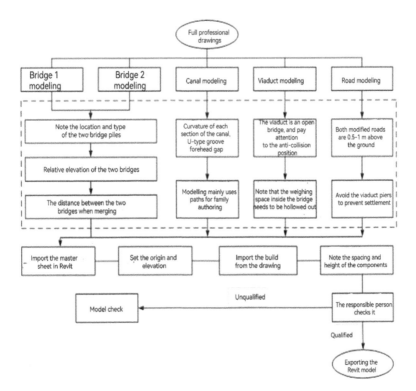

Figure 1. Schematic diagram of the modeling process.

5 APPLICATION OF BIM+AR

5.1 Program design link

Today, with the rapid development of the modern engineering field, many advanced technical means gradually appear in the engineering field and BIM and AR technology are one of the more widely used and well-applied ones. The application of these two technologies provides strong support for the development of engineering design work. First of all, with modeling software such as Revit Bentley as the main tool, the corresponding parameter data is used to build the corresponding three-dimensional model, after that, the model is transferred to Nacisworks, Lumion, and other software content, to carry out deep processing of the three-dimensional model, such as roaming processing, to get a better three-dimensional model. Finally, under the role of AR technology, the various 3D models constructed are imported into the same scene at the same time, to get a building model close to the real situation (Huang et al. 2019). The implementation of AR technology requires the help of software devices and hardware devices, and the working implementation steps of R technology are shown in Table 2 below.

Table 2. Steps of AR technology implementation.

Steps	Content
1	Capture real-scene information with mobile device cameras
2	Track and analyze camera location information and real-environment target locations
3	Interactively generate virtual scenes
4	Seamlessly fuse virtual scene information with real scenes
5	AR scene rendering

For a project, located under a railroad, the roadway consisted of two sections, one from east to west and the other from west to east, to support the movement of vehicles in both directions. In the middle of the channel, a fill section was established for the extension of the east-to-west section of the tunnel, and at the same time, the original channel section was used as the basis and appropriately re-shifted to obtain a new channel. Specifically, the BIM model of a project is shown in Figure 2 to Figure 3 (He 2021).

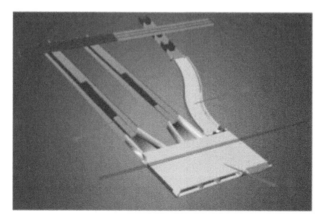

Figure 2. A project overall BIM model diagram.

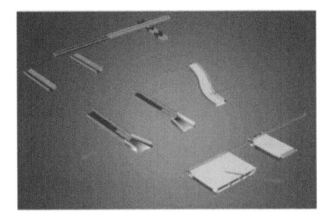

Figure 3. A project section BIM model diagram.

5.2 *Quality control link*

When municipal engineering is constructed, it may be affected by many factors, such as personnel factors, material factors, technical factors, etc. If it fails to manage personnel, materials, and technology effectively, it may interfere with the construction quality of the whole project, cause the problem of engineering changes, prolong the construction period, and increase the construction cost. Therefore, to ensure that the quality of construction meets the requirements, municipal engineering construction, based on the conventional construction plan, has strengthened the application of BIM + AR technology to build a more intuitive and effective three-dimensional model to guide the construction of the whole project. After the project is finished, it can also guide the acceptance of the project through the analysis of the 3D model, so that the quality of the whole project can be guaranteed (Wang et al. 2022).

5.3 *Site management link*

There is a lot of municipal engineering management, and site management is one of the most critical, effective implementations of the work, which can ensure that the whole site is good to provide help for the construction activities. And in the construction process, the construction site is easily affected by many factors and various problems, which are not conducive to the construction of the project. Therefore, engineering site management can be applied to BIM technology + AR technology. Through the various sensors installed at the construction site and the data collected by the personnel, which are automatically transmitted to the cloud and BIM software, according to the analysis and calculation of this engineering site information, BIM software can draw the corresponding site model map, and the model map can show in detail the details of the location of the site. When the construction site security and other aspects of the problem are considered, the relevant personnel can be observed through the site model map. When the safety and other aspects of the construction site problems are considered, the relevant personnel can understand the specific situation of the site accurately so that the site problems can be found in time, and make the corresponding rectification at the first time to ensure the goodness of the construction site (Niu 2022).

5.4 *Follow-up inspection link*

After the construction, by scanning the completed project and transmitting the scanning data to BIM software, BIM software can use these data to build the corresponding completed project model, and later, based on this, compare it with the expected engineering model to judge whether the construction situation meets the requirements and whether the settlement and offset of the building meet the requirements (Shi 2022).

6 CONCLUSION

In summary, in modern municipal engineering construction, to improve the quality of construction, shorten the construction cycle, and reduce the construction cost, BIM+AR technology has been gradually applied and has achieved relatively good results. Project construction lays a solid foundation.

REFERENCES

Cao Zhengguo, Gu Jian, Li Zhi. Application of AR+BIM+GIS in the External Survey of Railroad Construction Organization Design [J]. *Railroad Technology Innovation*, 2022(01):01–06.

Cao ZQ, He Yuelei, Lu Hongyao, *et al.* Research on the application of BIM+AR Technology in High-speed Rail Ballastless Track Monitoring [J]. *Railway Standard Design*, 2022, 66(10): 64–68.

Gao Zhaoxiang, Lv Ning, Su Jinyao, *et al.* Application of BIM+AR Intelligent Construction Technology in Municipal Engineering Projects[J]. *Urban Housing*, 2020, 27(03):197–198.

He X. Study on the Application of BIM+AR Multi-user Interaction and Collaboration Technology in Highway Engineering [J]. *Shanxi Transportation Science and Technology*, 2021(05):46–48+73.

Huang Jinsen, Zhang A-Q, Fang W-G. Research on the Application of BIM+AR Technology based on Building Safety Management [J]. *Fujian Construction*, 2019(09): 119–121.

Niu Lei. Application of BIM Technology in Bid Price of Urban Rail Transit Tunnel Engineering [J]. *Modern Urban Rail Transit*, 2022 (11): 92–94.

Shang Dunjiang, Liu Meirui. Research on the Visualization of BIM-AR Technology in Architectural Heritage Data Under "Internet" - The Development of the Gulangyu Huifeng Mansion project[J]. *Urban Architecture*, 2020, 17(07): 163–169.

Shi Weiguang. The Application of BIM Technology and Assembly Industrialization in Interior Decoration Design [J]. *Footwear Technology and Design*, 2022,2 (16): 141–143.

Wang Xingwen, Yu Hao, Yang Ping. The Rational Application of BIM Technology in Building Structure Design [J]. *Smart Building and Smart City*, 2022(11):111–113. DOI: 10.13655/j.CNKI. ibci. 2022.11.034.

Xu Zhanpeng, Zhang Zhongyu, Che Yongfu. The Intelligent Use of BIM + AR & amp 3D Scanning in Rail Transportation [J]. *Science and Technology Innovation*, 2021(34): 134–136.

Water Conservancy and Civil Construction – Oke & Ahmad (Eds)
© 2024 The Author(s), ISBN: 978-1-032-58618-2

Two-Dimensional numerical analysis of the Nanjiang gate water diversion to improve rivernet water environment

Di Hao
School of Naval Architecture and Maritime Zhejiang Ocean University, Zhoushan, China
School of Water Conservancy and Environmental Engineering Zhejiang University of Water Resources and Electric Power Hangzhou, China

Dongfeng Li*
School of Water Conservancy and Environmental Engineering Zhejiang University of Water Resources and Electric Power Hangzhou, China

Zihao Li
Qingtian County Organization Department, Qingtian, Zhejiang, China

Donghui Hu & Aijun Sun
Yuyao Water Conservancy Bureau, Yuyao, Zhejiang, China

Zhenghao Li
Department of Water Conservancy and Engineering Henan Vocational College of Water Conservancy and Environment, Zhengzhou, Henan, China

ABSTRACT: The development of the economy and society urgently requires flood control and water environment improvement. The two-dimensional water quality mathematical model of the riverside area of Shaoxing Binhai New City eliminates the irrelevant province. According to the diversion gates and pump stations of the Sanhui areas in the riverside area, the changes in water level elevation, flow, BOD concentration, and DO concentration at each point of the water system on the route are analyzed according to the indicators of newly built and existing sluice stations. The water diversion mode of the Nanjiang sluice is studied.

1 INTRODUCTION

The water area of the Sanhui area is 1,159,200 m^2, and the water volume is about 2,086 m^3. This plan adopts the new Nanjiang sluice gate and four culverts to divert water from the Huayang sluice gate, Xinlian sluice gate, Lianyi sluice gate, and Weimin sluice gate. According to the "Shaoxing Cao'e river comprehensive improvement project preliminary design report (draft)", the new Nanflowjiang sluice is a single hole of 4.5 m with a maximum flow rate of 27 m^3/s net width, and four culvert gate single holes of 1.6 m net width, a flow capacity of 5.13 m^3/s. It is important to understand the impact of local control on the whole when designing spatial control and overall collaborative schemes and regional evaluation (Lin 2016). The analysis of the hydrodynamics and the changes of rivers and lakes in certain boundary conditions is the basis of research problems, and a two-dimensional mathematical model as an important and convenient tool is one of the important means to solve the hydrodynamic problems. The analysis of a two-dimensional mathematical model of the hydrodynamics of the river network in the coastal Datian Plain (Li 2016) clarifies the role of reasonable gate control

---

*Corresponding Author: lidf@zjweu.edu.cn

DOI: 10.1201/9781003450832-42

in the regulation of the water system, which accelerates the exchange of water in the urban river network and also contributes to the improvement of the urban water environment. Chen (2014) used a two-dimensional hydrodynamic model to calculate the hydrodynamic force of the river network and the local area of rivers and lakes and obtained the relevant data. In this paper, the environmental improvement effect of the Nanjiang sluice diversion scheme is analyzed based on the data calculated by the two-dimensional mathematical model of the hydrodynamic water environment of the river and lake. A reasonable scheme cannot effectively improve the overall hydrodynamic state of the water system but also improve the water quality and promote the overall water environment (Ding 2016).

2 TWO-DIMENSIONAL MATHEMATICAL MODEL OF HYDRODYNAMIC WATER ENVIRONMENT OF RIVERS AND LAKES

2.1 *Basic theory*

The modeling theory of the hydrodynamic water environment of river networks and lakes includes the basic equations and solution conditions for the water flow movement and water quality movement under its action (Tong & Li 2016).

The two-dimensional system of non-constant equations is as follows.

$$\frac{\partial h}{\partial t} + \frac{\partial h\bar{u}}{\partial x} + \frac{\partial h\bar{v}}{\partial y} = hS \frac{\partial h\bar{u}}{\partial t} + \frac{\partial h\bar{u}^2}{\partial x} + \frac{\partial h\overline{uv}}{\partial y} = f\bar{v}h - gh\frac{\partial \eta}{\partial x} - \frac{h}{\rho_0}\frac{\partial p_a}{\partial x} - \frac{gh^2}{2\rho_0}\frac{\partial \rho}{\partial x} + \frac{\tau_{sx}}{\rho_0} -$$

$$\frac{\tau_{bx}}{\rho_0} - \frac{1}{\rho_0}\left(\frac{\partial S_{xx}}{\partial x} + \frac{\partial S_{xy}}{\partial y}\right) + \frac{\partial}{\partial x}(hT_{xx}) + \frac{\partial}{\partial x}(hT_{xy}) + hu_sS \frac{\partial h\overline{\bar{v}}}{\partial t} + \frac{\partial h\bar{\bar{v}}^2}{\partial x} + \frac{\partial h\overline{uv}}{\partial y} = -f\bar{u}h -$$

$$gh\frac{\partial \eta}{\partial y} - \frac{h}{\rho_0}\frac{\partial p_a}{\partial y} - \frac{gh^2}{2\rho_0}\frac{\partial \rho}{\partial y} + \frac{\tau_{sy}}{\rho_0}\frac{\tau_{by}}{\rho_0} - \frac{1}{\rho_0}\left(\frac{\partial S_{yx}}{\partial x} + \frac{\partial S_{yy}}{\partial y}\right) + \frac{\partial}{\partial x}(hT_{xy}) + \frac{\partial}{\partial y}(hT_{yy}) + hv_sS$$

$$(1)$$

Where t is time (s); x and y are the three axis directions of the Cartesian coordinate system; η is the water level (m); C is the Xiecai coefficient ($m^{1/2}/s$); h is the total water depth (m), $h = \eta + d$, and d is the static water depth; u and v are the velocity components (m / s) along the water depth in the x and y directions; μ is the viscous force coefficient (pa·s); g is the acceleration of gravity (m/s²); ρ_0 is the fluid density (kg/m^3); f is the Coriolis force coefficient, $f = 2\omega \sin \varphi$; P_a is atmospheric pressure (pa); S_{yy}, S_{xx}, and S_{xy} are the radiation stress components; S is the source item; v_s, u_s are the flow rates. $\overline{uv}$ is the average velocity along the water depth, which is defined by the following equation:

$$h\bar{u} = \int_{-d}^{\eta} u dz, \quad h\bar{v} = \int_{-d}^{\eta} v dz \qquad (2)$$

T_{ij} is the horizontal viscous stress, including viscous force and horizontal convection and is derived from the eddy current equation based on the velocity gradient averaged along the water depth:

$$T_{xx} = 2A\frac{\partial \bar{u}}{\partial x}, T_{xy} = A\left(\frac{\partial \bar{u}}{\partial y} + \frac{\partial \bar{v}}{\partial x}\right), T_{yy} = 2A\frac{\partial \bar{v}}{\partial y} \qquad (3)$$

The water quality equation is as follows.

$$\frac{\partial C}{\partial t} + u\frac{\partial C}{\partial x} + v\frac{\partial C}{\partial y} + w\frac{\partial C}{\partial z} = D_x\frac{\partial^2 C}{\partial z^2} + D_y\frac{\partial^2 C}{\partial z^2} + D_z\frac{\partial^2 C}{\partial z^2} + S_C + P_C \qquad (4)$$

326

Where C is the concentration of the state variable; D_x, D_y and D_z are the diffusion coefficients; P_C is the process item of the ECO Lab; S_C is the source sink; $D_x \frac{\partial^2 C}{\partial z^2} + D_y \frac{\partial^2 C}{\partial z^2} + D_z \frac{\partial^2 C}{\partial z^2}$ is a diffusion term; $u \frac{\partial C}{\partial x} + v \frac{\partial C}{\partial y} + w \frac{\partial C}{\partial z}$ is a convection term. Each variable is coupled by P_C nonlinear or linear.

2.2 *Definition of the solution conditions*

(1) Initial:
    The initial values of the mathematical model show that the water level at the initial moment of the river network is 2.8 m the normal water level and the flow rate is 0. The initial water quality condition is Class V water, that is, the BOD is 10 mg/l and the DO is 2 mg/l.
(2) Boundary conditions

In the hydrodynamic mathematical model, there are six different boundary conditions (Chen 2020), namely the land boundary conditions (vertical flow velocity is zero), land boundary conditions (flow velocity is zero), velocity boundary conditions, flux boundary conditions, water level boundary conditions, and flow boundary conditions. Each boundary must contain at least two nodes.

The initial boundary conditions use interpolation. When the boundary requires time and space, linear interpolation or piecewise cubic interpolation can be used.

This calculation is designed by introducing clean water as Class III water, that is, BOD is 4 mg/l, DO is 5 mg/l, and a given flow rate at the inlet. The water quality standards for BOD and DO of surface water are shown in Table 1.

Table 1. Surface water BOD and DO water quality standards are shown in table.

The serial number	Water quality index		Class I	Class II	Class III	Class IV	Class IV
1	Dissolved oxygen DO	$\geq$	7.5	6	5	3	2
2	Biochemical oxygen demand BOD	$\leq$	3	3	4	6	10

2.3 *Two-dimensional mathematical model verification analysis*

The study of this watershed activation scheme is validated using the previous mathematical model of hydrodynamic water quality in the Shangyu River Lake, which is described in the validation and application report of the model in Shaoxing Plain River Network, Shangyu Plain River Network, Hangzhou, Jiaxing, Huzhou, and Ningbo Plain River Network (Cui et al., 2017). The calculations are carried out using the Hydrodynamics and Water Quality Module (ECO-LAB) of the mike21 software.

Table 2. Nanjiang gate water diversion and revitalization program.

Gate	Nanjiang Gate	Huayang Gate	Xinlian Gate	Lianyi Gate	Kuo Weimin Gate	Yuwei Gate	Pumping Station
Flow rate (m^3/s)	27	5.13	5.13	5.13	5.13	Open	5

3 SANHUI PIECE NANJIANG GATE SEPARATE DIVERSION AND REVITALIZATION SCHEME

3.1 *Calculate boundary conditions*

3.2 *Water diversion line one*

Water diversion Line 1 is shown in Figure 1.

It can be seen from Figure 2 that the water level elevation gradually increases within 14 hours after the diversion, but it gradually decreases along the route because there are

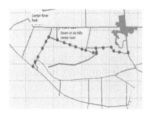 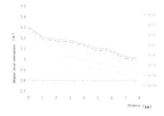 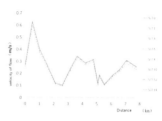

Figure 1. Line-location.   Figure 2. Water level elevation change plot.   Figure 3. Flow velocity varies.

multiple diversions on the line. It can be seen from Figure 3 that the flow velocity fluctuates greatly. The increase is the increase in flow after confluence, and the decrease is the larger turning angle of the river and the larger local head loss.

It can be seen from Figure 4 that the BOD concentration fluctuates greatly with time. After 14 hours, the BOD concentration reaches about 4.5 mg/l at 0-5 km from the water inlet, which is the standard of four types of water. The BOD concentration is low at 5.5 km, which is at the end of the central river. The reason is that the pumping station of the central river pumps water so that the living water first flows through the whole central river. The BOD concentration is between 3 mg/l to 4 mg/l after 14 hours at a distance of 6-8 km. It is the standard of three types of water. Due to the relatively small calculation time (14 hours), the increase in water quality calculation time will continue to improve the water quality. It can be seen from Figure 5 that the DO concentration generally increases. After 14 hours, the DO concentration in most of the river reaches 6.5 mg/l, which is the standard for Class II water. The DO concentration increases greatly at 5.5 km for the same reason as in Figure 4.

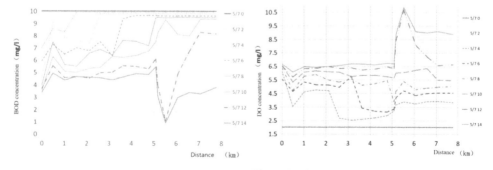

Figure 4. BOD concentration varies along the way.   Figure 5. DO concentration varies along the way.

3.3 Water diversion line two

The second diversion line is shown in Figure 6.

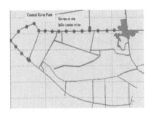 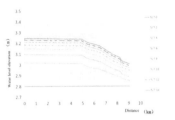 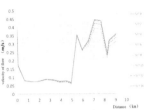

Figure 6. Line - location.   Figure 7. Water level elevation varies.   Figure 8. The flow rate varies.

It can be seen from Figure 7 that the water level elevation gradually increases within 14 hours after water diversion, the water level elevation along the way first stabilizes, and then decreases. The reason is that there is more river diversion after 5 kilometers. From Figure 8, it can be seen that the flow velocity fluctuates greatly. The flow velocity is low at 1-5 km from the diversion outlet and increases sharply after 5 km. The reason for the subsequent fluctuation is the decrease in flow velocity caused by the large loss of the local head.

It can be seen from Figure 9 that the BOD concentration fluctuates greatly with time. After 14 hours, the BOD concentration of 0-5 km away from the water inlet reaches 1 mg/l, which is the standard of Class I water. The BOD concentration is higher at 5 km, which is the end of the central river. Due to the small calculation time (14 hours), the living water pumped by the central river pumping station does not reach the place, so the BOD concentration is relatively high. However, it decreases later. The BOD concentration is between 4mg/l to 5mg/l after 14 hours, which is the standard of four types of water. As can be seen from Figure 10, the DO concentration also increases significantly at 1-5 km, increases at 6-8 km, and increases at about 7.5 mg/l. The oxygen concentration in the water increases, the water quality is improved, and a sharp decrease occurs at 5 km. The reason is the same as in Figure 9.

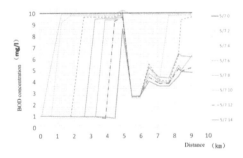

Figure 9. BOD concentration varies along the way.

Figure 10. DO concentration varies along the way.

3.4 BOD concentration plane distribution analysis

As can be seen from Figure 11, when the diversion flow of the Nanjiang Gate is 27 m³/s, the Yu Fence Gate is opened, the diversion time is 14 hours, and the BOD concentration value of most rivers in Sanhui is about 3.6-4.2 mg/l, which meets the standard of three types of water. The BOD concentration of the connecting river section of Huantang South River and Qiliuqiu Central River reaches 1.2 mg/l water quality improvement is very significant, and the BOD concentration of the central river near the central lake is 3 mg/l, reaching the standard of Class II water, the water quality of this section has also been greatly improved. The BOD concentration of the central river has been reduced from 6 mg/l to 1.2 mg/l, reaching the standard of Class I water, and the water quality of the central river has also been greatly improved (Wu 2018). Although the water is only changed for 14 hours, the water quality improves significantly.

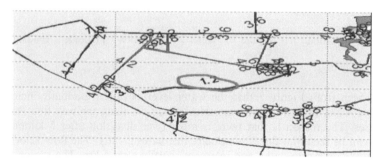

Figure 11. Overall BOD concentration distribution.

4 CONCLUSION

In this paper, by analyzing the changes in water level elevation, velocity, BOD concentration, and DO concentration at each point on the route of the Nanjiang Gate diversion scheme, it is finally determined that the flow discharge of the Nanjiang Gate is 27 m^3/s, the scheme is the optimal scheme under the three conditions of opening the Yuwei Gate and adding a pumping station in the Central River, and the scheme has played a significant role in improving the water environment of Sanhui. From the local area to the overall, to achieve the purpose of improving the water system environment of the riverside area of Shaoxing Binhai New City, it provides experience for the establishment of "three vertical, six horizontal, and four lakes" in the future river system planning of the riverside area, and provides experience for the improvement of the water environment of the "three vertical, six horizontal, and four lakes" in the water system planning.

ACKNOWLEDGMENT

This research was supported by the Funds Key Laboratory for Technology in Rural Water Management of Zhejiang Province (ZJWEU-RWM-202101), the Funds of Water Resources of Science and Technology of Zhejiang Provincial Water Resources Department, China (No.RC2239, No.RB2115, and No.RC2040), the Joint Funds of the Zhejiang Provincial Natural Science Foundation of China (No. LZJWZ22C030001, No. LZJWZ22E090004), the National Key Research and Development Program of China (No.2016YFC0402502), the National Natural Science Foundation of China (51979249), the University Student Innovation and Entrepreneurship Project of Zhejiang University of Water Resources and Hydropower (S202211481017, S202111481001).

REFERENCES

Cui, Chen, Xiang, Zhang, and Xu (2017). Evaluation of Water Environment Improvement Effect of River Network Connectivity in the Plain River Network Area. *Journal of Water Resources* (12), 1429–1437. doi:10.13243/j.cnki.slxb.20170576.

Chen Qingjiang River, Ding Rui & Zhao Hai. (2020). Improvement Effect of Smooth Flow of Living Water on Hydrodynamics and Water Quality in the Plain River Network Area. *Advances in Water Conservancy and Hydropower Technology* (03), 8–13.

Chen Weijin, Li Dongfeng & Zhang Hongwu. (2014). Analysis of Control Conditions for Hydrodynamic Model of Flood Control and Drainage in Shaoxing Plain River Network. *Journal of Zhejiang University of Water Resources and Electric Power* (03), 38–41.

Ding Yi, Jia Haifeng, Ding Yongwei & Sun Chaoxia. (2016). Hydrodynamic Optimization Control of Water Network in Water Towns Based on EFDC Model. *Journal of Environmental Sciences* (04), 1440–1446. doi:10.13671/j.hjkxxb.2015.0472.

Huang Guoru, Luo Haiwan, Chen Wenjie & Pan Jian. (2019). Scenario Simulation and Risk Assessment of Urban Flood Disaster in Donghaochong Basin, Guangzhou. *Progress in Water Science* (05), 643–652. doi: 10.14042 / j.cnki.32.1309.2019.05.004.

Li Dongfeng, Xie Feng, Bai Fuqing & Hu Jianyong. (2016). Impact Analysis of River Network Water System Adjustment in Linhai Datian Plain. *Journal of Zhejiang University of Water Resources and Electric Power* (06), 12–15.

Li Dongfeng, Pan Jie, Chen Haixiong & Zhang Hongwu. (2016). Two-dimensional Numerical Simulation of Flood Control and Drainage Effects of River Connection Projects. *Renmin Yellow River* (11), 31–33.

Lin Lan. (2016). The Current Situation of Water Pollution in the Yangtze River Delta Region Evaluation and Management Ideas. *Environmental Protection* (17), 41–45. doi : 10.14026/j.cnki.0253-9705.2016.17.008.

Tong Yunying, LI Dongfeng & NIE Hui. (2016). Research on a Hydrodynamic and Two-dimensional Mathematical Model of the River Network in Datian Plain. *Journal of Zhejiang University of Water Resources and Electric Power* (01), 14–17.

Wu Shiqiang, Dai Jiangyu & Shi Sha. (2018). Research Progress on Ecological Effect Assessment of Water Diversion Project Lakes. *Journal of Nanchang Institute of Technology* (06), 14–26.

Water Conservancy and Civil Construction – Oke & Ahmad (Eds)
© 2024 The Author(s), ISBN: 978-1-032-58618-2

Study on the urban design guidelines of waterfront space in Pearl River Delta—taking Foshan city as an example

Liang Liyun*
Guangzhou Academy of Fine Arts, Guangzhou, China

Liu Hui
South China University of Technology, Guangzhou, China

ABSTRACT: With the increasing attention to urban water environment problems, the compilation of urban design guidelines for water bodies has become an extremely important task in current urban planning and management. In this paper, we summarize the important role of current waterfront space design guidelines, which is no longer partial and shallow, but should be expanded horizontally and vertically from serving for design to serving for both refined management and implementation. We also propose a basic technical route for establishing such guidelines, including establishing goals and principles, rationally classifying water bodies, defining the spatial composition of waterfronts, identifying controlling elements, and highlighting the important points. This paper also takes the water system planning project of Foshan city as an example to demonstrate how to establish a comprehensive and complete framework of urban design guidelines, especially in the Pearl River Delta, where the water network is dense.

1 OVERVIEW OF WATERFRONT URBAN DESIGN GUIDELINES

On the one hand, people's demands for quality of life and living environment have been increasing in recent years. On the other hand, the problems of urban water environment, water ecology, and water safety have emerged from time to time. Therefore, urban ecological landscapes have received more and more attention, especially in the development and construction of waterfront areas. Projects, such as river ecological restoration, waterfront development, and waterfront landscape construction, are common in many cities. The content of these projects is not only limited to beautifying the city but also integrates multiple disciplines, such as water conservancy and flood control, social economy, urban planning, landscape ecology, historical and cultural protection, and human behavior psychology. Corresponding to the refined management of the city, the role of waterfront urban design guidelines is also changing.

1.1 *From partial design to exhaustive control and guidance*

Different disciplines are currently conducting research on the planning and construction of waterfront areas from their perspectives. These studies are mainly focused on the following aspects. (i) Urban economics studies the rational models of waterfront land development to maximize economic benefits. (ii) The landscape architecture proposes a landscape

---

*Corresponding Author: liangliyun@tom.com

DOI: 10.1201/9781003450832-43

renovation plan for a specific water or waterfront area to enhance the image of the city and improve the quality of life. (iii) The ecology focuses on restoring rivers and their surrounding ecosystems to balance the needs of residents for recreation and ecological protection.

The research results of the above disciplines have played a positive role in promoting the development and design of urban waterfronts. However, most of their research methods determine that they can only target a few major rivers or some key points of waterfronts in the city. Therefore, it is difficult to directly extend the results to the waterfront space of the whole city, especially in the river network cities with complex and changeable water systems. We should develop a set of design guidelines that will be of wide use in the planning of water systems throughout the city. In these guidelines, we can summarize a series of design requirements to comprehensively guide and coordinate various types of waterfronts in urban planning and management.

1.2 *Depth expansion from the shoreline to inland*

In the usual waterfront landscape design, designers only focus on the location of the land and water junction. Moreover, the red line of waterfront planning, which is the basis for waterfront design, often includes a thin layer of the waterfront. This also leads designers to ignore the space beyond the red line and lack the rationality to consider the construction of the waterfront area on a larger scale.

In fact, the elements that make up the river landscape are very diverse. As far as the concept of landscape ecology is concerned, urban landscape planning should pursue the coordination of large-scale textures. As an important part of the urban ecological environment, the waterfront area should be expanded more widely in space. In addition, another important motivation for the in-depth expansion comes from the planning administrative department, which is the initiator of the organization and compilation of urban design guidelines. This is because its scope of authority covers all kinds of waterfront land, not the shoreline.

1.3 *From design to refined management and implementation*

The urban design guidelines for waterfront spaces have three main functions:

For the planning management departments. They can master the basic construction control standards according to the guidelines to make scientific and reasonable decisions on waterfront land construction projects and control the development and design of these spaces in the entire city.

For Landscape designers and architects. When planning and designing, they can use the guidelines as design manuals and references to facilitate the coordination and unification of waterfront spaces with the other parts of the city, which truly completes the implementation of the guidelines.

For the public. With the publicity and promotion of the design guidelines, their concise, vivid, and easy-to-understand forms can stimulate public interest. Furthermore, they can encourage the general public as well as the users of the waterfront area to deepen their understanding of the planning and design of the waterfront area, thus forming spontaneous public participation in the design.

2 A TECHNICAL ROUTE TO CONSTRUCT A FRAMEWORK FOR WATERFRONT URBAN DESIGN GUIDELINES

Although the waterfront urban design guidelines are only a kind of special guideline for urban water bodies, they involve a wide range of factors and contents. From the functional division of the urban water system to the planning of coastal land, or, from the landscape characteristics of the entire river to the detailed design of important nodes, these guidelines must provide control suggestions. Therefore, to integrate the numerous design factors as concisely as possible in the guidelines, a structured and coherent framework must be established.

2.1 *Establishing overall goals and principles*

Principles of Ecological Protection and Sustainability. To give full play to the ecological effects of rivers on the urban environment, we can define some specific principles under this general principle, which includes protecting the existing river grids, water patches, and wetlands, connecting them with the urban green network to form an overall continuous ecological corridor, advocating the use of methods that have the least impact on the ecological environment, reducing damage to nature, creating multiple types of complex habitats to maintain the quality of waterfront plant and animal habitats and strengthening the circulation and regeneration function of the ecosystem.

Principles of humanity. The waterfront areas should be accessible and encouraged for public use. We should ensure that there are adequate sight corridors. This means that the roads to the river should be dense enough to ensure that the water body is perceptible in the city. We can make full use of the waterway resources to organize public transportation and excursions reasonably on the rivers. We can arrange appropriate facilities within the waterfront areas to meet the diverse needs of the public.

Principles of historical and cultural protection. The protection objects mentioned here should include all material and intangible cultural heritage related to rivers, such as waterfront traditional settlements, historical buildings, traditional customs, festivals, etc. We can integrate them into urban design to show the human landscape with local characteristics and keep them alive in the new urban life.

2.2 *Reasonable classification of water bodies*

Classification according to the area where the river flows. According to the urban construction intensity, functional requirements, and resource conditions in the flowing area, we divide the design objects into urban sections, rural sections, and suburban sections.

The "urban section" refers to the water bodies that flow through the urban settlements and their corresponding land areas. Such areas have high population density, heavy construction pressure, heavily polluted water bodies, and concentrated cultural and historical resources. The design guidelines for such waterfront spaces should emphasize the compactness of land use layout, the humanization and ecological benefits of landscape design, and the beautification of the city.

The "rural section" refers to the water bodies that flow through the rural settlements. In this section, the population density is low, and the traditional features are well preserved. However, under the influence of urban development, the intensity of development and water pollution is increasing day by day. The design guidelines should emphasize the excavation of local characteristics, the reproduction of historical features, and the restoration of the ecological environment.

The "suburban section" refers to the water body and corresponding land area flowing through non-residential areas, usually has a good natural environment and water quality but lacks ecological planning and has a fragile ecological environment. Therefore, the design guidelines should focus on protection and minimize manual intervention.

Classification according to the width of rivers. Generally, the narrower the river surface, the more intimate it is to users. However, in such rivers, the land on both sides is tense and the building volume may oppress the river space. With the widening of the river, the width of the land on both sides, the scale of the river space, and the way that people use it all change. The design guidelines must also be adjusted. In the water system planning of Foshan, the rivers are divided into 3 types with 10 m and 30 m as standard lines. (Table 1)

Table 1. Classification of rivers in the water system planning of Foshan.

Area that flows through	Width		Urban design principles of waterfront
Urban section	For living	≤ 10 m	Emphasize Humanized design principles;
		10-30 m	Create a livable urban environment;
		≥ 30 m	Meet people's psychological needs of being
	For production	Ports and Industrial Shorelines	close to water.
Rural section	≤ 10 m		Emphasize the ecologicalization of the
	10-30 m		landscape;
	≥ 30 m		Highlight the traditional style of Lingnan;
			Promote the development of agricultural ecology.
Suburban section	≤ 10 m		Protect as an important water source protection
	>10 m		and conservation area;
			Emphasize ecological design principles;
			Focus on ecological protection and restoration.

2.3 Defining the spatial composition of the waterfronts

Depending on the characteristics of use and design priorities, the waterfront can be divided into three interconnected parts. The order from water to land is a riverside area, waterfront green belt area, and construction area.

Riverside area, with a depth range generally about 1-5 m, is the shoreline directly adjacent to the body of water. It integrates the functions of water conservancy, flood control, ecological edge effect, and waterside activities.

The waterfront green belt area is an important urban public area that can form the waterfront landscape and provide leisure and entertainment for the citizens. It is often incorporated into urban parks and integrated with landscape design to form strip-shaped waterfront parks. The depth range of these areas varies greatly.

The construction area refers to all kinds of urban construction land with a certain depth along the water. Since such an area is farther from the water body than others, it is rarely designed in previous waterfront planning. However, it has been proved that the proper nature of land use, convenient transportation, and the volume and form of buildings along the river all have a decisive impact on the overall image of the waterfront area.

2.4 Identifying the controlling elements

According to the respective morphological characteristics of the above 3 areas, and summarizing the experience of waterfront design at home and abroad at the same time, we propose the following 12 main design control elements.

For the riverside area, 5 elements can be controlled, which include the form of the revetment, the material of the revetment, the hydraulic construction, the safety measures, and the plant design. For the waterfront green belt area, there are 4 elements, which include the paving design, the Landscape Facilities, the illumination, and the plant design. For the construction area, the building concessions, the building height control, and the building facade design are involved. These 12 elements are universal and necessary in the design of waterfront areas in most cities. Other corresponding control elements can also be added in each area according to actual needs.

To facilitate access or indexing by designers and managers, the design method of each element can be expressed in the form of diagrams, lists, and numbered indexes. Then, according to the type of river, spatial composition, and control elements, a general table is listed (Table 2).

Table 2. Index of urban design elements of waterfront areas in urban sections.

	≤ 10 m	10-30 m	≥ 30 m
Revetment form	A1, A2, A5	A4, A5, A6, A7, A8	A7, A8, A9, A10
Revetment material	B1	B2, B4	B3
Hydraulic construction	C1	C2, C5	C3, C4
Safety measures	E1	E2, E3	E4
Plant design	F1	F2	F3
Paving design	G1	G3	G4
Landscape facilities	H1	H3	H2, H1
Illumination	J1	J1	J2
Plant design	K1, K2	K3, K5	K4, K6
Building concessions	L1	L1	L2, L3
Building height control	M1	M2	M3
Building facade design	N1	N2	N3

3 AN EXAMPLE–URBAN DESIGN GUIDELINES FOR WATERFRONT AREAS IN URBAN WATER SYSTEM PLANNING IN FOSHAN

Foshan City in Guangdong Province is located in the Pearl River Delta District with dense water networks. It has a large urban area and scattered built-up areas, and the current situation of waterfront land is extremely complex. At the same time, the city is in a state of rapid construction and high-intensity development, which makes the waterfront land and shoreline resources all-around tight.

According to the main functions of the existing rivers in Foshan City, the rivers are firstly divided into two categories, outer rivers and inner rivers. The wide outer river with a total length of about 650 kilometers is mainly responsible for flood control and water transportation. They are controlled by the laws of water conservancy and the general design principles of waterfront areas. A large number of inner rivers no longer undertake the function of water transportation with the development of land transportation. The landscape function of inland rivers is brought to a prominent position and becomes a key planning object in the urban design guidelines for waterfront areas.

Through on-the-spot investigation and analysis of more than 2,700 inner rivers in the city, we have made a more detailed classification of them according to their characteristics and various factors involved in urban design. According to Table 1 and 12 design control elements, the design index table for various rivers is listed in Table 2. Figure 1 shows how to access and use the design guidelines.

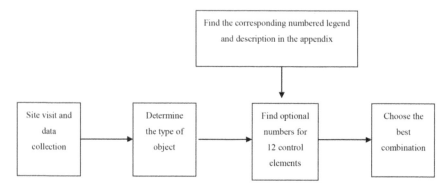

Figure 1. How to access and use the design guidelines.

4 CONCLUSION

A set of comprehensive, detailed, systematic, and practical urban design guidelines is an indispensable workbook and directional marker for designers to carry out detailed planning and design in the future. At the same time, it is also an important decision basis for government departments to conduct urban management. Compiling such a set of design guidelines is more important for cities located in the river network area, where the water system is widely distributed and the water system planning involves a lot of content and management departments. The construction of the framework of the guidelines, the classification of planning water bodies, and the selection of control elements are the keys to compiling this guideline. By studying the urban design guidelines for the waterfront area in Foshan, the experience and methods summarized in this paper can also be used for reference in other cities. This study also aims to create safe, ecological, prosperous, and sustainable cities in the future.

ACKNOWLEDGMENTS

Liang acknowledges the support from the Characteristic Innovation Project of the Guangdong Educational Commission (Grant No. 2020WTSCX045) and the Special Project of the Guangzhou Academy of Fine Arts (20XSB12).

REFERENCES

Foshan Planning Bureau. (2018) *Urban Blue Line Planning of Foshan.* http://www.foshan.gov.cn/zwgk/zfgb/rmzfbgshj/content/post_1740150.html, 2022/1/5

He, Yuxiao. Research on The Idea of Compiling Urban Design Guidelines-Ttaking the Qianhai Mawan Area of Shenzhen As An Example. *City House* 27(12), 121–122+125(2020).

Li, Minzhi., Ye, Weikang. Rational Construction of Controllable System of Urban Design Guidelines. *South Architecture* 04, 15–19(2012).

Liang, Liyun. Urban Water System Planning In A Densely Populated Area-Taking Foshan City As An Example. *In: 5TH International Conference on Traffic Engineering and Transportation System*, vol. 12058, 04–1. Spie(2021).

Mohammed, M. R. *Handbook of Waterfront Cities and Urbanism.* Routledge, New York (2023).

Xu, Hui., Xu, Xiangyang., Cui, Guangbo. Application of Landscape Spatial Structure Analysis in Urban Water System Planning. *Advances in Water Science* 18(1), 108–113(2007).

Water Conservancy and Civil Construction – Oke & Ahmad (Eds)
© 2024 The Author(s), ISBN: 978-1-032-58618-2

Effect analysis of water environment improvement in Sanhui District based on mike21 ECO-LAB modular

He Huan
School of Naval Architecture and Maritime Zhejiang Ocean University Zhoushan, China

Yanfen Yu*
Zhejiang Province Qiantang River Basin Center, Hangzhou, China

Zihao Li
Qingtian County Organization Department, Qingtian, Zhejiang, China

Donghui Hu & Aijun Sun
Yuyao Water Conservancy Bureau, Yuyao, Zhejiang, China

Zhenghao Li
Department of Water Conservancy and Engineering Henan Vocational College of Water Conservancy and Environment, Zhengzhou, Henan, China

ABSTRACT: The study of water diversion schemes is significant to improve the water environment. The development of the economy and society urgently needs the improvement of the water environment and the water diversion scheme is an important countermeasure. The Nanjiang Gate is an important gate for regulating water flow in the Sanhui District. To fully study the living water effect of the Nanjiang Gate, based on the two-dimensional mathematical model of the hydrodynamic and water environment of the Sanhui River network, the overall river network diversion water environment improvement scheme is made and the reasonable water level boundary conditions are calculated and analyzed. Then the separate diversion line scheme of Nanjing Gate is analyzed. By analyzing the changes in water level elevation, flow rate, BOD concentration, and DO concentration at important points on the line, the influence of the scheme on the line is analyzed. Finally, this scheme can effectively improve the water environment and provide a reference for improving the water environment in the Sanhui District.

1 INTRODUCTION

Currently, the main rivers in Sanhui District include the Qiliuqiu Beitang River, Zhongxin River, Sanhui Central River, Nanhuantang River, etc. The regional water District is 1,159,200 m^2 and the water volume is about 102,086 m^3. The water diversion eliminates most of the dead water District and creates a good flow field, thus reducing turbidity and improving water transparency, which is conducive to improving water quality. A reasonable water diversion scheme can effectively improve water quality. Economic and social development urgently needs flood control, drainage, and water environment improvement. A living water scheme is an important countermeasure. Nanjiang Gate is the main water

---

*Corresponding Author: yuyanfen@zjwater.gov.cn

source of Sanhui District. Based on the two-dimensional hydrodynamic model, the water diversion scheme of Nanjiang Gate in Sanhui District is designed.

2 TWO-DIMENSIONAL MATHEMATICAL MODEL AND VERIFICATION OF DYNAMIC WATER ENVIRONMENT OF SANHUI DISTRICT

2.1 Basic equation of two-dimensional mathematical model of the dynamic water environment of Sanhui District

(1) Fundamental equation

The two-dimensional unsteady equations are as follows.

$$
\begin{aligned}
\frac{\partial h}{\partial t} + \frac{\partial h\bar{u}}{\partial x} + \frac{\partial h\bar{v}}{\partial y} &= hS \frac{\partial h\bar{u}}{\partial t} + \frac{\partial h\bar{u}^2}{\partial x} + \frac{\partial h\overline{uv}}{\partial y} \\
&= f\bar{v}h - gh\frac{\partial \eta}{\partial x} - \frac{h}{\rho_0}\frac{\partial p_a}{\partial x} - \frac{gh^2}{2\rho_0}\frac{\partial \rho}{\partial x} + \frac{\tau_{sx}}{\rho_0} - \frac{\tau_{bx}}{\rho_0} - \frac{1}{\rho_0}\left(\frac{\partial S_{xx}}{\partial x} + \frac{\partial S_{xy}}{\partial y}\right) \\
&\quad + \frac{\partial}{\partial x}(hT_{xx}) + \frac{\partial}{\partial x}(hT_{xy}) + hu_sS \frac{\partial h\bar{v}}{\partial t} + \frac{\partial h\bar{v}^2}{\partial x} + \frac{\partial h\overline{uv}}{\partial y} \\
&= -f\bar{u}h - gh\frac{\partial \eta}{\partial y} - \frac{h}{\rho_0}\frac{\partial p_a}{\partial y} - \frac{gh^2}{2\rho_0}\frac{\partial \rho}{\partial y} + \frac{\tau_{sy}}{\rho_0}\frac{\tau_{by}}{\rho_0} - \frac{1}{\rho_0}\left(\frac{\partial S_{yx}}{\partial x} + \frac{\partial S_{yy}}{\partial y}\right) \\
&\quad + \frac{\partial}{\partial x}(hT_{xy}) + \frac{\partial}{\partial y}(hT_{yy}) + hv_sS
\end{aligned}
\tag{1}
$$

Where t is time (s); x and y are the three axis directions of the Cartesian coordinate system; η is the water level (m); C is Xiecai coefficient ($m^{\frac{1}{2}}/s$); h is the total water depth (m), $h = \eta + d$; d is the static water depth; uandv are the velocity component (m/s) along the water depth in the x and y directions; μ is the viscous force coefficient (pa·s); g is the acceleration of gravity (m/s²); ρ_0 is fluid density (kg/m^3); f is the Coriolis force coefficient, $f = 2\omega\sin\varphi$; P_a is atmospheric pressure (pa); S_{yy}, S_{xx}, and S_{xy} are the radiation stress components; s is the source item; v_s and u_s are the flow rates; $\overline{uv}$ is the average velocity along the water depth, which is defined by the following formula:

$$
h\bar{u} = \int_{-d}^{\eta} u\, dz, \quad h\bar{v} = \int_{-d}^{\eta} v\, dz
\tag{2}
$$

Where T_{ij} is the horizontal viscous stress, including viscous force and horizontal convection, which is derived from the eddy current equation based on the velocity gradient averaged along the water depth.

$$
T_{xx} = 2A\frac{\partial \bar{u}}{\partial x}, \quad T_{xy} = A\left(\frac{\partial \bar{u}}{\partial y} + \frac{\partial \bar{v}}{\partial x}\right), \quad T_{yy} = 2A\frac{\partial \bar{v}}{\partial y}
\tag{3}
$$

The water quality equation is as follows.

$$
\frac{\partial C}{\partial t} + u\frac{\partial C}{\partial x} + v\frac{\partial C}{\partial y} + w\frac{\partial C}{\partial z} = D_x\frac{\partial^2 C}{\partial z^2} + D_y\frac{\partial^2 C}{\partial z^2} + D_z\frac{\partial^2 C}{\partial z^2} + S_C + P_C
\tag{4}
$$

Where C is the concentration of the state variable; D_x, D_y and D_z are the diffusion coefficients; P_C is the process item of the ECO Lab; S_C is the source sink; $D_x\frac{\partial^2 C}{\partial z^2} + D_y\frac{\partial^2 C}{\partial z^2} + D_z\frac{\partial^2 C}{\partial z^2}$ is a diffusion term; $u\frac{\partial C}{\partial x} + v\frac{\partial C}{\partial y} + w\frac{\partial C}{\partial z}$ is a convection term. Each variable is coupled by P_C nonlinear or linear.

(2) Definite solution condition

The definite solution conditions include initial conditions and boundary conditions. The boundary conditions of model construction are shown in Figure 1.

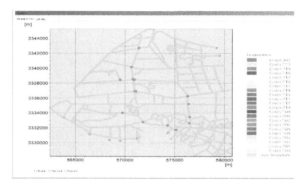

Figure 1. Model boundary conditions construction diagram.

Initial conditions: These refer to the initial value of the mathematical model. The water level at the given initial moment of the river network is 2.8 m and the flow rate is 0. The initial water quality conditions are Class V water, that is, BOD is 10 mg/l and DO is 2 mg/l.

Boundary conditions: In hydrodynamic mathematical models, there are six different boundary conditions, that is, the land boundary conditions (zero vertical velocity). land boundary conditions (zero velocity), velocity boundary conditions, flux boundary conditions, water level boundary conditions, and flow boundary conditions. Each boundary must contain at least two nodes. (Li et al. 2022).

Velocity boundary conditions: The velocity boundary needs to set the velocity in the x-direction and the y-direction. There are three kinds of input. The first is the constant that does not change with time; The second is that the velocity is constant along the boundary but the velocity changes with time. If the boundary conditions that change with time and change with space are to be used on the boundary, a data file with input containing x and y direction velocity (m/s) should be prepared before model calculation, and a file with the same time range should be prepared as the model. The input file should include the entire model cycle. However, the time step number does not need to be consistent with the time step of the hydrodynamic model. If the time steps do not match, the model will be interpolated automatically. The third is that the velocity varies with time and space on the boundary. If one uses boundary conditions on the boundary that vary with time and space, then he or she must prepare an input file with the same scope as the model before setting up the hydrodynamic module. The input file must contain the entire model cycle. (Mu & Liu 2022) However, the time step number does not need to be consistent with the time step of the hydrodynamic model. If the time steps do not match, the model will be interpolated automatically.

Water level boundary conditions: If the boundary is selected as water level (surface elevation, m), there are three ways to set water level boundary conditions. The first is the constant boundary conditions in which the water level does not change with time. The second is that the water level is constant along the boundary and changes with time. The third is that the boundary conditions of water level change with time and space. The principle of boundary input and calculation is the same as that of velocity boundary conditions.

Flow boundary conditions: In the flow boundary condition, the flow value is the total flow. The flow boundary has two forms. The first is that the flow rate that does not change with time is the constant boundary condition. The second is that the flow boundary conditions of variables change with time. The principle of boundary input and calculation is the same as that of velocity boundary conditions.

This calculation is to introduce clear water as Class III water, that is, BOD is 4 mg/l, DO is 5 mg/l, with a given planning design flow rate at the diversion. The water quality standards of surface water BOD and DO are shown in Table 1.

Table 1. Surface water BOD and DO water quality standards.

Water quality index		I	II	III	IV	V
Dissolved oxygen DO	<	7.5	6	5	3	2
Biochemical oxygen demand BOD	≤	3	3	4	6	10

2.2 *Verification analysis of two-dimensional mathematical model*

The dynamic water quality mathematical model of the lake water of Shangyu River is used in this study. The verification and application report of the model are obtained in the Shaoxing Plain River Network, Shangyu Plain River network, Hangzhou Plain River network, Jiaxing, Huzhou, and Ningbo Plain River networks. Calculations are made using the Hydrodynamic and Water Quality Module (ECO-LAB) of mike21 software.

3 SANHUI DISTRICT RUNNING WATER SCHEME

3.1 *Calculation of boundary conditions*

Table 2 shows the boundary conditions for the calculation of the Sanhui District.

Table 2. Calculation of boundary conditions for the Sanhui District.

Gate	Flow rate (m/s)
Nanjiang Gate	13.5
Huayang Gate	5.13
New coupling Gate	5.13
Power coupling Gate	5.13
Weimin Gate	5.13

The inlet position and flow velocity vector are shown in Figure 2.

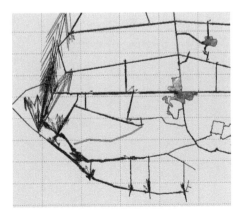

Figure 2. Flow vector diagram.

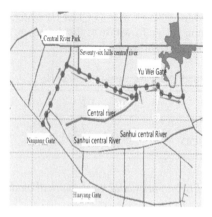

Figure 3. Location map of the water diversion line of Nanjiang gate.

3.2 *Separating diversion line of Nanjiang gate*

We select the first river line in the Sanhui piece, as shown in Figure 4. The influence of this scheme on the line is analyzed by analyzing the changes in water level elevation, flow rate, BOD concentration, and DO concentration at each point on the line (Li & Pan 2016).

From the changes in water level elevation along the route in Figure 3, the water level elevation at all points increases significantly within 6 hours after water diversion. This is because the Nanjiang Gate opens the inflow of fresh water. After 12 hours, the elevation of the water level at each point begins to be fixed and the water level stabilizes. As the distance from the inlet to the outlet changes, the elevation of the water level decreases continuously. Nanjiang Gate, which is 6.8 km away from the water diversion, with the coordinate of (569555.3013, 3333966.897), shows a sudden decrease in water level because the location of the Yu Wai Gate is closed and no water enters the place, and there is an increase due to the entry of water from the connecting section of the central lake (Li & Chen 2019). However, the increase is less than in other places.

As can be seen from Figure 5, the overall fluctuation of flow velocity after water diversion is relatively large. There are three reasons for the analysis. First, the route has multiple large corners, which increases the local head loss and the total head loss of water flow (Liang & Li 2013). Second, the diversion of the tributaries results in a decrease in water velocity at the first two locations. Third, the closure of the Yu Wei gate results in a flow velocity of 0 after

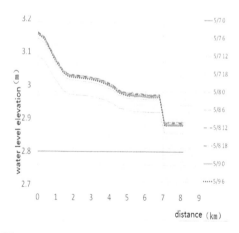
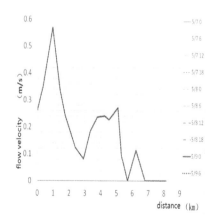

Figure 4. Water level elevation variation.   Figure 5. Velocity variation.

6.8 km. However, the overlapping of the lines at each period indicates that the flow velocity at each point is very stable over time.

As can be seen from the change of BOD concentration along the route in Figure 6, the overall BOD concentration decreases with the increase in time from May 7 to May 9 at 06:00, and most of the BOD concentration dropped to the standard of the Class III water. However, the BOD concentration or high water quality does not improve at a distance of 5.7 km, that is, t34 point, with the coordinate of (569079.9262, 3333622.423) and another coordinate (569555.3013, 3333966.897) at 6.8 km. The former is because the water is not circulated in the central river and the latter is because the Yuwai Gate is not opened and the water is not circulated.

As can be seen from Figure 7, the DO concentration in water increases significantly 6 hours after water diversion. One day after the diversion of water, DO concentration reaches more than 5 mg/l, which is the standard for Class III water, but after the distance of 6.8 kilometers, DO concentration decreases and is Class IV water. The reason is that the Yuwai Gate is not opened, and there is less running water flow and less water quality improvement.

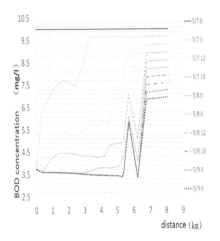

Figure 6. BOD concentration variation.　　Figure 7. DO concentration variation.

4 CONCLUSION

Based on the establishment and preliminary verification of the two-dimensional mathematical model of the dynamic water environment in Sanhui, reasonable water level boundary conditions are analyzed. The opening of the Nanjiang Gate results in a significant increase in water level elevation. The BOD concentration decreases as a whole and the DO concentration increases. Therefore, the use of this water diversion scheme can effectively improve the water environment. Yuwei Gate is not opened and the central river water does not flow, which affects the hydrodynamic force due to the lack of control. This paper preliminarily analyzes the separate diversion line of Nanjiang Gate and provides a reference for improving the water environment of Sanhui District. The influencing factors are complex and there are many calculated data. Due to the lack of comparison, the improvement of the water environment needs further analysis.

ACKNOWLEDGMENT

This research was supported by the Funds Key Laboratory for Technology in Rural Water Management of Zhejiang Province (ZJWEU-RWM-202101), the Funds of Water Resources of Science and Technology of Zhejiang Provincial Water Resources Department, China (No.RC2239, No.RB2115, No.RC2040), the Joint Funds of the Zhejiang Provincial Natural Science Foundation of China (No. LZJWZ22C030001, No. LZJWZ22E090004), the National Key Research and Development Program of China (No.2016YFC0402502), the National Natural Science Foundation of China (51979249), the University Student Innovation and Entrepreneurship Project of Zhejiang University of Water Resources and Hydropower (S202211481017, S202111481001).

REFERENCES

Li Dongfeng, Zhang Hongwu, Zhong Deyu, Hu Jianyong, Ma Liangchao & Jiang Caodong. (2021). Two-dimensional Numerical Model Analysis of the Influence of Sediment on the Lower Reaches of the Yellow River Estuary by Different Flow Channels. *Yellow River of the People* (05), 17–23+29.

Li Huiling, Chen Jing, Jin Qiu, Chen Liming & Xu Yifan. (2019). Study on the Influence of Gate Scale on the Hydrodynamic Water Environment of the River. Changjiang *People's River* (02), 181–185+196. doi:10.16232/j.cnki.1001-4179.2019.02.032.

Li Dongfeng, Xie Feng, Bai Fuqing & Hu Jianyong. (2016). Analysis of the Influence of River Network Adjustment in Linhai Datian Plain. *Journal of Zhejiang Institute of Water Resources and Hydropower* (06), 12–15.

Li Dongfeng, Pan Jie, Chen Haixiong & Zhang Hongwu. (2016). Two-dimensional Numerical Simulation of Flood Control and Drainage Effects of River Connection Projects. *The People's Yellow River* (11), 31–33.

Liang Baijun, Li Dongfeng & Zhang Hongwu. (2013). Numerical Calculation and Analysis of the Influence of Water Diversion from the Direct River of Guazhu Lake on Navigation. *Journal of Zhejiang Water Resources and Hydropower College* (04), 1–3.

Li Dongdong, Wei Yuehua, Hu Donghui, Luo Jiamin & Li Dongfeng. (2022). Two-dimensional Numerical Analysis of Water Environment Improvement in Diversion Lake. *Journal of Zhejiang Institute of Water Resources and Hydropower* (04), 37–41.

Mu Shousheng, Liu Yang, Wu Jingxiu, Fan Ziwu, Liu Guoqing & Xie Chen. (2022). Simulation Comparison and Field Test Study of Free Flow Fresh Water Scheme in Changzhou's Main Urban District. *Journal of Water Resources and Water Transport Engineering* (05), 148–156.

Zhou F & Chen M B. (2020). Comprehensive Impact of Port Construction on Hydrodynamics and Water Environment. *Water Transport Engineering* (09), 91–96. Doi:10.16233/j.carolCarrollnkiissn1002-4972.20200820.006.

Water Conservancy and Civil Construction – Oke & Ahmad (Eds)
© 2024 The Author(s), ISBN: 978-1-032-58618-2

Study on the optimal scheduling of the Jiangsu section of the south to north water transfer project east line in ten days

Ke Xu
College of Water Resources and Hydropower, Sichuan University, Chengdu, China

Chao Wang*
Department of Water Resource, China Institute of Water Resources and Hydropower Research, Beijing, China

Linan Xue & Chen Ji
College of Water Resources and Hydropower, Sichuan University, Chengdu, China

Hao Wang
College of Water Resources and Hydropower, Sichuan University, Chengdu, China
Department of Water Resource, China Institute of Water Resources and Hydropower Research, Beijing, China

ABSTRACT: The Jiangsu section of the South to North Water Transfer Project (JSNWTP) is an oversize inter-basin water transfer project (IWTP) consisting of different engineering units, such as rivers, lakes, gates, and pumps, which has a large-scale, complex structure, comprehensive function, and many influencing factors. At present, the ten-day scale scheduling cost of the JSNWTP is high and it is urgent to optimize the holistic scheduling on the ten-day scale. In this paper, based on the results of ten-day water level forecasts of impounded lakes, combined with constraints, the study adopts the differential evolutionary algorithms (DE) to analyze and calculate water sources, sectors, routes, and pumping stations, and generates a joint scheduling scheme of multi-water sources and double lines on a ten-day scale. On the premise of guaranteeing the completion of water transfer tasks, the cost of water transfer is reduced to achieve better project scheduling operation benefits. The results indicate that the model built in this paper can support the scientific scheduling decision of the JSNWTP to improve the scheduling and operation benefits on a ten-day scale and provide a new method for the benefit improvement of other water transfer projects.

1 INTRODUCTION

The IWTP is an important way to redress the contradiction between the supply and demand of interregional water resources and to realize rational allocation and effective utilization of water resources (Xu 1997; Zheng 2003). The JSNWTP is an important strategic infrastructure to alleviate the serious shortage of water resources in the eastern part of northern China (Li 2021), which is the world's largest modern large pump station group (Liu & Du 1986). Since the JSNWTP has been officially put into operation, the Eastern Route of JSNWTP has effectively improved the water shortage in the northern region, while playing a vital role in flood control, drought relief, drainage, and shipping in the areas along the JSNWTP (Lu et al. 2007). The JSNWTP is a very mega IWTP, including different

---

*Corresponding Author: wangchao@iwhr.com

DOI: 10.1201/9781003450832-45

engineering units, such as rivers, lakes, gates, and pump stations (Guo et al. 2018; Fang et al. 2018), with the characteristics of a huge scale, complex structure, comprehensive function, and other factors affecting the characteristics. Scientific and efficient scheduling decisions for the project are the key to giving full play to the comprehensive benefits of the project. The paper adopts the differential evolutionary algorithms (DE) to analyze and calculate water sources, sectors, routes, and pumping stations based on the results of ten-day water level forecasts of impounded lakes, combined with constraints, such as river water conveying capacity, the operation efficiency of the pump station, and water conveyance loss. It generates a joint scheduling scheme of multi-water sources and double lines on a ten-day scale. On the premise of guaranteeing the completion of water transfer tasks, the cost of water transfer is reduced to achieve better project scheduling operation benefits and support the scientific and efficient scheduling decision of the JSNWTP.

2 STUDY AREA

The JSNWTP relies on the nine cascade pumping stations formed in the Yangtze River to North Water Transfer Project in the province, and forms a double-line water transfer pattern of the canal line and the canal west line by construction and reconstruction of 18 pumping stations and new excavation and dredging of about 100 km of river channels, as shown in Figure 1.

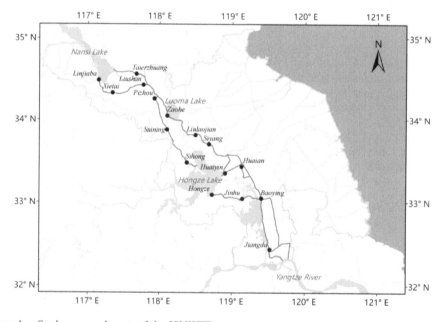

Figure 1. Study area and route of the JSNWTP.

The JSNWTP forms a complex multi-level hybrid relationship by rivers, lakes, gate pumps, and other engineering units. The general idea is to divide the whole project into three sections, including the Yangtze River to Hongze Lake, Hongze Lake to Luoma Lake, and Luoma Lake to out of the Jiangsu Province (OOJP) Section, with two impounded lakes as the nodes. And each section has two water conveyance routes, which are the canal route and the canal west route. At the same time, the JSNWTP is composed of pumping stations and several river sections along the routes.

3 OPTIMAL SCHEDULING METHODS AND CONSTRAINTS

3.1 *Optimal scheduling methods*

The scheduling objective of the paper is to meet the dispatching task OOJP in the next ten days and to make the JSNWTP the station with the lowest cost of scheduling operations for the whole line. Luoma Lake, Hongze Lake, and Yangtze River supply water for the OOJP dispatch. Therefore, it is necessary to calculate the transferable water of each water source in the next ten days.

The Luoma Lake Transferable Water. The Luoma Lake transferable water is the portion of the lake water storage corresponding to the average water level in the next ten days that exceed the lake water storage corresponding to the water level of the Luoma Lake in the JSNWTP, which can be described as follows.

$$\sum\nolimits_{t=1}^{p} Wav(t) = \sum\nolimits_{t=1}^{p} (Wsave(t) - Wb(t)) \tag{1}$$

Where p is the length of the period, which is one ten-day in this model. $Wav(t)$ is the adjustable water of Luoma Lake at time t. $Wsave(t)$ is the Storage water of Luoma Lake at time t. $Wb(t)$ is the water corresponding to the water level of Luoma Lake in JSNWTP at time t.

The ten-day water transfer object OOJP should be less than the adjustable water volume of Luoma Lake in the next ten days and the water volume transferred to Luoma Lake in the next ten days in the Jiangsu section, which can be described as follows.

Hongze Lake Adjustable Water. The water volume of 13.2 meters above the water level of Hongze Lake can be used for the regulation and storage of water for the ten-day dispatch. The adjustable water of Hongze Lake can be described as follows.

$$\sum\nolimits_{t=1}^{p} Wh(t) = \sum\nolimits_{t=1}^{p} (Ws(t) - Wp(t)) \tag{2}$$

Where $Wh(t)$ is the adjustable water volume of Hongze Lake at time t. $Ws(t)$ is the Storage water of Hongze Lake at time t. $Wp(t)$ is the corresponding water at time t when the water level of Hongze Lake is 13.2 m.

And if the water level in Hongze Lake is below 13.2 m, the water out of the lake is equal to the water into the lake, then the water is transferred northward from the Yangtze River.

Minimum Scheduling Cost. Based on the hydraulic connection between the engineering units, the impounded lakes, water transmission routes, and pump stations are connected in series. The dispatching cost is calculated according to the developed water dispatching scheme. By predicting the water levels Zh_t and Zl_t of Hongze Lake and Luoma Lake in period t, and the daily flow $Q_{n,t}$ of pumping into the pumping station in the future ten days, the scheduling operation cost $f(*)$ of the project in the future decade can be described as follows:

$$f(*) = \sum\nolimits_{t=1}^{p} f_t(Zh_t, Zl_t, Q_{n,t}) = \sum\nolimits_{t=1}^{p} \sum\nolimits_{n=1}^{N} M_{n,t} \tag{3}$$

Where N is the total number of pumping stations scheduled to be put into dispatch operation in ten days. $M_{n,t}$ is the operating cost of the pump station numbered n in period t.

The dispatching cost of the JSNWTP is calculated by using the cost of electricity for pumping station operation as the first cost.

$$M_{n,t} = Q_{n,t} \cdot t \cdot E(Q_{n,t}) \cdot hi \tag{4}$$

Where $Q_{n,t}$ is the flow rate of the pump station with number n in period t. t is the operation time of the pump station. $E(Q_{n,t})$ is the energy unit consumption of the pump station when the pumping flow is $Q_{n,t}$ (using historical data). hi is the pumping station head.

River Water Balance. The river channel between the two pumping stations forms a river section. The river connection is calculated using the water balance equation, taking into account the water transfer losses in the river segment.

Scheduling Decision of Pumping Stations. Based on the real-time rolling forecast results of the ten-day water level forecast model of Hongze Lake and Luoma Lake, the adjustable water of the two lakes in the next ten days and the dispatching plan OOJP is calculated based on the lake water level-storage capacity curve. The water volume $Wjs(t)$ transferred into Luoma Lake and the water volume $Whin(t)$ transferred into Hongze Lake are calculated. The water volume $Wpiz(t)$ transferred into Luoma Lake from Pizhou Station, the water volume $Wzaoh(t)$ transferred into Luoma Lake from Zaohe Station, the water volume $Whongz(t)$ transferred into Hongze Lake from Hongze Station, and the water volume $Whuaiy(t)$ transferred into Hongze Lake from Huaiyin Station are calculated as the main decision variables.

$$Wjs(t) = Wpiz(t) + Wzaoh(t) \tag{5}$$

After calculating the water to be transferred northward in the next ten days from the four stations for the two lakes, the water to be pumped out of Hongze Lake $Whout(t)$ and the water to be transferred northward from other pumping stations on each route in the system is calculated according to the river transmission losses. Based on the water conveyance capacity of the pumping stations of the JSNWTP and the water transferred north from each pumping station, the daily pumping flow is calculated for the next ten days. Finally, the daily pumping flow of 12 pumping stations on the two water transmission routes from the Yangtze River to Hongze Lake and Hongze Lake to Luoma Lake in the next ten days is obtained.

3.2 *Constraints*

Water Level Constraints of Impounded Lakes

$$Zt < Zn \tag{6}$$

Where Zn represents the normal storage level of Luoma Lake. Zt represents the water level of Lake Loma at time t.

Water Carrying Capacity Constraint of Pumping Station

$$Q_{\min} \leq Q_{i,t} \leq Q_{\max} \tag{7}$$

Where $Q_{\min}$ is the minimum water-carrying capacity. $Q_{\max}$ is the maximum pumping capacity of the pumping station. $Q_{i,t}$ is the flow of the pump station at time t.

Water Carrying Capacity Constraint of River

$$Q_{i,\min} \leq Q_{i,t} \leq Q_{i,\max} \tag{8}$$

Where $Q_{i,\min}$ is the minimum water transfer flow of river i. $Q_{i,\max}$ is the maximum water transfer flow of river i. $Q_{i,t}$ is the water transfer flow of river i at the t-th time.

River Conveyance Losses. The river water conveyance loss scenario is considered the South-North water transfer planning scheme.

4 DIFFERENTIAL EVOLUTION ALGORITHM

For the characteristics of multi-objective, multi-decision variables, and high operation frequency in the ten-day optimal scheduling model of the JSNWTP, DE is selected for optimization calculation. DE is a population-based global search optimization algorithm with simple principles and easy implementation, which contains initialization population, mutation, crossover, selection, and other steps (Storn & Price 1995; Wang et al. 2014).

4.1 *Initialization population*

We assume the individual dimension is D, the population size is NP, the evolutionary generations is t, and $X(t)$ represents the t-th generation population. Then the generation 0 population is randomly generated in the decision space of the problem.

$$X(0) = \{x_1(0), x_2(0), x_3(0), x_4(0)\} \tag{9}$$

Where $x_i(t)$ denotes the ith individual in the population of the t-th generation. The value of each dimension of the individual is initialized by the following equation.

$$x_{i,j}(t) = L_j + rand_{i,j}[0,1](U_j - L_j) \; 1 \le i \le NP, \; 1 \le j \le D \tag{10}$$

Where $[L_j, U_j]$ is the range of values in the jth dimension. $rand_{i,j}[0,1]$ is a uniformly distributed random number between 0 and 1.

4.2 *Mutation*

The mutation operation is performed on each vector in the population using the mutation operator. First of all, the three vectors, $x_{r1}(t)$, $x_{r2}(t)$, and $x_{r3}(t)$, are randomly selected from the current population, which requires r_1, r_2, and r_3 to be different integers randomly selected from the set $\{1,..., NP\}\backslash\{i\}$. Secondly, the difference between two vectors is scaled and added to the third vector to get a variation vector $V_i(t)$. The equation is described as follows.

$$V_i(t) = x_{r1}(t) + F[x_{r2}(t) - x_{r3}(t)] \tag{11}$$

Where F is the scaling factor of DE and the value range is [0,1].

4.3 *Crossover*

The crossover operation is performed on the individual $x_i(t)$ in the parent population and the newly generated mutant $v_i(t)$ to generate the experimental individual $u_i(t)$. The formula is described as follows.

$$u_i(t) = \begin{cases} v_i(t), if \quad rand_j[0,1] < Cr \quad or \quad j = j_{rand} \\ x_i(t), otherwise \end{cases} \tag{12}$$

Where j is a random number between [0, 1]. Cr is the crossover frequency between 0 and 1. In this model, the Cr value is 0.5. j_{rand} is an integer randomly selected in the interval [1, D] to ensure that $u_i(t)$ is selected at least one value from $v_i(t)$, so that the crossover result is different from both the parent and the mutant individuals to avoid invalid crossover.

4.4 *Selection*

Comparing the parent individual with the crossover individual, the better one is selected to enter the next generation population.

5 RESULTS AND DISCUSSION

To reduce the loss of water transmission, this paper uses the reservoirs along the water transmission line for water storage and takes into account factors such as water use, shipping, and the situation of water transmission projects along the line. Centralized water diversion is used as much as possible without affecting the safety of the water supply. The reservoir's real-time operation can be appropriately adjusted and revised on a rolling basis according to water conditions, lake and reservoir storage, and other conditions. Table 1 shows the specific monthly water scheduling plan.

Table 1. Water supply scheduling schedule for the Jiangsu section of the East route of the South-to-North water transfer project from 2018 to 2019 (million/m^3).

Item	Dec.	Jan.	Feb.	Mar.	Apr.	May.	Total
From the Yangtze River	168	120	104	110	246	263	1014
From Luoma Lake	140	100	87	91	207	219	844

The actual scheduling operation of the year is divided into two stages. The first stage of the scheduling operation starts on December 25th, 2018, and ends on January 27th, 2019. The second stage of the scheduling operation starts on February 20th, 2019, and ends on May 28th, 2019. In summary, the water transfer time is 130 days, the net water supply is 1092 million m^3, and the inter-provincial water exchange is 844 million m^3. Therefore, while meeting the annual water plan OOJP, the actual water pumped by the JSNWTP is more than the planned water pumped from the Yangtze River by 78 million m^3.

According to the scheduling plan for water supply from Luoma Lake to Shandong province for each month in Table 1, the scheduling objective is to satisfy the scheduling task OOJP and the lowest scheduling cost. The paper takes February as an example of optimization. The average water level of Hongze Lake in February is 13.39 m, which is above the available water level, the average water level of Luoma Lake in February is 22.83 m, and the water level of JSNWTP is 22.7 m. Therefore, Hongze Lake can be considered as the water source to save the scheduling cost of pumping water from the Yangtze River into Hongze Lake. The model recommended optimal water scheduling scheme for JSNWTP in February is shown in Table 2.

Table 2. Water dispatching optimization scheme for Jiangsu section of East line project in february 2018-2019 (m^3/s).

| | Hongze Lake-Luoma Lake | | | | | | The Yangtze River-Hongze Lake | |
| | Canal west route | | | Canal route | | | | |
	Pizhou	Suining	Sihong	Zaohe	Liulaojian	Siyang	Canal west route	Canal route
Feb.20	100	102.8	105	0	0	0	0	0
Feb.21	100	102.8	105	0	0	0	0	0
Feb.22	100	102.8	105	0	0	0	0	0
Feb.23	100	102.8	105	0	0	0	0	0
Feb.24	100	102.8	105	0	0	0	0	0
Feb.25	100	102.8	105	0	0	0	0	0
Feb.26	50	52.8	55	0	0	0	0	0
Feb.27	0	0	0	0	0	0	0	0
Feb.28	0	0	0	0	0	0	0	0

Table 2 shows the optimized scheduling scheme. The optimal scheduling only opens from the Hongze Lake to Luoma Lake section of the canal west line to operate for 7 days and without pumping water from the Yangtze River, which makes the future adjustable water of Luoma Lake sufficient to meet the task of water supply to Shandong province. When the

Hongze Lake is at a high water level, Hongze Lake is used as the water source, which effectively reduces the scheduling cost of drawing water from the Yangtze River while avoiding water abandonment. The water transfer process for each month of 2018-2019 is optimized and compared with the actual water transfer process, as shown in Table 3.

Table 3. Optimization effect of water supply scheduling plan for Jiangsu section of East line phase I project to Shandong in 2018-2019 (million m^3).

Item	Dec.	Jan.	Feb.	Mar.	Apr.	May.	Total
From Luoma Lake	1400	100	87	91	207	219	844
Planned from the Yangtze River	168	120	104	110	246	263	1014
Optimized from the Yangtze River	102	60	0	0	227	239	628
Optimized operating time (days)	10	23	7	25	30	30	125

The rolling calculation of the ten-day optimal scheduling model in this paper generates the monthly scheme. Compared with the original annual plan, the optimized pumping water from the Yangtze River is 386 million m^3 less than planned and 464 million m^3 less than actual. After the rational utilization of the impounded lakes for water source switching, the pumping of water from the Yangtze River can be significantly reduced. At the same time, the optimized operating days are 5 days less than the actual operating days.

6 CONCLUSION

In this paper, the units of pumping stations, routes, lakes, and water sources in the whole engineering system of the JSNWTP are represented and connected in series to form the optimal scheduling method of the JSNWTP. Then the paper forms the optimal scheduling logic according to the actual scheduling rules and business requirements. In addition, according to the simplified scheduling objectives, each project and each objective is expressed and connected with mathematical equations based on the hydraulic connection and water balance, and an optimal scheduling model based on the differential evolutionary algorithm is proposed. Finally, the study optimizes and analyzes the historical water scheduling plan and scheme of the first phase of the East Route of the JSNWTP from 2018 to 2019. When the water storage lake is rationally used for water source conversion, the pumping of water from the Yangtze River can be significantly reduced. At the same time, the optimized operation days are also 5 days less than the actual operation days, saving the scheduling operation cost and supporting the scientific and efficient scheduling decision of the JSNWTP.

REFERENCES

Fang, G. H., Guo, Y. X., Wen X. Application of Improved Multi-objective Quantum Genetic Algorithm in Optimizing Water Resources Scheduling in Jiangsu Section of South-North Water Transfer Project. *Water Resources Protection* 34(2), 34–41(2018).

Guo, Y. X., Zhang, J. S., Zheng, Z. Z. Study on Multi-objective Optimal Scheduling of Jiangsu Section of South-North Water Transfer Project. *Journal of Water Resources* 49(11), 1313–1327(2018).

Li, G. Y. Promote High-quality Development of the South-North Water Transfer Follow-up Project. *People's Daily* 13, 2021/07/29.

Liu, C. M., Du, W. Geographical System Analysis of the Water Balance of the South-North Water Diversion East Line - taking the East Line Phase I Project as an example. *Journal of Water Resources* 2, 3–14(1986).

Lu, H., Geng, L. H., Pei, Y. S. Research on Key Technical Issues of South-North Water Transfer Operational Risk Management. *South-to-North Water Transfers and Water Science & Technology* 5(5), 4–7(2007).

Storn, R., Price, K. *Differential Evolution—A Simple and Efficient Adaptive Scheme for Global Optimization Over Continuous Spaces.* University of California, Berkeley: ICSI, (1995).

Wang, S. W., Ding, L. X., Zhang, W. S. Advances in Differential Evolutionary Algorithm Research. *Journal of Wuhan University (Science Edition)* 60(4), 283–292(2014).

Xu, M. Y. Review of Foreign Inter-basin Water Transfer Project Construction and Management. *Yangtze River* 3(28), 11–13(1997).

Zheng, L. D. Inter-basin Water Transfer Projects in Chinese History. *South-to-North Water Transfers and Water Science & Technology.* 1, 8–9(2003).

Evaluation and analysis of low carbon traffic in Jiaozuo city based on comprehensive empowerment method

Zhezhe Zhang*

School of Energy Science and Engineering, Henan Polytechnic University, Jiaozuo, Henan, China

ABSTRACT: Based on the analysis of the current situation of low-carbon transportation development in Jiaozuo City, this paper constructs a low-carbon transportation evaluation index system for Jiaozuo City by selecting multi-dimensional indicators for the composition of the urban low-carbon transportation system based on the principles of index system establishment. The comprehensive weighting method combining hierarchical analysis (AHP) and coefficient of variation method is selected, taking into account both subjective and objective. The consistency of the test judgment matrix is improved, the nonlinear optimization problem is solved, and more stable and accurate index weights are obtained. The constructed index system is comprehensively evaluated by the gray correlation analysis method and the MATLAB calculation program is used in the process of calculating the gray correlation degree.to reduce the calculation amount. The qualitative and quantitative analysis of infrastructure, transportation, policy management and low-carbon, environmental protection, and energy consumption can be conducted to evaluate the development status of low-carbon transportation in Jiaozuo. This study will provide countermeasures for the development of low-carbon transportation in Jiaozuo city.

1 INTRODUCTION

Experience shows that with the continuous improvement of social and economic development levels, the proportion of transportation in total energy consumption is increasing [Shi 2011]. The concept of a low-carbon economy was first put forward in the UK article "The Future of Our Energy: Creating a Low-carbon Economy". Since then, it has opened the prelude to the development of the "low-carbon" concept in other countries around the world [DTI 2003]. Zhang Taoxin et al. believed that transportation demand is the basis of urban low-carbon transportation, and reducing energy consumption and pollution gas emissions in the urban transportation system, and building a convenient, safe, fair, and comfortable transportation system is an effective way [Zhang 2011]. Li Ye, Zou Diming, and others elaborated on the objectives, concepts, bases, and measures of low-carbon transportation in detail, taking meeting the transportation needs of urban citizens as the starting point, actively developing urban low-carbon transportation, promoting the development of slow traffic and new energy transportation, and building a harmonious urban low-carbon transportation system. Sheffield et al. deeply studied the basic calculation formula and influencing factors of urban carbon emission and established the carbon emission model of urban traffic. Taking Beijing as an example for empirical analysis, they put forward some practical measures for the current situation and actual situation of low-carbon transportation development in Beijing [Xie 2013]. Guo Jie conducted a comprehensive evaluation and

---

*Corresponding Author: zhe991027@163.com

DOI: 10.1201/9781003450832-46

research on the importance of a low-carbon transportation system and established a low-carbon transportation evaluation index system from four dimensions: transportation infrastructure, transportation technology and equipment, transportation organization and management, and highways. 18 evaluation indexes were selected and the arithmetic average assignment method was used for comprehensive evaluation [Guo 2012]. In the study of urban green transportation evaluation indexes and methods, Wang Zhenzhen established a green public transportation evaluation system for urban transportation in China from four dimensions of transportation function, environmental impact, resource-saving, and applicability to social and national economic development, proposed the main indexes for urban public transportation evaluation, selected 27 key transportation evaluation indexes, and used multidimensional synthesis and fuzzy method to comprehensively analyze the evaluation results [Wang 2003]. Based on an in-depth study of the development of low-carbon tourism passenger transportation in various cities in China, Lu Huifen conducted a comprehensive study on six dimensions of its energy consumption, economic efficiency, coordination, functionality, service, and environmental protection, determined 26 evaluation indicators, constructed a set of comprehensive evaluation index framework, and applied the gray-scale fuzzy evaluation method and data envelopment analysis to evaluate the two index frameworks separately to judge the accuracy of the index system.

This paper analyzes the development status of Jiaozuo's low-carbon traffic. First, the domestic study is discussed to determine the evaluation method of low-carbon transportation in Jiaozuo City based on the comprehensive empowerment method. Then, based on the urban low-carbon transportation system, the low-carbon transportation evaluation index system of Jiaozuo City is constructed by selecting indicators in four dimensions: transportation, infrastructure, policy management, and low-carbon environmental energy consumption. Finally, through the gray index analysis, countermeasures and suggestions are provided for the development of a comprehensive evaluation of low-carbon transportation in Jiaozuo City.

2 LOW-CARBON TRANSPORTATION EVALUATION MODEL CONSTRUCTION

In this paper, the gray correlation degree evaluation method based on hierarchical analysis is used to comprehensively evaluate urban low-carbon traffic. In calculating the weights, the comprehensive empowerment method is used and combined with the hierarchical analysis method (subjective method) and the coefficient of variation method (objective method).

2.1 Index weight calculation method

(1) AHP hierarchical analysis method
    The method of calculating index weights by hierarchical analysis method is shown below.

A. According to the above-established hierarchy, a hierarchy level is further established. The hierarchical class established in this paper mainly includes the target layer, standard layer, and index layer.
B. A judgment matrix is constructed based on the pairwise comparison of elements in the same hierarchy.

$$A = (a_{ij})_{n \times n} \qquad (1)$$

C. Based on the target layer above the standard layer, AHP hierarchical analysis is used to determine the criteria [Hua 2012]
    The properties of the judgment matrix are as follows. 1) $a_{ij} = 1$; 2) $a_{ij} > 0$; 3) $a_{ij} = \frac{1}{a_{ji}}$.
D. The weight vector is calculated as follows $W = (w_1, \ w_2, \ \cdots w_n)^T$

1) The geometric mean of the elements of each row is calculated.

$$\bar{W}_i = \sqrt[n]{\prod_{i=1}^{n} a_{ij}} \quad \text{Where } (i = 1, 2\ldots\ldots n) \text{ n is the matrix order} \tag{2}$$

2) Normalization is processed as follows.

$$W_1 = \frac{\bar{W}_i}{\sum\limits_{j=1}^{n} \bar{W}_j} \text{Where } (j = 1, 2\ldots\ldots n) \text{ n is the matrix order} \tag{3}$$

E. Consistency check.

The maximum eigenvalue of the matrix is judged.

$$\lambda_{\max} = \frac{1}{n} \sum_{i=1}^{n} \frac{(BW)_i}{W_i} \tag{4}$$

$$B = \begin{bmatrix} \beta_1(1) & \beta_1(2) & \cdots & \beta_1(m) \\ \beta_2(1) & \beta_2(2) & \cdots & \beta_2(m) \\ \cdots & \cdots & \cdots & \cdots \\ \beta_n(1) & \beta_n(2) & \cdots & \beta_n(m) \end{bmatrix} \tag{5}$$

The association coefficient in the matrix $\beta_j(i)$ between the value of the i-th index in the j-th evaluation object and the i-th optimal index is as follows.

$$CR = \frac{CI}{RI} \quad CI = \frac{\lambda_{\max} - n}{n - 1} \tag{6}$$

RI is the average random consistency index (n is the order), which is shown in Table 1.

Table 1. The values table of the R I.

Order n	1	2	3	4	5	6	7	8	9	10
RI	/	0	0.52	0.89	1.12	1.26	1.36	1.41	1.46	1.49

CR is called a judgment matrix random consistency index. If CR is 0.1, it indicates consistency, and if CR is 0.1, the matrix should be adjusted until the consistency is satisfied.

(2) Coefficient of variation method

The coefficient of variation method refers to the direct use of the information contained in each secondary index and the weight value of the index through calculation is obtained.

A. Coefficient of variation of the secondary index is as follows.

$$V_i = \frac{\sigma_i}{\bar{X}_i} \ (i = 1, 2\ldots\ldots n). \tag{7}$$

Where

V_i—Coefficient of variation in term i; σ_i—Standard deviation of the indicator in item i; X_i—Mean of the indicator in item i.

B. The calculation method of the weights is as follows.

$$W_2 = \frac{V_i}{\sum\limits_{i=1}^{n} V_i} \tag{8}$$

(3) Comprehensive empowerment law

The paper adopts the method of comprehensive integration empowerment. The weight coefficient vector of the decision target obtained by the AHP method is assumed to be W_1, and the vector of weight coefficients obtained by the coefficient of variation method is W_2. The weight vector of the subjective and objective comprehensive method is as follows.

$$W = \alpha W_1 + (1 - \alpha) W_2 \tag{9}$$

The weight value obtained by the comprehensive assignment method can change with α, when $\alpha = 0$ corresponds to the weight determined by the coefficient of variation method. When $\alpha=1$, the comprehensive assignment is obtained.

2.2 Gray correlation degree evaluation model

This paper evaluates the index system of urban low-carbon transportation based on the gray analysis method [Lu 2011].

$S = B \cdot W$ is the evaluation model for the gray correlation degree

Where S — The comprehensive evaluation result vector of each evaluation object;

W — The weight allocation vectors of each evaluation index

The specific calculation process is as follows:

(1) Determining the evaluation index data set

Through investigation, data inquiry, and other ways of data collection, then quantitative analysis and processing of relevant indicators can be obtained.

$$R' = \begin{bmatrix} r_{11}' & r_{12}' & \cdots & r_{1n}' \\ r_{21}' & r_{22}' & \cdots & r_{2n}' \\ \cdots & \cdots & \cdots & \cdots \\ r_{m1}' & r_{m2}' & \cdots & r_{mn}' \end{bmatrix} \tag{10}$$

(2) Determining the optimal index set

According to the optimal index set matrix constructed by the original data matrix, the largest of the n evaluation objects is the optimal value and the reverse index is the smallest is the optimal value. Thus, the optimal sequence can be obtained.

$$A^* = (r_1^*, r_2^*, \cdots, r_m^*) \tag{11}$$

The build matrix is:

$$A = \begin{bmatrix} r_{11}{}' & \cdots & r_{1n}{}' & r_1{}^* \\ r_{21}{}' & \cdots & r_{2n}{}' & r_2{}^* \\ \cdots & \cdots & \cdots & \cdots \\ r_{m1}{}' & \cdots & r_{mn}{}' & r_m{}^* \end{bmatrix} \tag{12}$$

(3) Unstructured quantification of index data
The common formulas are as follows:
Forward pointers:

$$r_{ij} = \frac{r_{ij}{}' - \min r_{ij}{}'}{\max r_{ij}{}' - \min r_{ij}{}'} \tag{13}$$

Reverse indicators:

$$r_{ij}{}' = \frac{\max r_{ij}{}' - r_{ij}{}'}{\max r_{ij}{}' - \min r_{ij}{}'} \tag{14}$$

The matrix A is dimensionless, and the standardized matrix R is obtained

$$R = \begin{bmatrix} r_{11} & \cdots & r_{1n} & 1 \\ r_{21} & \cdots & r_{2n} & 1 \\ \cdots & \cdots & \cdots & \cdots \\ r_{m1} & \cdots & r_{mn} & 1 \end{bmatrix} \tag{15}$$

(4) Calculation of the gray correlation coefficient
The correlation coefficient is as follows.

$$\beta_j(i) = \frac{minj\ mini\ |r_i{}^* - r_{ij}| + \alpha\ maxj\ maxi\ |r_i{}^* - r_{ij}|}{|r_i{}^* - r_{ij}| + \alpha\ maxj\ maxi\ |r_i{}^* - r_{ij}|} \tag{16}$$

Where $\alpha\alpha\ [0, 1]$, the resolution coefficient is indicated and the value range is set to reduce the influence of the extreme value on the calculation results. This paper takes $\alpha = 0.5$.

(5) The gray-weighted correlation degree is calculated to establish the gray correlation degree.

$$S_j = \sum_{i-1}^{m} \beta_j(i) \cdot w_i \tag{17}$$

(6) Comprehensive evaluation and analysis
Gray correlation of an evaluated object nS_j (where j = 1,2,3,.... the n) in descending order, the ranking result of an evaluated object can be obtained. The results show that the better-fitting indexes can better reflect the relationship between the evaluated objects and the optimal set.

3 CASE ANALYSIS OF LOW-CARBON TRAFFIC IN JIAOZUO CITY

3.1 *Data collection of the comprehensive evaluation index of low-carbon transportation in Jiaozuo*

Table 2. Comprehensive evaluation index data of low-carbon traffic in Jiaozuo City (the data are derived from the survey).

Evaluating indicator	A particular year			
	2017	2018	2019	2020
Public transport travel share rate (%) C11	27.36	31.15	36.67	41.37
Slow traffic travel sharing rate (%) C12	29.47	33.50	35.12	37.26
New energy vehicle use ratio (%) C13	11.43	15.60	21.37	26.97
Bus ownership of ten thousand people (ten thousand people / standard platform) C14	9.28	11.52	10.08	9.76
Average age of bus vehicles (year): C15	3.14	3.14	3.36	3.36
Road network density (km / km^2) C21	11.60	12.20	12.90	13.50
Per capita road area (m^2/ Man) C22	14.40	15.10	15.60	16.10
Bus lane setting rate (%) C23	9.70	12.40	19.20	24.50
Bus station: 500 m coverage rate (%) C24	96.00	100.00	100.00	100.00
Bicycle lane setting rate (%) C25	94.80	95.30	96.42	96.58
The Walking and Bicycle Transportation System C26	82.00	85.00	87.00	89.00
Low carbon Standard Specification Implementation Procedure (%) C31	60.00	62.00	67.00	69.00
Improvement of carbon emission detection mechanism (%) C32	15.00	19.00	21.00	23.00
Improvement of traffic management policies (%) C33	55.00	58.00	62.00	68.00
Low-carbon awareness penetration rate (%) C34	46.10	52.32	58.33	63.50
Traffic information degree: C35	47.61	51.38	56.20	63.72
Vehicle carbon dioxide emissions (10,000 tons) C41	273.56	259.28	232.74	221.60
Air quality rate of road section (%) C42	43.50	39.70	32.60	28.40
Road greening rate (%) C43	21.00	29.50	34.70	36.90
Vehicle transportation energy consumption per unit (ton of standard coal, ten Thousand person kilometers) C44	0.66	0.61	0.58	0.55
New energy utilization rate (%) C45	27.30	29.60	33.40	38.00

3.2 *Calculation of weight by comprehensive empowerment method*

3.3 *Evaluation based on the gray correlation degree*

(1) Determination of the evaluation index data set

According to the characteristics of the evaluation system, the data is collected through investigation, data inquiry, and other ways. Relevant indicators are quantitatively analyzed and processed.

$$R = \begin{bmatrix} 27.36 & 29.47 & 11.43 & 9.28 & 3.14 & 11.6 & 14.4 & 9.7 & 96 & 94.80 & 82 & 60 & 15 & 55 & 46.10 & 47.61 & 273.56 & 43.5 & 21.0 & 0.66 & 27.3 \\ 31.15 & 33.50 & 15.6 & 11.52 & 3.14 & 12.2 & 15.1 & 12.4 & 100 & 95.30 & 85 & 62 & 19 & 58 & 52.32 & 51.38 & 259.28 & 39.7 & 29.5 & 0.61 & 29.6 \\ 36.67 & 35.12 & 21.37 & 10.08 & 3.36 & 12.9 & 15.6 & 19.2 & 100 & 96.42 & 87 & 67 & 21 & 62 & 58.33 & 56.20 & 232.74 & 32.6 & 34.7 & 0.58 & 33.4 \\ 41.37 & 37.26 & 26.97 & 9.76 & 3.36 & 13.5 & 16.1 & 24.5 & 100 & 96.58 & 89 & 69 & 23 & 68 & 63.50 & 63.72 & 221.60 & 28.4 & 36.9 & 0.55 & 38.0 \end{bmatrix}^T$$

(2) Determination of the optimal index set

The optimal index set matrix is constructed according to the selection principle of the optimal index set. The negative indicators include C15, C41, C42, C44, and the rest are positive indicators.

Table 3. Results of calculation weight of comprehensive empowerment method.

Target layer A	Criterion layer B	The standard layer weight	Criterion layer C	Hierarchical analysis method for calculation Weight results		Coefficient of variation method calculation Weight results		Comprehensive empowerment method weight
				Corresponding index layer weight	Index layer weight	Variation coefficient	weighted value	
Burnt do market low carbon hand over Through finger treetop counterpoise repeat	Transport B1	0.2274	Public transport travel share rate is C11	0.4164	0.0959	0.1041	0.0598	0.0779
			Slow traffic travel sharing rate is C12	0.1612	0.0354	0.0711	0.0411	0.0383
			Use ratio of new energy vehicles C13	0.2633	0.0604	0.1460	0.0841	0.0723
			The number of ten million buses owned is C14	0.0985	0.0273	0.0764	0.0440	0.0357
			The average age of buses is C15	0.0625	0.0338	0.0698	0.0402	0.0370
	Infrastructure B2	0.2272	Road network density: C21	0.2526	0.0573	0.0660	0.0381	0.0477
			Road area per capita is C22	0.3816	0.0864	0.0475	0.0274	0.0569
			Bus lane setting rate is C23	0.1592	0.0369	0.0968	0.0558	0.0464
			Bus station of 500 meters coverage rate C24	0.0653	0.0188	0.0116	0.0066	0.0127
			Bicycle lane setting rate: C25	0.0415	0.0097	0.0090	0.0052	0.0075
			The Walking and Bicycle Transportation System C26	0.1019	0.0230	0.0347	0.0201	0.0216
	Policy Management B3	0.1222	Low carbon standard specification Implementation Procedure C31	0.1362	0.0168	0.0653	0.0374	0.0271
			Improvement degree of carbon emission detection mechanism C32	0.1362	0.0168	0.1753	0.1008	0.0588
			Improvement degree of traffic management policy: C33	0.4031	0.0532	0.0924	0.0534	0.0533
			Low-carbon awareness penetration rate: C34	0.0778	0.0098	0.1875	0.1080	0.0589
			Traffic information degree: C35	0.2456	0.0301	0.1067	0.0613	0.0457
	Low-carbon-environmental protection-Energy consumption B4	0.4232	Vehicle CO 2 emissions of C41	0.4161	0.1760	0.0190	0.0109	0.0935
			Air quality exceeding the standard rate of the road section C42	0.1075	0.0497	0.0351	0.0201	0.0349
			Road greening rate: C43	0.0813	0.0345	0.0886	0.0510	0.0428
			Vehicle energy consumption per unit of transport C44	0.1421	0.0603	0.0603	0.0347	0.0475
			New energy utilization rate: C45	0.1531	0.0678	0.1725	0.0996	0.0837

Table 4. Unclassified quantification and gray correlation coefficient results of the index data.

Evaluating indicator	Indicator nature	Standardization				Gray association coefficient results			
		R1	R2	R3	R4	R1	R2	R3	R4
C11	Forward pointer	0.6	0.75	0.89	1.00	0.33	0.36	0.77	1
C12	Forward pointer	0.79	0.90	0.94	1.00	0.33	0.37	0.46	1
C13	Forward pointer	0.42	0.58	0.79	1.00	0.33	0.36	0.46	1
C14	Forward pointer	0.81	1.00	0.88	0.85	0.33	0.37	0.55	1
C15	Reverse indicators	1.00	1.00	1.07	1.07	0.33	0.35	0.57	1
C21	Forward pointer	0.86	0.90	0.96	1.00	0.33	0.42	0.61	1
C22	Forward pointer	0.89	0.94	0.97	1.00	0.33	0.46	0.62	1
C23	Forward pointer	0.40	0.51	0.78	1.00	0.33	0.39	0.50	1
C24	Forward pointer	0.96	1.00	1.00	1.00	0.33	0.40	0.57	1
C25	Forward pointer	0.98	0.99	0.99	1.00	0.33	0.41	0.85	1
C26	Forward pointer	0.92	0.96	0.98	1.00	0.33	0.47	0.63	1
C31	Forward pointer	0.87	0.90	0.97	1.00	0.33	0.39	0.69	1
C32	Forward pointer	0.65	0.83	0.91	1.00	0.33	0.50	0.67	1
C33	Forward pointer	0.81	0.85	0.91	1.00	0.33	0.39	0.52	1
C34	Forward pointer	0.73	0.82	0.92	1.00	0.33	0.36	0.43	1
C35	Forward pointer	0.75	0.81	0.88	1.00	0.33	0.43	0.59	1
C41	Reverse indicators	1.23	1.17	1.05	1.00	0.33	0.44	0.75	1
C42	Reverse indicators	1.53	1.40	1.15	1.00	0.33	0.45	0.68	1
C43	Forward pointer	0.57	0.80	0.94	1.00	0.33	0.46	0.56	1
C44	Reverse indicators	1.20	1.11	1.05	1.00	0.33	0.45	0.62	1
C45	Forward pointer	0.72	0.78	0.88	1.00	0.33	0.42	0.56	1

The optimal sequence is $A^* = (r_1^*, r_2^*, \cdots, r_m^*)$

$$A^* = \begin{pmatrix} 41.37, 37.26, 26.97, 11.52, 3.14, 13.5, 16.1, 24.5, 100, 96.58, \\ 89, 69, 23, 68, 63.5, 63.72, 221.6, 28.4, 36.9, 0.55, 38 \end{pmatrix}$$

$$R = \begin{bmatrix} 41.37 & 37.26 & 26.97 & 11.52 & 3.14 & 13.5 & 16.1 & 24.5 & 100 & 96.58 & 89 & 69 & 23 & 68 & 63.50 & 63.72 & 221.60 & 28.4 & 36.9 & 0.55 & 38.0 \\ 27.36 & 29.47 & 11.43 & 9.28 & 3.14 & 11.6 & 14.4 & 9.7 & 96 & 94.80 & 82 & 60 & 15 & 55 & 46.10 & 47.61 & 273.56 & 43.5 & 21.0 & 0.66 & 27.3 \\ 31.15 & 33.50 & 15.6 & 11.52 & 3.14 & 12.2 & 15.1 & 12.4 & 100 & 95.30 & 85 & 62 & 19 & 58 & 52.32 & 51.38 & 259.28 & 39.7 & 29.5 & 0.61 & 29.6 \\ 36.67 & 35.12 & 21.37 & 10.08 & 3.36 & 12.9 & 15.6 & 19.2 & 100 & 96.42 & 87 & 67 & 21 & 62 & 58.33 & 56.20 & 232.74 & 32.6 & 34.7 & 0.58 & 33.4 \\ 41.37 & 37.26 & 26.97 & 9.76 & 3.36 & 13.5 & 16.1 & 24.5 & 100 & 96.58 & 89 & 69 & 23 & 68 & 63.50 & 63.72 & 221.60 & 28.4 & 36.9 & 0.55 & 38.0 \end{bmatrix}^T$$

(3) Dimensionless index data and calculation of gray correlation degree of index data

The evaluation matrix, R= (R1, R2, R3, R4), which is composed of the gray correlation coefficient, is obtained. The greater the gray correlation, the closer to the optimal value. Based on MATLAB software programming, the process of calculating the gray correlation degree is simplified. The results are shown in Figure 1.

Figure 1. Line plot of the gray correlation operation of each indicator.

(4) Calculation of Gray correlation degree

The grayscale correlation is calculated using the formula (8) in Chapter 2, and the specific calculation results are shown in Table 5. The evaluation objects are ranked according to the calculation results.

Table 5. Comprehensive evaluation results.

Evaluation object	2017	2018	2019	2020
Bear fruit	0.3720	0.4536	0.5427	0.7841
Sort	4	3	2	1

4 CONCLUSION

(1) Urban low-carbon transportation is an important part of low-carbon city construction. In this paper, by studying the current situation of low-carbon transportation in Jiaozuo city, the evaluation index system of low-carbon transportation in Jiaozuo city is determined and the weights of each index are determined based on the comprehensive assignment method. The constructed index system is comprehensively evaluated by the gray correlation analysis.

(2) In recent years, the degree of the Jiaozuo urban transportation system has been continuously improved, including the sharing rate of public transport, bicycle lanes, road area, low carbon awareness penetration rate, walking and bicycle transportation system, and the amount of carbon dioxide emissions.

(3) Jiaozuo City needs to implement a public transportation-oriented urban development strategy to reduce traffic with unreasonable demand for high carbon emissions by improving local walking conditions, building comprehensive bicycle facilities, and using traffic behavior control tools, such as land use planning, economic leverage, policies and regulations, and information dissemination and publicity and education to effectively guide and adjust people's traffic behavior and consumption concepts. We reform the transportation planning system, accelerate the legalization of comprehensive transportation planning and special public transportation plans, and further improve existing policies and regulations on land management, urban and rural planning, the automobile industry, energy conservation and emission reduction, public transportation, transportation management, and scientific and technological innovation. We establish a fiscal and tax incentive policy system to encourage the production and consumption of clean energy vehicles. The development of methods and technologies for energy saving and emission reduction in transportation according to local conditions is something to be strengthened in Jiaozuo City in the future.

REFERENCES

Analysis and Application of Hierarchical Analysis [J], South China University of Technology, 2012 (7).

Department of Trade and Industry (DTI). *UK Energy White Paper: Our Energy Future—Creating a Low Carbon Economy*. London: TSO, 2003.

Guo Jie, Chen Jianying, Ouyang Bin. Research on China Regional Low Carbon Transportation Evaluation Index System [J], *Integrated Transportation*, 2012 (6).

Lu Huifen, *Research on the Evaluation Index System of Low Carbon Passenger Transport System in Big Cities* [D], Beijing Jiaotong University, 2011 (6).

Sheffield, *Research on Influencing Factors of Urban Transportation Carbon Emission and Low-carbon Transportation Development* [D], Beijing Jiaotong University, 2013 (6): 1–12.

Shi Lixin, Huang Yin, Yu Juan. *Transportation Energy Consumption and Carbon emission research* [M]. Beijing: China Economic Press. 2011(10).

Wang Zhenzhen, *The Theory and Method of Sustainable Development Planning of Urban Transportation System* [D], Southeast University, 2003 (5).

Zhang Taoxin, Zhou Yueyun, Zhao Xianchao. The Current Situation of Low-carbon Urban Transportation Construction in China is also Analyzed [J], *Urban Transportation*, 2011 (1).

Water Conservancy and Civil Construction – Oke & Ahmad (Eds)
© 2024 The Author(s), ISBN: 978-1-032-58618-2

Structural design and mechanical properties of assembled right-of-way in ecologically sensitive areas

Jing Guo & Hongzhang Wang
CCCC Second Navigation Bureau Third Engineering Co., Ltd, Zhenjiang, China

Wei Qiao
Research Institute of Highway Ministry of Transport, Beijing, China

Xinglin Lin*
Ningde Sandu'ao Expressway Co., Ltd., Ningde, China

Yuan Tian
Transportation Technology Development Promotion Center, China Academy of Transportation Sciences, Beijing, China

Shanjun Yang
CCCC Second Navigation Bureau Third Engineering Co., Ltd, Zhenjiang, China

Fei Li
Suqian Highway Development Center, Suqian, China

Libin Zhang
Ningde Sandu'ao Expressway Co., Ltd., Ningde, China

ABSTRACT: Ecologically sensitive areas have sensitive and fragile characteristics and the construction of construction access roads in ecologically sensitive areas should avoid disturbing and damaging the ecological environment as much as possible. Based on the Yanluo Expressway under construction, this paper conducts a study on the form of assembled reinforced concrete pavement with steel pipe piles in the ecologically sensitive area, with the reinforced concrete panel size of 3.5 m * 2 m * 0.3 m and steel pipe piles ($\Phi 800 \times 10$ mm) entering the soil at a depth of 3.5 m. The concrete slab reinforcement and pile length calculation, combined with the field experimental data, show that the structural design of the right-of-way can meet the load-bearing capacity requirements while reducing the disturbance to the environment and achieving rapid recovery after the completion of the project.

1 INTRODUCTION

The construction conditions of the red line in the ecologically sensitive area are harsh and have high requirements for environmental protection. The assembled construction right-of-way can meet the normal passage and use needs of engineering construction vehicles, for a short construction period, and can be reused, with green, efficient, environmental protection, energy saving, and other social benefits (Chen 2017; Yu 2020). The assembled construction right-of-way in ecologically sensitive areas is mostly used for small vehicle roads with light loads and the cast-in-place method is still used for construction right-of-way with large loads.

---

*Corresponding Author: 358623193@qq.com

DOI: 10.1201/9781003450832-47

For the structural study of assembled right-of-way, Liu Weidong et al (Liu et al. 2015; Song 2018; Zhao 2011) found that assembled cement pavement slabs and traditional cement pavement slabs have a planar size effect, the size of assembled concrete pavement slabs should not be larger than 45 m^2, and the aspect ratio should not be larger than 2.5. Wang Huoming and Wang Xuan et al (Wang & Zhao 2011; Wang & Hu 2021; Wang et al. 2012) found that the factors affecting the bearing capacity of the right-of-way through mechanical simulation are: strength and modulus of subgrade > thickNess of sand bedding layer > width of joints > block size. The current research on assembled right-of-way mainly focuses on the dimensions of assembled road panels and the force characteristics of the panels themselves (Liu 2021; Pei & Ma 2022), while there is almost no research on the application of assembled pile slab structure in the construction right-of-way in ecologically sensitive areas.

For the construction of right-of-way in ecologically sensitive areas, the construction environment is complex, environmental requirements are high, and there are technical defects such as the uneven settlement of the road surface. Based on the construction project of the right-of-way of the section from Sucheng to Sihong, Jiangsu Province of the Yancheng-Luoyang National Expressway, this paper investigates the applicability of pile-slab construction right-of-way in ecologically sensitive areas by using Midas software for mechanical analysis and calculation of reinforcement and pile length and provides a reference for similar projects.

2 PROJECT OVERVIEW

Yancheng-Luoyang National Expressway from Sucheng to Sihong, Jiangsu Province ("Yanluo Expressway"), the starting point is located at the intersection with the Huai-Xu Expressway at the Cangji junction, with a total line length of 8.1 km. Important control points are the Cangji junction, the Tuyuan interchange, the Hongze Lake Wetland Special Bridge (3.28 km), and the S268 mainline overpass. The project line is located on the northwest side of Hongze Lake, near Chengzi Lake, through the Hongze Lake (Figure 1) Wetland National Nature Reserve.

Figure 1. Hongze lake (Suicheng District) wetland.

Hongze Lake wetland has formed its unique ecosystem under the interaction of its unique hydrology, soil, and climate. The wetland is rich in biological resources and the ecosystem is sensitive, fragile, and changeable. The traditional construction of soft foundation right-of-way requires the laying of a lime soil layer, which not only disturbs the soil layer in the ecologically sensitive area but also causes pollution to the soil and water sources around the roadbed during the water cycle in the wetland. The post-work reclamation construction is complicated and the wetland ecosystem suffers irreversible damage. Therefore, it is necessary to study the design and mechanical properties of pile-slab right-of-way structures in ecologically sensitive areas.

The construction right-of-way is designed as a 3.5 m wide one-way lane, which is mainly used for concrete tanker and dump truck transportation. Referring to the studies of Liu Zexin and Duan Zhengjun (Duan 2007; Liu 2021), the size of the concrete panel is appropriate to be 2 m × 3.5 m × 0.28 m under the design load of the right-of-way of this project, considering the planar size effect, the design width of the construction right-of-way, and the convenience of construction.

3 STRUCTURAL ANALYSIS AND INTERNAL FORCE CALCULATION OF ASSEMBLED RIGHT-OF-WAY

3.1 *Structure and material properties*

The pile-slab structure is mainly composed of precast reinforced concrete road slab and steel pipe piles, which are driven into the roadbed according to the design position, and the road slab is placed on the piles (Figure 2). The top of the slab is set at 0.5 m from the edge of the slab. The bottom of the slab is set in the same circular position as the slab with a slightly larger diameter than the steel pile and does not penetrate the thickness of the groove. The steel pile is inserted into the groove to restrain the displacement of the slab. The connection between the plate and the plate is made in the form of pre-buried round steel, such as setting the transmission rod. The pavement structure and prefabricated pavement prefabricated plate splicing situation and interface are shown in Figure 3. After the construction of the pile plate type structure, the road base surface clearing table according to the design position uses a vibration sinking pile party to lay steel pipe piles, and finally complete the lifting of prefabricated road panels can be put into use.

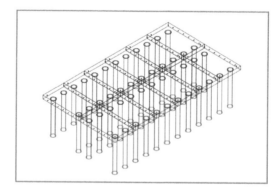

Figure 2. Pile-and-slab construction access road.

For the precast of assembled concrete road panels, materials are used the same construction materials as the mainline project, not for separate procurement. The road panel is cast with C30 concrete, the steel pipe piles are $\Phi 800 \times 10$ mm (Q235B), the road panel is double reinforced with HRB400 hot-rolled ribbed reinforcement, the lifting point reinforcement, as well as the transfer bar reinforcement, are all made of HPB300 bright round reinforcement. The mechanical properties of the materials are shown in Table 1.

Component stiffness requirements according to "steel design code" requirements, the stiffness of the structure is by the standard combination of load calculation, and the stiffness allowable values are shown in Table 2.

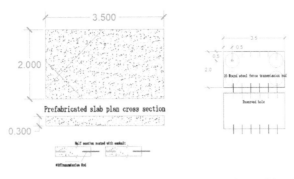

Figure 3. Prefabricated pavement prefabricated slabs splicing situation and interface details.

Table 1. Table of mechanical properties of major materials.

Serial No.	Component name	Material Type	Design value of tensile, compressive, and bending strength /MPa	Shear strength design value /MPa
1	Steel pipe piles	Q235B	f = 205 (16 < t < 40 mm) f = 215 (t ≤ 16 mm)	fv = 120 fv = 25
2	Concrete	C30	fc = 14.30 N/mm^2	ft = 1.43 N/mm^2
3	Rebar	HRB400 HPB300	fy = 360 N/mm^2 fy = 270 N/mm^2	fy' = 360 N/mm^2 fy' = 270 N/mm^2

Table 2. Allowable deflection of bending members.

Serial No.	Displacement absolute value	Allowable stiffness value	Notes
1	Absolute value of vertical displacement	L/250	L indicates the span
2	Absolute value of lateral displacement	H/150	H indicates the height of the bracket from the ground at the maximum value of displacement

3.2 Overall structural force simulation calculation

The calculated load of the structure is mainly composed of structural self-weight and live load. The self-weight of the structure is mainly the self-weight of the concrete panels and the self-weight of the precast concrete single slab is taken as 3.75 tons (the density of reinforced concrete ρ is taken as 2.5 t/m^3). The main load of the live load is a 70 t concrete tanker (expansion factor has been considered), the wheelbase is 1.8 m, the wheelbase is 4.0+1.4 m, and the single wheel landing area is 0.3 m × 0.2 m. There are 3 axles in total, with 10 wheels. The front axle gravity is 110 kN, and the rear axle gravity is 2×120 kN. The combination of working condition one (normal operation) considering the self-weight and live load is shown in Table 3.

Table 3. Load combination.

Serial No.	Load combination	Calculation formula
1	Standard load combination	1.0×Self-weight +1.0×Active load
2	Calculation of load combinations	1.3×Self-weight +1.5×Active load

In this paper, Midas finite element software is used to establish the analytical model of the pile-slab structure right-of-way, the model is established by the beam lattice method, and the model is shown in Figure 4. The calculation process is simulated strictly according to the construction process and considers the effect of effectiveness during normal use, the main material mechanical parameters are shown in Table 1. The calculation model unit characteristics, loads, and constraints are taken according to the relevant design drawings and current specifications. The main boundary condition is the common node connection between the steel pipe pile and the panel. The bottom of the steel pipe pile is the constraint conditions DZ, RX, RY, and RZ. The vehicle loads are simulated as follows according to the lane surface loads.

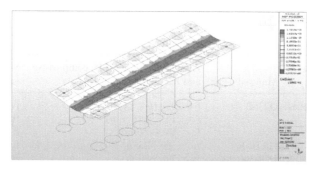

Figure 4. Calculation model and plate unit stress schematic.

The results of the structural simulation are shown in Table 4. Combined with Figures 4 and 5, it can be seen that when the pavement panel structure model is calculated using the plate unit, the maximum deformation of a single plate is located in the middle of the like, the maximum tensile stress of the plate is 2.2 MPa, and the compressive stress is 2.7 MPa. In the whole right-of-way panel structure, the load of the first and last concrete slab is significantly larger than the middle part, and this area needs to be reinforced in practical engineering applications. When calculating the stress of the connecting reinforcement of the right-of-way slab using the beam unit, the maximum integrated stress of the beam unit is 90.4 MPa and the maximum stress appears at the intersection of the splice between the slab and the central axis of the steel pipe pile.

Table 4. Pile and plate structure calculation results.

Component	Type	Working condition I	Notes
Concrete Panel	Tensile stress /MPa	2.2	Standard value combinations
	Compressive stress /MPa	-2.7	
	Bending moment /kN•m	51.4	
	Shear force /kN	±137.4	
Steel pipe piles	Combined stress /MPa	-0.3	Design value combinations
	Support reaction force /kN	161.0	Design value combinations
Connecting reinforcing steel	Combined stress /MPa	-90.4	Design value combinations

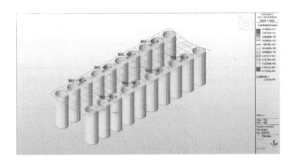

Figure 5. Stress diagram of beam unit.

3.3 Concrete slab reinforcement and pile length calculation

According to the settlement results in Table 4, it can be seen that the stress of connecting reinforcement is 90.4 MPa $\leq$ f = 270 MPa, and the strength of connecting reinforcement with HPB300Φ16 smooth round steel bar meets the load-bearing requirements. The precast reinforced concrete slab is subjected to the compressive stress of 2.7 MPa $\leq$ fc = 14.3 MPa, which meets the requirements. The concrete panel tensile stress is 2.2 MPa $\geq$ ft = 1.43 MPa and the reinforcement of the panel at the time of modeling does not meet the requirements. Therefore, the reinforcement verification of the plate is required.

According to the model division, the section 500 mm * 300 mm is taken for calculation and the bending bearing capacity calculation of the positive section and the shear bearing capacity calculation of the oblique section of the panel are carried out respectively. The maximum internal bending moment of the precast concrete road panel can be obtained through simulation as M_{max} = 51.4 kN, the maximum shear stress V_{max} = 137.4 kN, and the reinforced concrete panel material mechanical properties are shown in Table 1.

Flexural calculations refer to the "concrete structure design code", according to Formula (1) to get the tensile area As = 570 mm^2. The longitudinal reinforcement rate of the longitudinal stress reinforcement is calculated as ρ = 0.38% > ρ_{smin} = 0.002, that is, the reinforcement rate meets the tensile requirements.

$$A_s = \frac{\xi \alpha_1 f_c b h_0}{f_y} \tag{1}$$

Where As is the reinforcement area of the tensile reinforcement, ξ is the height of the relative pressure zone, α_1 is the calculation factor, f_c is the design value of axial compressive strength of concrete, b is the width of the interface, h_0 is the effective height of the interface, and f_y is the design value of the tensile strength of the steel bar. ξ takes the value of 0.108 and α_1 takes the value of 1.

The cross-sectional test is performed with Formula (2) to find the cross-sectional shear stress V = 473.69 kN > V_{max} = 137.40 kN, that is, the hoop section size meets the requirements. The calculated hoop reinforcement rate ρ_{sv} 0.01% < ρ_{svmin} = 0.13% does not meet the minimum hoop rate, then the shear hoop reinforcement is taken as $\rho_{svmin} \times$ b=635.56 mm^2/m according to the structural reinforcement.

$$V = 0.25 \beta_c f_c b h_0 \tag{2}$$

Where V is the section shear stress and β_c is the section shear stress. The rest of the symbols are the same as in Formula (1) and β_c takes the value of 1.

According to the calculation of reinforcement configuration in Table 5, it can be seen that the upper part of the plate with reinforcement Φ16@150 mm (HRB400) is arranged along

Table 5. Configuration steel calculation table.

Parts	Constructions	The actual configuration of reinforcing steel	Results
Upper longitudinal bar	As = 300 mm²	2Φ12+1Φ16 (427mm², ρ = 0.28%) HRB400	Reinforcement meets
Lower longitudinal reinforcement	As = 570 mm²	2Φ16+1Φ20 (716mm², ρ = 0.48%) HRB400	Reinforcement meets
Hoop reinforcement	Av/s = 636 mm²/m	Φ8@150 (3) (1005mm²/m, ρsv = 0.20%) HPB300	Reinforcement meets

the 3.5 m direction and the lower part of the plate with reinforcement Φ20@150mm (HRB400) is arranged along the 3.5 m direction. To disperse the stress in the plate as much as possible and avoid stress concentration, the reinforcement can adopt the same diameter and spacing as 3.5m along the 2m direction.

As can be seen from Figure 6, the pile structure uses the column unit for the steel pipe pile Φ800 × 10 mm compressive bearing capacity calculation. According to the simulation results, it can be seen that the maximum pile end reaction design value compressive force is 161.0kN.

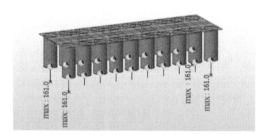

Figure 6. Schematic diagram of column unit support reaction force.

According to the geological data (Table 6), it is known that the bearing layer is deep and the piling depth is too large by using an end-bearing pile. Therefore, the steel pipe column is calculated in the form of a friction pile for the force. According to the simulation results, the maximum pile end reaction force design value is 161.0 kN. According to the soil parameters in Table 6, the ultimate bearing capacity design value of a single pile is calculated according to the "Specification for Port Engineering Pile Foundation" using Formula 3.

$$Q_d = \frac{1}{\gamma_R} \left(U \sum q_{fi} l_i + \eta q_R A \right) \quad (3)$$

Table 6. Geological conditions.

Layer number	Soil layers	Elevation /m	Layer thickNess /m	Pile end bearing capacity /kPa	Pile side friction resistance /kPa
A-3	Vegetative fill	14.8~14.4	0.4	–	–
B-1	Clay	14.4~13.2	1.2	–	30
C-1	Clay	13.2~11.8	1.4	–	35
C-2	Clay	11.8~-6.6	18.4	–	40

Where Q_d is the design value of the ultimate bearing capacity of a single pile, γ_R is the bearing capacity sub-factor, U is the perimeter of the pile section, q_{fi} is the standard value of the ultimate lateral friction resistance of layer i soil of a single pile, l_i is the length of the pile through the i-th layer of soil, η is the load-bearing capacity discount factor, q_R is the standard value of the ultimate pile end resistance of a single pile, and A is the cross-sectional area of the pile. γ_R and η refer to the specification "Code for Port Engineering Pile Foundation" to take the value.

After calculating the depth of 3.5 m, Q_d = 170.25 kN > 161.0 kN to meet the requirements, that is, the pile into the soil is 3.5m (0.4 m of A-3 plain fill layer, 1.2 m of B-1 clay layer, 1.4 m of C-1 clay layer, 0.5 m of C-2 clay layer), which meets the bearing capacity requirements.

4 DISCUSSION

The current engineering application of the assembled construction roadbed and subgrade is almost no different from the traditional roadbed, its construction occupies the same land resources as the conventional right-of-way, shortening the construction period is only the pavement laying maintenance period, the road panel is also prone to breakage and misalignment after the uneven settlement of the local foundation.

For the construction characteristics of the Yantai Expressway construction roadbed, two solutions are also proposed at the early stage of design, one is the use of assembled expanded base plus pavement plate structure, and the other is the use of soil curing agent directly on the pavement soil curing treatment. Both of the above solutions have less disturbance to the soil layer of the wetland reserve, occupy fewer land resources, and are conducive to post-work reclamation and rapid ecological restoration. After preliminary numerical simulation, the deflection of the center of the panel is 26 mm when using the same road panel as the pile slab structure, which is much larger than the deflection value specified in the "Concrete Structure Design Code". The prefabrication and on-site construction of the expanded foundation are more complicated than the steel pile construction. The construction of program two is simple but the hardened soil road construction right-of-way bearing capacity is less than 1/10 of the cement concrete pavement. The water stability is poor, and settlement deformation occurs under heavy loads. Therefore, the pile-slab structure is finally selected for the construction application on site after comparison.

The pile-slab structure has high stiffness and minimal settlement and its force structure is simple and transparent. The design of this structure is relatively simple, and the reinforcement conversion of precast pavement slabs and the pile length can be calculated according to the actual engineering geological conditions. Compared with the traditional flexible or semi-rigid fill roadbed, this structure can greatly improve the bearing capacity of the construction right-of-way by using a rigid roadbed.

The main nodes of the pile-slab structure are the inter-slab connection and pile-slab connection, which is also the weak link of the pile-slab structure and is most prone to damage and deformation. According to the simulation study of this paper, the installation accuracy should be controlled strictly according to the design requirements during construction, and the gap between the panels should not be larger than 1.5 cm. A groove should be set at the bottom of the panels without penetrating the panels to restrain the displacement of the panels.

5 CONCLUSION

In this paper, the structural design and mechanical properties of the construction right-of-way of Yanluo Expressway are studied and the following conclusions are obtained by numerical simulation and field construction testing.

(1) The pile-slab structure is fully applicable to the load-bearing and environmental protection requirements of the right-of-way in the ecologically sensitive area. The concrete panel size is 3.5 m * 2 m * 0.3 m with double reinforcement, and the steel pipe pile ($\Phi 800 \times 10$ mm) needs to enter the soil at a depth of 3.5 m to meet the load-bearing capacity requirements.

(2) In the first and last concrete panels outside the right-of-way, as the load is greater than the middle part, the foundation needs to be hardened here in the construction of the right-of-way to avoid breaking the ring at the end. The top of the steel pipe pile needs to be reliably connected to the precast panel, and a pile cap can be added to the steel pipe pile to make its contact area with the concrete panel larger and avoid local concentration of force.

(3) In the actual engineering application, the pile slab structure construction right-of-way covers 60% and 40% of the traditional construction right-of-way in terms of floor area and construction duration, respectively. After self-inspection by the construction unit, the construction right-of-way has good levelness, no obvious uneven settlement, and no broken and misplaced panels. Under the action of vehicle load, the panel corners at the joints are broken and peeled off due to the long-term local concentrated force, and the concrete panel corners need to be wrapped with steel to ensure their durability and aesthetics. The recovery rate of concrete panels can reach 95% and steel piles 98%, while the recovery rate of pile cap deformation is 64%.

REFERENCES

Chen, Z. J. Study on the Application Technology of Prefabricated Assembled Cement Concrete Pavement. Highway and Transportation *Technology (Applied Technology Edition)*, 2017. 13(04): pp. 63–65.

Duan Zhengjun. *Study on Reasonable Plan Dimensions of Cement Concrete Pavement Panels*, 2007, Chang'an University. Page 77.

Liu Weidong, Li Limin, and Hou Ziyi. Structural Analysis of Precast Concrete Assembled Pavement Slabs. *Journal of Heilongjiang Engineering College*, 2015. 29(05): pp. 24–30.

Liu Zexin. *Structural Design and Numerical Simulation Optimization of Assembled Cement Concrete Pavement Slabs*, 2021, Chang'an University. Page 118.

Pei, Y. F., Ma, Y. Z. Analysis of Energy-saving and Environmentally Friendly Temporary Road Practices on Construction Sites. *Jushe*, 2022(05): pp. 175–177+180.

Song Wa Li *et al*. Research on Key Technology of Large-size Precast Concrete Slab Pavement. *Journal of Chongqing Jiaotong University (Natural Science Edition)*, 2018. 37(09): pp. 28–33.

Wang Huoming and Zhao Jun. Finite Element Computational Analysis of Structural Mechanical Properties of Precast Block Pavement. *Highway Traffic Technology*, 2011(06): pp. 16–20.

Wang Huoming *et al*. Experimental Study on the Load-bearing Characteristics of Precast Concrete Block Pavement. *Highway Traffic Technology*, 2012(02): pp. 24–30.

Wang Xuan and Hu Chunhua. Mechanical Response Analysis and Joint Optimization of Assembled Cement Concrete Pavement Panels. *Technology and Markets*, 2021. 28(01): pp. 64–66+69.

Yu Wanyou. Application of Assembled Concrete Roads in Urban Subway Construction. *Railway Construction Technology*, 2020(02): pp. 111–114.

Zhao Junjun. *Application of Cement Concrete Precast Blocks in an Emergency Stopping Lane of Highway*, 2011, Chongqing Jiaotong University. Page 89.

Water Conservancy and Civil Construction – Oke & Ahmad (Eds)
© 2024 The Author(s), ISBN: 978-1-032-58618-2

Research on the strategic choice of logistics development in Urumqi based on AHP-SWOT

Xuegang Liang*
Department of Traffic Engineering, Henan Polytechnic University, Jiaozuo, China

Lan Li*
Department of Logistics Engineering, Xinjiang Agricultural University, Urumqi, China

Chengming Zhu*
Department of Traffic Engineering, Henan Polytechnic University, Jiaozuo, China

ABSTRACT: As the capital city of Xinjiang Uygur Autonomous Region and the international trade center for Central and West Asia, Urumqi is also an important node of the "One Belt, One Road" strategy. The study of its development in logistics can help further promote the rapid development of the logistics industry in Xinjiang. In this paper, we analyze the key factors affecting the development of logistics in Urumqi through a series of problems in the current situation of logistics development in Urumqi and establish a complete system framework combining quantitative and qualitative analysis by using hierarchical analysis and SWOT analysis. By using the Delphi method to score each factor, it is concluded that the "Belt and Road" strategy is an important opportunity for logistics development and the high logistics cost is an indispensable problem to solve the logistics development in Urumqi. The development strategy of Urumqi logistics should choose the WO strategy. Finally, on this basis, we propose the development of a logistics industry in Urumqi.

1 INTRODUCTION

Urumqi is located in the middle of Xinjiang, which is adjacent to Central Asian countries. It is the second bridgehead of the Asia-Europe Continental Bridge in western China, an important gateway for China to open up to the west, an important central city in northwest China, and an international trade center in Central and West Asia. In the context of comprehensively deepening reform, opening up, and implementing the "Belt and Road" high-quality development strategy, the transportation infrastructure in Xinjiang is completed and its functions are constantly improved. With location advantages and superior transportation conditions, Urumqi is a must-go destination for the three major economic corridors of "One Belt, One Road": New Asia-Europe Continental Bridge, China-Central Asia-West Asia, and China-Pakistan. Urumqi city has opened several foreign opening projects, such as the China-Europe Classical Train, allowing access to many countries and regions in Central Asia and Europe and realizing the penetration of the western land and sea corridor under the "One Belt and One Road" strategy. At present, the logistics industry in Urumqi is still in the initial stage and there are some problems in the management system. For example, the managers of logistics enterprises lack sufficient understanding of logistics, the logistics technology,

---

*Corresponding Authors: 1919516343@qq.com, 1057666390@qq.com and zhuchengming@hpu.edu.cn

DOI: 10.1201/9781003450832-48

equipment, and infrastructure are relatively backward, and logistics-related enterprises lack front-line high-quality talent. The study of the development of the logistics industry in Urumqi plays an important role in promoting the development of the regional economy.

2 LITERATURE REVIEW

Hierarchical analysis can decompose a complex decision problem into objective, criterion, and solution levels layer by layer (Guo 2013). The importance of each influencing factor in each layer is quantified by establishing a vector matrix and the weight values of each different element are determined according to the calculation results. Finally, the solution is derived from the analysis of the relationship between the weights and the magnitudes. In the 1980s, Heinz Weihrich, an American professor of management, proposed the SWOT analysis. By analyzing the internal conditions and external environment of the company, it is concluded that the strengths, weaknesses, opportunities, and threats of the company are the four factors that must be considered to determine the strategic choices of the company (Lv 2020). Based on the AHP-SWOT analysis method, many scholars have conducted more studies on the development of regional logistics and made their contributions from different perspectives. Feng used the AHP-SWOT analysis method to propose that the development of China's space security should seize opportunistic strategies, rely on the space power strategy, seize international space development opportunities, and build a space power infrastructure to support the national space development strategy (Feng et al. 2019). Ma and others believed that the development of green logistics in Hefei must build on its strengths, avoid its weaknesses, avoid external threats, overcome internal weaknesses, and fully combine external opportunities with local advantages. All enterprises and government departments at all levels must work together to put the development of green logistics on the agenda (Ma et al. 2016). Li and Yang proposed that the transportation advantages of Shaybak District, the imperfection of related supporting industries of logistics in Shaybak District, the constraints of the logistics industry becoming a new economic growth point in Shaybak District, and the small scale and low grade of existing enterprises are the key elements affecting the development of the logistics industry in Shaybak District (Li & Yang 2022). Wang believed logistics requires a higher quality for logistics practitioners (Wang 2022). The logistics industry is not simply the loading, unloading, and handling of various goods, packaging, and transportation. In the current logistics industry, most of the employees are recruited from the labor market with insufficient knowledge and skill reserves. The development of the logistics industry needs a greater number of qualified personnel. Various relevant research institutes, colleges, and universities should undertake targeted training to promote the development of the entire logistics industry. Zi proposed that with the vigorous development of the economy, the cooperation between regions, cities, and enterprises is increasingly strengthened and the development of regional logistics connects the production enterprises and consumers at all levels into a whole structure, which can greatly promote the socio-economic development of a region (Zi 2021). The development of regional logistics can improve the quality of logistics supply and reduce the costs in the logistics chain, thus promoting social and economic development.

3 STRATEGIC ANALYSIS OF LOGISTICS DEVELOPMENT IN URUMQI BASED ON AHP-SWOT

3.1 *Analysis of the current situation of logistics development in Urumqi*

The transportation industry in Urumqi is developing rapidly. With an integrated transportation system, which includes highway, railroad, air, and pipeline transportation systems, Urumqi is an important integrated transportation framework linking the north and south of the country, as well as at home and abroad. In 2019, Urumqi became one of the 23 national cities designated for the construction of logistics hubs. It is developed to promote the economic integration between the logistics hub and the surrounding areas to drive various industrial parks in Urumqi and realize various functions, such as cargo distribution, warehousing, and transit between various regions in the territory.

The logistics costs in Urumqi are high and the technical level and management system are not perfect. Most of the logistics companies in Urumqi are developed from consignment companies and there is a lack of technical communication among the various enterprises and a lack of a public information platform for logistics to provide information exchange services. The uneven distribution of logistics parks and warehouses, most of which are located in the suburbs of the city, has led to the dispersion of logistics services, which has affected the further development of the logistics industry.

The northwest region is an important area of the "Belt and Road" initiative, an important node for cargo transportation and opening up to the outside world. The logistics industry in Urumqi is bound to receive unprecedented development due to the construction of "One Belt, One Road." With the continuous increase in opening up to the outside world, the development of economic construction in cities along the route, and especially with the current far-reaching impact of the new epidemic in various countries and regions of the world, the demand for materials related to epidemic prevention and control is increasing. The demand for logistics services from the general public is also increasing. In addition, by promoting the "One Belt and One Road" initiative, the western region is benefiting more from national logistics policies, such as transportation and highway infrastructure construction, as well as expanding the economic development momentum in the west, thus increasing the potential for logistics development in Xinjiang.

3.2 *Establishing a SWOT analysis model of logistics development strategy in Urumqi*

For the internal and external environment of logistics development in Urumqi, a SWOT analysis model of logistics development strategy in Urumqi is listed, focusing on spatial location, resources and technology, transportation, enterprise development, and policy promotion.

Table 1. SWOT matrix of logistics development strategy in Urumqi.

	Strengths (S)	Weaknesses (W)
Internal condition	S1: Gradual and steady development of logistics in Urumqi S2: Urumqi has a good industrial base and strong open dynamics S3: Rich in resources, unique in nature S4: Transportation industry in Urumqi is developing rapidly S5: National policy support drives the development of logistics in Urumqi	W1: Logistics technology is backward W2: Logistics cost is still high W3: Xinjiang logistics management system is not perfect W4: Logistics center construction is still lagging behind W5: Logistics enterprises are small in scale and low in professionalism
	Opportunity (O)	Threat (T)
External condition	O1: The construction of "One Belt, One Road O2: Western development strategy brings opportunities for the logistics development O3: Xinjiang port logistics continues to develop O4: Regional logistics is increasingly becoming an international trend	T1: The expansion of domestic and foreign logistics enterprises has intensified market competition T2: Insufficient demand in the logistics market T3: Increasingly prominent environmental problems T4: Insufficient high-end talents in logistics

Table 2. Judgment matrix of logistics development strategy in Urumqi.

A	Strengths	Weaknesses	Opportunities	Threats
Strengths	1	2/3	1/3	1/2
Weaknesses	3/2	1	1/2	2
Opportunities	3	2	1	3
Threats	2	1/2	1/3	1

Table 3. Judgment matrix of the advantages of logistics development in Urumqi.

Strengths	S1	S2	S3	S4	S5
S1	1	4	1/2	2	3
S2	1/4	1	1/5	1/3	1/2
S3	2	5	1	3	4
S4	1/2	3	1/3	1	2
S5	1/3	2	1/4	1/2	1

Table 4. Judgment matrix of the disadvantages of logistics development in Urumqi.

Weakness	W1	W2	W3	W4	W5
W1	1	1/2	4	3	3
W2	2	1	7	5	5
W3	1/4	1/4	1	1/2	1/3
W4	1/3	1/3	2	1	1
W5	1/3	1/3	3	1	1

Table 5. Judgment matrix of logistics development opportunities in Urumqi.

Opportunities	O1	O2	O3	O4
O1	1	8	5	3
O2	1/8	1	1/2	1/6
O3	1/5	2	1	1/3
O4	1/3	6	3	1

Table 6. Judgment matrix of threats to logistics development in Urumqi.

Threats	T1	T2	T3	T4
T1	1	2	3	3
T2	1/2	1	2	2
T3	1/3	1/2	1	1
T4	1/3	1/2	1	1

3.3 *Quantitative analysis of key factors in the SWOT model*

Using the scale method from 1 to 9, the decision makers involved in the evaluation compare the relative importance of each factor between any two pairs. The judgment matrix A is obtained by using the hierarchical analysis method, where the maximum characteristic root value of the judgment matrix A is calculated and tested for consistency. Similarly, the relative importance of internal advantage S, disadvantage W, external opportunity O, and threat T are compared between two pairs, and the judgment matrix S, W, O, and T are obtained in turn.

The eigenvalues and eigenvectors of the above judgment matrices A, S, W, O, and T can be used as the basis for weight calculation through the consistency test of CI, RI, and CR indicators.

According to the results of the stratified single ranking in Table 7, the stratified total ranking of all the factors that can influence the development of logistics in Urumqi for the overall goal can be ranked hierarchically in total. It can be seen from the calculation results that the hierarchical total ranking S3, W2, O1, and T1 account for a larger weight relative to the weights of each element in the group.

Table 7. Hierarchical single alignment and consistency tests.

Matrix	λmax	CI	RI	CR	Normalized Eigenvectors
A	4.0968	0.0323	0.8820	0.0366	0.5774, 1.1067, 2.0598, 0.7598
S	5.0681	0.0170	1.1100	0.0153	1.6438, 0.3839, 2.6052, 1.0000, 0.6084
W	5.0721	0.0180	1.1100	0.0162	1.7826, 3.2271, 0.3589, 0.6683, 0.7248
O	4.0665	0.0222	0.8820	0.0251	3.3098, 0.3195, 0.6043, 1.5651
T	4.0104	0.0035	0.8820	0.0039	2.0598, 1.1892, 0.6389, 0.6389

Table 8. Hierarchical general arrangement and consistency test.

	X1	X2	X3	X4	X5
S	0.0338	0.0079	0.0535	0.0205	0.0125
W	0.0648	0.1173	0.0130	0.0243	0.0263
O	0.2611	0.0252	0.0477	0.1235	-
T	0.0768	0.0443	0.0238	0.0238	-

3.4 Construction of the strategic quadrilateral

According to the above calculation and analysis, after determining the total ranking of each influencing factor of SWOT strategy selection, the SWOT quadrilateral can be used to make strategy selection and construct the strategy quadrilateral.

According to the strategic quadrilateral, it can be judged that triangle O1AW2 has the largest area. Therefore, the development of logistics in Urumqi should choose the WO strategy, that is, the reversal type strategy. The development of the logistics industry in Urumqi needs to overcome the disadvantages and grasp external opportunities, such as the national economic development strategy.

3.5 Suggestions for the development of logistics in Urumqi

The logistics development in Urumqi should overcome internal disadvantages and grasp external opportunities. Making full use of the local advantages and opportunities for opening up to the outside world, the government and local logistics enterprises should work together to actively develop the logistics industry and make full use of the western development strategy and the "Belt and Road" strategy to promote the economic and social development of Xinjiang.

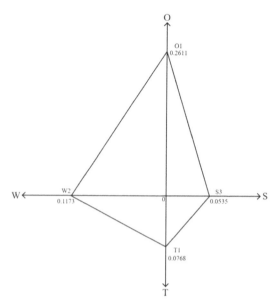

Figure 1. Strategic quadrilateral.

First of all, to accelerate the planning and construction of modern logistics centers, Urumqi should rely on the existing urban transportation infrastructure and various industrial parks to strengthen the existing infrastructure and market environment of the industrial parks. With the correct guidance from the government, as well as feedback from the market, a modern logistics center should be built. The construction of a logistics center can realize the integration of the logistics system and promote the modernization of logistics by relying on the existing transportation infrastructure. Secondly, it can actively introduce talents and cultivate middle and senior logistics talents. Modern logistics requires a high level of logistics practitioners. Combining theory and practice, only through the close cooperation between relevant universities and various logistics enterprises and targeted training for students in Xinjiang universities, relevant research institutions and universities can provide more qualified researchers and technical workers for the logistics industry, thus promoting the rapid development of the modern logistics industry in Urumqi.

4 CONCLUSION

This paper uses the AHP-SWOT method to analyze the logistics development strategy of Urumqi and concludes that the disadvantages and opportunities are the important reasons affecting the logistics development of Urumqi. The "Belt and Road" strategy is an important opportunity for the logistics development of Urumqi and the high logistics cost is a solution to the logistics development of Urumqi. The high cost of logistics is an indispensable problem for the development of logistics in Urumqi. However, it must be noted that the correlation between the factors needs to be further considered. And the development strategy of the logistics industry is complex, systematic, and limited by many realistic factors. Therefore, further research is needed to ensure the comprehensiveness of the strategic analysis and the accuracy of the calculation.

REFERENCES

Feng, S., Su, X., Wang, Y. *Research on Development Strategy of Space Security Based on AHP-SWOT* [C]. Dalian, Liaoning, China: 2019.

Guo Haiyan. *Research on Fuzzy Evaluation of Enterprise Strategic Decision Risk Based on AHP [Master Thesis]* [D] Shandong University of Technology, 2013.

Lv Rui. *Research on the Development Strategy of Maishou E-commerce Company Based on SWOT-AHP [Master's Thesis]* [D] Lanzhou University of Technology, 2020.

Li Yanan, Yang Shuxin. Strategic Choice of Logistics Industry Development in Shaybak District of Urumqi City [J]. *Journal of Urumqi Vocational University*. 2015, 24(03): 15–20.

Ma Zhengqi, Wang Jiacai, Zhu Ying, *et al.* AHP-SWOT Analysis on the Development of Green Logistics in Hefei City [J]. *Journal of Hefei University (Comprehensive Edition)*. 2016, 33(04): 41–46.

Wang Huizhen. Analysis of the Logistics Personnel Training System of International Land Ports under the Belt and Road Initiative [J]. *Logistics Science and Technology*. 2022, 45(01): 168–170.

Zi Pei jade. The Impact of Regional Logistics on the Regional Economy and its Approach [J]. *Modernization of shopping malls*. 2021(13): 47–49.

Analysis of impact indicators for dynamic management of on-street parking spaces based on hierarchical analysis

Chunmei Hu, Yang Ming*, Shuting Deng & Qianrong Tu
School of Civil Engineering and Architecture, Nanchang Hangkong University, Nanchang, Jiangxi, China

ABSTRACT: An index system of parking space usage for dynamic management is established by taking into account three factors: road conditions, dynamic traffic changes, and the demand for parking space utilization, which is to lessen the traffic impact of on-street parking on road sections and realize the dynamic management of on-street parking space usage. An evaluation model for the dynamic management of on-street parking use based on the hierarchical analysis approach was built by the qualitative and quantitative analysis of the indicators, and the reliability of the model was examined through MATLAB simulation. Research results show that the index scores have a significant influence on the realization of the dynamic management of on-street parking spaces, with the dynamic traffic impact accounting for 0.73, indicating that it is closely related to the use of on-street parking spaces. A theoretical basis from the results of the study is provided for realizing the dynamic management of on-street parking spaces.

1 INTRODUCTION

On-street parking, which is crucial to resolve the paradox between parking supply and demand, is the practice of designating several temporary parking spaces for vehicles on both sides or one side of a city road using traffic signs and markings. Additionally, on-street parking restricts effective lane widths and lateral clearances, resulting in traffic delays and accidents. Therefore, some scholars believe that an in-depth study of the impact of on-street parking on the dynamic traffic of road sections is particularly necessary.

Scholars both domestically and internationally have conducted studies on this topic. Among scholars at home and abroad, the research idea is to make coefficient corrections based on the basic capacity model proposed in the United States, which has lane width, lateral clearance, and parking for vehicles as indispensable correction coefficients in various scholars' studies (Zeng 2003; Jiang 2008; He 2012; Shao 2018; Ye 2018). Zhan (2020) incorporated correction coefficients of intersection impact which were based on this consideration of the road network capacity. Guo (2011, 2012) established a different basic capacity model based on conflict analysis techniques with corrections, and the results of this study were later applied to the traffic flow modeling under on-street parking by Wei (2018). As regards journey delay, Scholars divided the motor vehicle movement in the on-street parking section into the process of deceleration-following-acceleration leaving and calculated the corresponding delay model by refining the parking process (Feng 2008; Jie 2016; Mei 2009; Pei 2020). Wei *et al.* (2017) studied the journey delay under in-stopping based on the model of meta-cellular automata about the energy consumption of the traffic flow, and showed that the larger the proportion coefficient of vehicles is to be stopped, the longer the

---

*Corresponding Author: mingyang@nchu.edu.cn

obstruction time to the following vehicles will be caused by in-stopping behaviors. Krzysztof *et al.* (2018) first added driver behavior to the study of the impact of on-street parking, based on the meta-cellular automaton and its extended model studying the impact of travel time under on-street parking. VISSIM simulations have also been used to analyze the changes in journey delay under on-street parking, and the results showed that the simulation results were more closely matched with the change in actual road journey delay (Gao 2010; Liu 2009; Li 2016; Shi 2020; Yang 2007). Some scholars (Li 2015) have used the hierarchical analysis to discuss the impact of on-street parking on vehicle delays. Others (Lin 2016; Zhou 2017) have evaluated and analyzed the technology and policy of on-street parking charges based on hierarchical analysis to select the optimal way to charge for on-street parking. Scholars have used fuzzy hierarchical analysis for the choices of parking behavior and the evaluation of the level of operation of on-street parking systems in terms of four aspects: people, vehicles, roads, and the traffic environment (Zheng 2017; Manville 2021). To determine the choice of on-street parking management, some scholars have also conducted studies on on-street parking fees (Shoup 2021; Granger 2021).

Most of the studies on on-street parking are based on theoretical models and microsimulation analyses of roads and traffic, which quantify the impact of on-street parking, but do not provide specific suggestions or measures for improvement, especially when it comes to the management of parking usage. Given this, based on the perspective of traffic managers, this paper selects road conditions, dynamic traffic impacts, and parking utilization as factors relevant to the dynamic management of on-street parking use. Based on the research of a large number of scholars, a qualitative and quantitative analysis of the relevant factors is conducted to lay a preliminary and theoretical foundation for the dynamic management of on-street parking use.

2 DETERMINATION OF A SYSTEM OF DYNAMIC MANAGEMENT INDICATORS FOR ON-STREET PARKING

As a complement to the allocated parking and public parking, on-street parking spaces are set up within the red line of the road, sharing the pressure of urban parking. Furthermore, they take over the driving functions of the road, which naturally impacts the operation of the road as well as traffic flow dynamics. A hierarchical structure is used in this paper to construct a model of on-street parking management evaluation, which is divided into three parts: the uppermost target layer (F layer), the middle criterion layer (S layer), and the indicator layer (T layer), as shown in Figure 1.

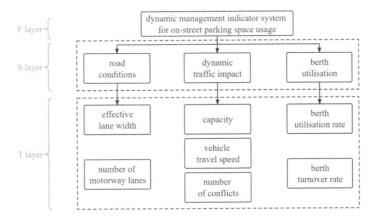

Figure 1. The system of dynamic management indicator for on-street parking space usage.

2.1 Road conditions S_1

Road conditions include effective lane width T_1 and the number of motorway lanes T_2. Effective lane width T_1 refers to the actual width of the road that can be used for driving after the installation of on-street parking. The larger the effective lane width is, the smaller the impact of on-street parking is on the dynamic traffic on the road. Following the installation of on-street parking, the number of motorway lanes T_2 are lanes that can be used by motor vehicles. Added on-street parking to a motorway will directly reduce the capacity of the roadway for vehicular traffic. Generally, roads with on-street parking are better as the effective travel lane width T_1 and motorway lanes T_2 are wider.

2.2 Dynamic traffic impact S_2

Dynamic traffic impacts consist of three aspects: traffic volume T_3, vehicle speed T_4, and the number of conflicts T_5. Capacity T_3 refers to the maximum number of vehicles passing through a section of road per unit of time under both on-street parking and off-street parking conditions. An increase in capacity reduces the impact of on-street parking on the road. In particular, on-street parking in and out of the roadway section will cause significant changes in vehicle speed T_4. A traffic conflict occurs when two or more road users park in the same lanes at the same time, which is close to each other. If one of the parties takes an abnormal driving behavior, unless the other party takes the corresponding avoidance behavior, there may be a collision. Traffic conflicts T_5 include conflicts between motor vehicles and motor vehicles, motor vehicles and non-motor vehicles, and non-motor vehicles and non-motor vehicles.

2.3 Berth utilization S_3

Berth utilization consists of two components, berth utilization rate T_6 and berth turnover rate T_7. The berth utilization rate T_6 is the percentage of the total time of a vehicle occupying a parking space in a year. The berth turnover rate T_7 is the average number of vehicles parking in each parking space in a given period, which is often expressed as a ratio of the cumulative number of vehicles parking in a given period to the capacity of the parking facility. Berth utilization rate T6 and berth turnover rate T7 represent efficiency in on-street parking concerning time and space, respectively, which are critical indicators of on-street parking's use.

3 MODEL BUILDING

3.1 Establishment of the recursive hierarchy

When using the hierarchical analysis to solve a problem, it is critical to organize and hierarchize the factors related to the decision, therefore a structural model with a hierarchy can be constructed. A hierarchical model consists of three main layers. The top layer is called the objective layer, the bottom layer is called the solution layer, and the middle layer is called the criterion layer. Generally, there will be a sub-criteria layer below the criterion layer, also known as the indicator layer. The specific number of layers should be determined according to the actual problem. In this paper, we take the dynamism of management of on-street parking as the objective layer, and road conditions, dynamic traffic impact, and parking utilization as the criterion layer, and establish a progressive hierarchy model. Among these, on-street parking has a more significant impact on road conditions and dynamic traffic on the road segment and should be considered as a key evaluation indicator, while parking utilization can be considered as a supplementary evaluation indicator. The factors of the indicator layer corresponding to road conditions are effective lane width and the number of motor vehicle lanes. The factors of the indicator layer corresponding to dynamic traffic impact are capacity, vehicle speed, and the number of conflicts. The factors of the indicator layer corresponding to berth utilization are berth utilization rate and berth turnover rate.

3.2 *Establishment of the judgment matrix*

There must be a judgment matrix for each criterion level, since the importance of each element may differ. As indicated in Table 1, the importance of factors in the criterion layer is often quantified by using the numbers 1 through 9, with aij standing for the comparative outcome of the i-th factor relative to the j-th component.

Table 1. Definition of judgment matrix scales.

Scales	Meanings
1	The same effect of factor i relative to factor j
2	Slightly stronger effect of factor i relative to factor j
3	A strong influence of factor i relative to factor j
4	A very strong influence of factor i relative to factor j
5	An extremely strong influence of factor i relative to factor j

Note: 2, 4, 6, and 8 indicate that the effect of factor i relative to factor j is between the two adjacent levels above.

3.3 *Hierarchical single ranking and consistency test*

(1) Consistency indicator CI

$$CI = \frac{\lambda_{\max} - n}{n - 1} \tag{1}$$

In Equation (1), where $\lambda_{\max}$ is the judgment matrix's maximum eigenvalue and n is the judgment matrix's order.
(2) Consistency test

$$CR = \frac{CI}{RI} \tag{2}$$

In Equation (2), CR is the random consistency ratio of the judgment matrix, and RI is the random consistency index. In general, the greater the order of the matrix is, the greater the possibility of random deviation of consistency. The correspondence is seen in Table 2. When CR = 0, it is not necessary to test the consistency of the judgment matrix. When CR < 0.1, the consistency of the judgment matrix reconstructs the judgment matrix and then tests.

Table 2. Average random consistency indicators.

N	1	2	3	4	5	6	7	8	9	10	11	12	13	14
RI	0	0	0.52	0.89	1.12	1.24	1.36	1.41	1.46	1.49	1.52	1.54	1.56	1.58

3.4 *Hierarchical total ranking and consistency test*

The entire hierarchical ranking also needs to be checked for consistency, and the synthetic weight of each element to the target must be determined as stated in Equation (3).

$$CR = \frac{\sum_{i=1}^{n} b_i CI_i}{\sum_{i=1}^{n} b_i RI_i} \tag{3}$$

4 MATLAB SIMULATION ANALYSIS

4.1 *Construction of the judgment matrix*

Hierarchical relationships are created by comparing two factors on the hierarchy diagram. The judgment matrix between the target and criterion layers is F-S, and the matrices between the criterion and indicator layers are S_1-T, S_2-T, and S_3-T, a total of four judgment matrices.

As shown in Table 3, the F-S matrix involves comparing factors S_1, S_2, and S_3 in the criterion layer to obtain the F-S judgment matrix.

Table 3. F-S judgment matrix.

	S_1	S_2	S_3
S_1	1	1/5	3
S_2	5	1	7
S_3	1/3	1/7	1

The only factors relevant to the first factor "road conditions" in the criterion layer are T_1 and T_2 in the indicator layer, so only these two factors are compared in the S_1-T judgment matrix shown in Table 4. The factors related to the second factor "dynamic traffic impact" in the criterion layer are T_3, T_4, and T_5, so only the three factors are compared with each other in the S_2-T judgment matrix. The S_2-T judgment matrix is presented in Table 5. The factors related to the third factor "berth utilization" in the criterion layer are only T_6 and T_7, therefore only two factors, T_6 and T_7, are evaluated with each other in the S_3-T judgment matrix. The S_3-T judgment matrix is shown in Table 6.

Table 4. S_1-T judgment matrix.

	T_1	T_2
T_1	1	5
T_2	1/5	1

Table 5. S_2-T judgment matrix.

	T_3	T_4	T_5
T_3	1	1/2	3
T_4	2	1	5
T_5	1/3	1/5	1

Table 6. S₃-T judgment matrix.

	T₆	T₇
T₆	1	3
T₇	1/3	1

4.2 Criterion-level factor weights and consistency tests

As can be seen from Table 3, in the criterion layer, the factors related to the target layer are S_1, S_2, and S_3, and their relative importance is determined by comparing them with each other by using a 9-level scale method two by two. Referring to Table 1, by quantifying the relative importance, the F-S judgment matrix is then established $\begin{bmatrix} 1 & 1/5 & 3 \\ 5 & 1 & 7 \\ 1/3 & 1/7 & 1 \end{bmatrix}$. The maximum eigenvalue $\lambda_{max} = 3.0649$ of the matrix F-S is found using MATLAB, and the normalized weight vector of the related eigenvector is as follows:

W=(0.188, 0.731, 0.081)

This matrix's consistency test produced the estimated findings CI=0.03245, RI=0.52, and CR=0.0624<0.1, indicating that the consistency of the matrix F-S was satisfactory.

As shown in Figure 2, the weights of the factors in the criterion layer and the target layer are represented. In the criterion layer, the dynamic traffic impact factor has the highest weight relative to the target layer, which is at 0.731, indicating that the dynamic traffic impact factor is the most significant in the dynamic management index system of on-street parking use, which is followed by the road condition factor, with a weight of 0.188. The berth utilization factor has the lowest weight, which is at 0.081.

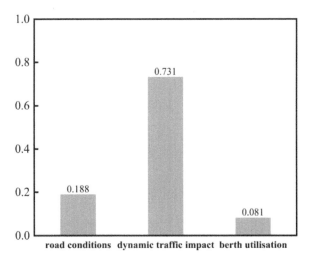

Figure 2. Weights of the factors in the criterion layer to the target layer.

4.3 Indicator layer factor weights and consistency tests

Weights in the indicator layer are solved in the same way as those in the criterion layer, and the eigenvalues and eigenvectors of the S_1-T, S_2-T, and S_3-T matrices and the F-S matrix are also solved similarly. Because of space limitations, they are not listed one by one. The weights obtained after the weighting process of each factor in the indicator layer are shown in Table 7. The consistency tests were conducted one by one, and the test results all met the requirements.

Table 7. The weight of the indicator layer for the criterion layer.

Indicator weights	T_1	T_2	T_3	T_4	T_5	T_6	T_7
	0.833	0.167	0.309	0.582	0.109	0.75	0.25
Consistency check	$\lambda_{max}=2$		$\lambda_{max}=3.0037$			$\lambda_{max}=2$	
	CI=0		CI=0.00185			CI=0	
	RI=0		RI=0.52			RI=0	
			CR=0.0036<0.1				

Figure 3 shows the weights of the factors in the criterion layer and the target layer. The factors in the indicator layer weight are 0.833, followed by the number of motorway lanes with a weight of 0.167. The factors in the indicator layer weigh 0.582 with the dynamic traffic impact factors in the guideline layer. The reason for this is that vehicle speed takes precedence over capacity and the number of conflicts. The factors in the indicator layer are 0.75 and 0.25 for berth utilization and berth turnover, respectively, about the factors in the criterion layer for berth utilization.

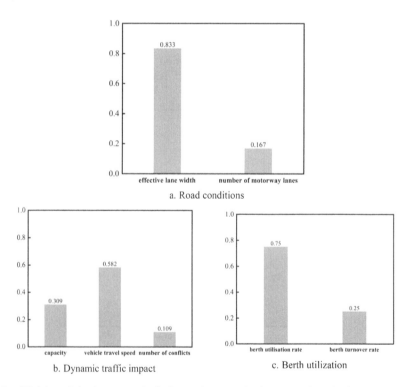

Figure 3. Weights of the factors at the indicator layer to the factors at the criterion layer.

4.4 Integrated impact function for dynamic management of parking spaces

All elements were ranked, and their relative weights were calculated, based on the criterion and weights of indicator layers. Equation (3) serves as the formula for determining the total weights and displays the results of that calculation in Table 8. By using Equation (2), the consistency test of the total ranking of the hierarchy was conducted and the calculated result CR=0.0036<0.1 satisfied the consistency test.

Table 8. Weights of all indicator factors for the target layer.

Criterion layer	Weight	Indicator layer	Weight
S_1	0.188	T_1	0.157
		T_2	0.031
S_2	0.731	T_3	0.226
		T_4	0.425
		T_5	0.08
S_3	0.081	T_6	0.061
		T_7	0.02

Figure 4 shows the weights of the factors for the target layer. Of the seven influencing factors, the vehicle speed is the most influenced with a weight of 0.425; effective lane width and capacity are not significantly different with weights of 0.157 and 0.226 respectively; the number of conflicts in the dynamic traffic impact weights 0.08; the number of motorways and parking berth utilization rate weights 0.031 and 0.061 respectively; and the berth turn-over rate has the smallest impact with a weight of 0.02.

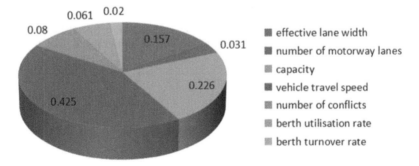

Figure 4. Weights of the factors for the target layer.

Various factors must be evaluated when managing on-street parking dynamically, and the weights assigned to them vary accordingly. After determining the weights of each evaluation factor, seven factors, such as vehicle speed, capacity, and parking utilization rate, are used as independent variables to evaluate the dynamic management of on-street parking, and a comprehensive impact function is established for the dynamic management of on-street parking:

$$Q = \sum_{i=1}^{n} a_i T_i \qquad (4)$$

where:
a is the weight of influence accounted for by each indicator factor.
T is the value of the change corresponding to each indicator factor.
n is the number of indicator factors, of which there are 7.

In the subsequent study, the change values T corresponding to the seven indicator factors can be obtained by using road cameras and field surveys. The weight values calculated in this study and the change values T of the indicators obtained from the surveys at different periods are substituted into Equation (4) to find the Q values within different periods. Equation (4) aims to quantify the link between on-street parking and the traffic impact of road sections, and the quantified value Q can be used as an effective basis for developing a dynamic management scheme for on-street parking use.

5 CONCLUSION

This article quantifies the evaluation of the dynamic management of on-street parking spaces based on hierarchical analysis, constructs an evaluation system and multi-level structure evaluation model for dynamic management of on-street parking spaces, and verifies the reliability of the model through simulation cases. The study shows that the multi-level model of structure evaluation can evaluate the impact scores of different indicators on the realization of dynamic management of on-street parking spaces, and establishes a comprehensive impact function for dynamic management of on-street parking space usage. In further research, actual road cases will be taken for analysis to provide an effective decision basis for traffic managers in developing dynamic management solutions for the use of on-street parking.

ACKNOWLEDGMENT

This research was supported by the 2022 Graduate Innovation Special Fund Project of Nanchang Hangkong University (YC2022-s755). The authors would like to thank the anonymous reviewers for their constructive comments.

REFERENCES

Feng W. *"Study on the Impact of On-street Parking on Traffic Flow on Road Sections"*, Beijing Jiaotong University, (2008).

Gao L. P., Sun Q. X., Liu M. J., Liang X., and Mao B. H. "On-street Parking Based Delay Model and Simulation Study of Mixed Machine-off-street Traffic", *Journal of System Simulation*, 804–808 (2010).

Granger A., Hybel J., Madsen E. and Mulalic I. "A Model for Estimation of the Demand for On-Street Parking", *Economics of Transportation*, (2021).

Guo H. W. *"Modelling and Characterization of the Impact of On-street Urban Parking"*, Beijing Jiaotong University, (2011).

Guo H. W., Gao Z. Y., and Yang X. B. "Modeling Travel Time Under Influence of On-street Parking", *Journal of Transportation Engineering-ASCE*, 229–235 (2012).

He Y. Q. and Li J. "On-street Parking Based Roadway Capacity Study", *Journal of Civil Engineering and Management*, 44–47 (2012).

Jiang B. and Ge X. "A Study of the Impact of On-street Parking on Road Delays and Capacity", *Journal of Transportation Engineering*, 20–23 (2008).

Jie D. L. "Quantitative Analysis of the Impact of On-street Parking on Traffic Flow Delays", *China Engineering Consultants*, 49–52 (2016).

Krzysztof M. "A Computer Simulation of Traffic Flow with On-street Parking and Drivers' Behavior Based on Cellular Automata and a Multi-Agent System", *Journal of Computational Science*, 32–42 (2018).

Li Z. Q. *"Analysis of the Impact of On-street Parking on Road Traffic Flow"*, Chongqing Jiaotong University, (2015).

Li Y., Gong X., Wang Y. X., and Zu Y. "VISSIM-based Simulation to Set Traffic Volume Conditions for On-street Parking Spaces", *Urban Transport of China*, 55–59 (2016).

Lin T., Lv G. L., Tian F., and Lu Y. *"Evaluation of On-Street Parking Charging Policies in Pilot Areas in Shenzhen"*, *Urban Transport of China*, 30–39 (2016).

Liu H. and Yan K. F. *"Analysis of the Form and Delays of On-street Parking Provision"*, Traffic & Transportation, 132–134 (2009).

Manville M. and Pinski M. "The Causes and Consequences of Curb Parking Management", *Transportation Research Part A: Policy and Practice*, 295–307 (2021).

Mei Z. Y., Chen J. and Wang W. "Set up a Delay Model for On-street Parking on Traffic Flow Formation on a Road Section", *Journal of Harbin Institute of Technology*, 164–168 (2009).

Pei Y. L. and Yang Y. F. "Stalling Delay Model Under the Influence of Two-way, Two-vehicle in-Road Parking", *Urban Transport of China*, 119–126 (2020).

Shao C. Q. and Yang Q. Q. "Analysis of the Impact of On-street Parking Behavior on the Capacity of Essential Road Sections", *Journal of Chongqing Jiaotong University (Natural Science)*, 80–84 (2018).

Shi W. L. *"Study on the Impact of On-street Parking on the Operational Characteristics of Traffic Flow on Secondary Roads"*, Fujian Agriculture and Forestry University, (2020).

Shoup D. "Pricing Curb Parking", *Transportation Research Part A: Policy and Practice*, 399–412 (2021).

Wei L. X. and Liang Y. J. "A Meta-automaton Traffic Flow Energy Model Considering On-street Parking", *Bulletin of Science and Technology*, 216–220 (2017).

Wei J. R. *"Simulation Modeling Study of the Impact of On-street Parking Behavior on Road Traffic Flow"* Beijing Jiaotong University, (2018).

Yang W. L. *"Study of the Impact of On-street Parking on Dynamic Traffic"*, Jilin University, (2007).

Ye X. F., Yan X. C., Chen J., Wang T., and Yang Z. "Impact of Curbside Parking on Bicycle Lane Capacity in Nanjing", *Transportation Research Record Journal of the Transportation Research Board*, 1–10 (2018).

Zeng X. M., Li C. H., Luo Q. Z. and Zhang Y. H. "Analysis of the Impact of on-street Parking on Road Capacity", *Highway Engineering*, 58–67 (2003).

Zhan Y. X. *"Study on the Impact of On-street Parking on the Capacity of Urban Road Networks"*, Northern Polytechnic University, (2020).

Zheng H. B. *"Analysis of Parking Behavior Based on Fuzzy Hierarchy Analysis"*, Changsha University of Technology, (2017).

Zhou X. B. "Analysis of the Technical Model of On-street Parking Charging Based on Hierarchical Analysis", *Engineering and Technological Research*, 77–78 (2017).

Research and application of new ecological slope protection technology for highways in retention areas

Danxuan Xue*, Xiaodong Zhu & Xingyu Zhang
China Municipal Engineering North China Design and Research Institute Ltd., Tianjin, China

ABSTRACT: Most of the existing slope protection methods of subgrade slopes in retention areas were mortar rubble masonry pavement, which had a poor ecological effect. Under the action of long-term wind and sun, local damage was easy to occur and even led to the overall instability of the slope. Based on the design principles of environmental protection, economy, landscape, and safety and durability, a new ecological slope protection technology was developed. The new technology was resistant to erosion, soak, and salinity, which not only ensured the safety of the roadbed slope but also focused on ecological green, achieving the dual objectives of safety and ecology. The technology was applied to the Jinshi Expressway in Tianjin. By establishing the numerical model of the slope with or without protection, the variation rule of the stability of the subgrade slope and the distribution of the plastic zone inside the subgrade during the rise and fall of water level on both sides were obtained. The stability coefficient of the roadbed with protection was higher than that of the unprotected state, which showed that the new ecological slope protection not only could improve the overall stability of the roadbed but also reduce the range of the plastic zone and the displacement at the foot of the slope, therefore the protective effect was good.

1 INTRODUCTION

As the national and regional organization of highway networks becomes clearer, more and more riverine highways are being built in floodplains, and riverine highway projects are taking shape (Gao 2019). The slopes of highway foundations in floodplains are subject to river erosion all year round, and if slope protection is not in place, the safe use of the highway will be seriously affected when the strength of the soil base is low. At present, the slope protection of highways in floodplains mostly adopts the traditional slurry stone slope protection (Wang 2004), which leads to the destruction of the plant growth environment and makes it difficult to restore the original ecological environment. In addition, after the completion of the highway project, the exposed rock, concrete, and other masonry materials will seriously affect the visual effect of the overall environment.

Ecological environment and soil and water conservation are gradually being paid attention to, and some ecological restoration measures are beginning to be preferred to complete the process of highway slope management. There are many common ecological protection methods, such as common grass planting (Wu 2011), the geotechnical bag (He 2012), concrete chain block planting (Wang 2013), plant fiber blanket (Li *et al.* 2016), spray seeding (Mickovski & Beek 2006), planted concrete (Wang 2019), etc. However, the slope protection in the retention area needs to have the characteristics of scouring resistance, soak resistance, and salinity resistance, so it is difficult for a single ecological protection measure to meet the

---

*Corresponding Author: 541564705@qq.com

demand. To ensure the safety of the slope of the roadbed and focus on ecological green, it is urgent to develop a new ecological slope protection technology that is resistant to scouring, soaking, and salinity, which can achieve the dual objectives of safety and ecology.

2 NEW ECOLOGICAL PROTECTION TECHNOLOGIES

Based on the design principles of environmental friendliness, economy, landscape, safety, and durability, a new type of protection technology is developed, which is resistant to scouring, soaking, and salinity and is more ecological in comparison with common ecological protection measures.

2.1 Concept design

The ecological protection scheme is designed as follows, provided that the slope of the roadbed itself can be kept stable.

A layer of three-way composite geomaterials is first laid on the slope, anchored at the top and bottom. The ecological bags are then placed on top by using a staggered seam. Next, to prevent the ecological bags from falling off, a three-way geogrid is laid on the outer partition and tied to the bottom layer of geomaterials to make the protection system an integral whole and to ensure that it does not collapse under the effect of flooding and scouring. At the same time, to enhance stability, slurry sheet rock footings are installed for reinforcement. After the slope is completed, the outside of the three-way geogrid is greened by spray seeding with a thickness of not less than 10 cm of guest soil, as shown in Figure 1.

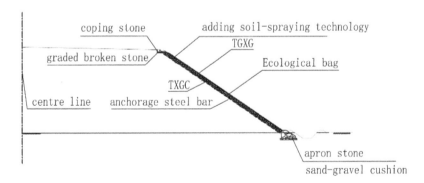

Figure 1. Design of the new ecological slope protection scheme.

2.2 Material requirements

2.2.1 Geotechnical bags
a) Durability
The geotechnical bags are made of polymeric, an environmentally friendly material. Longitudinal strength retention remains at 50% and lateral strength retention remains at 50% after 2000 h of irradiation. The geotechnical bag has a service life of up to 50 years in this design solution.
b) Hydrolysis resistance
The ecological bag has no obvious discoloration, is non-toxic and non-degradation, and has no corrosion. The plant is easy to grow, and the bag body is permeable to water but impervious to the soil.

c) Scouring resistance
At a constant flow velocity of 0.5-1.5 m/sec, the overall shape of the ecological bag is well maintained without deformation and collapse.
d) Specifications
Length is not less than 80 cm longitudinally and 40 cm horizontally. Length is not less than 60 cm; width is not less than 30 cm; height is not less than 15 cm; weight is not less than 40 kg.

2.2.2 *Composite geomaterials*
The following table shows the material requirements for composite geomaterials.

Table 1. List of composite geomaterials indicators.

				TX150-GC	
Serial number	Indicators	Test Method	Unit	Nominal value	Allowable deviation
1	Rib section shape			Rectangular	
2	Hexagonal median distance	TR 041 B.4	mm	≥ 76	
3	Radial stiffness at 2% strain	TR 041 B.1	kN/m	≥ 185	
4	Static top breaking strength (CBR)	EN ISO 12236	kN	≥ 1.5	
5	Dynamic perforation (drop cone test)	EN ISO 13433	mm	≥ 31	
6	Effective pore size	EN ISO 12956	μm	110	± 40
7	Vertical permeability coefficient (flow rate index)	EN ISO 11058	$(VI_{H50})\ ms^{-1}$	≥ 0.11	
8	Product grammage	TR 041 B.3	kg/m^2	≥ 0.36	

2.2.3 *Three-way geogrids*
a) Technical specifications
A monolithic tensile grating made by direct punching and stretching of polypropylene sheets with equilateral triangular mesh holes.
b) Geometric requirements
Hexagonal center spacing ≥ 76 mm; rib cross-sectional shape is rectangular.
c) Mechanical indicators
Quality control tensile modulus is greater than or equal to 295 KN/m/2%.
The radial tensile modulus ratio is greater than or equal to 0.65.
d) Light aging
The minimum carbon black content is not less than 2%.

2.2.4 *Tie down straps*
Ultraviolet resistant; anti-aging; anti-abrasion; high strength polymer composite.

2.2.5 *Filled soil for ecological medium bags*
Native soil suitable for plant growth is chosen to fill and add organic fertilizer, medium, coarse sand, and other admixtures.
General slope mix ratio: (55-65%) native soil + (20-30%) medium coarse sand + (10-15%) mushroom fertilizer or other organic fertilizer. The fullness is filled to 80%.

2.2.6 *Main guest soil materials*
Greening material: organic content >80%; The contents of N, P, and K are not less than 5%; PH value is 4.5-6.0.

Special greening agents: mainly composed of water retention agents (100 times more), polymer coagulants, plant growth agents, etc.

Long-lasting compound greening fertilizer: fertility effectiveness generally is up to 2-3 years.

Local soil material: local fertile or cooked soil is used as much as possible.

2.2.7 *Ratio of tree material*

The seeds for spraying plants should be selected from native salinity-tolerant plants, in the form of herbaceous seeding or grass-irrigation mixes, with species and ratios referring to the following table.

Table 2. Material ratios for slope guest soil spraying plants (grass-irrigation mix).

Name of material	Purple locust	Torch tree	Bearded seeds	Dogtooth root	Tall fescue	Perennial ryegrass
Quantity g/m^2	6	4	10	2	5	3

2.3 *C. Mechanistic analysis*

The new eco-slope meets the requirements of scour resistance, soak resistance, salinity resistance, and ecology in four aspects: scour resistance, the slope protection system consisting of geomaterials, geobags, and three-way grids, which can avoid direct scouring of the slope surface by water, and prevent local scouring and erosion. In terms of salt and alkali resistance, firstly, the selected geotechnical materials have undergone acid and alkali tests. And secondly, we adopt the method of mixed sowing of salt and alkali-resistant grass and irrigation to improve the survival rate of plants. In terms of ecology, after calculation, the cost of the new slope protection is the same as that of slurry slate, and its ecology is much higher than that of slurry slate, which can beautify the environment and have a good ecological effect.

3 ENGINEERING APPLICATIONS

The whole project is located in Jinghai District, Tianjin, starting from the interchange of Hu Xinzhuang, and ending at the border between Tianjin and Hebei Province, with a total length of 12.508 km, a two-way 6-lane motorway, a design speed of 120 km/h, and a roadbed width of 34.5 m. The whole project is located in the Jiakouwa detention basin, with a flood prevention standard of 1 in 100 years and a flood level of 5.94 m.

A. On-site construction

To improve the effect of landscape greening on the slopes of the Jinshi Expressway, the subject group selected the position of the bridgehead (K6+415.370~K6+515.370), the position of the bridge tail (K7+514.230~K7+614.230), and the position of the bridgehead (K8+928.570~K9+028.570) of the Wangkou Drainage Canal on both sides of the Jinshi Expressway for a total of 600 m as the test section for engineering application, and the construction process is as follows:

B. Stability analysis

For the characteristics of the highway project in the stagnant area, a typical road section of the Jinshi Expressway is selected for modeling. As shown in the figure below, the mainline two-way six-lane highway has a roadbed width of 34.5 m, a roadbed slope of 1:1.75, and a height of 6 m roadbed model. The traffic situation is also considered, and the surface of the

Figure 2. Photographs of the construction process of the Jinshi Expressway.

roadbed is evenly distributed 30 kPa of the traffic load. Flood water can seep from the ground and slope into the interior of the roadbed model. The model with or without protection is established, as shown in Figures 3 and 4. The model calculation range from the foot of the slope to the boundary of the distance is not less than 1.5 times the slope height; the top of the slope to the boundary of the distance is not less than 2.5 times the slope height; the total height of the upper and lower boundary is not less than 2 times of the slope height. Figure 4 sets up concrete protection at the foot of the slope on both sides, and also uses truss units to simulate the protection of the slope surface by grating on both sides of the slope.

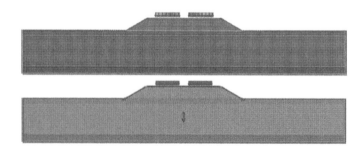

Figure 3. Stability analysis model for unprotected and protected subgrade slope.

As shown in Figure 5, in the course of flood level rise-hold-fall, the change of stability coefficient is manifested in 3 stages, namely rapid increase-slow decrease-rapid decrease. The reasons for the above 3 stages can be explained as follows: in the first stage, the water level rises so that the water pressure has an inward thrust on the slope of the roadbed, thus increasing the anti-slip capacity of the slope, which is manifested as a rapid increase in the stability coefficient; in the water level maintenance stage, due to the continuous infiltration of water into the soil within the slope, the internal soil mechanical parameters decrease, while the weight of the soil increases, resulting in a slow decrease in the stability coefficient; in the water level reduction stage, the water level decreases at a faster rate than the infiltration rate of the water body. As the water level decreases faster than the infiltration rate of the water body, the thrust of the external water body acting on the slope of the roadbed disappears rapidly, the weight of the roadbed soil body is larger and the nature of the soil body further deteriorates, causing the stability coefficient to decrease rapidly after the water level decreases. From the plastic zone distribution map (Figures 4 and 5), the plastic zone is located below the bottom of the roadbed, deep into the foundation. The foot of the slope deformation is the largest, and no berm slope foot plastic zone range is larger.

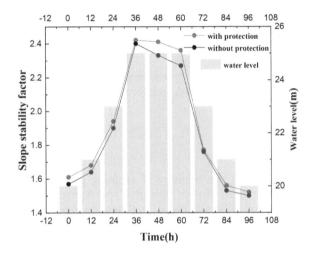

Figure 4. The trend of the coefficient of stability with and without protection.

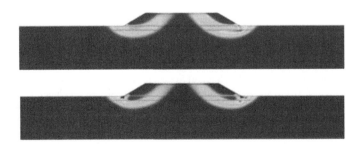

Figure 5. Distribution of the plastic zone of unprotected and protected subgrade slope.

By establishing numerical models for slopes with and without protection respectively, the stability change pattern of the slope of the roadbed during the rise and fall of the water level on both sides is obtained, and the stability coefficient with protection is higher than that without protection.

4 CONCLUSION

This project takes the slope of the roadbed in the Jinshi Expressway (one of the third batch of typical demonstration projects for green highway construction by the Ministry of Transport) in Tianjin as the research object, investigates and analyses the slope section size, hydrological conditions, and nature of fill in the stagnant flood zone. It also collects the basic parameters required for the numerical simulation of the slope of the roadbed in the stagnant flood zone, develops a new protection technology, and establishes a slope stability model for the slope of the roadbed during the change of flood water level to explore the dynamic change trend of slope stability. The main conclusions are as follows:

(1) Based on the design principles of environmental friendliness, economy, landscape, safety, and durability, the new protection technology developed for roadbed slopes in stagnant floodplains is resistant to scouring, immersion, salinity, and ecology.
(2) Through the finite element simulation of unprotected slope in the process of rise and fall of water level on both sides of the roadbed slope respectively, the new ecological slope has a certain protection effect by integrating the pore water pressure distribution field inside the roadbed, the dynamic characteristics of stability coefficient, and the distribution of plastic zone inside the roadbed.
(3) The next step is to study the green environmental protection technology of unstable slopes, which has both slope reinforcement and ecological slope protection effects.

ACKNOWLEDGMENT

This project was supported by the Ministry of Housing and Urban-rural Development Science and Technology Program Project "Research and Application of Key Technologies for Ecological Protection of Subgrade Slope in Retention Area of Green Expressway " (2020-K-030).

REFERENCES

Gao P. *Research on Settlement Deformation and Stability of Road Embankment in the Stagnant Floodplain.* D. Chang'an University (2019).

He Y.B. Construction and Quality Control of Geotechnical Mold Bags in Highway Slope Protection. *J. China Construction*, 8:198–199 (2012).

Li H.J, Kong Y.P, and Zhang Y. A Review of the Environmental Protection Benefits of Plant Fiber Blankets. *J. China Soil and Water Conservation Science*, 14:146–154 (2016).

Mickovski S. B. and Beek L.P.H. A Decision Support System for the Evaluation of Eco-engineering Strategies for Slope Protection. *J. Geotechnical and Geological Engineering*, 243: 28–29 (2006).

Wu Q. Research on Reinforcement Mechanisms of Expressway Side Slope Protection with Greensward. *J. Journal of Central South University of Forestry&Technology*, 31:106–109 (2011).

Wang S. Experiments on the Effect of Different Ratios of Deep and Shallow-rooted Plants on the Impact and Shear Resistance of Slopes. *J. Road traffic science and technology*, 30:39–44 (2013).

Wang K. *Research on Ecological Protection Technology of Roadbed Slope*. D. Southwest Jiaotong University (2019).

Wang Z.D. Roadbed Design in Floodplain. *J. Chinese and Foreign Highways*, 05:32–34 (2004).

Water Conservancy and Civil Construction – Oke & Ahmad (Eds)
© 2024 The Author(s), ISBN: 978-1-032-58618-2

Analysis of parking choice behavior in commercial areas based on hybrid logit model

Shuting Deng, Yang Ming*, Chunmei Hu & Qianrong Tu
School of Civil Engineering and Architecture, Nanchang Hangkong University, Nanchang, Jiangxi, China

ABSTRACT: The commercial areas of city centers are characterized by dense buildings and high traffic flow. The law of parking selection behavior in commercial areas is closely related to the psychological factors of drivers. A structural equation model is established to explore the behavior pattern of the parking choice in the urban central business district. Exogenous latent variables and endogenous latent variables are applied to the model. They are personal basic information, actual travel information, travel purpose, time perception, and attitude perception respectively. Taking the Causeway Bay commercial area in Nanchang City as an example, the elements that significantly affect drivers' choice behavior are investigated to obtain the significant degree of influence of various variables on parking choice behavior according to the survey data. Finally, it is found that the SEM fit indices are close to the exponential range and the model fit is good. Driver's age and driving age significantly affect the parking choice behavior, with loading coefficients of 0.63 and 0.93 respectively. A strong positive influence exists in the potential variables time perception and attitude perception, and their load factors are both more than 0.7, indicating that drivers place a high value on time management and parking security in the parking process.

1 INTRODUCTION

City centers typically have problems with "parking difficulty", including the imbalance between the parking supply and demand, the structure of parking facilities, and parking fees. Adding more parking spaces, limiting or altering parking demand, and increasing the rates of parking facility utilization can resolve this problem [1]. The Already established central cities can alleviate parking problems by increasing the utilization of parking facilities. A lack of parking lots, excessive wait times for a car, dissatisfaction with parking management in the central business district, and excessive parking fees are the main obstacles for drivers. The research on the influence of latent variables in the parking process can enhance our understanding of parking choice behavior and adjust the parking strategy, making parking more enjoyable.

Latent variables are variables that cannot be measured directly in fields like economics, sociology, and psychology [2]. Parking-related studies can utilize latent variables by including them in structural equation models (SEM) [3,4], Logit models [5,6], MIMIC-RF models [7], and other models for analyzing drivers' behavior when switching from cars to buses, subways [8], bicycles [9], and other transportation modes. People's intrinsic psychological factors need to be taken into consideration when analyzing latent variable factors. Psychological factors can be well explained by the structural equation model. Parking choice behavior and parking demand are influenced by parking pricing policies [10], individual

---

*Corresponding Author: mingyang@nchu.edu.cn

DOI: 10.1201/9781003450832-51

socioeconomic attributes [11], trip purpose, and parking space availability [12]. Parking choice behavior in commercial areas has been examined in terms of parking allocation, sharing and prediction [13,14], and parking reservation [15]. Based on a polynomial logit model, Wei Liang et al. [16] analyzed the preferences of social attributes, driving attributes, and consumption attributes. Many studies on parking choice behavior were based on the non-set meter model [17-21], prospect theory [22], and other approaches. Using psychological perception factors, Qin Huanmei et al. [23] studied the effect of travel purpose on travelers' parking choice behavior. According to Sana Ben Hassine et al. [24], Tunisian drivers value parking safety, parking cost, and parking duration when making parking choices. Studies on parking choice behavior include shared parking, street parking, tourist attraction parking, and campus parking. Study variables include walking distance, travel time, parking cost, and latent variables. Human psychological factors need to be considered when designing latent variables. Parking choice behavior can be analyzed and generalized by using appropriate latent variables.

A structural equation model is developed by analyzing the factors that influence parking choice behavior by using personal basic information, actual travel information, travel purpose, time perception, and attitude perception as latent variables. The target area of this study is the city's central business district. An example of analyzing parking choice behavior in a commercial area is using the survey data from Causeway Bay commercial area in Nanchang City.

2 MODEL BUILDING

2.1 *The principle of the hybrid logit model*

The hybrid Logit model is constructed on the basis of the random utility theory of the discrete choice model (non-set counting model). A hybrid logit model combines SEM with a discrete selection model. A hybrid logit model is formed by integrating latent, explicit, and measured variables into the discrete choice model, as shown in Figure 1.

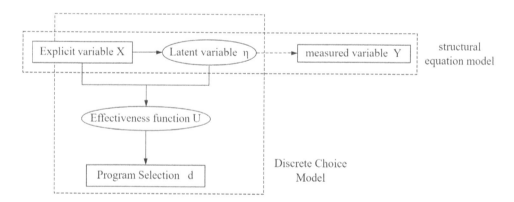

Figure 1. Schematic diagram of the hybrid logit model.

Two parts make up SEM: the structural model and the measurement model. A measurement model reflects the measured variables of a latent variable, while a structural model reflects their causal relationships. The SEM establishes the connection between the measured variables and the latent variables they measure, and uses data testing to verify the hypothesis. SEM is an effective method for describing the relationship between latent and measured variables. SEM transforms latent variables into directly observable measured variables. SEM is mathematically defined as follows:

$$\eta = B\eta + \Gamma\xi + \zeta \qquad (1)$$

$$y = \Lambda_y\eta + \varepsilon \qquad (2)$$

$$x = \Lambda_x\xi + \delta \qquad (3)$$

Where η is "the endogenous latent variable"; ξ is "the exogenous latent variable"; ζ is a random term vector. Equation (1) is the structural model, representing the linear relationship between latent variables. y and x are both measurable variables, which are the measured variables of exogenous and endogenous latent variables respectively. ε and δ are random vectors, so there's no correlation between them. Equations (2) and (3) are measurement equations.

In the hybrid logit model, there are

$$U_{in} = \sum a_{il}x_{iln} + \sum b_{ik}\eta_{ikn} + \varepsilon_{in} \qquad (4)$$

$$P_{in} = \frac{\exp(\sum a_{il}x_{i\,ln} + \sum b_{ik}\eta_{ikn})}{\sum_{j \in A_n} \exp(\sum a_{jl}x_{j\,ln} + \sum b_{jk}\eta_{jkn})} \qquad (5)$$

Where $0 \leq P_{in} \leq 1$; U_{in} is the utility function of scheme i for the traveler n; a and b are the undetermined coefficient; x_{iln} is the 1 th variable eigenvalue of traveler n's choice of the scheme i; η_{ikn} is the k th variable eigenvalue of scheme i for the traveler n; ε_{in} is the random term of the utility function.

2.2 Model assumptions

Personal basic information, actual travel information, travel purpose, time perception, and attitude perception are considered the influencing factors of parking choice behavior. SEM is developed with personal basic information, actual travel information, and travel purpose as exogenous latent variables, and time perception and attitude perception as endogenous latent variables. As shown in Tables 1 and 2, each latent variable is reflected by several measured variables.

Table 1. Exogenous latent variable hypothesis.

Exogenous latent variable	Symbol of variable	Exogenous measured variable
Personal basic information	X1	Gender
	X2	Age
	X3	Age of experience in motor vehicle operation
	X4	Level of education
Actual travel information	X5	Duration of this stop
	X6	Parking fee payment method at this time
	X7	Whether this experience will affect the next parking choice.
Purpose of travel	X8	Work
	X9	Education
	X10	Shopping
	X11	Entertainment
	X12	Job demand

Table 2. Endogenous latent variable hypothesis.

Endogenous latent variable	Symbol of variable	Endogenous measured variable
Perception of time	Y1	I think the search for a berth is a waste of time.
	Y2	I tend to choose berths with short walking times.
	Y3	I tend to choose parking lots with high vacancy rates.
Perception of attitude	Y4	I think the quality of parking lot service is very important.
	Y5	I think it's important for an easy park.
	Y6	I think it's important to have monitored parking locations.
	Y7	I think it's important to have someone patrol the parking lot.

3 DATA SURVEY AND INSPECTION

3.1 *Variable selection*

The Causeway Bay business district in Nanchang City is chosen as the study area to further examine the behavior pattern of parking choice. A total of 199 questionnaires are collected by using behavioral and willingness surveys, with 196 valid questionnaires and a 98% valid recovery rate. Table 3 shows the sample variables in the questionnaires. Most drivers depend on their cell phones to find out where to park at their destinations according to the survey results. An acceptable maximum parking fee of 3 RMB/hour was accepted by 43% of the parkers. An acceptable walking time after getting out of the car was 5 minutes for 54% of the parkers.

3.2 *Reliability and validity analysis*

Reliability and validity refer to the reliability and accuracy of the questionnaire results, respectively. Only the latent variables are used for reliability and validity analyses. Depending on the degree of preference, latent variables are divided into five levels. Table 4 below shows the results of the reliability and validity analyses by using SPSS. Based on the reliability analysis, all latent variables have coefficients greater than 0.7, indicating an acceptable level of reliability. As a rule of thumb, the higher the KMO value is, the more common and suitable the factors among the variables are for the factor analysis. As shown in Table 4, all latent variables have KMO greater than 0.5, and all factor loadings are greater than 0.8, suggesting that the questionnaire results are accurate.

4 MODEL EMPIRICAL ANALYSIS

4.1 *Structural equation model*

AMOS is used to perform a maximum likelihood estimation based on the basic assumptions of the model. There is a positive correlation between the loading factors of the measured variables Y1 to Y7 in Figure 2, indicating that the latent variables can be measured by using the measurement problem. As shown in the AMOS output, the chi-square value is 2.482,

Table 3. Description of questionnaire variables.

Name of variable	Symbol of variable	Value	Name of variable	Symbol of variable	Value
Gender	X1	1: Male 2: Female	I think the search for a berth is a waste of time.	Y1	It is graded by degree of preference.
Age	X2	1: 18-25 years old 2: 26-35 years old 3: 36-45 years old 4: 46-55 years old 5: Over 55 years old	I tend to choose berths with short walking times.	Y2	It is graded by degree of preference.
Age of experience in motor vehicle operation	X3	1: Less than 1 year 2: 1 year-6 years 3: More than 6 years	I tend to choose parking lots with high vacancy rates.	Y3	It is graded by degree of preference.
Level of education	X4	1: High school and below	I think the quality of parking lot service is very important.	Y4	It is graded by degree of preference.
		2: Junior College	I think it's important to have an easy park.	Y5	It is graded by degree of preference.
		3: Undergraduate Degree	I think it's important to have monitored parking locations.	Y6	It is graded by degree of preference.
		4: Master's degree or above	I think it's important to have someone patrol the parking lot.	Y7	It is graded by degree of preference.
Duration of this stop	X5	1:<1 h 2: 1 h–3 h 3: 3 h–6 h 4: >6 h	Purpose of travel	X8 (work) X9 (education) X10 (Shopping) X1 (entertainment) X12 (job demand)	1: Yes 2: No
Parking fee payment method at this time	X6	1: Payment by working unit 2: Individual payment			
Whether this experience will affect the next parking choice.	X7	1: Yes 2: No			

Table 4. Test results of sample data.

Latent variable	Symbol of variable	Cronbach's α	KMO	Factor load
Perception of time	Y1			0.934
	Y2	0.906	0.738	0.933
	Y3			0.889
Perception of attitude	Y4			0.924
	Y5			0.908
	Y6	0.922	0.821	0.887
	Y7			0.885

which meets the evaluation criterion of less than 3. The value of RMSEA (root mean square and square of asymptotic residuals) is 0.087, which is slightly above the optimal range of values (<0.05). Data analysis may be affected by the fact that the majority of drivers in the sample size chose underground parking lots.

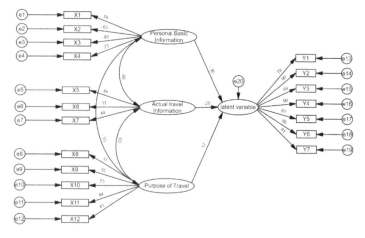

Figure 2. Model standardization fitting path.

4.2 Hybrid logit model

According to the survey results, 3,724 pieces of valid data are selected. SPSS is used to test the fit of the hybrid logit model, and the results are presented in the table below. Table 5 shows that the model significance is 0.000, which is less than 0.05, and the independent variable is well matched with the dependent variable. Table 6 shows the results of the R^2 test of the model. Values are within the range of 0.2 to 0.4, and the model fits well.

Table 5. Model fit test.

Model	Model fitting condition Minus 2-time logarithmic likelihood	Likelihood ratio test		
		χ^2	DF	P
Only intercept	254.868	–	–	–
Ultimate results	196.259	58.610	20	0.000

Table 6. Model test results.

R^2	Cox & Snell	Nagelkerke	Mcfadden
Value	0.258	0.352	0.226

4.3 Analysis of results

The following conclusions can be drawn from the parameter estimation results for the structural equation model.

(1) Gender, age, driving age, and education have standardized loading coefficients of 0.18, -0.63, -0.93, and 0.23, respectively, on personal basic information. Personal basic information is positively influenced by gender and education, and negatively influenced by age and driving age, with age and driving age having a significant impact.

(2) There are standard loading coefficients of 0.49, -0.11, and 0.44 for three exogenous variables, namely the length of this parking, the payment method, and whether this parking influences the next parking choice. Travel information is significantly influenced by the duration of this parking and whether it affects the next parking choice. Parking payment methods have a weak negative impact on actual travel information.

(3) Commuting, school, shopping, recreation, and work need all have standardized loading coefficients of 0.37, 0.35, 0.73, 0.44, and 0.43, respectively. All five exogenous measurements are positively correlated with trip purpose. As a commercial district, the survey area has the characteristics of being influenced most positively by shopping.

(4) There are standardized loading coefficients of 0.79, 0.90, 0.89, 0.90, 0.81, 0.86, and 0.81 for the seven measured variables corresponding to the latent variables. Positive effects are found for all measured variables on the latent variables, especially for Y2, Y3, and Y4.

5 CONCLUSION

Two latent variables, time perception, and attitude perception are introduced to analyze drivers' parking choice behavior towards commercial areas. Scale reliability is tested by using structural equation modeling. As a final step, a mixed logit model is constructed by using the personal basic information, actual travel information, travel purpose, and latent variables as independent variables and parking choice results as dependent variables. The main results of the research are listed below.

(1) A study of the factors that influence parking choice behavior in commercial areas is conducted. The exogenous latent variables are personal basic information, actual travel information, and parking purpose, while the endogenous latent variables are time perception and attitude perception. Using a structural equation model, the relationship between the different influences on parking choice behavior is examined.

(2) The parking lot of Causeway Bay in Nanchang City is considered an example to investigate parking choice behavior in the commercial area. Structure equation models reflect the importance of each influencing factor.

(3) Time perception and attitude perception of motor vehicle drivers show strong positive effects. In other words, drivers care about the control of parking time and the level of security in parking lots.

(4) Using a questionnaire survey process, with most respondents selecting parking lots underground, may lead to bias in the results of data analysis. This can be improved in subsequent studies.

(5) The variables used in the survey do not account for personal factors such as income and occupation. In future studies, multiple variables can be considered together to refine the influencing factors. The next step of this study is planned to propose corresponding parking recommendations for more significant influencing factors.

ACKNOWLEDGMENT

This research was supported by the 2022 Graduate Innovation Special Fund Project of Nanchang Hangkong University (YC2022-135). The authors would like to thank the anonymous reviewers for their constructive comments.

REFERENCES

[1] Han H. and Xie T. "Research on the Transfer Behavior of Car Travel Mode to Public Transport during Commuting Peak Period Based on Latent Variable RF", *Journal of Inner Mongolia University (Natural Science Edition)*, 211–218 (2020).

[2] Han X and Wang D. "Behavior Analysis on Choices of Urban Residents on Shared Parking", *Journal of Hebei University of Water Resources and Electric Engineering*, 57–60 (2018).

[3] Hassine S. B., Mraihi R., and Kooli E. "Drivers' Parking Choice Behavior", 2019 *International Colloquium on Logistics and Supply Chain Management (LOGISTIQUA)*, 2019.

[4] Hu X. H., Yang Y. Z., and Zhang L. P. "Research on the Time Value of Parking Choice Behavior based on Multiple Factors", *Technology & Economy in Areas of Communications*, 26–31 (2019).

[5] Huo Y. Y., Guo C., Zhu Y., and Feng C. X. "Use Intention Model of Shared Autonomous Vehicles and Its Impact Factors", *Journal of Northeastern University (Natural Science)*, 1057–1064 (2021).

[6] Liang W., Hu J. M., Zhang Y., and Wang Z. W. "Multinomial Logit Model-based Parking Choice in a Mall at City", 2016 *Chinese Control and Decision Conference (CCDC)*, 325–328 (2016).

[7] Liu C. C. and Yan L. "The Central Business District Parking Improvement Research", *Logistics Sci-Tech*, 82–85 (2017).

[8] Liu Z. W., Liu J. R., and Deng W. "Travelers' Choice Behavior of Autonomous Vehicles Based on Latent Class", *Journal of Jilin University (Engineering and Technology Edition)"*, 1261–1268 (2021).

[9] Ma X. L., Sun X. D., He Y. L., and Chen Y. X. "Parking Choice Behavior Investigation: A Case Study at Beijing Lama Temple", *Procedia - Social and Behavioral Sciences*, 2635–2642 (2013).

[10] Meng F. Y., Du Y. C., Chong Y. E., and Wong S. C. "Modeling Heterogeneous Parking Choice Behavior on University Campuses", *Transportation Planning and Technology*, 154–169 (2018).

[11] Mo B. C., Kong H., Wang H., Wang X. K., and Li R. M. "Impact of Pricing Policy Change on On-street Parking Demand and User Satisfaction: A Case Study in Nanning", *Transportation Research Part A*, 445–469 (2021).

[12] Ning Y., Yan M., Xu S. X., Li Y. N., and Li L. X. "Shared Parking Acceptance Under Perceived Network Externality and Risks: Theory and Evidence", *Transportation Research Part A*, 150 (2021).

[13] Pan S. L., Zheng G. J., and Chen Q., "The Psychological Decision-Making Process Model of Giving up Driving under Parking Constraints from the Perspective of Sustainable Traffic", *Sustainability*, 1–19 (2020).

[14] Qin H. M., Pang Q. Q., and Wang J. "Analysis of Parking Choice Behavior Based on Psychological Perception", *Technology & Economy in Areas of Communications*, 44–48+73 (2020).

[15] Shu S. N., Bian Y., Rong J., Li S., and Ma J. L., "Staged Approach from Driving to Bicycling based on the Transtheoretical Model", *Journal of Southeast University (Natural Science Edition)*, 373–380 (2018).

[16] Wang J., Fu X. S., Hu X. W., and Wang Z. R. "Choice Behavior Analysis of Residential Shared Parking Incorporating Psychological Latent Variable", *Journal of Dalian Jiaotong University*, 1–5+12 (2018).

[17] Wang W. "*Research on the Methods for Evaluating Policy Effects Based on SEM*", *Beijing Jiaotong University*, 1–72 (2021).

[18] Xiang H. Y., He S. Z. and Xu T. "Modeling of Park & Ride Behaviors of Commuting Corridors Based on Structural Equation Model", *Journal of Chongqing Jiaotong University (Natural Science)*, 90–95 (2018).

[19] Yang X. F., Jin Y., and Fu Q. "Analysis of the Influence of Different Factors on the Selection of Shared Parking", *Logistics Sci-Tech*, 55–58+62 (2018).

[20] Yu J. W., Wan Q., Yang J. Q., Zhou W. Z., and Li Z. B. "Distribution Model of Shared Parking Space in Urban Business District at Night", *Forest Engineering*, 91–95+100 (2018).

[21] Zhang j. and Feng H. "Research on Public Parking Demand Forecast Model in a Commercial Zone", *Modern Electronics Technique*, 123–125+129 (2020).

[22] Zhang Y., Sun S. R., and Zhang W. H. "Parking Space Sharing Willingness Research in a Residential Area Based on Structural Equation Model", *Forest Engineering*, 143–150+158 (2021).

[23] Zheng S. D., Li J., Shao C. F., and Liu B. H. "Holiday Parking Choice Behavior Model Based on the Prospect Theory", *Shandong Science*, 89–97 (2019).

[24] Zong F., Yu P., Tang J. J., and Sun X. "Understanding Parking Decisions with Structural Equation Modeling", *Physica A: Statistical Mechanics and its Applications*, 408–417 (2019).

Analysis of college students' bus travel behavior in the post-epidemic era

Congling Zheng* & Xianghong Li*
School of Energy Science and Engineering, Henan Polytechnic University, Jiaozuo, Henan, China

ABSTRACT: To explore college students' public transportation travel behavior in the post-epidemic era, two factors were incorporated into the traditional theory of planned behavior, namely, risk perception and epidemic prevention strategies, therefore a model of college students' public transportation travel behavior was constructed. The travel intention survey was used to obtain seven dimensions of the travel influence information, and the structural equation modeling was combined to realize the analysis of the influence factors of college students' public transportation travel. The path coefficient quantified the degree of influence of different factors on college students' intention to travel by public transportation, and it was found that travel attitude, perceptual behavior control, epidemic prevention strategies, and subjective norms all had significant positive effects on travel intention, among which the travel attitude had the greatest effect on travel intention, and risk perception had a significant negative effect on travel intention, and epidemic prevention strategies had a significant negative effect on risk perception. Suggestions are made to protect the safe travel of college students by adopting customized buses for schools, ensuring the disinfection and ventilation of vehicles, and publicizing the health status of drivers.

1 INTRODUCTION

After the reform of the higher education system, China's higher education has gained the opportunity to develop rapidly, and the number of university students in school has grown year by year and the group size is huge. Students are an important part of the social composition, and they are equally valued by society as the general citizens in urban travel, so it is important to study the travel behavior of college students for urban transportation development. The urban transportation system under an epidemic can be divided into five phases: the usual phase, the latent phase of the epidemic, the outbreak phase, the post-epidemic phase, and the recovery and enhancement phase (Li *et al.* 2020). Under the strict control of the state, the epidemic was successfully controlled, which allows people to work and live normally and gradually enter the post-epidemic stage. Investigating the college students' views on epidemic prevention strategies for public transportation will help the management of each link to understand the applicability and effectiveness of epidemic prevention strategies in people's minds, and provide data reference and opinion reference for each department to formulate epidemic prevention strategies.

In the contemporary society that promotes low-carbon living, the promotion of public transportation can effectively reduce the number of private car trips, which in turn reduces the emission of harmful gases, such as carbon dioxide, and the country has vigorously promoted the mode of public transportation. This includes bus lanes, special signals, BRT

---

*Corresponding Authors: 212202020089@home.hpu.edu.cn and lixianghong001@126.com

DOI: 10.1201/9781003450832-52

systems, and other powerful measures aimed at improving the competitiveness of buses. The bus system in Jiaozuo City is well developed, with the proportion of green buses and air-conditioned buses in Jiaozuo reaching 75% and 65%, respectively, by 2018 (Zhang 2018). Over 2,690 tons of carbon dioxide emissions were reduced throughout the year, effectively contributing to the improvement of people's travel environment and air quality in many ways. The new coronavirus pneumonia epidemic had a huge impact on public transportation, with 44,236 million public buses and tram passengers nationwide in 2020 decreased by 36.1%. Public transport travel becomes less safe and people's resistance to this situation where they can easily have close contact with others may reduce their choice of using public transport. This can increase urban traffic congestion, which harms the efficiency of public resource allocation and climate.

The theory of planned behaviors is a continuation of the Theory of Rational Behavior proposed by Ajzen and Fishbein (Fishbein & Ajzen 1875). Based on the Theory of Rational Behavior theory, Ajzen combined the multi-attribute attitude theory and TRA to create a theoretical model of social psychology (Ajzen 2012). The Theory of Planned Behavior has been applied in various fields, for example, in the fields of travelers' travel behavior, unsafe transportation behavior, residents' travel mode choice behavior, travel behavior of medium and long-distance travel passengers, and consumers' shopping behavior (Ding 2013; Lu 2014; Yu 2015; Yin et al. 2008; Zhang et al. 2010), but there are relatively few studies with college students as the object. In this paper, from the perspective of travel behavior, based on the theory of planned behavior, we analyze the influencing factors of college students' behavior toward public transportation travel in the post-epidemic era context. Based on this, the proposed public transportation prevention strategies are analyzed, expecting to maintain the positive attitude of college students' public transportation travel while effectively preventing and controlling the epidemic.

2 UNDERLYING THEORY AND MODEL ASSUMPTIONS

2.1 *Fundamental theory*

The Theory of Planned Behavior contains five factors, namely Attitude towards the Behavior (AB), Perceived Behavior Control (PBC), Subjective Norm (SN), Behavior Intention (BI), and its framework diagram, as shown in Figure 1. Attitude towards the Behavior is an individual's positive or negative evaluation of adopting a behavior; the subjective norm is some social pressure that an individual feels when performing a behavior; perceived behavioral control is an individual's determination of the ability to perform the behavior and bear the consequences of the behavior; and behavioral intention indicates a person's tendency to perform the behavior.

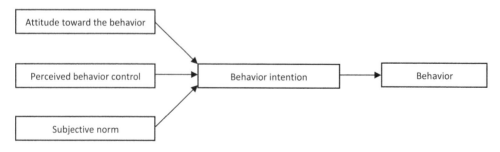

Figure 1. Framework diagram of the theory of planned behavior.

2.2 Model assumptions

By incorporating the two factors of risk perception and epidemic prevention strategy into the traditional behavioral theory, a model of college students' bus travel behavior was constructed, including the following five factors: attitude, which refers to the positive or negative feelings of residents about the possible scenarios of bus travel; perceptual behavioral control, which refers to the influence of past experiences and acquired knowledge on bus travel behavior; subjective norms, which refer to the influence of important external individuals or groups on bus travel behavior; behavioral intention, which refers to college students' willingness to travel by bus; and behavior, which refers to the choice of bus travel during the epidemic. The relationship framework between the influencing factors is shown in Figure 2.

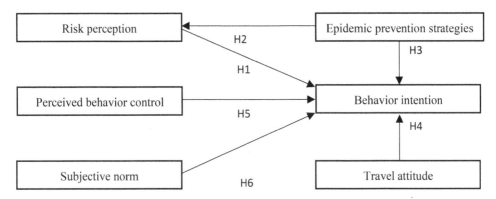

Figure 2. Assumptions for a model of the behavior of college students' public transportation travel.

(1) The relationship among risk perception, travel attitude, and travel intention

Risk perception is a person's subjective assessment of the characteristics and severity of the existence of risk. In the context of an epidemic, a student who has a higher risk perception will be more afraid of the negative effects caused by the epidemic, so the lower his travel intention will be and the more negative his travel attitude will be. In summary, the following hypotheses are proposed.

Hypothesis one (H1): there is an inverse effect of college students' risk perception of the epidemic on travel intention.

(2) The relationship among prevention strategies, risk perception, and travel intention

If college students have a higher awareness of various prevention strategies, their risk perception of risks will decrease, which will directly or indirectly affect their attitudes toward travel behavior. In summary, the following hypotheses were proposed.

Hypothesis two (H2): there is an inverse effect of college students' satisfaction with epidemic prevention strategies on risk perception.

Hypothesis three (H3): there is a positive influence of college students' satisfaction with the prevention strategy on their travel intention.

(3) According to TPB theory, we can know the relationship between the influence of AB, SN, and PBC on BI. The following hypotheses can be formulated.

Hypothesis four (H4): there is a positive influence on college students' attitude toward the tool and their intention to choose it.

Hypothesis V (H5): there is a positive influence from the environment and peer influence on college students' willingness to choose a certain travel tool.

Hypothesis 6 (H6): there is a positive effect of college students' perceived ability to make their own decisions and their willingness to choose a travel tool.

3 ANALYSIS OF COLLEGE STUDENTS' BEHAVIOR REGARDING PUBLIC TRANSPORTATION TRAVEL DURING THE EPIDEMIC

The data obtained from this survey came from a survey on the travel behavior of Henan University of Technology students in the city. 430 questionnaires were distributed and collected, of which the number of valid questionnaires was 414, with an effective value of 96.30%. In the data screening, the questionnaires with malicious all-choice 1 were removed, and the average was used to fill in the gaps of some individual questions in the questionnaire. The following data results were obtained after the statistics.

3.1 *Questionnaire design*

In the scale design section, a Likert scale was used, as shown in Table 1. It is a common type of scale used to measure the attitude of the subject under investigation toward something, and the scale selected on a 5-point scale can achieve good results in terms of questionnaire

Table 1. Questionnaire design content for TPB variables.

Variable name	Observed variables	Variable number
Risk perception	The virus poses a serious threat to life safety.	RC1
	The virus has a serious impact on daily study life.	RC2
	The virus is very psychologically taxing.	RC3
	You fully understand the symptoms of a virus infection.	R 4
Epidemic prevention strategies	Contact distance with fellow passengers affects travel intentions.	S1
	Are you aware that the status of staff vaccination training affects travel intentions?	S2
	The degree of disinfection and ventilation of the riding environment will affect the intention to travel.	S3
	Are you aware that the physical condition of your fellow travelers and the level of protection will affect your travel intentions?	S4
Travel attitude	Direct access to your destination	AT1
	Reasonable fares	AT2
	Ability to get to your destination quickly	AT3
	Short waiting time for trains	AT4
	I feel safe inside (emergency safety facilities, safety of travel process, and disinfection situation).	AT5
	The interior of the car makes me feel comfortable.	AT6
	I am less influenced by weather conditions.	AT7
Subject norm	Policy support and transportation conditions will affect my choice of this transportation.	SN1
	The opinions of my peers have a great influence on my choice of this means of travel.	SN2
	The choice of travel partners will influence whether I choose this tool or not.	SN3
Perceived behavior control	The choice of this method is entirely up to your own will.	PBC1
	Ride with confidence after doing self-protection measures during the epidemic	PB 2
	You may know about the protection and are willing to use it to travel.	PBC3
Behavior intention	I accept the use of its travel in the event of an outbreak.	BI1
	I encourage others to use it for travel in the event of an outbreak.	BI2
	I prefer to choose this mode of transportation when traveling in a group	BI3

reliability. According to the purpose of this research paper, the following scale questions were set.

3.2 Reliability and validity tests

The KMO measure of Bartlett's sphericity test was performed on the collected data. The passing indicator of KMO was 0.5, and the range between 0.5 and 0.6 was indicated very poor: between 0.6 and 0.7 indicated fair; between 0.7 and 0.8 indicated suitable; between 0.8 and 0.9 indicated very suitable; above 0.9 indicated very suitable. The closer it is to 1, the more suitable the factor analysis is. According to Table 2, the KMO values of each variable are greater than 0.6, and although some of the values do not reach 0.7, they are greater than 0.5, which meets the relevant requirements. The total reliability value of 0.935 exceeds 0.90, indicating that the reliability is very excellent and good. After testing the reliability of the latent variables to which they belonged, the results of the alpha values were all greater than 0.6, indicating that they were reliable.

Table 2. Results of the reliability and validity analysis.

Latent variable	α value	Overall α value	Measurement question items	KMO metric	Bartlett's sphericity test		
					Approximate cardinality	Degree of freedom	Significance
Travel attitude	0.883	0.935	AT1, AT2, AT3, AT4, AT5, AT6, AT7	0.719	165.664	28	0.000
Subject norm	0.777		SN1, SN2, SN3	0.675	185.995	3	0.000
Perceived behavior control	0.767		PBC1, PBC2, PBC3	0.708	149.578	15	0.000
Behavior intention	0.686		BI1, BI2, BI3	0.653	144.962	3	0.000
Risk perception	0.837		RC1, RC2, RC3, RC4	0.781	153.169	6	0.000
Epidemic prevention strategies	0.820		S1, S2, S3, S4	0.736	140.854	6	0.000

3.3 Model construction and analysis

A preliminary structural equation model of college students' bus trips was constructed by using AMOS24.0 software, as shown in Figure 3. The potential variables are represented by using oval boxes and the observed variables are represented by using rectangular boxes. The measurement errors of RC1~BI3 are expressed by e1~e26. A judgment is required regarding the evaluation of whether the fit between the designed model plots and the collected data is appropriate. The fit index cardinality of freedom ratio (CMIN/DF), fit index (GFI), gauge fit index (NFI), relative fit index (RFI), comparative fit index (CFI), and asymptotic residual mean square and root square (RMSEA) were used to evaluate the model fit. The results of each test index are shown in Table 3, and it can be learned that the fitness between them is good.

The results of the model parameters calculation are shown in Table 4. When it is significant (p-value less than 0.1), it verifies the hypothesis proposed in our paper, indicating a significant effect between the two. As for whether the significant effect is positive or negative, it can be determined based on the positive or negative estimate value. The absolute values of CR in the relationship are all greater than 1.96, indicating that the significance level is reached. The model is judged to comply, indicating that the explanatory power of the model is good.

The path relationships among the factors show that, in terms of the degree of influence on travel intention, travel attitude, perceived behavioral control, epidemic prevention policy,

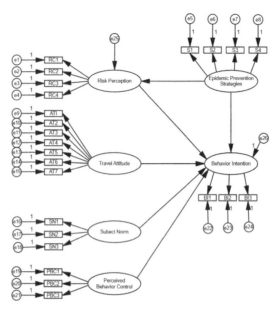

Figure 3. The structural equation model diagram of university student travel.

Table 3. Output of adaptation values.

Indicators	Adaptation indicators	Check output value	Whether to adopt
CMIN/DF	1~3	2.961	Adaptation
GFI	>0.9	0.859	Adaptation
NFI	>0.9	0.903	Adaptation
RFI	>0.9	0.928	Adaptation
CFI	>0.9	0.901	Adaptation
RMSEA	>0.08	0.063	Adaptation

and subjective norms all have significant positive effects, but risk perception has a significant negative effect on them. The epidemic prevention policy also had a significant negative effect on risk perception. This also verifies the six hypotheses in the previous paper.

(1) The coefficient of 0.366 between immunization policy and intention to travel indicates that when college students are more satisfied with the immunization policy of the public transportation mode, they are more confident about the safety of travel. The intention to use it for travel is also greater. This is in line with the previous hypothesis H3 and the four observed variables under epidemic prevention strategy, and their path correlation coefficients with epidemic prevention strategy are 0.986, 0.961, 0.979, and 1.000, respectively, where the highest level of influence is known that the health condition of the people around them is good and they have protected themselves from the outbreak. The next most influential factor is that people care about the distance between them and the passengers in the same vehicle when they travel. The third ranking is whether the disinfection and ventilation measures are in place for the passenger environment. The last ranking is the degree of knowledge about the staff's training in epidemic prevention. In

Table 4. The output of model coefficients.

			Estimate	S.E.	C.R.	P
Risk perception	<—	Epidemic prevention strategies	-.586	.148	-3.959	.001
Behavior intention	<—	Risk perception	-.219	.123	-17.805	.003
Behavior intention	<—	Travel attitude	.387	.101	3.831	***
Behavior intention	<—	Subject norm	.270	.102	2.647	***
Behavior intention	<—	Perceived behavior control	.386	.096	4.021	.003
Behavior intention	<—	Epidemic prevention strategies	.366	.115	3.182	.002
RC4	<—	Risk perception	1.000	–	–	–
RC3	<—	Risk perception	1.003	.192	5.224	***
RC2	<—	Risk perception	1.008	.241	4.182	***
RC1	<—	Risk perception	.977	.259	3.772	***
AT1	<—	Travel attitude	.869	.126	6.897	***
AT2	<—	Travel attitude	1.000	–	–	–
AT3	<—	Travel attitude	.563	.133	4.233	***
AT4	<—	Travel attitude	.583	.134	4.351	***
AT5	<—	Travel attitude	.675	.131	5.152	***
AT6	<—	Travel attitude	.706	.144	4.903	***
AT7	<—	Travel attitude	.762	.139	5.482	***
SN3	<—	Subject norm	1.000	–	–	–
SN2	<—	Subject norm	1.129	.109	10.358	***
SN1	<—	Subject norm	1.346	.129	10.434	***
PBC3	<—	Perceived behavior control	1.000	–	–	–
PBC2	<—	Perceived behavior control	1.036	.128	8.094	***
PBC1	<—	Perceived behavior control	.717	.115	6.235	.002
S1	<—	Epidemic prevention strategies	.986	.241	4.091	***
S2	<—	Epidemic prevention strategies	.961	.172	5.587	***
S3	<—	Epidemic Prevention strategies	.979	.223	4.390	***
S4	<—	Epidemic prevention strategies	1.000	–	–	–
BI1	<—	Behavior intention	.726	.141	5.149	***
BI2	<—	Behavior intention	.627	.096	6.531	***
BI3	<—	Behavior intention	.753	.168	4.482	***

conclusion, the degree of influence of the above four is less different, but all have a great influence on travel intention and can be based on the conclusion of this analysis to make relevant recommendations for the bus company later in the paper.

(2) The path coefficient relationship between travel attitudes and travel intentions is 0.387, which is a positive relationship, in line with hypothesis H4. This part of the question reflects the degree of influence of attitudes toward travel with this mode of transportation on travel intentions. The most influential factor is AT3 with a path coefficient of 0.957, indicating that people value the low fare of the bus more. The path coefficients of both the risk perception factor and its observed variables were high, indicating that people were well aware of the dangers of the epidemic.

(3) The path coefficient between perceived behavioral control and travel intention was 0.386, indicating a positive effect between the two. Among the three observed variables, the path coefficients of RC2 and RC3 are larger, which indicates that people are willing to travel by public transportation when they know enough about epidemic prevention and can protect themselves well. The relationship between the two has a positive effect, which is consistent with hypothesis H6 in the previous paper.

The above analysis shows that when making a trip, there are several influencing factors of travel intention at the same time, under their combined effect, people make travel choices.

Among the more influential factors are attitudes, prevention strategies, and perceptual behavioral control. This can also be shown from the results. During the epidemic period, college students' willingness to use public transportation to travel depends mainly on their attitudes towards this mode. In addition, the intention to travel will enhance if the public transportation companies' epidemic prevention measures are accepted by college students or if they have extensive knowledge and protective equipment for epidemic prevention themselves. Subjective norms represent social pressure felt by individuals, and their perception of risk has relatively less influence on travel intentions.

4 RECOMMENDATIONS FOR SAFE TRAVEL

4.1 Suggestions for schools

After the statistics on travel choices by period, we can learn that people prefer to travel during the daytime on double holidays (35.7%), on double holidays, and Friday evenings (28.5%). This indicates that people are more likely to travel out of school and within the city on double breaks. Since there are no classroom constraints on double days, people prefer to go off-campus to relax and have fun. As for the specific time of the day, when people prefer to travel, the results show that the proportion is 35.7% for noon (11:00-13:00) and 32.4% for the afternoon (16:00-18:00).

With the normalization of the epidemic, schools can set the time slots available for travel to meet students' needs for time away from home based on everyone's preference for travel time. Students enjoy a journey where they can easily come into contact with outside people and there is a significant shortfall in safety. Schools may consider offering customized buses to serve students on their trips. This will reduce the likelihood of catching a virus during the journey.

4.2 Suggestions for college students to travel

You should learn to protect yourself when you travel and be sure to wear a mask carefully and regularly throughout and wash your hands regularly. You can carry disinfectant. When making physical contact with the outside world, for example, when riding shared bicycles outside the school, you should try to disinfect the seats and handlebars that come in contact with them.

When traveling, you can take the same cab with your classmates, or members of the school, to avoid the presence of people with unknown health conditions in the riding environment. In addition, we can't dismiss the bus in its entirety, in that the bus companies can guarantee the implementation of epidemic prevention measures with high quality. For example, strict disinfection and ventilation of the vehicle, as well as health assurance of the passengers and crew members can make us travel with confidence. When taking public transportation, it is important to be careful to spread out and increase the distance between you and others. When traveling, you should also follow the school's requirements strictly and return to campus within the specified period.

4.3 Recommendations for related transportation companies

In the prevention and control of the bus travel system, the most important thing that should be looked at is the health of the driver. The driver's working environment does not have good ventilation in the confined environment of the bus, and passengers from different places are staying in the same environment as the driver. If there is an accident in the driver's health, it will cause a huge impact on virus transmission. Therefore treating drivers should require them to take their temperature several times a day, and wear masks correctly, and for the health of the driver, they should be shown to the passengers. We should focus on training the driver group on epidemic prevention, requiring them to understand the general knowledge related to the epidemic and to have the right way to handle the situation when they encounter red-coded people and so on.

At the level of the riding environment, the bus company should do the disinfection and ventilation the vehicle. The observed variable S3, the effect of vehicle disinfection, and ventilation on travel intention are set in this paper. The coefficient between them is 0.91, which indicates a significant positive relationship between them. After the extermination work, the work record sheet can be displayed to passengers, which not only enhances the psychological satisfaction of travel safety but also allows people to participate in supervision. The seats in the car can be set at a good safety distance, and by dismantling some seats, passengers must keep a safe distance. The frequency can be increased and the departure time can be reduced to avoid the situation of people gathering at the bus stop. At the stop signs, there can be propaganda and display of epidemic prevention, using loudspeaker broadcasting and banners to call people to wear masks and show green codes, which shows people the countermeasures made by the bus for epidemic prevention and publicizes the health status of drivers.

5 CONCLUSION

This paper extends the traditional theory of planned behavior by introducing two factors, risk perception, and epidemic prevention strategies, and combines structural equation modeling to realize the analysis of factors influencing college students' public transportation trips. The data on college students' public transportation trips obtained from the survey were applied to the model and the corresponding path coefficient maps were obtained. The results show that travel attitude, perceptual behavior control, epidemic prevention strategies, and subjective norms all have significant positive effects on travel intention, with travel attitude having the strongest effect on travel intention. Path coefficient quantified the degree of influence of different factors on college students' intention to travel by public transportation and obtained the degree of importance for college students to attach to each epidemic prevention strategy. Based on this, relevant suggestions were made to the related transportation companies for their prevention efforts, including the disclosure of driver health information and a safe distance from each other for the occupants. With the increasing maturity of people's work on epidemic prevention, how to achieve sustainable development of urban transportation and keep the bus attractive for college students to travel is a topic worth thinking about while ensuring safety.

REFERENCES

Ajzen I. *The Theory of Planned Behavior* [M]. New York: Handbook of Theories of Social Psychology. 2012.

Ding Liying. A Study of Fuzhou Residents' Travel Behavior Intentions to Taiwan Based on Theory of Planned Behavior [J]. *Journal of Jilin Normal University (Natural Science Edition)*, 2013, 34 (1): 117–119.

Fishbein M. and Ajzen I. Belief, Attitude, Intention and Behaviour: an introduction to theory and research [J]. *Philosophy & Rhetoric*, 1975, 41 (4): 842–844.

Li Ye, Liu Xinghua, and He Qing. Urban Transportation System Resilience During the COVID-19 Pandemic [J]. *Urban Transport of China*, 2020, 18 (03): 80–87+10.

Lu Xin. *A Study on Travelers' Medium and Long Distance Travel Mode Choice Behavior Based on Planned Behavior* [D]: [Master's degree thesis]. Cheng Du: Southwest Jiaotong University, 2014.

Statistical Bulletin on the Development of the Transport Industry in 2020 [J]. Finance & Accounting for Transport, 2021 (06): 92–97.

Yin Shijiu, Wu Linhai, and Du Lili. A Study of Consumers' Willingness to Shop Online Based on the Theory of Planned Behavior [J]. *Consumer Economics*, 2008, 24 (4): 35–39.

Yu Xiaolong. A Study of Travel Mode Selection Based on the Theory of Planned Behavior [D]: [Master's degree thesis]. Zi Bo: Shandong Polytechnic University, 2015.

Zhang Lei, Ren Gang, Wang Weijie. A Study of Bicycle Unsafe Behavior Model Based on the Theory of Planned Behavior [J]. *China Safety Science Journal*, 2010, 20 (7): 43.

Zhang Ming. "Bus Priority" to Lead the Acceleration of Jiaozuo Bus [J]. *People's Public Transportation*, 2018 (11): 50–52.

The design of revenue adjustment method for highway public-private-partnership (PPP) projects under inflation

Zhenyao Wu*
Guizhou Communications Polytechnic, Guiyang, China

ABSTRACT: Inflation affects the costs and benefits of highway PPP projects. By analyzing the impact of inflation on the project and the problem of insufficient or excessive income of the project caused by regular price adjustment, this paper puts forward the government's income adjustment method, and constructs a decision-making model based on the fuzzy multi-objective decision-making of the price adjustment range and the proportion of government income adjustment, taking into account the interests of social capital, road users, and government departments. Through case analysis, it is found that the model can decide the optimal combination scheme of price adjustment range and government share ratio, which verifies the feasibility of the model. The decided scheme can increase the net present value of the project, reduce the uncertainty of its value, and help to enhance the investment confidence of social capital. At the same time, the government departments can share part of the inflation risk through income adjustment, or avoid the excess income of the project caused by excessive price adjustment, which is conducive to the benign development of the project and the protection of social public interests.

1 INTRODUCTION

Increasing investment in infrastructure construction has become an important way to ease the downward pressure on the economy, and transportation infrastructure is one of its main investment areas. The PPP mode is widely used in infrastructure investment due to its advantages of relieving government financial pressure, stimulating market vitality, and improving operational efficiency (Carbonara & Pellegrino 2018). As far as the highway is concerned, this kind of project has become one of the typical projects built in the PPP mode due to its relatively complete user payment system and operation process. However, its construction cost is high, and the period of investment payback and concession is long (Buyukyoran & Gundes 2018). During the concession period lasting for several decades, inflation has led to price increase year by year, currency devaluation, increased project operating costs, reduced revenues, and even losses (Mirzadeh & Birgisson 2015). How to mitigate and reasonably allocate the inflation risk in PPP projects is one of the problems to be solved.

Ye and Tiong (2003) think that adjusting the toll price of the project according to the inflation situation is an effective method to mitigate the inflation risk, but they do not provide a quantitative calculation method.

Wibowo *et al.* (2012) and others theoretically discussed the impact of price adjustment delay on social capital due to external factors and established a compensation model for price adjustment delay. Mirzadeh and Birgisson (2015) proposed to reduce the inflation risk borne by social capital and cost compensation rather than price adjustment because the most direct impact of inflation is the increase in project cost. Under this method, the risk is completely borne by government departments and social capital. Compared with a price

---

*Corresponding Author: zhenyao2512@163.com

DOI: 10.1201/9781003450832-53

adjustment, the government needs a lot of fiscal expenditure to compensate for the increase in cost. Cai Xiaoyan and Zhou Guoguang (2017) put forward a method of income adjustment based on a price adjustment. When the price adjustment is lower than the inflation rate, the government will compensate for the income, while the social capital will share the excess income. This research concept opens up a new idea for the sharing of inflation risks. The government departments are included in the risk sharing. However, due to the single factors considered in the model, the cycle and changes of price adjustment in the time value of money under inflation are ignored, which is easy to make wrong decisions.

The impact of inflation is widespread in PPP projects, but there are relatively few quantitative studies. Based on the influence of inflation on project parameters, this paper describes the change of traffic volume under inflation through the method of traffic route selection in traffic engineering, puts forward the method of government revenue adjustment based on considering the impact of the price adjustment cycle and inflation on the operating cost, project income, and net present value, and meets the interests of social capital, users, and government departments. A decision model of the range of price adjustment and the proportion of government revenue adjustment is constructed. This study is conducive to quantifying the inflation risk in PPP projects and providing theoretical guidance for the government to formulate sharing policies of inflation risk.

2 IMPACT OF INFLATION ON PROJECT PARAMETERS

2.1 *Project traffic volume under inflation*

Generally speaking, in addition to the toll highway, there are many other free roads to the OD connected by one highway, but these roads are long and the road conditions are relatively poor. In order to be closer to reality, from the perspective of path selection, we shall analyze the path selection of road users with or without inflation, that is, the change in traffic volume.

2.1.1 *Project traffic volume under general conditions*

Without considering the impact of inflation and toll price adjustment, there are multiple roads connecting the two OD regions. Path A is the highway; toll price is P; average travel time is t_a; path B is the free road with the shortest distance and the best road condition except for highway; and average travel time is t_b. If the user does not choose the highway, he/she will give priority to route B instead of other roads, and the traffic path of OD can be simplified in Figure 1. Let the total traffic demand between the two places be Q, in which the traffic volume of route A is Q_a. The traffic volume of route B is Q_b, and there is a relationship $Q = Q_a + Q_b$.

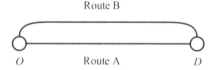

Figure 1. Routes from O to D.

Each road user has its independent time value x (unit: yuan/minute). It is generally considered that its value follows the normal distribution (Chan et al. 2015), and $f(x)$ and $F(x)$ represent the probability distribution function and cumulative probability function of the user's time value respectively. Assuming that the road users are rational, they will choose two routes according to the minimum generalized travel cost (Chen et al. 2018). It x_0 is the time value that makes the generalized travel costs tc_a tc_b of routes A and B equal, then

$$tc_a = tc_b, \text{ i.e, } P + t_a x_0 = t_b x_0 \tag{1}$$

$$x_0 = \frac{P}{t_b - t_a} \quad (t_b > t_a) \tag{2}$$

$$t_a = t_0 \left[1 + 0.15 \left(\frac{Q_a}{c_a} \right)^4 \right] \tag{3}$$

Where t_0 is the travel time of free flow on path A, and c_a is the actual carrying capacity of path A. Similarly, t_b can be calculated by this method.

Where Q_b can be expressed as:

$$Q_b = Q \int_0^{x_0} f(x)dx \tag{4}$$

$$\frac{Q_b}{Q} = F(x_0) - F(0) = F(x_0) = F \left[\frac{P}{t_b(Q_b) - t_a(Q - Q_b)} \right] \tag{5}$$

2.1.2 Project traffic volume considering inflation

The time value x is anchored by currency. When inflation occurs, the currency depreciates, and the time value expressed in the currency should also rise with the inflation rate. If f_j is the inflation rate of the jth year, the unit time value x_j of the year will be:

$$x_j = x_{j-1} \cdot (1 + f_j) \tag{6}$$

$$x_j^0 = \frac{P}{t_b(Q_j^b) - t_a(Q_j - Q_j^b)} \tag{7}$$

$$\frac{Q_j^b}{Q_j} = F(x_j^0) = F \left[\frac{P}{t_b(Q_j^b) - t_a(Q_j - Q_j^b)} \right] \tag{8}$$

2.2 Impact of inflation on other parameters

1) Project operating costs. The most direct impact of inflation is the increase in project cost. The operating cost C_j of the jth year needs to take into account the growth rate and inflation rate of equipment α_C and other operation and maintenance costs f_j.

$$C_j = C_{j-1} \cdot (1 + \alpha_C^j) \cdot (1 + f_j) \tag{9}$$

2) Project revenue. The annual income of the project is determined by the annual traffic volume and toll price, both of which are affected by inflation, so the projected income is also a variable affected by inflation. The projected income R_j for the jth year is expressed as:

$$R_j = Q_j^a \cdot P = (Q_j - Q_j^b) \cdot P = Q_j \cdot (1 - F(x_j^0(f_j))) \cdot P \tag{10}$$

3) The Net present value of the project. Generally, the NPV of the project is calculated by considering the time value of money without inflation. In practice, the existence of inflation not only affects the price level of each period, making the cash flow of the project change but also changes the demarcation of the time value of money because of the devaluation of the currency. The relationship between the actual cash flow RCF without considering inflation and the nominal cash flow NCF without considering inflation is as follows:

$$RCF_j = \frac{NCF_j}{\prod_{i=1}^{j}(1 + f_i)} \tag{11}$$

$$NPV = \sum_{j=tc}^{T} \frac{RCF_j}{(1+r)^j} = \sum_{j=tc}^{T} \frac{RCF_j \cdot \prod_{i=1}^{j}(1+f_j)}{(1+r)^j \cdot \prod_{i=1}^{j}(1+f_j)} = \sum_{j=tc}^{T} \frac{NCF_j}{(1+r)^j \cdot \prod_{i=1}^{j}(1+f_j)} = \sum_{j=tc}^{T} \frac{R_j(f_j) - C_j(f_j)}{(1+r)^j \cdot \prod_{i=1}^{j}(1+f_i)}$$

$$(12)$$

Where t_C is the project construction period and T is the project concession period.

3 ADJUSTMENT METHOD OF GOVERNMENT REVENUE CONSIDERING REGULAR PRICE ADJUSTMENT

Due to inflation, the price level has risen, and the operating cost has increased year by year but the profit has decreased. To mitigate the impact of inflation, the price adjustment cycle π and the price adjustment range η are usually agreed upon at the negotiation stage. The toll price P_j of the jth year and the time y_m of the mth price adjustment are respectively expressed as:

$$P_j = \begin{cases} P_{j-1} & if \ y_{m-1} \leq j < y_m \\ P_{j-1} \cdot (1+\eta) & if \ j = y_m \end{cases} \tag{13}$$

$$y_m = y_{m-1} + \pi \ (m = 1, 2, ...) \tag{14}$$

Income regulation consists of income compensation and excess income sharing. The income compensation is to share the inflation risk of the part not covered by the price adjustment. If the total actual income $\sum_{y_{m-1}}^{y_m} R_j$ within the price adjustment interval π year (i.e., from year y_{m-1} to year y_m) is less than the total income $\sum_{y_{m-1}}^{y_m} R_j^f$ adjusted at the inflation rate each year, the government department will compensate the part of the income difference and FTG_j will be used to represent the income compensation of year j. Among them, the proportion of government revenue adjustment is $\delta\%(\delta \leq 1)$.

$$FTG_j = \begin{cases} \delta \cdot (\sum_{y_{m-1}}^{y_m} R_j^f - \sum_{y_{m-1}}^{y_m} R_j), & if \ j = y_m \ and \ \sum_{y_{m-1}}^{y_m} R_j < \sum_{y_{m-1}}^{y_m} R_j^f, \\ 0, & otherwise. \end{cases} \tag{15}$$

$$R_j^f = P_j^f \cdot Q_j^{af} \tag{16}$$

$$P_j^f = P_{j-1}^f \cdot (1+f_j) \tag{17}$$

$$R_j = P_j \cdot Q_j^a \tag{18}$$

In the formula, Q_j^{af} and Q_j^a are the traffic volume of route A in the jth year when the price is adjusted, according to the inflation rate and the agreed price adjustment range respectively.

The sharing of excess income refers to what when $\sum_{y_{m-1}}^{y_m} R_j > \sum_{y_{m-1}}^{y_m} R_j^f$ occurs, the government department obtains the excess income $\delta\%(\delta \leq 1)$ shared by the social capital party. As the government provides income compensation to bear some risks, when the project obtains excess income, the government departments have the right to share part of the income, which is also to avoid damage to the public interests caused by excessive prices.

$$SER_j = \begin{cases} b \cdot (\sum_{y_{m-1}}^{y_m} R_j - \sum_{y_{m-1}}^{y_m} R_j^f), & if \ j = y_m \ and \ \sum_{y_{m-1}}^{y_m} R_j > \sum_{y_{m-1}}^{y_m} R_j^f, \\ 0, & otherwise \end{cases} \tag{19}$$

Under the income regulation policy, the government's net expenditure NGP is:

$$NGP = \sum_{j=1}^{T} \frac{FTG_j - SER_j}{(1+r)^j \cdot \prod_{i=1}^{j}(1+f_i)} \tag{20}$$

4 DECISION ON THE RANGE OF PRICE ADJUSTMENT AND THE PROPORTION OF INCOME ADJUSTMENT

The decision-making model mainly considers the interests of the social capital party, users, and government departments. The social capital party wants the maximum net present value of the project, the users want the price adjustment to be as small as possible, and the government departments want the minimum net expenditure, which is expressed as:

$$\begin{cases} \max NPV \\ \min \eta \\ \min NGP \end{cases} \tag{21}$$

Due to the mutual influence and conflict of the three parties' interests, it is difficult to directly select an optimal or non-inferior adjustment scheme for the price adjustment range and the government's share proportion. Therefore, the fuzzy set theory is used to solve the problem of multi-objective decision-making. Let f_i be the vector composed of the ith objective value of m schemes, namely $f_i = f_i(A_j) = (f_i(A_1), f_i(A_2), ..., f_i(A_m))$. The three target values of the jth scheme are expressed by the vector f_j as $f_j = (f_1(A_j), f_2(A_j), f_3(A_j))^T$. The decision matrix F can be expressed as:

$$F = (f_{ij})_{3 \times m} = \begin{array}{c} \\ f_1 \\ f_2 \\ f_3 \end{array} \begin{bmatrix} A_1 A_2 \cdots A_m \\ f_{11} f_{12} \cdots f_{1m} \\ f_{21} f_{22} \cdots f_{2m} \\ f_{31} f_{32} \cdots f_{3m} \end{bmatrix} \tag{22}$$

For the objective of maximizing NPV, the membership calculation formula is

$$\mu_{ij} = \frac{f_{ij} - f_{i\,\min}}{f_{i\,\max} - f_{i\,\min}} \tag{23}$$

The degree of membership of the target of price adjustment and minimization of government net expenditure is:

$$\mu_{ij} = \frac{f_{i\,\max} - f_{ij}}{f_{i\,\max} - f_{i\,\min}} \tag{24}$$

The final scheme is obtained by using the max-min combination formula:

$$\mu_{ij}^* = \max_{1 \le j \le m} \min_{1 \le i \le n} \left\{ \mu_{ij} \right\} \tag{27}$$

5 CASE ANALYSIS

5.1 *Case background and simulation parameters*

A highway was built in a province, and the PPP mode was adopted for public bidding. The construction period was 3 years, and it was put into operation in 2014. The construction investment was discounted to the end of year 0, that is, 384,638,400 yuan at the beginning of the construction period. A highway group company that wins the bid gains revenue by charging fees for passing vehicles. The concession period is 33 years, including 30 years of operation. According to the feasibility study report of the project, the initial average toll price of the project is 55.14 yuan/vehicle, the free flow travel time is 102 minutes, and the road traffic capacity is 130656 vehicles/day. At the same time, there is another Class III highway between the two places. The free flow travel time of this section is 208 minutes, and the road traffic capacity is 5000 vehicles/day. The time value x of the local road users follows the normal distribution $N(2, 0.67^2)$. The government stipulates that the price adjustment cycle of the project is 5 years, and the inflation rate follows the normal distribution $N(0.03, 0.006^2)$. The total traffic volume of the project and the annual operating cost without inflation are calculated in the feasibility study report of the project. MATLAB R2015a is used for simulation, and the number of iterations is 2000.

5.2 Model simulation

The relatively optimal scheme of government share proportion under each price adjustment proportion is selected, and then the scheme set A of price adjustment range and income adjustment share proportion is formed, as shown in Table 1. The value ranges of the three decision objectives are $NPV \in [109.01, 332.29]$, $\eta \in [12\%, 18\%]$ $NGP \in [9.47, 331.64]$.

Table 1. Scheme set of price adjustment range and government share ratio.

Scheme A	A1	A2	A3	A4	A5	A6	A7	A8
Share proportion δ	60%	50%	50%	20%	50%	30%	20%	50%
Price adjustment range η	12%	13%	14%	15%	16%	17%	18%	18%
NPV (Million yuan)	134.88	119.41	160.66	109.01	244.46	261.49	318.12	332.29
NGP (Million yuan)	331.64	236.57	159.32	61.75	111.55	40.51	9.47	23.72

$$\mu_{ij}^* = \max_{1 \le j \le 8} \min_{1 \le i \le 3} \{\mu_{ij}\} = \max_{1 \le j \le 8} \min_{1 \le i \le 3} \begin{bmatrix} 0.116 & 0.047 & 0.231 & 0 & 0.607 & 0.683 & 0.937 & 1 \\ 1 & 0.833 & 0.667 & 0.5 & 0.333 & 0.167 & 0 & 0 \\ 0 & 0.295 & 0.535 & 0.813 & 0.683 & 0.904 & 1 & 0.956 \end{bmatrix} = \mu_{25}$$

5.3 Discussion on simulation results

If the project does not adjust the price during the operation period and the government does not provide income regulation, the inflation risk will be solely borne by the social capital party. At this time, the average net present value of the project is -1021.93 million yuan, which is a serious loss and difficult to attract investment. After the price adjustment, the

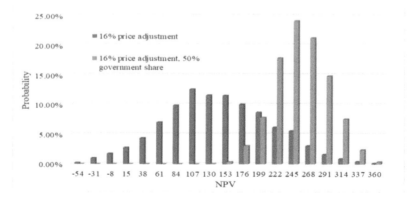

Figure 2. Probability distribution of project NPV with or without income adjustment.

average net present value of the project has increased to 165.99 million yuan, avoiding the loss of the project, but it is still significantly lower than the net present value of 340.72 million yuan adjusted at the annual inflation rate, which is caused by the income loss of the unadjusted price within the interval of five-year price adjustment. After the government provides income adjustment, as shown in Figure 2, the average net present value increases to 281.34 million yuan, and the standard deviation decreases by 49.99%. At this time, the government provides income adjustment to increase the net present value of the project, reduce the uncertainty of its value, help attract social capital and increase its enthusiasm for participation, because the 16% price adjustment is not enough to cover the income loss caused by inflation. Through income adjustment, the government departments share part of the inflation risk, and the net present value of the project increases.

6 CONCLUSION

Given the inflation risk of highway PPP projects, this paper proposes a regulation method of government revenue based on a regular price adjustment. Through case simulation, considering the interests of social capital, users, and government departments, the optimal scheme of price adjustment range and government revenue regulation share proportion is determined. Further analysis of the scheme shows that, compared with the two methods of no price adjustment and only price adjustment without income adjustment, under the risk of inflation, the linkage adjustment of price adjustment and income adjustment can increase the net present value of the project, reduce the uncertainty of its value, and help to enhance the investment confidence of social capital. At the same time, the government departments can share part of the inflation risk through income adjustment, or avoid the excess income of the project caused by excessive price adjustment, which is conducive to the benign development of the project and the protection of social public interests. The methods and models proposed in this paper are not only applicable to highway projects but also can be applied to other transport infrastructure projects.

REFERENCES

Buyukyoran F., Gundes S. Optimized Real Options-based Approach for Government Guarantees in PPP Toll Road Projects [J]. *Construction Management and Economics*, 2018, 36 (4): 203–216.

Cai Xiaoyan and Zhou Guoguang. The Real Option Value Analysis of Income Adjustment During the Operating Period in the Toll Road Public Private Partnership projects [J]. *Journal of Transportation Systems Engineering and Information Technology*, 2017, 17 (3): 7–11.

Carbonara N. and Pellegrino R. Revenue Guarantee in Public-private Partnerships: A Win-win Model. *Construction Management and Economics*, 2018, 36 (10–12): 584–598.

Chan A. P. C., Lam P. T. L., and Wen Y. Cross-sectional Analysis of Critical Risk Factors for PPP Water Projects in China [J]. *Journal of Infrastructure Systems*. 2015, 21 (1): 4014031.

Chen Q., Shen G., Xue F., and Xia B. Real Options Model of Toll-Adjustment Mechanism in Concession Contracts of Toll Road Projects [J]. *Journal of Management in Engineering*, 2018, DOI: 10.1061/(ASCE) ME.1943–5479.0000558.

Mirzadeh I. and Birgisson B. Evaluation of Highway Projects under Government Support Mechanisms Based on an Option-Pricing Framework [J]. *Journal of Construction Engineering and Management*, 2015:04015094.

Wibowo A., Permana A., Kochend Rfer B., et al. Modeling Contingent Liabilities Arising from Government Guarantees in Indonesian BOT/PPP Toll Roads [J]. *Journal of Construction Engineering and Management*, 2012, 138 (12): 1403–1410.

Ye S. and Tiong R. L. K. Effects of Tariff Design in Risk Management of Privately Financed Infrastructure Projects [J]. *Journal of Construction Engineering and Management*, 2003, 129 (6): 610–618.

Water Conservancy and Civil Construction – Oke & Ahmad (Eds)
© 2024 The Author(s), ISBN: 978-1-032-58618-2

Impact analysis on dynamic response of asphalt pavement structure with semi-rigid base effected by interlayer evolution properties under different service conditions

YuBin Zhang & Xiangbiao Wang
Achui Transport Consulting & Design Institute Co., Ltd, Anhui, China

Rui Zhang, Wenzhi Yuan* & Qun Yang
The Key Laboratory of Road and Traffic Engineering, Ministry of Education Tongji University, Shanghai, China

ABSTRACT: In order to explore the relationship between the bonding state exiting in different pavement layers and the mechanical response, a finite element analysis model of the pavement was established with three preset interlayer contact interfaces: the upper asphalt layer, middle layer-lower layer, and lower layer-base layer. And a half-sine load was used to simulate the vehicle driving process. Seven equations for the decay of the friction coefficient with service time were constructed and divided into three types of working conditions according to the decay rate to characterize the interlayer bonding under different service years. For the friction coefficients predicted by different equations under different years, the mechanical response of the pavement structure was calculated. It was found that for the selected typical pavement structures: (1) Poor interlayer bonding conditions lead to peak shear stresses at the interface location, and the maximum shear stress occurred at the bottom of the upper surface layer; (2) As the service time increases (the interlayer bonding condition deteriorated), the maximum shear stress in the asphalt layer increased by 30% and the maximum tensile strain at the bottom of the asphalt layer increased 5.00 times; (3) The trends of the maximum shear stress and the tensile strain of the asphalt layer were the same under different working conditions. The maximum shear stress and the tensile strain of the asphalt layer grew relatively slowly when μ=0.4-1.0. With the decay of the interlayer bond state, when $\mu<0.4$, the maximum shear stress and the tensile strain of the asphalt layer began to grow rapidly.

1 INTRODUCTION

When analyzing pavement structures, it is commonly assumed that the interface state of the pavement structure is completely continuous, i.e., "full friction" or "full bond" between the layers [1]. However, differences in material properties and construction quality can lead to a change in the interface bonding conditions (such as the contact condition of the pavement structure layers) from the full bonding state to the partial bonding state, significantly affecting the mechanical response of the pavement [2]-[5]. In addition, for constructed roads, the interlayer bonding conditions will inevitably change under external environmental factors such as traffic loads, temperature, and humidity. And the bonding performance will be impaired. Therefore, treating them as full bonds would overestimate pavement performance [6].

---

*Corresponding Author: 2011380@tongji.edu.cn

420

DOI: 10.1201/9781003450832-54

To gain a deeper understanding of the effects of bonding state on pavement structures, the characterization of interlayer bonding, the evaluation methods, and the influencing factors have been well researched [7]-[11]. It has been shown that poor interlayer contact conditions lead to stress and strain redistribution, which significantly affects the mechanical responses of the pavement structure [12]-[15], and predisposes it to shear damage at weak interlayer bonding conditions [16]-[18]. This can lead to premature failure of the pavement structure and shorten the service life of the pavement [19]-[23]. Shuhua Wu [24] found that the maximum interlayer shear stress changed most when the interlayer contact condition changed from full bonding to partial bonding or to fully sliding, with a value of 0.48 MPa. Ashraful Alam [25] *et al.* found that the tensile stress in partially bonding systems increased by about 15% compared to fully bonding pavement systems by finite element simulations. Manik Barman [26] *et al.* found that debonding can increase the magnitude of stress by 40-55%. Mariana R. Kruntcheva [27] *et al.* considered different degrees of interface bonding conditions between pavement layers and found that poor interlayer bonding conditions could reduce the life of the pavement structure by 80%. Chun S. [28] *et al.* also demonstrated that interlayer bonding conditions play an important role in the structural response characteristics of a layered flexible pavement system as well as the pavement service life.

As interlayer bonding conditions have become a more important subject and it has been studied in routine pavement construction, the question of the long-term performance of the bonding state has arisen [29]. However, current domestic and international research on interlayer contact was more concerned with the bonding state between the subgrade and the surface layer and mostly did not consider the actual laying process of the surface layer. Moreover, it was generally only a comparative study of a specific bonding state without considering the long-term performance of the bonding state as it decayed over time. Therefore, to explore the influence of the decay of the interlayer bonding conditions on the pavement structure, this paper characterized the interlayer bonding conditions by the friction coefficient, constructed the decay equation of the interlayer bonding conditions during the service life of the pavement, and analyzed the influence of the decay of the interlayer bonding conditions on the mechanical responses of the pavement during the service time.

2 THE THREE-DIMENSIONAL DYNAMIC FINITE ELEMENT MODEL

2.1 *Pavement structure and material parameters*

Referring to the pavement structure of several domestic highways and the recommended pavement structure in the design specification of asphalt concrete pavement, the values of the structural parameters of asphalt concrete pavement at the room temperature of 20°C were selected and shown in Table 1. In this paper, the contact conditions were set at three interfaces in the calculation model, as shown in Figure 1. The first position was between the upper and middle surface layers. The second position was between the middle layer and

Table 1. The structure and material parameters of semi-rigid base asphalt pavement.

Structural layer	Modulus (MPa)	Poisson's ratio	Density $(kg \cdot m^{-3})$	Rayleigh damping	Thickness (cm)
AC–13	12,000	0.25	2,400	0.9	4
AC–16	11,000	0.25	2,400	0.9	5
AC–25	10,000	0.25	2,400	0.9	7
Base	13,000	0.25	2,300	0.8	34
Subbase	11,000	0.25	2,300	0.8	20

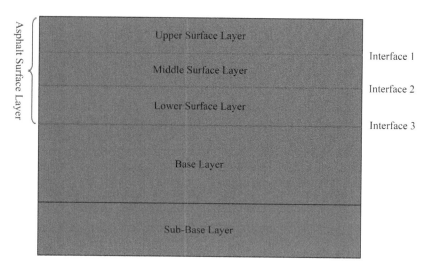

Figure 1. The preset contact conditions.

lower layer, and the third position was between the lower layer and base layer. And the contact condition between the base layer and the subgrade was set as continuous contact.

2.2 Model size and cell division

In the structural analysis of pavements, it is generally assumed that the layers of the pavement are elastic layers with infinite planes, and the roadbed is an elastic half-space body. When using finite element calculations, the size of the model cannot be taken to infinity but can only be as large as possible when building the model. However, if the size is too large, it will greatly increase the calculation workload. If the size is too small, it will affect the accuracy of the calculation. Therefore, the size of the model should be reasonably determined so that it does not increase the computational workload too much while ensuring computational accuracy. In this paper, the geometry of the model was set to 3.80 m×3.80 m×3.80 m. The mesh cells of the pavement structure were C3D8R (three-dimensional 8-node reduced integral cell body), which can avoid the simulation grid bending seriously. In order to save calculation time, a non-uniform meshing method was used, with 0.02 m for the surface part, 0.05 m for the base part, and a global seeding density of 0.20 m for the soil base part. The model and model division are shown in Figure 2.

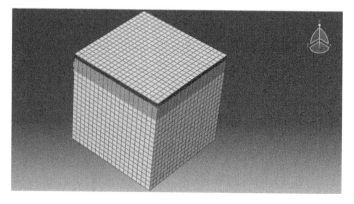

Figure 2. Grid model division.

2.3 Boundary conditions

The bottom surface of the model was constrained by U1=U2=UR3=0, the left and right sides of the road surface were constrained by U1=UR2=UR3=0, and no other surfaces were constrained.

2.4 Load contact areas and load models

In traditional calculations for pavement mechanics and our current design codes, the distribution of the tire-pavement contact area is reduced to a circular form. The interaction between the tire and the road is very complex and influenced by the tire structure. At present, radial tires are the mainstream of the market. According to the statistics of the Tire Branch of the China Rubber Industry Association, 38 key enterprises (including overseas factories) in China will produce 50.15 million radial tires in 2021, accounting for 94.6% of the total output of integrated outer tires. Given that radial tires have a harder tread and softer walls, the larger the grounding area is, the more rectangular shape of its grounding area is, which gets closer to a rectangle as the load increases. So, this paper used a rectangle as the shape of the tire-road contact. The standard load of the BZZ-100 single-axle twin wheel set was used. The tire inflation pressure was 0.7 MPa. The approximate size of the individual tire area was determined to be 18.9 cm × 18.9 cm. And the spacing between the twin wheels was 34 cm. The spacing between the wheel gaps on both sides was 180 cm, as shown in Figure 3.

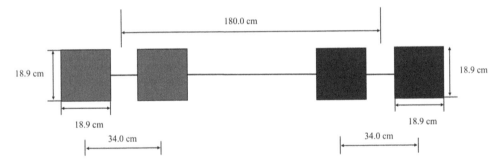

Figure 3. Rectangular load form.

To better simulate the moving process of vehicle movement, a half-sine load action was used, as shown in Equation (1).

$$P(t) = 0.7 \sin(10\pi t) \text{MPa}, 0 \leq t \leq 0.1s \tag{1}$$

2.5 Interlayer contact

Interlayer contact simulations were carried out by using the penalty function formulation in ABAQUS/Explicit. And the kinetic contact algorithm, the balanced master-slave algorithm, and the small slip formulation used by default in ABAQUS/Explicit were used for the interlayer contact simulations. When simulating the normal behavior of the contact surfaces, it was assumed that the two contact surfaces would not separate and remain bonded, ensuring that vertical stresses and displacements at the contact surfaces were transmitted continuously. To simulate the tangential behavior of the contact surfaces, the Coulomb friction model was used to describe the interaction between the contact surfaces, and the friction behavior between the two surfaces was modeled by the friction coefficient μ.

2.6 Indicators for analysis of interlayer bonding conditions

The main modes of structural damage to pavements are fatigue cracking, permanent rutting deformation, low-temperature cracking, and shear damage, which should be taken into account in the structural design. Therefore, the main analysis indicators that can be selected for pavement structural analysis are: (1) the deformation value; (2) the maximum tensile stress within each structural layer; (3) the maximum shear stress within the asphalt surface layer. The magnitude of the interlayer shear stress and the tensile strain at the bottom of the asphalt layer is directly correlated with the corresponding interlayer contact state. So, the maximum shear stress and the tensile strain of the asphalt layer were used in this paper as evaluation indicators for the overall performance of the pavement structure.

The maximum shear stress at the bottom of each surface layer and the tensile strain at the bottom of the asphalt layer was taken as representative values under certain contact conditions. The change of the representative values under different contact conditions was analyzed to study the effect of the contact conditions between layers on the maximum shear stress and the tensile strain.

3 THE EQUATION CONSTRUCTION OF INTERLAYER BONDING CONDITIONS DECAY

Under the combined effect of traffic loads and environmental factors, the bond between layers of the pavement structure gradually decays over time. Typically, asphalt pavements have a design life of around 15 years. To simulate the decay of the structural interlayer bonding state, the friction coefficient μ was considered to be 1 for the new pavement and 0 at the design service life of 15 years, based on which the friction coefficient-service life (μ-t) equation was constructed. In order to better describe the variation of the friction coefficient with time, seven sets of equations were constructed as follows: Equations (2)-(8), involving linear, quadratic, exponential, triangular, and logarithmic forms, as shown in Figure 4.

$$y_1 = -\frac{1}{15}x + 1 \tag{2}$$

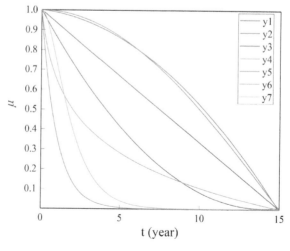

Figure 4. Friction coefficient decay with service time.

$$y_2 = -0.0039x^2 - 0.0086x + 1 \tag{3}$$

$$y_3 = 0.0051x^2 - 0.1426x + 1 \tag{4}$$

$$y_4 = e^{-x} \tag{5}$$

$$y_5 = \cos\left(\frac{\pi}{30}x\right) \tag{6}$$

$$y_6 = \frac{-\ln\left(x + \frac{1}{e}\right) + \ln\left(15 + \frac{1}{e}\right)}{1 + \ln\left(15 + \frac{1}{e}\right)} \tag{7}$$

$$y_7 = (x + 1)e^{-x} \tag{8}$$

According to the different decay of the friction coefficient with service life in Figure 4, they were divided into three categories, where y1, y2, and y5 were set to category I, y3 and y6 to category II, and y4 and y7 to category III. Three different working conditions were represented by the three categories of equations separately. Based on the constructed μ-t equation, the friction coefficients were calculated for each year in turn, and the results are shown in Table 2.

Table 2. The estimated friction coefficient for the corresponding service life.

t (year)	μ						
	y1	y2	y3	y4	y5	y6	y7
0	1.00	1.00	1.00	1.00	1.00	1.00	1.00
1	0.93	0.99	0.86	0.37	0.99	0.65	0.74
2	0.87	0.97	0.74	0.14	0.98	0.50	0.41
3	0.80	0.94	0.62	0.05	0.95	0.41	0.20
4	0.73	0.90	0.51	0.02	0.91	0.34	0.09
5	0.67	0.86	0.41	0.01	0.87	0.28	0.04
6	0.60	0.81	0.33	0.00	0.81	0.24	0.02
7	0.53	0.75	0.25	0.00	0.74	0.20	0.01
8	0.47	0.68	0.19	0.00	0.67	0.16	0.00
9	0.40	0.61	0.13	0.00	0.59	0.13	0.00
10	0.33	0.52	0.08	0.00	0.50	0.11	0.00
11	0.27	0.43	0.05	0.00	0.41	0.08	0.00
12	0.20	0.34	0.02	0.00	0.31	0.06	0.00
13	0.13	0.23	0.01	0.00	0.21	0.04	0.00
14	0.07	0.12	0.00	0.00	0.10	0.02	0.00
15	0.00	0.00	0.00	0.00	0.00	0.00	0.00

4 MODEL CALCULATION AND ANALYSIS

Based on Table 2, the predicted friction coefficient μ for each equation at different service years was simulated and analyzed for each mechanical response of the pavement structure.

4.1 *Maximum shear stress within the surface layer*

Shear stress is an important factor in the shear deformation (such as pushing and packeting) and rutting of asphalt surfaces. The higher the shear stress is, the more severe the shear deformation and rutting are. The unit group with the highest shear stress directly below the wheel load was selected to analyze the variation of shear along the depth, as shown in Figure 5.

It can be seen from Figure 5 that:

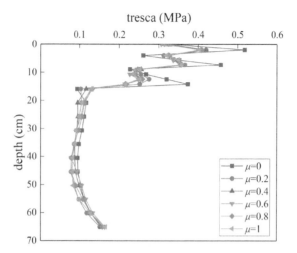

Figure 5. Variation of shear stress with depth for different friction coefficients.

Along the depth direction of the road surface structure, the shear stress values at all three interfaces, such as interface 1, interface 2, and interface 3, showed a peak value within the surface layer, with the maximum peak value occurring at interface 1, i.e., at the bottom of the upper surface layer. At the same time, as μ increases, the bond between the layers becomes better, and the peak value at each layer decreases. In the three interfaces, the peak value decreased by 22.25%, 21.73%, and 42.39%, respectively. The largest decrease in peak value was found at interface 3 due to the difference in pavement materials between the two layers at the interface.

As service life increased, the bonding state between the layers of the pavement structure decayed under traffic loads and environmental effects, resulting in greater maximum shear stress at the bottom of the upper surface layer, as shown in Figure 6.

It can be seen from Figure 6 that:

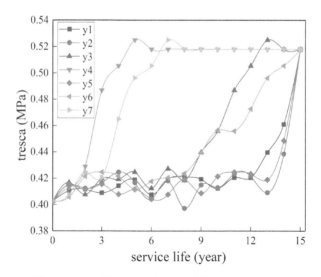

Figure 6. Variation of shear stress with service time for different equations.

During the 15 years of service, the maximum shear stress in the surface layer increased from 0.40 MPa to 0.52 MPa, which increases by 30%. The trend of the maximum shear stress with service time was consistent with the friction coefficient service life (μ-t) equation for each of the three categories. For Category I, the maximum shear stress of the surface layer showed a small fluctuating increase of about 4.41% between 0 and 12 years and a sharp increase of 23.18% between 12 and 15 years. For Category II, the maximum shear stress of the asphalt surface layer showed a small fluctuating increase of about 5.20% between 0 and 8 years and a sharp increase of 23.18% between 8 and 15 years. For Category III, the maximum shear stress in the asphalt surface layer increased sharply by 30.34% between 0 and 5 years due to the rapid decay of the contact condition, reaching the maximum shear stress. Corresponding to μ values, it could be noticed that when $\mu < 0.4$ the maximum shear stress changed most dramatically.

4.2 Tensile strain at the bottom of the asphalt layer

The level of tensile strain at the bottom of the asphalt layer affected the fatigue life of the semi-rigid base asphalt pavement. So, the tensile strain of the asphalt layer was analyzed for different friction coefficients. Figure 7 shows the response of the horizontal tensile strain at the bottom of the asphalt layer for the entire time course of the semi-sinusoidal load.

The half-sine load reached its peak $t = 0.05$s. Observation of Figure 7 showed that there was a lag in the peak tensile strain relative to the peak load, with the peak strain occurring at around $t = 0.055$s. Therefore, the strain response $t = 0.055$s was selected for analysis.

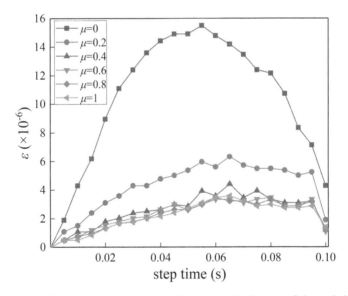

Figure 7. Variation of the time course of the tensile strain at the bottom of the asphalt layer for different friction coefficients.

It can be seen from Figure 8 that:

Over the 15 years of service, the maximum tensile strain at the bottom of the asphalt layer increased from 3.10×10^{-6} to 15.49×10^{-6}, about 5.00 times the initial value. The trend of the maximum tensile strain with service time was also consistent with the friction coefficient service life (μ-t) equation for the three categories. For category, I, the maximum tensile strain

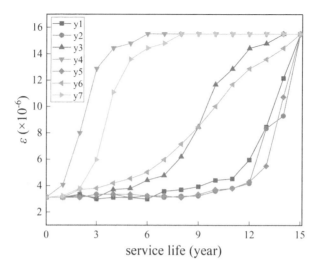

Figure 8. Variation of tensile strain at the base of asphalt layers with service time for different equations.

at the bottom of the asphalt layer increased slowly, about 1.92 times between 0 and 12 years, and increased dramatically between 12 and 15 years, about 2.60 times. For category II, the maximum tensile strain at the bottom of the asphalt layer tended to increase slightly more, about 1.20 times between 0 and 8 years, and increased sharply by a factor of 2.50 between 8 and 15 years. For Category III, between years 0 and 6, the maximum tensile strain at the bottom of the asphalt layer increased sharply and reached a peak tensile strain. According to the μ values for the corresponding years, it could be seen that when $\mu<0.4$, the maximum tensile strain increased rapidly. Compared with Chapter 3.1, it could be found that the variation law of maximum tensile strain at the bottom of the asphalt layer under each equation was consistent with the variation law of maximum shear stress.

5 CONCLUSION

(1) Seven friction coefficient service life (μ-t) equations were constructed to characterize the decay of the interlayer bonding conditions over time by means of the friction coefficient μ, which were divided into three types to characterize three different working conditions.
(2) Due to the poor contact condition, the dispersion of traffic load propagation was poor, resulting in peak shear stress at the interface location. The maximum shear stress occurred at the bottom of the upper layer of the asphalt layer. It indicated that attention should be paid to the interlayer treatment, especially the upper and middle surface layer interface.
(3) Under different conditions, the maximum shear stress in the asphalt layer grew at different rates with time due to the different decay rates of the friction coefficients. The faster the decay of the friction coefficient μ is, the higher the maximum shear stress in the asphalt layer is. The trend of tensile strain at the bottom of the asphalt layer was consistent with the maximum shear stress of the asphalt layer.
(4) When $\mu=0.4$-1.0, the maximum shear stress in the asphalt layer and the tensile strain at the bottom of the asphalt layer grew relatively slowly. With the decay of the interlayer bonding conditions, when $\mu<0.4$, the maximum shear stress in the asphalt layer and the tensile strain at the bottom of the asphalt layer started to grow rapidly.

ACKNOWLEDGMENT

The paper was supported by Anhui Transportation Holding Group Technology Project (No. JKKJ-2020-29).

REFERENCES

[1] Alam A., Haselbach L., and Cofer W. F. Evaluation of Critical Stresses in Pervious Concrete Pavement Aystems Due to Partial to Full Bonding Between Subgrade, Subbase, and Surface layers [J]. *Int. J. Eng. Tech. Res*, 2014, 2(9): 323–327.

[2] Barman M., Vandenbossche J. M., and Li Z. Influence of Interface Bond on the Performance of Bonded Concrete Overlays on Asphalt Pavements [J]. *Journal of Transportation Engineering, Part B: Pavements*, 2017, 143 (3): 04017008.

[3] Canestrari F., Cardone F., Gaudenzi E., et al. Interlayer Bonding Characterization of Interfaces Reinforced with Geocomposites in Field Applications [J]. *Geotextiles and Geomembranes*, 2022, 50 (1): 154–162.

[4] Chen Hu A. P., Guo Z., and Zhou X. Stress Singularities of Contact Problems with a Frictional Interface in Anisotropic Biomaterials [J]. *Fatigue & Fracture of Engineering Materials & Structures*, 2012, 35 (8): 718–731.

[5] Chen S., Wang D., Yi J., et al. Analytical Calculations for Asphalt Pavement Considering Interlayer Performance [J]. *Journal of Transportation Engineering, Part B: Pavements*, 2022, 148 (2): 04022022.

[6] Chun S., Kim K., Greene J., et al. Evaluation of Interlayer Bonding Condition on Structural Response Characteristics of Asphalt Pavement Using Finite Element Analysis and Full-scale Field Tests [J]. *Construction and Building Materials*, 2015, 96: 307–318.

[7] Collop A. C., Sutanto M. H., Airey G. D., et al. Shear Bond Strength Between Asphalt Layers for Laboratory Prepared Samples and Field Cores [J]. *Construction and Building Materials*, 2009, 23 (6): 2251–2258.

[8] Graziani A., Canestrari F., Cardone F., et al. Time-temperature Superposition Principle for Interlayer Shear Strength of Bituminous Pavements [J]. *Road Materials and Pavement Design*, 2017, 18 (sup2): 12–25.

[9] Guo C., Wang F. and Zhong Y. Assessing Pavement Interfacial Bonding Condition [J]. *Construction and Building Materials*, 2016, 124: 85–94.

[10] Hristov B. Influence of Different Interface Properties on the Interlayer Bond Shear Stiffness [C]// *IOP Conference Series: Materials Science and Engineering. IOP Publishing*, 2018, 365 (3): 032056.

[11] Hu T., Zi J., Li J., et al. Laboratory and Field Investigation of Interlayer Bonding Between Asphalt Concrete Layer and Semi-rigid Base Constructed by Using Continuous Construction Method[J]. *Construction and Building Materials*, 2017, 150: 418–425.

[12] Hu X.and Walubita L. F. Effects of Layer Interfacial Bonding Conditions on the Mechanistic Responses in Asphalt Pavements [J]. *Journal of Transportation Engineering*, 2011, 137 (1): 28–36.

[13] Kim H., Arraigada M., Raab C., et al. Numerical and Experimental Analysis for the Interlayer Behavior of Double-layered Asphalt Pavement Apecimens [J]. *Journal of materials in civil Engineering*, 2011, 23 (1): 12–20.

[14] Kruntcheva M. R., Collop A. C. and Thom N. H. Properties of Asphalt Concrete Layer Interfaces [J]. *Journal of Materials in Civil Engineering*, 2006, 18 (3): 467–471.

[15] Kruntcheva M. R., Collop A. C., and Thom N. H. Effect of Bond Condition on Flexible Pavement Performance [J]. *Journal of Transportation Engineering*, 2005, 131 (11): 880–888.

[16] Li S., Huang Y., and Liu Z. Experimental Evaluation of Asphalt Material for Interlayer in Rigid–flexible Composite Pavement [J]. *Construction and Building Materials*, 2016, 102: 699–705.

[17] Malick Diakhate, Annabelle Phelipot, Anne Millien, and Christophe Petit (2006) Shear Fatigue Behaviour of Tack Coats in Pavements, *Road Materials, and Pavement Design*, 7:2, 201–222, DOI: 10.1080/14680629.2006.9690033

[18] Raab C.and Partl M. N. Interlayer Bonding of Binder, Base and Subbase Layers of Asphalt Pavements: Long-term Performance [J]. *Construction and Building Materials*, 2009, 23 (8): 2926–2931.

[19] Ragni D., Ferrotti G., Petit C., et al. Analysis of Shear-torque Fatigue Test for Bituminous Pavement Interlayers [J]. *Construction and Building Materials*, 2020, 254: 119309.

[20] Wang X., Su Z., Xu A., et al. Shear Fatigue Between Asphalt Pavement Layers and its Application in Design [J]. *Construction and Building Materials*, 2017, 135: 297–305.

[21] Wellner F. and Hristov B. Numerically Supported the Experimental Determination of the Behavior of the Interlayer Bond in Asphalt Pavement [J]. *Transportation Research Record*, 2015, 2506 (1): 116–125.

[22] Wu S., Chen H., Zhang J., et al. Effects of Interlayer Bonding Conditions Between the Semi-rigid Base Layer and Asphalt Layer on Mechanical Responses of Asphalt Pavement Structure [J]. *International Journal of Pavement Research and Technology*, 2017, 10 (3): 274–281.

[23] Yang K. and Li R. Characterization of Bonding Property in Asphalt Pavement Interlayer: A Review [J]. *Journal of Traffic and Transportation Engineering (English Edition)*, 2021, 8 (3): 374–387.

[24] Yang Kai, Li Rui, Yu Yi, Pei Jianzhong, and Liu Tao (2020). Evaluation of Interlayer Stability in Asphalt Pavements Based on Shear Fatigue Property. *Construction and Building Materials*, 258, 119628. DOI: 10.1016/j.conbuildmat.2020.119628

[25] Yang Y. H., Shen Y., and Gao X. X. Analysis Interlayer Shear Stress of Ultra-thin Wearing Course Considering Temperature and Different Interlayer Contact Conditions [C]// *Applied Mechanics and Materials. Trans Tech Publications Ltd*, 2014, 505: 102–105.

[26] Ye H., Wang X., Fang N., et al. Interlayer Working Conditions Classification and Treatment Measures of Airport Asphalt Pavement Overlay [J]. *Journal of Advanced Transportation*, 2019, 2019.

[27] You L., Yan K., and Liu N. Assessing Artificial Neural Network Performance for Predicting Interlayer Conditions and Layer Modulus of Multi-layered Flexible Pavement [J]. *Frontiers of Structural and Civil Engineering*, 2020, 14 (2): 487–500.

[28] You L., Yan K., Man J., et al. Anisotropy of Multi-layered Structure with Sliding and Bonded Interlayer Conditions [J]. *Frontiers of Structural and Civil Engineering*, 2020, 14 (3): 632–645.

[29] You L., Yan K., Shi T., et al. Analytical Solution for the Effect of Anisotropic Layers/Interlayers on an Elastic Multi-layered Medium Subjected to Moving Load.[J]. *International Journal of Solids and Structures*, 2019,172:10–20.

Water Conservancy and Civil Construction – Oke & Ahmad (Eds)
© 2024 The Author(s), ISBN: 978-1-032-58618-2

Research on the ultimate protection ability of grade A W-beam barrier based on the bus collision

Hao Wang
Shandong High-speed Co., Ltd, Jinan, China

Shuai Gong*
Beijing Hualuan Traffic Technology Co., Ltd, Beijing, China

Hong Guo
Shandong High-speed Co., Ltd, Jinan, China

Shuming Yan
Beijing Hualuan Traffic Technology Co., Ltd, Beijing, China

ABSTRACT: In order to understand the ability of the grade A w-beam barrier to protect the bus, the methods of the computer simulation and the full-scale impact test with real vehicles were used to research it. The results show that the grade A w-beam barrier can meet the requirements of the old standard, but can not satisfy the requirements of the new standard. All the buses have the phenomenon of riding over the barrier when the collision energy ranges from 145 kJ to 160 kJ, which does not meet the evaluation criteria. It can be seen that the buses all have a good driving posture and the driving track can meet the requirements of the exit box at the collision energy of 140 kJ, according to the simulation collision analysis and the real test data of the vehicle crash. Therefore, the grade A w-beam barrier can effectively protect the bus with 140 kJ collision energy, that is, the ultimate protection energy of the barrier to the bus is 140 kJ. The research can provide basic data for the rational application of the grade A w-beam barrier in practical engineering.

1 INTRODUCTION

On December 1, 2013, the *Standard for Safety Performance Evaluation of Highway Barriers* (JTG B05-01-2013) (hereinafter referred to as the "New Standard") was promulgated and implemented. The standard is more strict and perfect than the *Evaluation Specification for Highway Safety Barriers* (JTG/T F83-2004) (hereinafter referred to as the "Old Standard") promulgated and implemented on December 31, 2004. The grade A w-beam barrier is the main representative form of the semi-rigid barrier, which consists of a double wave plate with a center height of 600 mm from the ground, a column with a size of 140 mm×4.5 mm, and a hexagonal block with a thickness of 4.5 mm connected by bolts. The grade A w-beam barrier is widely used in Chinese expressways, but vicious accidents caused by the collision of the barrier are common. It is necessary to combine the provisions of the old and new standards to study the protective ability of this barrier on the bus.

---

*Corresponding Author: gongshuaizhen@163.com

DOI: 10.1201/9781003450832-55

2 NEW STANDARD AND OLD STANDARD

For the bus crash test of the grade A barrier, both the "New Standard" and the "Old Standard" requires a vehicle mass of 10 t, a collision speed of 60 km/h, and a collision angle of 20 ° as the collision conditions, but the "New Standard" and the "Old Standard" have different provisions on the tolerance error of collision condition and the evaluation index.

2.1 *The collision condition tolerance error*

Table 1 shows the comparison of the bus collision conditions and allowable errors in the new and old standards.

According to Table 1, the "Old Standard" has positive and negative provisions on the collision parameters, while the "New Standard" cancels the negative errors of the vehicle mass and the collision speed, reducing the negative error range of the collision angle. The "New Standard" requires that there should be no negative error in the collision energy, that is, the collision energy of the grade A w-beam barrier must be above 160 kJ, while the "Old Standard" does not have clear requirements for this. According to the error lower limits of the three collision factors, the "Old Standard" requires that the collision energy of the grade A barrier must be above 122 kJ.

Table 1. Comparison of the bus collision conditions and allowable errors of the new and old standard.

Impact condition	Normal value and tolerance	"Old Standard"	"New Standard"
Vehicle Quality	Standard (t)	10	
	Allow errors (kg)	-300~300	0~300
Impact speed	Standard value (km/h)	60	
	Allow errors (km/h)	-3~3	0~4
Impact angle	Standard value (°)	20	
	Allow errors (°)	-1.5~1.5	-1~1.5
Impact energy	Standard value (kg)	160	
	Allow errors (kg)	------	Positive error

2.2 *Evaluation index*

Table 2 shows the comparison of the evaluation indexes for safety performance between the "New Standard" and the "Old Standard". It can be concluded that the provisions of the "New Standard" and the "Old Standard" are the same for the containment function, but the language of the "New Standard" is more refined. As for the vehicle attitude in the guiding function, the "Old Standard" requires that the vehicle does not have the phenomenon of transverse turn and u-turn. Considering that transverse turn and u-turn can easily lead to a vehicle rollover, the "New Standard" directly stipulates that the vehicle shall not roll over after impact. For the guidance function, in order to investigate the impact of the accident vehicle on the normal running vehicle adjacent lane, the "Old Standard" requires that the exit angle is less than 60% of the impact angle, which can only evaluate the state of the vehicle at the time of departure, but the measurement error is larger. While the "New Standard" requires the vehicle to use the guidance exit frame, leaves the barrier after a certain distance, and shall not cross the straight line parallel to the barrier, so the evaluation is more comprehensive and operable. Through consecutive years of barrier development and application experience, the "New Standard" no longer takes the barrier deformation index as a control index, but it requires the record of the maximum dynamic vehicle incline-out distance VI, and calculates the normalized maximum dynamic vehicle incline-out VI_n, in order to provide a basis for the reasonable setting of the barrier and study a more reasonable way.

Table 2. Comparison of the evaluation indexes of safety performance for the new and old standard.

Evaluation item	"Old standard"	"New standard"
Containment function	It should be able to effectively block and guide vehicles, and it is prohibited to cross, climb over, ride over and wear the barrier in any form.	It should be able to prevent vehicles from crossing, climbing, and riding.
	In the event of a crash, the debris from the assembly, the barrier, or other objects on the barrier shall not invade the cab or obstruct the driver's view.	The barrier components and their detaches shall not invade the vehicle occupant compartment.
Orientation function	After the impact, vehicles should maintain the normal driving posture, and no sideways turn, u-turn, and other phenomena occur.	Vehicles must not roll over after the impact.
	It should have a good steering function, and the vehicle exit angle after the impact should be less than 60% of the impact angle.	The wheel track after the vehicle exits the departure point shall not cross straight line F when it passes through the guide exit frame in Figure 1.
Barrier deformation	The maximum dynamic deformation of the semi-rigid w-beam barrier is less than or equal to 100 cm.	Only the maximum dynamic lateral deflection of barriers (D) and the maximum dynamic widening distance of lateral deflection of barriers (W) shall be recorded.
Vehicle incline-out	No requirements.	The maximum dynamic vehicle incline-out distance (VI) shall be recorded, and the normalized maximum dynamic vehicle incline-out (VI_n) shall be calculated.

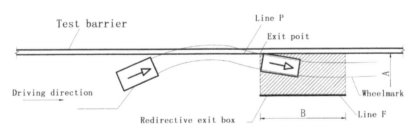

Figure 1. The guiding exit box of the wheel mark.

Table 3. The value of parameter A and parameter B.

Impact vehicle type	A	B
Medium and large truck	$4.4 + V_W + 0.16 V_L$	20

Note: V_W is the total vehicle width (m); V_L is the total vehicle length (m).

3 COLLISION MODEL AND COLLISION ANALYSIS

3.1 *Collision model*

In the special test site for the full-size impact of real vehicles, the test section was established according to the grade A corrugated beam barrier structure, and the two ends of the corrugated beam barrier were effectively anchored by steel wire ropes. According to the requirements of the "New Standard", the test bus was purchased, and the vehicle assembly is complete. The steering system, suspension system, wheels, front and rear axles, and tire pressure meet the technical requirements of normal driving. The vehicle loads are uniformly distributed and fixed with the vehicle body. The fuel of the fuel tank is replaced by water, and the mass is 90% of the fuel tank filled with the fuel. The test vehicle is clean inside and outside, and clear reference lines and reference points are set on the top and side according to the requirements of the image data acquisition. Figure 2 shows the barrier and vehicle test model.

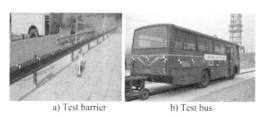

a) Test barrier   b) Test bus

Figure 2. The barrier and vehicle test model.

The 1:1 finite element simulation model is established according to the actual size of the guardrail and vehicle. The parts of the bus body are mainly connected by spot welding. The doors and the car body are connected by the hinge point units. The tire pressure is measured by a test at 0.8 MPa. The contact type between components based on the penalty function method is adopted to solve the boundary nonlinear problem (Jung *et al.* 2015; Yan 2011). Figure 3 shows the simulation model.

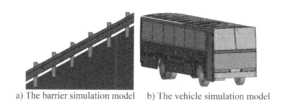

a) The barrier simulation model   b) The vehicle simulation model

Figure 3. FEA model.

3.2 *Collision analysis*

The standard collision conditions include a 10-ton bus, impact speed of 60 km/h, impact angle of 20°, and impact energy of 160 kJ. Under the standard crash, the simulation analysis of the A-grade corrugated beam barrier of the bus collision was carried out, and the full-size crash test of the grade A w-beam barrier of the bus collision was organized (Kovar *et al.* 2021; Ochoa & Ochoa 2015; Pajouh *et al.* 2018; Sennah & Mostafa 2018; Wei *et al.* 2017).

In Figure 4, it shows the comparison of the simulation and the test results of the bus collision with an A-grade w-beam barrier under standard collision conditions, which can be seen as follows: under the standard collision conditions, the vehicle attitude and barrier deformation obtained by the simulation calculation is the same as the test results, which proves that the simulation model of structure collision has high accuracy, and lays a foundation for the use of the simulation technology to research the collision simulation. The

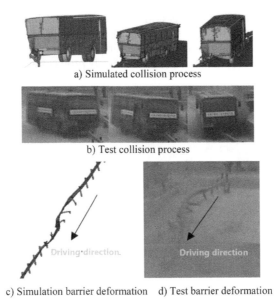

a) Simulated collision process

b) Test collision process

c) Simulation barrier deformation   d) Test barrier deformation

Figure 4. Comparison of the results between simulation and test.

riding phenomenon occurs after the bus collides with the A-grade corrugated beam barrier, which does not meet the requirements of the "New Standard". From the perspective of safety, it is necessary to further study the protective ability of the A-grade corrugated beam barrier to the bus and explore the limited protective ability of the A-grade corrugated beam barrier based on the bus collision.

4 EXPLORATION OF ULTIMATE PROTECTION CAPABILITY

Based on the simulation model of the bus and barrier in the previous Section 2.1, the gradual approximation method is adopted to explore the ultimate protection capability of the grade A w-beam barrier based on the bus collision, that is, exploration from the collision energy of 160 kJ, simulation calculations are performed at 5 kJ intervals until a collision energy that met the safety requirements of the "New Standard" is found.

The simulation results show that when the collision energy is between 145 kJ and 160 kJ, the bus rides, which does not meet the evaluation criteria. Figure 5 shows the simulation results at 145 kJ collision energy.

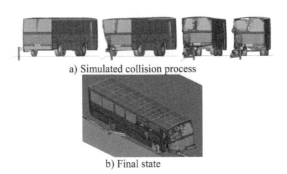

a) Simulated collision process

b) Final state

Figure 5. Simulation results with 145 kJ collision energy.

435

Figure 6 shows the simulation results of the grade A w-beam barrier for the bus collision when the collision energy is 140 kJ. It can be seen that the bus has a good driving attitude after the collision barrier, and the driving trajectory meets the requirements of the guiding exit frame. The Grade A w-beam barrier can effectively protect the bus with 140 kJ collision energy.

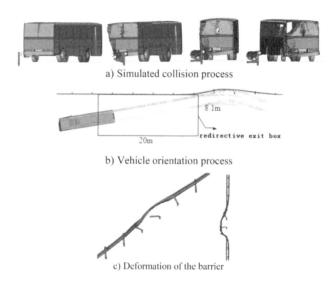

Figure 6. Simulation results with 140 kJ collision energy.

The simulation analysis results are summarized in Table 3, and it can be seen that all the indicators meet the requirements of the evaluation criteria 0.

Table 4. Simulation evaluation results.

Content of evaluation		Simulation result	Qualified or not
Containment function	Whether the vehicle traverses, climbs, and rides the barrier;	No	Yes
	Whether the barrier members and their detachments intrude into the passenger compartment of the vehicle;	No	Yes
Orientation function	Whether the vehicle overturns after the impact;	No	Yes
	Whether the wheel track of the vehicle after the impact meets the requirements of the guiding exit box.	Satisfy	Yes
Maximum dynamic lateral deflection of barriers (D)		1.119 m	
Maximum dynamic widening distance of lateral deflection of barriers (W)		1.212 m	
Maximum dynamic vehicle incline-out distance (VI)		1.520 m	
Maximum dynamic vehicle incline-out (VIn)		1.520 m	

The comprehensive simulation results show that the ultimate protection energy of the grade A w-beam barrier to the bus is 140 kJ.

5 ULTIMATE PROTECTION CAPACITY TEST VERIFICATION

A full-size collision test was carried out under the limit protection energy of 140 kJ, and the correctness of the simulation analysis results of the limit protection ability of the grade A w-beam barrier based on the bus collision was verified. Figure 7 shows the test vehicle and test barrier.

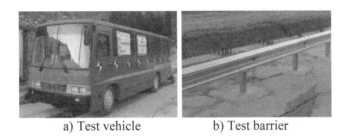

a) Test vehicle   b) Test barrier

Figure 7. Ultimate protection capability verification test.

Figure 8 shows the results of the collision test. It can be seen that the grade A w-beam barrier can effectively intercept and guide the bus under the impact energy of 140 kJ, and the indicators meet the requirements of the evaluation standards. However, the bus is seriously tilted and close to capsizing, indicating that the impact energy of 140 kJ is the limited protection ability of the w-beam barrier to the bus. The experiment provides data support for the correctness of the analysis results of computer simulation.

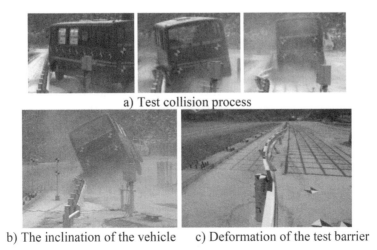

a) Test collision process

b) The inclination of the vehicle   c) Deformation of the test barrier

Figure 8. Results of the proof test of limit protective ability.

6 CONCLUSION

By using the comprehensive methods of the computer simulation analysis and the real vehicle full-size impact test, the ultimate protection ability of the grade A w-beam barrier based on the bus collision was studied. The main conclusions are as follows:

(1) The grade A w-beam barrier can meet the requirements of the "Old Standard", but can not satisfy the requirements of the "New Standard".
(2) The buses have the phenomenon of riding over the barrier when the collision energy ranges from 145 kJ to 160 kJ, which does not meet the evaluation criteria.
(3) According to the simulation collision analysis and the real vehicle crash test data, the buses all have a good driving posture and the wheel mark can meet the requirements of the guiding exit box at the collision energy of 140 kJ. The 140 kJ limit protection energy of the grade A w-beam barrier to the bus is verified.

The research results provide basic data for the reasonable application of the grade A w-beam barrier in practical highway engineering.

REFERENCES

Jung W. Y., Noh M. H., and Lee S. Y. "Estimation of THIV for Car Crash Against Attachable Roadside Barriers Made of the High Strength Steel". *Applied Mechanics and Materials*. Papers 751, 222–227 (2015).

Kovar J., Sheikh N., R. Bligh, et al. "Development and Testing of Structurally Independent Foundations for High-speed Containment Concrete Barrier". *Transportation Research Record Journal of the Transportation Research Board*. Papers 2675 (4), 297–307 (2021).

Ochoa C. M. and T. A. Ochoa. "Physics Models for Vehicle Crush During Frontal Fixed-barrier Impacts". *Transportation Research Record Journal of the Transportation Research Board*. Papers 2472: 220–235 (2015).

Pajouh M. A., Julin R.D., Stolle C. S., et al. "Rail Height Effects on the Safety Performance of Midwest Guardrail System". *Traffic Injury Prevention*. Papers 19 (1-4), 219–224 (2018).

Sennah K. and Mostafa A. "Performance of a Developed TL-5 Concrete Bridge Barrier Reinforced with GFRP Hooked Bars: Vehicle Crash Testing". *Journal of bridge engineering*. Papers 23 (2), 04017139.1–04017139.20 (2018).

The Ministry of Transport of the People's Republic of China: *Guidelines for Design of Highway Safety Facilities*. China Communications Press, Beijing, 2006. (In Chinese)

The Ministry of Transport of the People's Republic of China: *Standard for Safety Performance Evaluation of Highway Barriers*. China Communications Press, Beijing, 2013. (In Chinese)

Wei Z., Robbersmyr K. G., and Karimi H. R. "Data-based Modeling and Estimation of Vehicle Crash Processes in Frontal Fixed-barrier Crashes". *Journal of the Franklin Institute*. Papers 354 (12), 4896–4912 (2017).

Yan S. M. "Feasibility Analysis of Barrier Safety Evaluation with Finite Element Simulation Method". *Journal of Vibration and Shock*. Papers 1 (30), 152–156 (2011). (In Chinese)

Water Conservancy and Civil Construction – Oke & Ahmad (Eds)
© 2024 The Author(s), ISBN: 978-1-032-58618-2

Study on public transportation mode considering group heterogeneity

Jie Ji*, Yutong Li & Ruiqi Dong
College of Transportation Engineering, Chang'an University, Xi'an, China

ABSTRACT: In order to promote low-carbon travel for residents, this paper constructs a multivariate ordered logistic model based on the interaction effect between socioeconomic attributes and residents' attitudes towards travel modes and explores the influence of residents' attitudes towards public transportation. The results show that the elderly over 60 years old are less likely to choose public transportation than the young. Respondents who are tenants travel by bus more than those who own their homes. After adding the factor of travel attitude to model A, the pseudo-R square is significantly improved and the preference for bus travel will significantly increase. In model C, the pseudo-R square increased to 0.207, indicating the rationality of travel attitude and socioeconomic attributes. Among them, women with positive attitudes towards public transportation were less likely to choose to travel by bus than men. Both the significance level and the regression coefficient of subway attitude and bus attitude factors on the frequency of choosing subway travel are quite close, indicating that subway and bus are fungible. The results can provide a theoretical basis for formulating differentiated traffic route planning and guiding policies to promote the sustainable development of green transportation.

1 INTRODUCTION

According to the statistics of global carbon dioxide emissions, a quarter of the total carbon emissions comes from the transportation industry, among which motor vehicles account for half of the carbon emissions (International Energy Agency 2009). In China, with the energy consumption of the transportation industry growing year by year, it has become a major factor in carbon emissions (Zhu et al. 2015). Therefore, how to reduce the carbon emissions of motor vehicles and alleviate air pollution has become the primary problem to be solved by the government and the transportation sector. At present, many countries and regions have taken a series of measures to reduce carbon emissions, one of which is to encourage citizens to travel by bus. Based on the comprehensive consideration of the urban living environment, structural function, and spatial layout, professor Robert Cervero (Robert 1998) proposed the concept of a "bus city". He emphasized that urban public transport should actively guide urban development, and highlighted the symbiosis between urban public transport and citizens, realizing the human-centered urban development mode. On May 12, 2022, China's Ministry of Transport proposed the planning tasks of promoting urban sustainable development by providing convenient, efficient, green, and economic services under the "double carbon" goal (Political Bureau of the CPC Central Committee 2020). On the other hand, despite China's growing car ownership and the propensity for motor vehicle travel, most travelers are still at low and middle-income levels. Therefore, promoting public

---

*Corresponding Author: 862940776@qq.com

DOI: 10.1201/9781003450832-56

transportation to solve the problem of carbon emissions and meet the demand of most people is of great significance.

Public travel behavior is the result of both subjective and objective factors (He et al. 2017) and attitude is the most direct factor affecting behavioral intention (WU et al. 2017). Nowadays, domestic and foreign research on the impact of travel attitudes on travel modes has become mature in terms of theories and methods. Wu (2017) divided the subjective attitude into three categories and they found that the subjective attitude of residents on low-carbon commuting was closely related to the choice of individual travel mode and the composition of influencing factors of behavioral intention. The path of action and the intensity of action were different with different subjective attitudes held by groups. Cheng (2016) studied the influence of socioeconomic attributes, activity attributes, and attitude variables on travel mode choices for low-income urban commuters. Li (2021) took Beijing as an example and the study found that commuters' potential preferences or dislikes for travel modes significantly affected their travel mode choices. Fu (2016) believed that residents' subjective attitude towards a certain mode of travel not only affects this mode of travel but also affects other modes of travel. Existing studies have shown that the impact of attitudes on travel modes varies according to specific social population groups (Pucher et al. 2010). This group heterogeneity manifests as groups with different socioeconomic attributes, such as differences in gender, age, driver's license holding, household income, and car ownership, which will lead to differences in travel demand and behavioral characteristics (Waygood et al. 2015). By classifying travelers into heterogeneous categories and analyzing the influencing factors of travelers in each category, it can avoid blindly classifying travelers according to a certain characteristic of travelers (Liu et al. 2019). At the same time, it is helpful to more accurately grasp differences in travel satisfaction among different groups (Zhang et al. 2020). Therefore, considering group heterogeneity is of great significance for formulating differentiated bus travel optimization policies. In this paper, the study establishes a multivariate ordered logistic model based on the interaction effects between socioeconomic attributes and residents' attitudes towards travel modes. The results of this study are helpful for cities and traffic managers to understand the differentiated behavior characteristics and patterns of commuters and provide a reference for the government to guide and promote public travel from the carbon-intensive mode to the low-carbon mode represented by public transport.

2 DATA SURVEY AND STATISTICS

A total of 397 questionnaires are collected by random sampling, of which 27 pieces are deleted according to the needs of the research objectives and the remaining 370 pieces are valid questionnaires. The KMO test and Bartlett sphere validity test are conducted on the questionnaires and the results are shown in Table 1. According to Table 1, KMO=0.735, greater than 0.5. In addition, for the Bartlett sphericity test, the P value is 0, less than the significance level of 0.05, which rejects the null hypothesis, indicating that there is a strong correlation between variables and the sample data through the KMO test and Bartlett test.

Table 1. The KMO test and Bartlett test.

KMO sampling appropriateness quantity		0.735
Bartlett's test of sphericity	The approximate chi-square	1305.910
	Degrees of freedom	78
	Significance	0.000

The questionnaire mainly uses Likert 5-level scale to collect passengers' travel attitudes. In addition to summarizing the selection of evaluation indicators for each mode of public transport service quality in the relevant literature, the questionnaires also consider factors such as travel frequency, gender, age, driving license, and housing, and adjust the index selection of each dimension according to the actual traffic conditions. The descriptive statistics are shown in Table 2.

Table 2. The descriptive statistics (N = 260).

Variables	Classification	Proportion
Gender	Male	64.60%
	Female	35.40%
Age	18–29	31.20%
	30–39	28.80%
	40–49	16.20%
	50–59	11.50%
	≥ 60	12.30%
Driver's license	Yes	64.60%
	No	35.40%
Private car ownership	0	26.20%
	1	61.50%
	≥ 2	12.30%
The housing situation	Rental	20.00%
	Own-occupied	80.00%
Self-driving frequency	Never	62.70%
	<1 day per month	7.70%
	1–3 days per month	7.70%
	1 day per week	3.10%
	2–3 days per week	6.20%
	4–5 days per week	12.70%
Subway Frequency	Never	30.80%
	<1 day per month	6.90%
	1–3 days per month	15.80%
	1 day per week	8.50%
	2–3 days per week	15.80%
	4–5 days per week	22.30%
Bus frequency	Never	38.50%
	<1 day per month	5.00%
	1–3 days per month	12.30%
	1 day per week	10.40%
	2–3 days per week	20.00%
	4–5 days per week	13.80%
Bicycle frequency	Never	57.70%
	<1 day per month	10.00%
	1–3 days per month	7.70%
	1 day per week	5.80%
	2–3 days per week	8.10%
	4–5 days per week	10.80%
Walking frequency	Never	36.20%
	<1 day per month	5.00%
	1–3 days per month	3.80%
	1 day per week	5.80%
	2–3 days per week	6.50%
	4–5 days per week	42.70%

3 MODEL BUILDING

Multiple ordered logistic regression model is used in this experiment. The multivariate ordered logistic regression model is a kind of nonlinear regression analysis model, which is a common method to analyze the relationship between the dependent variable and multiple independent variables of ordered multiple categories.

$$p(y = j/x_i) = \frac{1}{1 + \exp(-(\alpha + \beta x_i))} \tag{1}$$

Where Y represents the travel frequency level and J (1, 2, 3, 4, 5, 6) is assigned to each level, x_i denotes the ith factor affecting travel frequency. The Logit model is established as follows:

$$\log_{it}(P_j) = \ln[P(y \le j)/p(y \ge j + 1) = \alpha_j + \beta x \tag{2}$$

Type: Pj is the probability that the travel frequency of travelers belongs to a certain level; $P_j = p(y=j)$ j=1, 2, 3, 4, 5, 6, and $X = (x_1, x_2, \cdots, x_i)^T$ represents the characteristic independent variable; β is the model regression coefficient; α_j is the model intercept term.

After getting α_j and β, the probability of a particular situation (y=j) can be obtained from the the the following equation:

$$P(y \le j/x) = \frac{\exp(-(dj + \beta x_i))}{1 + \exp(-(\alpha_j + \beta x))} \tag{3}$$

4 ORDINAL REGRESSION MODEL ANALYSIS OF BUS GROUPS

This study selects the frequency of driving, subway, bus, and bicycle trips as the dependent variables to set up four parallel models and the five factors of driving attitude, subway attitude, bus attitude, bicycle attitude, and walking attitude obtained by factor analysis, which are used as the explanatory variables of each model. Gender, age, driving license, family private car ownership, and housing status are used as control variables to ensure that the independent variables in each model are completely consistent. Model A represents results only under the action of the control variables and the explanatory variables are added to model A to form model B. Considering the interaction effect between travel mode attitude and some control variables, the interaction term is added to model B to form model C.

According to Model A, respondents [age =1] (18-29 years old) choose the bus more than respondents [age=5] ($\ge$ 60 years old), which is mainly because the comfort level of bus lines in Xi 'an may not meet the needs of elderly people over 60 years old. The respondents who rent a house choose the bus more than those who own a house, which may be related to the characteristics of renting a house that prioritizes the distance to the workplace. When the travel attitude factor is added to model A, the pseudo-R square is significantly increased, which indicates that the travel attitude factor has a meaningful role in explaining the choice of bus travel modes. Model B confirms that the preference for bus travel would significantly promote bus choices. Based on model C, the interaction terms between gender and public transport attitude are analyzed and it is found that women with positive attitudes towards public transport are more likely to choose public transport than men. This phenomenon can be explained from two aspects. First, women tend to take on more domestic obligations than men, such as delivering children to and from school and purchasing food and daily necessities. Second, men tend to work more and choose to take more long-distance travel modes such as driving by themselves. In addition, both the significance level and the regression

Table 3. Ordinal regression model of bus group.

Variables	Categories	Model A	Model B	Model C
Attitude	Subway		0.063 (0.123)	0.635 (0.596)
	Bus		0.447***(0.124)	1.611***(0.612)
	Bicycle		0.114 (0.118)	0.157 (0.753)
	Car		-0.138 (0.125)	-0.155 (0.515)
	Walk		-0.124 (0.122)	-0.548 (0.575)
Gender	Male [1]	-0.158 (0.245)	-0.168 (0.253)	-0.172 (0.285)
	Female [2]			
Age	18-29	0.775*(0.451)	0.993**(0.479)	1.271**(0.511)
	30-39	0.003 (0.444)	0.195 (0.461)	0.299 (0.496)
	40-49	0.513 (0.448)	0.575 (0.458)	0.705 (0.502)
	50-59	0.369 (0.469)	0.456 (0.481)	0.240 (0.537)
	≥ 60			
Driver's license	Yes	0.212 (0.308)	0.25 (0.313)	0.398 (0.328)
	No			
Private car	0	0.481 (0.439)	0.304 (0.475)	0.626 (0.501)
	1	0.32 (0.366)	0.182 (0.401)	0.472 (0.428)
	2			
Housing condition	Rental	0.577*(0.319)	0.675**(0.328)	0.633*(0.345)
	Own-occupied			
Gender*attitude	[Gender =1] * subway attitude			0.375 (0.271)
	[Gender =2] * subway attitude			
	[Gender =1] * bus attitude			-0.126022
	[Gender =2] * bus attitude			
	[Gender =1] * bicycle attitude			0.196 (0.287)
	[Gender =2] * bicycle attitude			
	[Gender =1] * car attitude			-0.008 (0.290)
	[Gender =2] * car attitude			
	[Gender =1] * walk attitude			-0.305 (0.269)
	[Gender =2] * walk attitude			
	Pseudo-R square	0.052	0.109	0.207

Level of significance: ***$p \leq 0.01$, **$p \leq 0.05$, *$p \leq 0.1$.

coefficient of the factors of subway attitude and bus attitude on the frequency of choosing subway travel are quite close, which indicates that modes of subway and bus are fungible.

5 CONCLUSION

This study investigates the attitude toward travel modes and social demographics for Xi'an residents, and it takes attitudes on travel way as explained variables, a variety of personal and social economic attributes as control variables, based on the driving, subway, bus, and bicycle, four kinds of modes, to construct four groups of parallel regression model and add the interaction effect analysis. By comparing the pseudo-R-square of the goodness of fit, the rationality of adding travel attitude and interaction term to the model is verified. The results show that: (1) compared with young people, the elderly over 60 years old are less likely to choose public transportation. It is suggested to strengthen the improvement of the aging adaptability of public transportation to improve the stability and comfort of the vehicles; (2) respondents who rent houses usually choose buses more than those who own houses. It is suggested that the planning and renovation of bus routes should reasonably adjust the location of stations to accurately serve residents along the routes by obtaining rental or owner-occupied conditions of nearby residential communities through investigation; (3) women who have a positive attitude towards public transportation are more likely to take public transportation than men, which has a positive effect on improving women's sense of security by considering gender factors in the process of road configuration and optimization; (4) the substitutability between subway and bus promotes the government to pay more attention to the coordinated development of subway and bus and solve the problem of poor connection between them.

REFERENCES

Cervero Robert. (1998) *The Transit Metropolis: a Global Inquiry*. London, Island Press.

Cheng Long, Chen Xuewu, and Yang Shuo. Low-income Commuters' Mode Choice Utilizing Attitude-behavior Model [J]. *Journal of Transportation Systems Engineering and Information Technology*,16 (1), 176–181, (2016).

Fu Xuemei and Juan Zhicai. Commuting Mode Choice Behavior Based on ICLV Model [J]. *Journal of Systems & Management*,25 (06), 1046–1050, (2016).

He Yao, Fang Xiaoping, and Yang Yang. The Research of Public Low-carbon Travel Behavior Decision Model [J]. *Journal of Railway Science and Engineering*, 14 (5), 1077–1085 (2017).

International Energy Agency. *Transport Energy and CO2: Moving Towards Sustainability [R]*. Paris: OECD Publishing, 2009.

Li Hongchang, Cui Jinli, and Pei Xinghua. Analysis of the Impacts of Beijing's Perceived Value on Shared Bicycle Choice [J]. *Science Technology and Engineering*,21 (05), 2034–2041, (2021).

Liu Jianrong and Wen Huiying. Incorporating Heterogeneity and Subjective Evaluation into Travel Mode Choice [J]. *Journal of South China University of Technology (Natural Science Edition)*,47 (10), 124–129, (2019).

Political Bureau of the CPC Central Committee. 14th Five-Year Plan [R]. *Beijing, Political Bureau of the CPC Central Committee*, 2020.

Pucher J., Dill J., and Handy S. Infrastructure, Programs, and Policies to Increase Bicycling: an International Review. *Prev. Med.* 50, S106–S125 (2010).

Waygood E. O. D., Sun Y., and Letarte L. Active Travel by Built Environment and Lifecycle Stage: Case Study of Osaka Metropolitan Area [J]. *International Journal of Environmental Research and Public Health*, 12 (12), 15900–15924, (2015).

Wu Wenjing, Sun Renchao, Zong Fang, and Jia Hongfei. Subjective Attitude Identification and Impact Analysis of Residents\Low-Carbon Commuting Travel [J]. *Journal of Chongqing Jiaotong University (Natural Science)*,40 (05), 53–58 (2021).

Zhang Xiaoqi, Miao Wang and Yang Jiankun. Statistical Test and Modeling Analysis of Group Heterogeneity in Commuting Satisfaction [J]. *Journal of Transportation Engineering*, 22 (1), 29–34, (2022).

Zhu Bangzhu, Wang Kefan, and Wang Ping. Research of China's Carbon Emissions Growth Drivers in Stages [J]. *Economic Perspectives*, (11), 79–89 (2015).

Water Conservancy and Civil Construction – Oke & Ahmad (Eds)
© 2024 The Author(s), ISBN: 978-1-032-58618-2

Traffic management with households under stochastic bottleneck capacity when school is near home

Boyu Lin*
Key Laboratory of Transport Industry of Big Data Application Technologies for Comprehensive Transport, Ministry of Transport, Beijing Jiaotong University, Beijing, China

ABSTRACT: This paper investigates household travel behaviors under stochastic bottleneck capacity when school is near home. From different studies, this paper considers the scenario that a school is close to home, and a traffic bottleneck exists between the school and work location. Furthermore, the bottleneck follows a binary distribution with a good capacity and a bad capacity. In user equilibrium, there are four possible equilibrium patterns. The end departure time should be the expected arrival time of the school in each equilibrium pattern under the assumption that free-flow travel time is zero. And the critical time points, the mean travel cost, and the boundary conditions of four equilibrium patterns are derived analytically. Due to the impacts of the school-work start time difference, the degraded ratio of capacity and the probability of degraded capacity on the mean travel cost are studied. The results show that commuters' travel choices can be regulated by adjusting the start time difference.

1 INTRODUCTION

Traffic congestion has become an important factor limiting urban development, and the traffic bottleneck has an exceeding impact on traffic congestion. The bottleneck model proposed by Vickrey describes the process by which travelers pass a single traffic bottleneck from home to work (Vickrey 1969). A lot of problems with individual trips are investigated based on Vickrey's model, such as mode choice, parking, stochasticity, and travel time reliability. Compared to individual trips, household trips are common in Asian countries. Jia *et al.* investigated the bottleneck model with household (Jia *et al.* 2016). Liu *et al.* extended the study by considering both household trips and individual trips (Liu *et al.* 2017). Vickrey's model is simple because of the certainty of all factors. But in the actual scenario, due to the occurrence of unpredictable events, travel time has great uncertainty. Zhu *et al.* introduced travel time reliability in general (Zhu *et al.* 2018). And Yu considered the variability of the number of departures per day (Yu 2015). Long *et al.* proposed the general model by considering the stochasticity in bottleneck capacity (Long *et al.* 2022). This paper investigated another case of household trips, i.e., the school is very close to home, and the departure time choice model is built under the stochasticity in bottleneck capacity.

2 THE CLASSICAL BOTTLENECK MODEL

In 1969, Vickrey proposed the classical bottleneck model by using the queuing theory. The departures with the constant number of N going to work from home by passing a traffic

---

*Corresponding Author: lby.studious@bjtu.edu.cn

DOI: 10.1201/9781003450832-57

bottleneck, and the capacity is s every day. This model measures social results by total travel cost. Since the departure rates during peak hours are greater than the bottleneck capacity s, there will be queues at the bottleneck. Therefore, some travelers will arrive early or late inevitably. To simplify the model, the free-flow time is assumed to be zero. The cumulative departures at t can be given as:

$$D(t) = \int_{t_s}^{t} r(x)dx \qquad (1)$$

where t_s is the departure time of the first traveler and $r(t)$ is the departure rate at t. The number of the queue at bottleneck can be expressed as:

$$Q(t) = \max\{D(t) - s(t - t_s), 0\} \qquad (2)$$

and the queuing time can be given as follows:

$$T(t) = \frac{Q(t)}{s} = \max\left\{\frac{D(t)}{s} - t + t_s, 0\right\} \qquad (3)$$

Assuming that the expected arrival time of work is t^*, the total travel cost of the traveler can be expressed as:

$$C(t) = \alpha T(t) + \beta \max\{t^* - t - T(t), 0\} + \gamma \max\{t + T(t) - t^*, 0\} \qquad (4)$$

where α is the unit queuing cost, β is the unit cost of schedule delay early, and γ is the unit cost of schedule delay late. According to the experimental verification of small, the condition $0 < \beta < \alpha < \gamma$ needs to be satisfied (Small 1982).

3 THE BOTTLENECK MODEL WITH HOUSEHOLDS WHEN SCHOOL IS NEAR MORE

3.1 Scenario description

For household trips, including one adult and one child, the adult first sends the child to school and then goes to work. In previous studies, the bottleneck is usually assumed to be between home and school, but commuters could not encounter the bottleneck from home to school when the school is close to home. And they encounter the bottleneck on the way to work. The network is shown in Figure 1. Following the study of Liu, et al., the bottleneck capacity is assumed to follow a binary distribution of poor and good capacity (Liu et al. 2020). The probability of the good capacity $\bar{s}1 - \pi$ is, and the probability of the poor one $\theta\bar{s}\pi$ is, where $0 < \theta \le 10 \le \pi \le 1$. Moreover, the capacity is assumed to be constant within a day and changes stochastically from day to day. The total travel cost for the household can be given as follows:

$$C(t) = \beta \max\{t_1^* - t - T(t), 0\} + \gamma \max\{t + T(t) - t_1^*, 0\} \\ + \alpha T(t) + \beta \max\{t_2^* - t - T(t), 0\} + \gamma \max\{t + T(t) - t_2^*, 0\} \qquad (5)$$

Figure 1. A network in the household trip when school is near home.

where t_1^* is the expected arrival time of school, and t_2^* is the expected arrival time of work. Following the study of Jia, et al. and Liu, et al., we consider the condition of $t_1^* \leq t_2^*$.

3.2 Equilibrium departure rates

In equilibrium patterns, children experience schedule delay early (SDE) or not. Children will avoid being late because the unit cost of a scheduled delay late is great than that of early. The end time of household departure should be t_1^* in equilibrium consequently. There will be three scheduled delay types for adults in equilibrium due to the stochastic bottleneck capacity. They may experience schedule delay early (SDE), schedule delay early or late (SDE/L), or schedule delay late (SDL). Similarly, there will be two types of queuing experienced by adults. They are always queuing (AQ) and possibly queuing (PQ), respectively. Combining the above types, there will be 4 situations for household trips. And in accordance with Wardrop's first principle to minimize the expected travel cost, the equilibrium condition is,

$$\frac{d(E[C(t)])}{dt} = 0, \quad \text{if } r(t) > 0. \tag{6}$$

where $E[C(t)]$ is the expected travel cost. Four situations and the equilibrium departure rates for each are shown in Table 1.

Table 1. Possible situations and the equilibrium departure rates.

Situation	The equilibrium departure rate
S1: SDE+SDE+AQ	$r_1 = \dfrac{(\alpha+\beta)\theta\bar{s}}{(\alpha-\beta)[\theta(1-\pi)+\pi]}$
S2: SDE+SDE/L+AQ	$r_2 = \dfrac{(\alpha+\beta)\theta\bar{s}}{(\alpha+\gamma)\pi + (\alpha-\beta)(1-\pi)\theta}$
S3: SDE+SDL+AQ	$r_3 = \dfrac{(\alpha+\beta)\theta\bar{s}}{(\alpha+\gamma)[\theta(1-\pi)+\pi]}$
S4: SDE+SDE/L+PQ	$r_4 = \dfrac{(\beta+\alpha\pi)\theta\bar{s}}{\pi(\alpha+\gamma)-(1-\pi)\beta\theta}$

3.3 Possible equilibrium departure patterns

The above four situations in Table 1 are arranged according to rules, and four equilibrium patterns are obtained as shown in Figure 2, where t_i^j denotes the i-th critical time points in Pattern j.

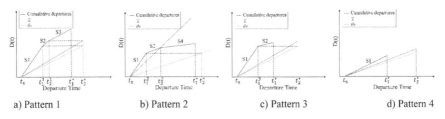

a) Pattern 1   b) Pattern 2   c) Pattern 3   d) Pattern 4

Figure 2. A network in the household trip when school is near home.

3.4 The critical time points and the mean travel cost

The critical time points as shown in Figure 2 can be obtained by referring to Vickrey and Long. The expressions can be obtained in Table 2, and the detailed derivations are omitted.

In equilibrium, the total travel cost is the same for all commuters. The first household only experiences the schedule delay early, i.e., $E[C(t_s)] = \beta(t_1^* - t_s) + \beta(t_2^* - t_s)$. Thus, corresponding to the t_s in Table 2, the mean travel cost of the four equilibrium patterns can be obtained as shown in Table 3.

3.5 The boundary conditions

According to the magnitude of critical time points and equilibrium departure rates in Figure 2, the boundary conditions of four equilibrium patterns can be derived. The

Table 2. The critical time points of equilibrium patterns.

Pattern	Time points
1	$t_s = t_1^* + \dfrac{\beta + \gamma}{\alpha + \gamma + 2\beta}(t_2^* - t_1^*) - \dfrac{\alpha + \gamma}{\alpha + \gamma + 2\beta} N\left(\dfrac{1 - \pi}{\bar{s}} + \dfrac{\pi}{\theta \bar{s}}\right)$
	$t_1^1 = t_s + \dfrac{\theta s(t_2^* - t_s)}{r_1}$
	$t_2^1 = t_1^1 + \dfrac{\bar{s}(1 - \theta)}{r_2}(t_2^* - t_s)$
2	$t_s = t_2^* - \dfrac{\pi \alpha + \beta}{(\alpha + \gamma + \beta)\pi + \beta}(t_2^* - t_1^*) + \dfrac{\beta}{(\alpha + \gamma + \beta)\pi + \beta}\dfrac{N}{\bar{s}}(1 - \pi) - \dfrac{(\alpha + \gamma)\pi}{(\alpha + \gamma + \beta)\pi + \beta}\dfrac{N}{\theta \bar{s}}$
	$t_1^2 = t_s + \dfrac{\theta \bar{s}(t_2^* - t_s)}{r_1}$
	$t_2^2 = t_1^2 + \dfrac{\bar{s}}{r_2 - \bar{s}}(t_1^2 - t_s) - \dfrac{\theta s}{r_2 - \bar{s}}(t_2^* - t_s)$
3	$t_s = t_1^* + \dfrac{\pi(\beta + \gamma)}{\alpha + \beta + \pi(\beta + \gamma)}(t_2^* - t_1^*) - \dfrac{(\alpha - \beta)(1 - \pi)}{\alpha + \beta + \pi(\beta + \gamma)}\dfrac{N}{\bar{s}} - \dfrac{\pi(\alpha + \gamma)}{\alpha + \beta + \pi(\beta + \gamma)}\dfrac{N}{\theta \bar{s}}$
	$t_1^3 = t_s + \dfrac{\theta \bar{s}(t_2^* - t_s)}{r_1}$
4	$t_s = t_1^* - \dfrac{\alpha - \beta}{\alpha + \beta}N\left(\dfrac{1 - \pi}{\bar{s}} + \dfrac{\pi}{\theta \bar{s}}\right)$

Table 3. The mean travel cost of equilibrium patterns.

Pattern	The mean travel cost
1	$\dfrac{\beta(\alpha - \gamma)}{\alpha + \gamma + 2\beta}(t_2^* - t_1^*) + \dfrac{2\beta(\alpha + \gamma)}{\alpha + \gamma + 2\beta} N\left(\dfrac{1 - \pi}{\bar{s}} + \dfrac{\pi}{\theta \bar{s}}\right)$
2	$\dfrac{\beta[(\alpha - \gamma - \beta)\pi + \beta]}{(\alpha + \gamma + \beta)\pi + \beta}(t_2^* - t_1^*) - \dfrac{2\beta\beta}{(\alpha + \gamma + \beta)\pi + \beta}\dfrac{N}{\bar{s}}(1 - \pi) + \dfrac{2(\alpha + \gamma)\beta\pi}{(\alpha + \gamma + \beta)\pi + \beta}\dfrac{N}{\theta \bar{s}}$
3	$\dfrac{\beta[\alpha + \beta - \pi(\beta + \gamma)]}{\alpha + \beta + \pi(\beta + \gamma)}(t_2^* - t_1^*) + \dfrac{2(\alpha - \beta)(1 - \pi)\beta}{\alpha + \beta + \pi(\beta + \gamma)}\dfrac{N}{\bar{s}} + \dfrac{2\pi(\alpha + \gamma)\beta}{\alpha + \beta + \pi(\beta + \gamma)}\dfrac{N}{\theta \bar{s}}$
4	$\beta(t_2^* - t_1^*) + 2\beta\dfrac{\alpha - \beta}{\alpha + \beta}N\left(\dfrac{1 - \pi}{\bar{s}} + \dfrac{\pi}{\theta \bar{s}}\right)$

448

boundary conditions of each equilibrium pattern are shown in Table 4. The specific derivations can refer to the previous research.

where $\Delta t_1 = \frac{2\beta\theta - (\alpha+\gamma)\pi(1-\theta)}{(\alpha+\beta)[\theta(1-\pi)+\pi]} N \left(\frac{1-\pi}{\bar{s}} + \frac{\pi}{\theta \bar{s}} \right)$, $\Delta t_2 = \frac{\alpha+\gamma}{\gamma+\beta} \frac{N}{\theta \bar{s}} - \frac{[(\alpha+\gamma)\pi+2\beta]}{(\gamma+\beta)\pi} \frac{N}{\bar{s}}$, and $\Delta t_3 = \frac{\alpha+\beta-(\alpha-\beta)[\theta(1-\pi)+\pi]}{(\alpha+\beta)[\pi(\beta+\gamma)+(\alpha-\beta)[\theta(1-\pi)+\pi]]} \left[(\alpha-\beta)(1-\pi)\frac{N}{\bar{s}} + \pi(\alpha+\gamma)\frac{N}{\theta \bar{s}} \right]$.

Table 4. The boundary conditions of equilibrium patterns.

Pattern	Boundary conditions
1	$t_2^* - t_1^* < \Delta t_1$
2	$\Delta t_1 < t_2^* - t_1^* < \min\{\Delta t_2, \Delta t_3\}$
3	$\max\{\Delta t_1, \Delta t_2\} < t_2^* - t_1^* < \Delta t_2$
4	$t_2^* - t_1^* > \Delta t_3$

4 NUMERICAL STUDIES

4.1 The equilibrium pattern diagrams

In this section, the values of Arnott et al. are adopted, i.e., $\alpha = 6.4\$/h$, $\beta = 3.9\$/h$, $\gamma = 15.21\$/h$. The good capacity is $\bar{s} = 3000$veh/h. And the total travel demand is $N = 5000$veh. The school-work start time difference $t_2^* - t_1^*$ is denoted as Δt.

Figure 3 shows the equilibrium patterns in the plane of θ-Δt and θ-π. In the plane of θ-Δt, Patten 1 - Pattern 4 exist simultaneously. Nevertheless, Pattern 1 and Pattern 4 do not exist in the same plane of θ-π. As shown in Figure 3(a), Pattern 1 only exists when Δt is small respectively, and Pattern 4 only exists when Δt is large. In Figures 3(b) and 3(c), this result is also reflected.

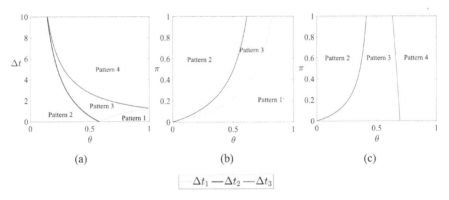

Figure 3. A network in the household trip when school is near home. Where (a) $\pi = 0.5$, (b) $\Delta t = 0.5$h, (c) $\Delta t = 2$h.

4.2 Impacts of Δt, θ and π on the mean travel cost

Figure 4 shows the isolines of mean travel costs in different patterns. The mean travel cost approaches infinity as θ approaches 0, therefore, the range θ is [0.4,1] to make the trend of isolines more obvious. The mean travel cost decreases with Δt in Pattern 1 and Pattern 2 as shown in Figure 4(a), but it increases with Δt in Pattern 3 and Pattern 4. Otherwise, the mean

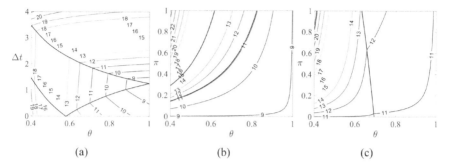

Figure 4. The isolines of mean travel cost in different patterns. Where (a) $\pi = 0.5$, (b) $\Delta t = 0.5$h, (c) $\Delta t = 2$h.

travel cost decreases θ, and the rate of cost reduction decreases θ. The mean travel cost increases with π expectedly.

5 CONCLUSIONS

The traffic management of household travel in morning commuting is investigated in this paper. The stochasticity of bottleneck and the scenario of the school near home are considered. It is worth noting that the traffic bottleneck is between the school and work location. In equilibrium patterns, the end departure time of the household should be t_1^* to ensure that children are not late for school. There are 4 situations considering the schedule delay and queuing for children and adults. And four possible equilibrium travel patterns are obtained. The theoretical boundaries of each equilibrium pattern have also been derived. Then, the equilibrium boundaries under different parameters are obtained by numerical studies. In addition, the influence of $\Delta t\theta$, and π on mean travel cost is analyzed by isolines. In the reality, it is difficult to adjust $\theta\pi$, which is usually at a great cost, hence the government can adjust the two expected arrival times to impact departure travel behavior. The pricing strategy can be considered in the following works. Time-varying tolls or step tolls can be examined to ease congestion.

REFERENCES

Arnott R., de Palma A., and Lindsey R. Economics of a Bottleneck. *J. Urban. Econ.* 27 (1), 111–130 (1990).
Jia Z., Wang D.Z., and Cai X. Traffic Managements for Household Travel in the Congested Morning Commute. *Transp. Res. Part E: Logist. Transp. Rev.* 91, 173–189 (2016).
Liu Q., Jiang R., Liu R., Zhao H., and Gao Z.Y. Travel Cost Budget Based User Equilibrium in a Bottleneck Model with Stochastic Capacity. *Transp. Res. Part B: Methodol.* 139, 1–37 (2020).
Liu W., Zhang F., and Yang H. Modeling and Managing Morning Commute with Both Household and Individual Travels. *Transp. Res. Part B: Methodol.* 103, 227–247 (2017).
Long J., Yang H., and Szeto W.Y. Departure Time Choice Equilibrium and Tolling Strategies for a Bottleneck with Stochastic Capacity. *Transp. Sci.* 56 (1), 79–102 (2022).
Small K.A. The Scheduling of Consumer Activities: Work Trips. *Am. Econ. Rev.* 72 (3): 467–479 (1982).
Vickrey W.S. Congestion Theory and Transport Investment. *Am. Econ. Rev.* 59 (2), 251–260 (1969).
Yu Xueyue. Travel Behavior and Tolling Strategies in a Bottleneck Model with Stochastic Demand. *Hefei university of technology* (2015).
Zhu S., Jiang G., and Lo H.K. Capturing Value of Reliability Through Road Pricing in Congested Traffic Under Uncertainty. *Transp. Res. Part C: Emerg. Technol.* 94, 236–249 (2018).

Water Conservancy and Civil Construction – Oke & Ahmad (Eds)
© 2024 The Author(s), ISBN: 978-1-032-58618-2

Analysis of research articles and international standards related to wastewater

Li Yefu & Wang Rong*
China Harbour Engineering Co., Ltd., Beijing, China

Hou Xiangyu
The Institute of Seawater Desalination and Multipurpose Utilization, MNR, Tianjin, China

Zhang Shoujia
China Harbour Engineering Co., Ltd., Beijing, China

ABSTRACT: Water is a key element of a sustainable future and a vital ingredient for renewable energy sources, food production, and the improvement of sanitation and health. Since the development of scientific research and standard for water provides tools for measurement of water use to realize optimization, and also offers best practices on the treatment and use of wastewater, many works focus on it. Understanding the status of scientific research and international standard related to wastewater treatment will be of great significance to the technology update and scientific management. In this paper, the field's distribution and the published trend of water-related literature were analyzed. And the current status of international standards, norms, and guidelines, related to the methods and management of water and wastewater treatment covered in ISO, were also analyzed. Results showed that more than 214,000 articles and 1,200 international standards were published, and the number has increased rapidly in recent years.

1 INTRODUCTION

Water is an essential ingredient in the sustainability of our planet (Clare 2021). Yet, nearly half of the world's population is living in areas lacking sufficient water, which could increase up to 5.7 billion in 2050 (Clare 2020). Therefore, how to use water and wastewater rationally is an important issue. Scientific research reports on water and water treatment can show the latest technological development in these fields. An accurate and timely grasp of the latest technological development can provide the necessary support for updating technologies, improving business management capabilities, and promoting the development of standards. The ISO standard covers almost every water issue and represents a consensus on practical solutions and best practices for sustainable water management. In addition, the ISO standard offers harmonized technology and terminology, allowing countries sharing the same water resources to work together. Regulators can rely on ISO standards as a solid base to create a public policy that addresses water-related challenges, to achieve their national and international water management commitments. Therefore, this article analyzes the current scientific literature and international standards related to water and wastewater treatment, based on the search and analysis in the SCOPUS database] and ISO official website.

---

*Corresponding Author: wangrong@chec.bj.cn

DOI: 10.1201/9781003450832-58

2 ANALYSIS METHODS

The SCOPUS database was used to analyze the scientific research literature related to wastewater treatment before September 2021. The Search includes the scopes of title, abstract, and keywords, with keywords of "wastewater and treatment". The search time is limited to 1960 to 2022. The search results were analyzed based on the article category, subject category, and years, which are taken by SCOPUS's analysis system. The current situation analysis of international standards is taken based on the ISO official website. The standards, pages, news, publications, and documents related to water and wastewater were searched and analyzed on the ISO official website, with search keywords of "water", "wastewater", "treatment", etc.

3 RESULTS AND DISCUSSION

3.1 *Scientific research*

The research literature related to wastewater treatment searched from the SCOPUS database was more than 214,000. The literature analysis results classified by article category are shown in Figure 1. Results showed that most of this literature is research articles, accounting for 77%. The conference paper is the second, accounting for 15%.

Figure 1. Literature analysis results are classified by article category.

Literature analysis results classified by subject category are shown in Figure 2. The number of literature in the fields of environmental science, engineering, and chemical engineering accounted for more than half. Especially environmental science has the most papers,

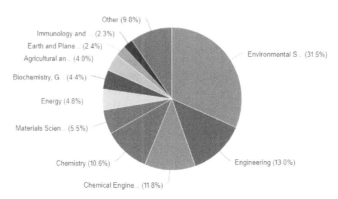

Figure 2. Literature analysis results are classified by subject category.

accounting for 31.5%. The next fields with many articles are chemistry, materials science, energy, biochemistry, genetics and molecular biology, agricultural and biological sciences, etc., the papers of which all exceed 10,000.

As shown in Figure 3, the research on wastewater treatment mainly started at the end of the 1960s, and then gradually increased. The research on wastewater treatment has received widespread attention since 2000. And much scientific research related to wastewater treatment was reported. The number of articles in the past 20 years has increased rapidly, and the rate of the increase was much faster than that of the last century. Especially in the past few years, the number of articles related to wastewater treatment has increased significantly. These research results have promoted the advancement of technology and management methods in water treatment, which will help promote the development of related industries.

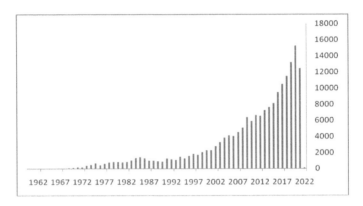

Figure 3. Literature analysis results classified by years.

3.2 *International standards*

The ISO has more than 1,200 standards related to water, with more standards in development. These international standards cover 10 water-related sectors (Figure 4), such as water quality, irrigation, footprinting, and pipes.

Figure 4. Water-related sectors covered in the ISO (Clare 2020).

The number of international standards for water issues was analyzed, as shown in Figure 5. Results showed that there are more than 500 standards for water quality covered in the ISO. These water quality standards are developed by ISO/TC 147 applicable to everything from plant treatment agents to natural mineral water. They provide common terminology, water sampling methods, and guidance on reporting and monitoring to determine a variety of properties and contaminants, from mineral content to the level of bacteria and impurities. There were more than 100 standards for piping and valves, drinking water, water treatment, water evaluation, irrigation, and sludge. Standards for irrigation are developed by ISO/TC 23/SC 18. They ensure greater efficiency by providing test methods and performance measurement. Giving a boost to sustainability, the ISO 16075 series of standards (developed by ISO/TC 282, water reuse) provides guidelines for treated wastewater use in irrigation projects. Standards about sludge recovery, recycling, treatment, and disposal are developed by ISO/TC 275. These guide defining the methods to characterize, categorize, prepare, treat, recycle, and manage sludge and products from urban wastewater collection systems, night soil, stormwater, and water and wastewater treatment plants for urban and similar industrial waters. Standards about drinking water supply and wastewater systems are developed by ISO/TC 224. These standards offer good-practice guidance for the quality criteria of the service and performance indicators relating to drinking water supply systems and wastewater systems. Standards in the area of water reuse are developed by ISO/TC 282. These standards provide a reference for water reclamation and reuse, which can address supply issues by creating new sources. Standards about Piping and valves are developed by ISO/TC 138, ISO/TC 5/SC 2, and ISO/TC 153. These standards improve the reliability of water supply systems, thus enabling efficient delivery of water.

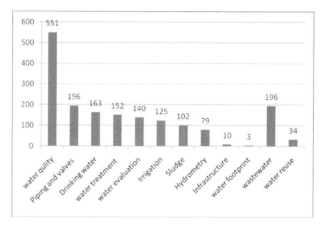

Figure 5. The number of international standards for water issues.

There are 152 international standards related to water treatment covered by ISO. The number of wastewater treatment international standards is 44, 3 of which are under development. The effective international standards related to wastewater treatment, ISO 7875-2:1984, were first published in 1984. It was applied to the determination of low concentrations of bismuth active substances. The method is suitable for the influents and effluents of sewage treatment plants and wastewater. When investigating surface water, it may be necessary to handle large sample volumes. Subsequently, a lot of work focused on the formulation of wastewater treatment standards, and the formulation of relevant standards accelerated after 2000. Especially since 2015, new international standards for wastewater treatment have been released every year.

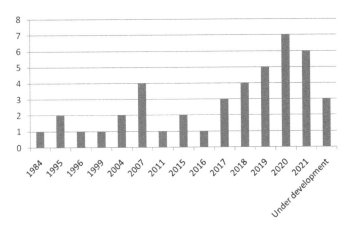

Figure 6. Developing time of International standards for wastewater treatment.

4 CONCLUSIONS

By continuously collecting relevant information, the research and standard on water and wastewater treatment are obtaining more and more attention. Related scientific research reports have increased year by year, especially in recent years. It indicates that wastewater treatment technologies are constantly evolving and updating. For enterprises, development opportunities and challenges coexist. It provides support for technology updates and improvement of the efficiency of the management department. But at the same time, it is also easy to eliminate some companies from the competition. In addition, the development of ISO international standards related to water and wastewater treatment has attracted much attention in these years. These standards cover 10 water-related sectors and help businesses in many ways, with tools for measuring water use to achieve optimization as well as offering best practices on the treatment and use of wastewater. Both businesses and consumers benefit from ISO standards when they are used by businesses and authorities to improve the quality of drinking water, water supply services, water quality, and wastewater treatment.

REFERENCES

Clare N. (2020). *Leaving No Drop to Waste: Standards to Support Treating Wastewater for Irrigation Projects Just Updated*, https://www.iso.org/news/ref 2598. html

Clare N. (2021). *World Water Day: Standards that Protect our Most Precious Resource*, iso.org, https://www.iso.org/news/ref 2644. html

https://www-scopus-com-443.webvpn.cams.cn/term/analyzer.uri

SCOPUS Database, http://www.scopus.com/

The ISO Official Website, https://www.iso.org/search.html

Water Conservancy and Civil Construction – Oke & Ahmad (Eds)
© 2024 The Author(s), ISBN: 978-1-032-58618-2

Structural optimization and practical engineering with the rise of modern technology application

Manli Boukari, Jianyong Pang*, Monkam Ngameni Huguette Maeva & Fatoumata Kir Kalissa
School of Civil Engineering and Architecture, Anhui University of Science and Technology, Huainan, China

ABSTRACT: Evolutionary Structural Optimization (ESO) is a design process that gradually eliminates wasteful material from a structure while it is being designed. The objective of this research is to understand the evolutionary algorithm family and its real-world applications. The Evolutionary Structural Optimization (ESO) technique was used to discover the best structural topologies for development and implementation. However, due to the ineffective removal criterion, its efficiency decreases. The bi-directional evolutionary structural optimization (BESO) technique is utilized to develop an evolutionary topology optimization approach for stress reduction strategy. The Windowed Evolutionary Structural Optimization (WESO) technique and an enhanced elimination parameter are presented in this study. The former uses the system's average compressive forces as an optimization criterion, with a configurable window setting the elimination frequency as a self-adaption condition. As a result, the traditional ESO's concerns of poor optimization efficiency and skewed optimization outcomes can be partially addressed. Many cutting-edge practical uses and variations of early evolutionary approaches are also demonstrated. After that, the WESO technique can be expanded and used with a sophisticated finite element model. The approach has proven to be effective at managing different components in large structures. The WESO approach outperforms the Michell theoretical solution in terms of computational effectiveness and topology stability. Finally, the WESO approach has proven to have a lot of possibilities in modern engineering applications after establishing the ideal topology for a double-deck construction.

1 INTRODUCTION

The purpose of structural optimization is to maximize the utility of a limited number of resources to achieve a certain goal. Shape, dimension, and topology are the three types of structural optimization [1]. One of the most particular of three categories, the structural topology approach, provides data on the location, number, dimension, and structure of "opportunities" in such a spectrum. An initial topological optimization problem approaches (Figure 1) was given. The Finite Element (FE) methodology [2], a relatively new concept, can be used to apply modern topology optimization algorithms to generalized issues.

Structural optimization could provide the least substantial use and charge for the structural strategy. It is becoming increasingly popular. Traditional structural optimizations concentrate on size optimization during the planning phase, resulting in few building

---

*Corresponding Author: jypang@aust.edu.cn

456

DOI: 10.1201/9781003450832-59

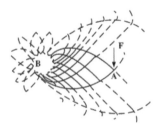

Figure 1. The first approach to the challenge of structural optimization.

improvements [3]. In the practice of engineering, since gravity advantages are considered less important owing to only one character of building structures, excitement for topology optimization is often seen to become more restrained [4]. However, in a period of sustainable and robust infrastructures, when redundancy is a key notion, should examine improving every single building to its maximum efficiency.

A popular solution to this problem is to utilize global measurements like the p-norm and the Kresselmeier-Steinhauser (KS) roles to estimate the maximum stress [5]. For sensitivity assessment, global stress measurements were computationally intensive, but at the cost of losing sufficient regulator of local pressure performance. The third obstacle arises from the fact that anxiety stages were extremely sensitive to topological variations, and it's a good problem in structural optimization systems [6]. Utilizing discrete structural analysis approaches like level-set approaches and ESO-type techniques, the "singularity" issue could be automatically addressed [7]. Beginnings utilizing a level-set approach to deal with stress restrictions in topology optimization were provided, followed by a large number of studies utilizing a level-set approach on another issue. Another major area of topology optimization seems to be the evolutionary structural optimization (ESO) approach. Because of its effectiveness and resilience, the convergence and mesh-independent bi-directional evolutionary structural optimization (BESO) approach, which allows either material removal or additions, has indeed been broadly applied to both academics and engineering studies [8]. To understand, the ESO approach, or its derivative BESO methodology, still has not been expanded for stress minimization design, even though it was originally built on a stress criterion by gradually discarding lightly stressed components. The continuous structure topological optimization technique avoids several mathematical principles by optimizing the organization topology simply utilizing physical concepts and computer-effective solution abilities [9]. Later, several researchers began to use regression analysis and filtration approaches to improve the BESO method's stability [10]. However, because of the difficulties of its intrinsic numerical solutions and the difficulty in detecting the optimization outcomes, structural topology optimization technologies are currently limited to the system level and simple situations of employment [11]. It encounters significant difficulties in more complex structural equipment or modules, including supporting walls, tubed buildings, basements, and structural steel. The finite element (FE) approach determination takes a period to solve the systems, and topology optimization normally needs numerous cycles to resolve to a moderately ideal outcome, both of which increase the computing charge [12]. The computation complexity of topology sensitivity is an applied difficulty that the ESO/BESO approach faces in three-dimensional (3-D) restricted topology optimization. To be more specific, for each restriction, a FE problem must be solved and the appropriate topological sensitivities must be computed [13]. Depending on the standard ESO approach, an assembly-free sensitivity is created in this study by employing the FE-programmed ANSYS as a device to efficiently and precisely address the topology optimization issues of two-dimensional (2-D) and three-dimensional (3-D) structures [14].

2 RELATED WORK

The design process suggested in this research was confirmed through a test case, and it represents an essential framework for such an industry's widespread adoption of additive manufacturing technology [15]. This paper proposes a unique multiscale optimization technique that optimizes designs at the macroscale and microscale. For verification and analysis of prospective benefits, a set of optimization issues minimizing adherence and complying systems are solved. [16]. This research improves BESO topology optimization techniques. Finite element analysis produces elemental sensitivity values, which are translated into node susceptibility numbers. The results demonstrate the novel BESO technique's effectiveness in producing convergence and mesh-independent algorithms [16]. The method is then illustrated on a wing box, demonstrating that the topology optimization for a mass reduction of the real-world structures may be explored utilizing the suggested organization [17]. Evolutionary Structural Optimization (ESO) is a computational structural optimization technique that incorporates finite element analysis (FEA). In contrast to ESO, which utilizes an initial large structure, bi-directional ESO (BESO) can begin with a minimal quantity of material (just that required to handle load and support cases. Two distinct procedures are employed to discover the optimal structure under any load conditions, and the outcomes of each are presented [18].

3 METHODOLOGY

3.1 *Topology optimization in the structural engineering applications*

Important work has indeed been conducted on the development of supporting mechanisms for high-rise structures utilizing topology optimization. Skidmore, Owings, and Merrill (SOM) engineers used theoretic studies on the high-rise stimulating strategy to create conceptual models for the high-rise structures that were together aesthetically attractive and physically effective [19]. Topology optimization has been used to determine the best number, placement, and form of openings in the outer strengthened concrete panels of a Japanese office complex nearby the Station. The walls were optimized for vertical and horizontal loading combinations by modeling them as basic rectangular plates. To specify the shape of the canopy support network, topology optimization investigations were performed. It was discovered that the final form had striking similar characteristics to a tree trunk [20].

3.2 *Difficulties for implementing the ESO approach in building structures*

Since development, the significance of this work requires a big volume of materials, a complicated construction technique, and a high cost, the decision of design and synthesis has long been a popular research area in architectural design. As a result, before construction, a solid structural design is required. Figure 2 depicts a schematic for architectural structure design [21]. Because of its implications for design and effectiveness, the early ESO technique was primarily focused on building structures [22].

(i) The Joint calculation across many platforms is essential. The ESO or BESO approach initially calculates a finite element model linear or nonlinear in the ANSYS before completing the optimization calculations. The target matrix is then optimized with the MATLAB program. Finally, the ANSYS software completes the unit's life-or-death processes. Manufacturer operability will be reduced by two software cooperative computation procedures.

(ii) Cost of high run time The classic ESO process utilizes longer to complete the feature deletion if utilizing ANSYS for structural optimization, implying reduced computational effectiveness [23]. The computational cost would indeed be tremendous if there were more components or higher-order components.

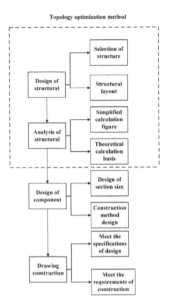

Figure 2. Design of building structure flowchart.

(iii) Although the early ESO approach is straightforward to implement, using fixed elements deletions or deletion frequency may result in an unequal spreading of component stress levels within the same generations' deletions. This raises the likelihood of some modification of the optimization results. In this, the organizational inventors want to employ an easy and actual optimization strategy to solve a strategy problem. Although the initial ESO approach is easy and extensively used, it is critical due to its grid reliance, checkerboard effect, and incapability to discover the best solution [24].

3.3 *Effective stress and stiffness models*

The designed domain is divided into several finite elements, each with its topological design parameter. $a = (a_1, a_2, \ldots, a_{nel})^T$ denotes the nel-dimensional vector holding the design parameters [25]. Within the scope of the BESO approach, the design parameters have either a 0 or a 1 number, in equation (1);

$$x_i \in \{0, 1\}, i = 1, 2, \ldots, nel. \tag{1}$$

The model parameters could be understood as a widespread mass density or with an indication parameter, with several ones corresponding to solid material and a value of 0 corresponding to invalid.

$$S_i = x_i S_o \tag{2}$$

S_o denotes the matrix of stiffness is such a solid substance. In training, the invalid is given a relatively low rigidity to minimize re-meshing and stiffness matrix singularities. Following that, the stress levels criteria are explored, which seeks to manage the maximum stress only at a microscopic scale in rank-2 layered compounds. The consistent description of the stress-strain vector for a power-law parameter interruption form is presented by

$$\sigma_i = \frac{S_i P_i v_i}{a_i} = S_o P_i v_i \tag{3}$$

where v_i is the dispersion vectors of a i-th element and P_i denotes the relationship between movement and strain. According to this concept, the element stress condition is considered to be independent of the related design parameter. It should be noted that, while the Ersatz materials approach is utilized to prevent remeshing, strains are only analyzed for solid materials and are reduced to 0 for voids.

3.4 *A Measure of global P-norm stress*

Concerning the local character of stress, the traditional p-norm global stress measurement is used to approximate the maximum stress.

$$\sigma_{PN} = \left[\sum_{i=1}^{nel} \sigma_{vm,i}^p \right]^{1/p} \tag{4}$$

Where $\sigma_{vm,i}$ is represented as von Mises stress at the i-th material's centroid and p is the stress norm component The p-norm is considered to yield the median stress when p = 1 and reaches the maximum stress σ_{max} when $p \to \infty$. A greater value of p appears to be preferable because it gives a better estimation of the maximum stress, but as p rises, the p-norm issue gets ill-conditioned [26]. The greater the score of p, the more intense the oscillation throughout the optimization method, because the algorithm only targets the maximum stress and ignores the others. As a result, the amount of a norm variable p must be carefully determined so that the issue smoothness is preserved whereas the maximum stress is suitably estimated. a comprehensive description of the p-norm global measures attribute

The component von Mises stress is expected to be generated at the centroid without loss of precision as

$$\sigma_{vm} = \left(\sigma^T V \sigma \right)^{1/2} \tag{5}$$

This equation signifies the stress in Voigt notation, and V has been the coefficient of stress matrices.

$$v = \begin{bmatrix} 1 & -1/2 & 0 \\ -1/2 & 1 & 0 \\ 0 & 0 & 3 \end{bmatrix} \tag{6}$$

in the 2D planar stress example and yields the expected result

$$\sigma_{vm} = \sqrt{\sigma_{aa}^2 + \sigma_{bb}^2 - \sigma_{aa}\sigma_{bb} + 3\sigma_{ab}^2} \tag{7}$$

In the 2D instance, σ_{aa}, σ_{bb} and σ_{ab} are the constituents of the stress vectors.

The mathematical formula for minimizing the p-norm globe stress measurement under a material consumption constraint would be as follows.

$$\begin{aligned} \min : &\ \sigma_{PN} \\ \textit{Subject to} \quad : &\ KU = F \\ : &\ V(a) = \sum a_i v_i = V_{req} \\ : &\ a_i = 0 \quad or \quad 1, \quad i = 1, \dots, nel, \end{aligned} \tag{8}$$

Where, K represents the global matrix of stiffness and U and F represents the global movement and power directions, respectively. The dimension of the i-th element is denoted by v_e, whereas the total and needed material volumes are denoted by $V(x)$ and V_{req}, correspondingly.

3.5 Windowed Evolutionary Structural Optimization (WESO) method

An examination says the ESO/BESO method's problem is inappropriate element cancellation. ESO deletion criteria depend on stress intensity sensitivity research, which can be improved in two ways: The mean stress distribution of design domain components is employed as the element removal criterion in iterative computations [28]. Standard deletion criteria use discrete components to reduce strain energy and avoid deletion frequency. Adaptive component removal criteria [29] avoid the typical removal level problem. Early in the optimization calculation, the number of inefficient components is high, and the deletion rate reduces control efficacy; subsequently, the number of components decreases.

3.6 WESO performance indicator

Measuring the advantages and downsides of topology findings is a topic in the optimization phase. As a result, a technique performance metric is required. This research utilizes the performance index (PI), which is described as

$$PI = \frac{L_i \overline{T_i}}{L_o \overline{T_o}} \tag{9}$$

where L_i seems to be the volume of the i-th iteration modeling environment, L_o would be the volume of a design phase area, and T_o is the first design domain's average strain power density. PI designates volume and stress. The i-th iteration's structure is superior to the design phase's. Smaller PI numbers also improve topology. Mean strain power density and system characteristics of mean strain energy content define the minimal stress component spectrum in the minimal period. It avoids ESO's sorting process [30]. Adaptive removal criteria increase optimization by decreasing deleted areas' strain power density. ESO's constant removal frequency causes non-ideal optimization challenges. The new technique provides good removal criteria. It can model complex structures with high-order components.

3.7 Comparison of basic supported beams

Figure 3 depicts the variation curves of a sensitivity threshold computation time with several iterations for such three approaches under identical conditions [27]. Figure 3 depicts the relationship between the RF/ER value and the WESO technique optimization variable within the same conditions. As indicated in Figure 3, the ESO/BESO approach takes a long time to calculate the sensitivities threshold, as well as its calculation production to keep as the number of optimization units reduces.

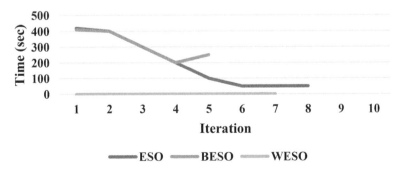

Figure 3. Sensitivity curve for calculation and iteration times.

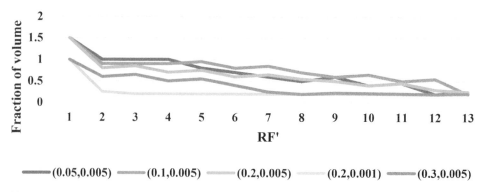

Figure 4a. Relationship between the fraction of volume and RF'.

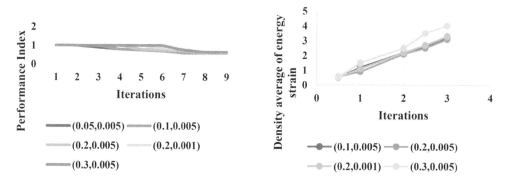

Figure 4b. PI variations and Density average of energy strain with the iteration cycle.

As shown in Figure 4, when the optimization goal is met, the size of EF' is equivalent, the PI and mean strain energy curvature shift gradually, and the total size is equivalent; if ER is low, EF' will require more repetitions to fulfill the target. The WESO method's sensitive adaptive adjustment structure allows architectural topology optimization solutions to be less influenced by EF/ER, improving its engineering usability. This is modern stiffness optimization. The technique and research above show that picking portions with more nodes can prevent organizational optimization distortions, and WESO has superior optimization efficacy. WESO gives you a lot of ways to optimize, and with the same parameter, you can make a stiffer topology design.

3.8 Topological construction of D-shape bridge finding structures

The WESO approach was long to three-dimensional interplanetary. The same flexible optimization criterion was utilized for both the 2-D and 3-D tasks. As a result, considerable changes to the approach for 3-D situations were unneeded. Furthermore, the findings of 3-D topology optimization could more intuitively illustrate the best topological shape of every structural element, providing designers with more exciting and instructional conceptual design options. To illustrate the resemblance between the $3-D$ topology optimization findings and the real structural engineering outcomes, a $3-D$ model with such a rising to-span proportion of $1/8$ and secure terminals was employed in this example. As a result, the WESO approach was used to determine the spatially structural shape. For modeling, the 3-D structural solid component (Solid65) was employed. The process structure dimension is $300m \times 40m \times 36m$, and the structure was discretized using standard hexahedron components using element lengths of 1 m. There was a total of 280,000 elements used. $E = 3.07 \times 10 N/mm^2$, $v = 0.2$, and initial non-optimization of the highest unit (i.e., Bridge plate); applied distribution heaviness of $1 GPa$; and the optimization complete transformation rate = 0.3 are the parameters. The driving plates component is not seen in the results obtained to make the optimization specifics of the bridges clearer. The optimization result is close to that of a deck arch bridge. At the same period, the network topology optimization findings diverge from the real design of a slanted arch bridge component [28]. This phenomenon demonstrates that, during the WESO technique's optimization procedure, the construction will develop towards the lowest load path direction, just as the load transmitting path in such an actual bridge does. Depending on the optimization findings, maximum stress does not occur on the major arch, and also the distribution of stress is well dispersed, suggesting that the optimization outcome will be in an even structural system [24]. Meanwhile, there are some disparities between the optimization values and the actual building of an arch bridge with inclined instead of vertical arch members. This behavior indicates that during the WESO technique optimization phase, the architecture will develop toward the lowest load path direction. The effectiveness of the WESO approach on the complicated construction with a great number of pieces is therefore validated by topology optimization study of a realistic bridge.

3.9 Computer-aided engineering: Grid filtering technique

The mesh-dependency problem caused by statistical instability is a problem that practically all topology optimization approaches face. For regularization, almost all approaches employ some form of filtering/smoothening. Estimating the value of a perimeter restriction for a fresh design challenge, on the other hand, is far from easy [29]. As a result, the WESO and BESO solutions are discussed in the published MATLAB program. BESO's evolutionary rate is 0.02 in this case; in WESO, $RF = 0.02$ and $ER = 0.01$. Both systems have a filtering

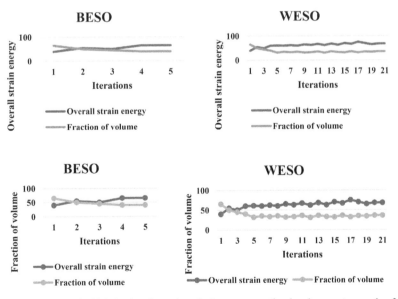

Figure 5. When WESO/BESO begins from the whole structure, the development records of total strain energy and volume fraction are shown.

radius of 6 mm. C denotes maximum stress capacity, which expresses the degree of homogeneity of specific structural applications. The smaller these numbers of C under the same settings, ever more equal the material's dispersion. WESO's sensitivity evaluation seems more stable. Each method yields distinct topological results. WESO is more strain-resistant than BESO. The WESO architecture approach, without filtering technologies, outperforms the competition. Figure 5 depicts the evolution of total strain energy and volume segments when the WESO/BESO approach is applied to a whole design using the same mesh filtering technique [13]. WESO has three times as many iterations as BESO, but its strain power at the same residue volume fraction is lower. WESO's topological stress solutions are more equal and offer superior material allocation. WESO performs many iteration simulations when the topology changes drastically, as in Figure 5b. BESO's filtering technologies cause the optimization technique to miss a superior topology.

Throughout the architecture of concrete structures buildings, steel bars were scattered within a concrete structure to increase the tensile properties of concrete, therefore the construction of steel bars became the subject of research. The WESO method's reinforcing optimization outcomes require filter technologies [30]. The arrangement of reinforcing bars is somewhat distinct, although it has little effect on that building's reinforcement architecture. Meshless filtering technology, for example, can be instantly turned into construction plans. If a mesh filtering method is built to smooth out the reinforcing pattern, it will not lengthen the program's creative development and have limited applicability.

4 RESULT AND DISCUSSION

4.1 *Comparison between WESO and solution*

Michell's theory illustrates systematic optimization approaches. The Michell truss has the lowest structural tensile stress when the desired volume is chosen in the initial design domain evaluation. The WESO technique aims to eliminate low-efficiency components regardless of a design's stress state, so the best solution is the lightest and stiffest. Only Michell's theory applies. So, the WESO technique optimization results can be evaluated by comparing the WESO technique explanation to the accessible Michell solution under identical boundary conditions. The early design domain for the beam was discretized using eight-node two-dimensional quadrilateral fundamentals (Plane82) using element sizes of $20mm \times 20mm$. Hemp attributed the gap to design limitations. WESO's sensitivity control precision was improved by using planar components with so many sensor nodes to create a thin mesh. Due to extensive calculation spans in design, the calculation shows that an employee's support may lose stability during the WESO methodology calculation procedure. Loads, clearances, manufacturing, and aesthetics for double-deck beams are challenging. We'll separate the research because it's long. The bridge deck web topology using WESO Web FE lacks apertures. Homogeneous computation architecture, Solid65, and bridge deck distance were also mentioned. WESO's double-deck concrete box girder under a curve transfers force perfectly. produced a double-deck girder. The web concept topological solution generates a strut-and-tie framework at primary nodes; 2) the framework's mechanical characteristics and operating principles are investigated; 3) an architectural system, such as structural bars, is completed, and designed documents are generated based on the ACI Code (2014) configuration. Design is informed by research. Topology was studied in experimental architecture. Both structures have identical diameters, decking, cross-girder size, concrete composition, substrate, beginning circumstances, and loads. All lower decks were strengthened vertically and laterally. Calculated are vertical and longitudinal stir-up frequencies. Double-deck WESO saves materials and enhances ventilation and lighting. Two-fold midspan topology Theoretical stress. Inefficiency and bending, topology beats the empirical model. Improve box girders. Design possibilities include WESO and double-deck box girder optimization. WESO compares complex structural designs. Normalize.

5 CONCLUSION

Topology optimization opens up new possibilities in civil/structural strategy and planning. It has been proposed as a method for increasing communication among engineers and designers during the conceptual construction process. Including the Windowed Evolutionary Structural Optimization (WESO) approach, an enhanced element elimination criterion depending on strain energy sensitivity is provided. To manage the correctness of the optimization calculation, the WESO technique establishes the reactive element removal criteria and includes the strain energy performance index. It can successfully avoid structural optimization distortions and the architecture from falling into such a local optimal solution, hence enhancing calculation performance yet further. A comparative analysis of the topological findings of simply supported beams demonstrates the WESO method's advantage. By implementing the WESO technique to the systemic discovery of a double-deck concrete box girder and trying to compare it to previous ideas, it was discovered that the suggested technique could provide a novel concept for the double-deck concrete box girder as well as enhance applications in various structural engineering fields, especially the complex structural system.

AUTHORSHIP CONTRIBUTION STATEMENT

Manli Boukari: Methodology, Conceptualization, Writing-original draft, Formal analysis. *Jianyong Pang*: Supervision, Validation, Resources, and Funding acquisition. *Manli Boukari and Monkam Ngameni Huguette Maeva*: Software, Analysis, *Fatoumata Kir Kalissa*: Writing-Reviewing, and editing.

CONFLICT OF INTEREST

The authors have no competing interests to declare that are relevant to the content of this article.

ACKNOWLEDGMENTS

Funding: Natural Science Foundation of Anhui Province, China (No. 2008085QD178).

REFERENCES

[1] Kingman JJ, Tsavdaridis KD, Toropov VV 2015 Applications of Topology Optimization in Structural Engineering: High-Rise Buildings and Steel Components. *Jordan Journal of Civil Engineering* 9:23
[2] Dar FH, Meakin JR, Aspden RM 2002 Statistical Methods in Finite Element Analysis. *Journal of biomechanics* 35:1155–1161
[3] Vanderplaats GN 2006 Structural Optimization for Statics, Dynamics and Beyond. *Journal of the Brazilian Society of Mechanical Sciences and Engineering* 28:316–322
[4] Sudhakaran G, Kandipati B, Surya GB, Shree VK, Sivaprasad M, Jayakumar M 2017 Evolutionary Algorithm based Structural Optimization for Patch Antenna Design and its Performance Analysis. In: 2017 International Conference on Advances in Computing, Communications and Informatics (ICACCI). *IEEE*, pp 2189–2192
[5] Xia L, Zhang L, Xia Q, Shi T 2018 Stress-based Topology Optimization Using Bi-Directional Evolutionary Structural Optimization Method. *Computer Methods in Applied Mechanics and Engineering* 333:356–370

[6]   Xia L, Xia Q, Huang X, Xie YM 2016 Bi-directional Evolutionary Structural Optimization on Advanced Structures and Materials: *A Comprehensive Review*. 43

[7]   Wang, L., Zhang, H., Zhu, M., & Chen, Y. F. 2020. A New Evolutionary Structural Optimization Method and Application for Aided Design to Reinforced Concrete Components. *Structural and Multidisciplinary Optimization, 62*(5), 2599–2613.

[8]   Hooshmand S, Behshameh M 2013 A Tabu Search Algorithm with Efficient Diversification Strategy for High School Timetabling Problem. *Information Technology* 5:15

[9]   Slowik A 2020 Evolutionary Algorithms and their Applications to Engineering Problems. *Neural Computing and Applications* 17

[10]   Osyczka, A., Krenich, S., Krzystek, J., & Habel, J. 2004. Evolutionary Optimization System (EOS) for Design Automation. In *IUTAM Symposium on Evolutionary Methods in Mechanics* (pp. 309–320). Springer, Dordrecht.

[11]   Plocher, J., & Panesar, A. 2019. Review on Design and Structural Optimisation in Additive Manufacturing: Towards Next-generation Lightweight Structures. *Materials & Design, 183*, 108164.

[12]   Andrey V 2018 The use of Methods of Structural Optimization at the Stage of Designing High-rise Buildings with Steel Construction. 7

[13]   Zhao F 2014 A Nodal Variable ESO (BESO) Method for Structural Topology Optimization. *Finite Elements in Analysis and Design* 86:34–40

[14]   Huang X, Xie Y 2008 A new look at ESO and BESO Optimization Methods. *Structural and Multidisciplinary Optimization* 35:89–92

[15]   Tang Y, Kurtz A, Zhao YF 2015 *Bidirectional Evolutionary Structural Optimization (BESO) Based Design Method for Lattice Structure to be Fabricated by Additive Manufacturing*. 23

[16]   Huang X, Xie Y 2007 Convergent and Mesh-independent Solutions for the Bi-Directional Evolutionary Structural Optimization Method. *Finite Elements in Analysis and Design* 43:1039–1049

[17]   Munk DJ 2006 *A Bi-directional Evolutionary Structural Optimization Algorithm for Mass Minimization with Multiple Structural Constraints*. 29

[18]   Young V, Querin OM, Steven GP, Xie YM 2000 3D and Multiple Load Case Bi-Directional Evolutionary Structural Optimization (BESO). 29

[19]   Kingman J, Tsavdaridis KD, Toropov VV 2014 Applications of Topology Optimization in Structural Engineering. *Civil Engineering for Sustainability and Resilience International Conference* (CESARE)

[20]   Huang X, Xie M 2010 *Evolutionary Topology Optimization of Continuum Structures: Methods and Applications*. John Wiley & Sons

[21]   Huang X, Xie Y 2008 Optimal Design of Periodic Structures using Evolutionary Topology Optimization. *Structural and Multidisciplinary Optimization* 36:597–606

[22]   Xie YM, Huang X 2010 Recent Developments in Evolutionary Structural Optimization (ESO) for Continuum Structures. In: *IOP Conference Series: Materials Science and Engineering*. IOP Publishing, p 012196

[23]   Komini L, Dyrmishi B, Pojani N 2016 *Implementation of Topology Optimization on Building Design-Study Case*.

[24]   Muc A, Muc-Wierzgoń M 2012 An Evolution Strategy in Structural Optimization Problems for Plates and Shells. *Composite Structures* 94:1461–1470

[25]   Yang X, Xie Y, Steven G, Querin O 1999 Bidirectional Evolutionary Method for Stiffness Optimization. *AIAA Journal* 37:1483–1488

[26]   Bindolino G, Ghiringhelli G, Ricci S, Terraneo M 2010 Multilevel Structural Optimization for Preliminary Wing-box Weight Estimation. *Journal of Aircraft* 47:475–489

[27]   Almeida VS, Simonetti HL, Neto LO 2013 Comparative Analysis of Strut-and-tie Models using Smooth Evolutionary Structural Optimization. *Engineering Structures* 56:1665–1675

[28]   Menna C, Mata-Falcón J, Bos FP, Vantyghem G, Ferrara L, Asprone D, Salet T, Kaufmann W 2020 Opportunities and Challenges for Structural Engineering of Digitally Fabricated Concrete. *Cement and Concrete Research* 133:106079

[29]   Li S, Zhao X, Zhou G 2019 Automatic Pixel-level Multiple Damage Detection of Concrete Structure using Fully Convolutional Network. *Computer-Aided Civil and Infrastructure Engineering* 34:616–634

[30]   Liang X 2019 Image-based Post-disaster Inspection of Reinforced Concrete Bridge Systems using Deep Learning with Bayesian Optimization. *Computer-Aided Civil and Infrastructure Engineering* 34:415–430

Water Conservancy and Civil Construction – Oke & Ahmad (Eds)
© 2024 The Author(s), ISBN: 978-1-032-58618-2

Analysis of the influencing factors of personal credit for practitioners in the field of engineering construction based on the fuzzy-DEMATEL model

Song Xue, Tong Su & Yi Zhou*
Business School, Hohai University, Nanjing, China

ABSTRACT: The construction industry is the pillar industry of China's national economy, but the research on the personal credit of practitioners is still lagging behind. Based on the grounded theory, this paper first conducts a qualitative analysis of 100 documents from the CNKI database and official government websites on Nvivo to establish the concept of personal credit of practitioners in this field from the perspective of institutionalization and identifies the three dimensions of influencing factors: individual, environment and governance. For multi-stage, multi-factor, and multi-expert decision-making situations, considering the fuzziness of expert statements, this paper introduces Fuzzy Logic and the Fuzzy-DEMATEL model that is established, extending the single value of the comparison matrix to a fuzzy interval and then giving the decision makers a proper judgment space. By standardizing the triangular fuzzy number matrix on the expert scoring results, a comprehensive influence matrix of personal credit is established to determine causal factors, outcome factors, and key influencing factors. The research on the personal credit of practitioners in the engineering construction field in this paper provides a conceptual theoretical basis for the subsequent research and also provides a theoretical basis for the construction of personal credit collection and evaluation systems of practitioners in this field.

1 INTRODUCTION

The construction industry is an important pillar industry of China's national economy. At present, infrastructure projects around the country are currently under construction, and the construction industry has ushered in a period of exuberant development. The latest data released by Jiyan Consulting shows that the total output value of China's engineering construction industry will be 26.4 trillion yuan in 2020, accounting for 25.88% of the gross domestic product. However, with the rapid development of the industry, the credit problems of market entities have become more and more prominent. Perfunctory supervision, jerry-building, accepting bribes, and other damage to personal credit behavior occurs, disrupting the order of the construction market, which is not conducive to the healthy development of China's construction industry. Present researches on personal credit are mainly concentrated in the field of financial lending, while researches in the engineering construction field mostly focus on enterprise credit and individual practicing credit. There is no clear and comprehensive definition of personal credit and few researches on the influencing factors in the engineering construction field. Therefore, it is particularly necessary to study the personal credit of practitioners in the field of engineering construction. This paper establishes the concept of personal credit of practitioners in the field of engineering construction from the perspective of institutionalization, identifies the influencing factors, and analyzes the relationship between them, so as to provide reference for subsequent research on personal credit of practitioners in similar fields.

---

*Corresponding Author: 1643080428@qq.com

DOI: 10.1201/9781003450832-60

2   LITERATURE REVIEW

The credit question belongs to the ethics morals research category earliest, the United States is recognized as the world's most developed country in credit management. Many scholars study credit, but the main research areas are concentrated in the field of financial bank lending.

In the field of engineering construction, Abramowitz [1] first proposed that in terms of individual professionals, architects need not only the ability to design and build buildings but also a high degree of ethical integrity. Fellows R [2] analyzed the immoral behaviors caused by different interest criteria in engineering projects through case studies, and their influence on personal credit was indirectly proved. Based on the analysis of the credit status of each subject in the construction industry, Wu Yuping [3] discussed the causal and influencing factors of the lack of construction enterprises' credit from several aspects, such as the core characteristics of enterprises, social system, credit evaluation system, and legal system, and then proposed countermeasures and measures to establish a credit system in this field. Fan Zhiqing [4] combined with the characteristics of the construction market, proposed an evaluation index system to measure the credit of practitioners and conducted an empirical analysis using the material element analysis method. Chen Yangyang [5] constructed a construction contractor credit evaluation index system, including basic information, financial capability, management capability, technical capability, and credit record, and proposed an evaluation model based on binary semantics. Huang Feina [6] constructed a professional credit evaluation system in the field of engineering construction that includes project management performance reflecting individual professional credit contribution and individual professional ethics reflecting individual professional credit contribution.

To sum up, the studies on personal credit in the engineering construction field are mostly based on the professional perspectives of registered constructors, cost engineers, and supervision engineers to propose the construction of their personal practice credit system respectively. Few studies can clearly construct the overall credit connotation of practitioners in the engineering construction field from multiple perspectives, and there are only a few studies on the identification of personal credit influencing factors.

3   PERSONAL CREDIT

Personal credit contains slightly different meanings from different perspectives. In a narrow sense, it refers to the economic capital that obtains the trust of the counterparties, mainly including the transaction ability and performance ability shown by the credit subject in financial lending, securities trading, commercial trading transactions, and other trading activities. In a broad sense, it refers to the extent to which individuals keep their promises and are willing to perform. From an institutionalized point of view, personal credit generally refers to a system in which an individual's credit rating is evaluated, recorded, and filed at any time, based on the resident's household income and assets, incurred borrowings and repayments, credit overdrafts, penalties and lawsuits in the event of bad credit, so that the supplier of credit can decide whether to lend to him or her and how much to lend, mainly referring to commercial and financial credit [7].

4   GROUNDED THEORY IN PERSONAL CREDIT CONCEPT ESTABLISHMENT

The concept of personal credit for practitioners in the field of engineering construction will be established based on the grounded theory in this paper. Grounded theory is the most scientific method of qualitative research recognized by the academic world, which requires three levels of coding, including open coding, axial coding, and selective coding, to discover and refine the scope of the data collected and organized, and finally, the reliability of the research results must be proven through a saturation test after the coding is completed.

4.1 Data collection and arrangement

The data for this paper is mainly from academic literature and policy documents.

4.1.1 Academic literature

The existing studies on personal credit evaluation, personal credit governance and system construction, and credit risk impact in the engineering construction field all reflect the personal credit of practitioners to different degrees. In this paper, we searched 25 documents with the highest relevance on the China Knowledge Network database by keywords for the four topics, among which 5 documents each were reserved for saturation test.

4.1.2 Policy documents

The keywords of "personal credit" were searched in the official government websites and public databases, and a total of 19 local government regulations and documents from 2001 to 2021 were finally compiled, 15 of which were used for initial coding and 4 for saturation testing.

4.2 Category mining and extraction

The data for this paper is mainly from academic literature and policy documents.

4.2.1 Open coding

By coding 95 documents on NVivo software, 32 subcategories were obtained.

Table 1. List of subcategories.

Subcategory	Primitive Resources (connotation)
Age	Demographic characteristics; common considerations for
Sex	personal credit
Education degree	
Marital status	
Professional period	
Professional title	
Project participation	The specific content of the professional work of the
Work unit	practitioner
Satisfaction rate of proprietor	
Working awards	
Working punishment	
Personal annual income	Reflect the practitioner's personal financial ability; The better the financial ability, the smaller the possibility of breach of trust and breach of contract due to interest
Monthly income per household	Personal income is not representative of a person's overall
Monthly household consumption	financial situation and should be combined with the average
per capita	monthly household income, pointing out
Financial assets	Reflect the consolidated asset condition
other assets	
misconduct record	Publication of bad work records on the platform
credit record	
industry and commerce	Comprehensive measurement of the social credit of the
administration punishment	practitioner
public prosecution credit record	
personal credit rating	Scoring based on personal credit indicators
personal credit risk judgment	Make judgments about personal credit risk

(continued)

Table 1. Continued

Subcategory	Primitive Resources (connotation)
credit risk early warning	Carry out early warning control when risk is judged to exist
credit record	Timely recording of every change in credit status
credit information preservation	Preservation and management of personal credit
credit information disclosure	information
credit information confidentiality	Part of the credit information is public, and personal information involving privacy is protected confidentially and visible only to some people
judgment of continuing employment	The work disposition of the practitioner's company based on personal credit and other factors
judgment of promotion	
project arrangement	
impact on project bidding	Personal credit level affects the tenderer selection of the winning bid

4.2.2 Axial coding

Although 32 subcategories have been formed in the open coding, they are relatively large in number and have loose boundaries between them, so it is necessary to sort out the organic relationships, including causality relationships, temporal relationships, similarity relationships, and so on, between the subcategories with the help of the axial coding. By clustering 32 subcategories using NVivo and asking experts' opinions, we finally summarized 9 main categories, such as basic personal information, professional work situation, and personal credit evaluation, as shown in the second column of Table 2.

4.2.3 Selective coding

NVivo was used to cluster the nine main categories again. According to the similarity of words and homogeneity of functions, the main categories were divided into three groups. Three core categories of "personal credit content", "personal credit management", and "personal credit influences" were derived after discussion, and finally, the concept of personal credit of practitioners in the engineering construction field was refined [8].

Table 2. Code table based on grounded theory.

Core Category	Main Category	Subcategory
Personal credit content	Personal basic information	A1 age
		A2 sex
		A3 education degree
		A4 marital status
	Financial condition	A6 income
		A7 expenditure
		A8 assets
	Professional work	A9 professional period
		A10 professional title
		A11 project participation
		A12 work unit
		A13 satisfaction rate of proprietor
		A14 working awards
		A15 working punishment

(continued)

Table 2. Continued

Core Category	Main Category	Subcategory
	Credit condition	A16 good behavior record
		A17 credit record
		A20 public prosecution record
		A21 industry and commerce administration punishment
Personal credit management	Personal credit evaluation	B1 personal credit rating
		B2 personal credit risk judgment
	Credit early warning	B3 credit risk early warning
	Credit information management	B4 credit record
		B5 credit information preservation
		B6 credit information disclosure
		B7 credit information confidentiality
Personal credit influences	Work disposition	C2 judgment of promotion
		C3 project arrangement
	Treatment of project tenderer	C4 impact on project bidding

4.2.4 *Saturation test*

Secondary coding was performed on the 29 previously reserved documents, and the results of the secondary coding did not have any missing or conflicting concepts from the primary coding. Therefore, the primary results were considered to pass the saturation test and the coding was correct and complete.

4.3 *Results and analysis*

This paper establishes the concept of personal credit based on the grounded theory and concludes that personal credit in the engineering construction field is a three-dimensional concept system including personal credit content, personal credit management, and personal credit influence. Based on the introduction of the personal credit definition in this paper, the research results of this paper are biased toward the personal credit concept of practitioners in the engineering construction field from the perspective of institutionalization.

The conceptual model shows an institutional system of personal credit cycles for practitioners in the engineering construction field. The supervisory department evaluates and

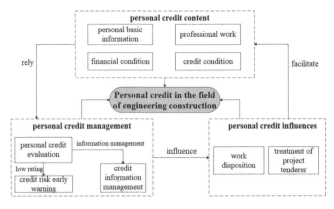

Figure 1. The conceptual model of personal credit of practitioners in the field of engineering construction.

manages the personal credit specifics of the practitioner, including basic information, etc., and the credit supplier decides the disposal of the practitioner based on the information provided by the supervisory department and thus has an impact on the practitioner personally, which in turn promotes the personal credit specifics of the practitioner in a good direction.

5 IDENTIFICATION OF FACTORS INFLUENCING PERSONAL CREDIT

Considering the relevant theories, the three dimensions of its conformation system correspond to the broader impact dimensions of personal credit in the engineering construction field respectively: personal credit contains content corresponding to individual practitioners, personal credit management corresponding to regulation, and personal credit impact corresponding to the environment. This paper will identify the secondary influencing factors in the engineering construction field from these three dimensions with relevant theoretical foundations.

5.1 *Individual dimension*

First, willingness to work is mainly related to the individual characteristics of the person and therefore is often characterized by basic personal information. Some practitioners have the workability to complete their work tasks with high quality, but their personal credit is affected by their low willingness to work due to their personal ethical qualities and other reasons, resulting in the impairment of the completion of the entire project. It is mainly related to the demographic characteristics of the practitioner, including education level, gender, education, age, and marital status.

Second, practitioners who do not have sufficient workability to complete their own tasks in a project will result in poor task completion, which will affect the quality of the project and eventually feed back to their own personal credit. When empirically analyzing the degree of influence of workability, indicators such as years of experience, job title, number of projects involved, work unit, owner satisfaction rate, work awards, and work penalties are often used to characterize the situation.

Third, an important manifestation of the impact of moral credit level on personal credit is a large material demand for money and a weak moral consciousness, so to maximize personal interests, there will be a breach of trust such as bribe-taking, which will seriously affect the quality of the project. However, there is no exact moral indicator to quantify a person's moral credit level, so it is usually characterized by some credit records such as credit records, tax records, records of industrial and commercial administrative penalties, and other indicators in credit assessment or empirical analysis.

Finally, if a person's financial condition is poor, such as carrying a large loan or in urgent need of money, coupled with the practitioner's weak sense of ethics and credit, then they are more likely to accept bribes, cut corners, and other acts of enrichment and breach of trust for profit, jeopardizing project security and hurting personal credit. The financial status of the practitioner can be measured by financial indicators such as annual income, asset situation, monthly household income per capita, and monthly household expenditure per capita.

5.2 *Environment dimension*

First, the social environment includes the construction of laws and regulations and the construction of social credit as a whole. The construction of laws and regulations restrains individuals from committing illegal crimes, such as accepting bribes, and other personal breaches of trust from the legal level, while the construction of social credit influences individuals to become a citizen with good personal credit from the moral level.

Second, practitioners in the field of engineering construction do not exist as individuals alone but are closely linked to the economic environment. Individual development is closely related to macroeconomic factors such as GDP growth, energy prices, exchange rates, and interest rates. These macroeconomic factors have a significant impact on the practitioner's own personal credit. Therefore, it is believed that the macroeconomic environment indirectly affects individual credit [9].

The working environment also matters and its typical aspect is the disposition of the practitioner's work. Another notable manifestation is credit contagion, including contagion among competitors, credit channel contagion, and corporate credit to personal credit. As to credit channel contagion, it means that practitioners engaged in engineering projects have business dealings with each other and credit dependence. For contagion between corporate and individuals, it means if a practitioner's firm has poor credit, there is a greater potential for personal credit to be impacted.

5.3 Governance dimension

Advanced supervision refers to the supervision of practitioners before they participate in project work. At present, the main way of advanced supervision for practitioners in the field of engineering construction in China is the management of practitioners' access qualifications. The professional degree of the personnel is assessed through the licensing system, consisting of the examination system, registration system, and practice system. Proper advanced supervision can effectively reduce the impact on personal credit brought about by insufficient professional competence.

For process supervision, the supervision department is entrusted during a project by the construction project legal person to supervise and manage the investment, schedule, and quality of the construction project according to the relevant laws, construction contracts, supervision contracts, and other legally binding documents. However, there are often many problems in the actual supervision process. Engineering supervision enterprises do not seriously fulfill their contractual obligations due to their inertia and low level of professional ethics. Some supervisors of local public construction fail to check the internal quality management of enterprises according to the current construction acceptance norms and carry out dynamic supervision of projects. If effective supervision cannot be achieved in the project, some practitioners may be emboldened to make breach of trust contract.

For the object of postmortem supervision, the personal data of practitioners published on the national construction market supervision public platform, a typical postmortem supervision platform, is lacking, only basic information such as name, education, engineering performance, and practice registration included, and a considerable number of individual data is missing. On the platform in Guangxi, the bad behaviors of practitioners could not be found, while immortal behaviors are publicized such as inadequate supervision and failure to carry out supervision according to the strategy in Guangdong. Under the same circumstances, the disposition of bad behavior in Guangdong has a greater deterrent effect on the personal credit of practitioners in the field of engineering construction.

After the above identification process, the final system of personal credit influencing factors for practitioners in the field of engineering construction is established.

Table 3. System of personal credit influencing factors.

Primary Influencing Factors	Secondary Influencing Factors	Content
Individual practitioner	Willingness to work	Whether the practitioner has a sufficient willingness to participate in the work
	Working ability	Whether the practitioner has the sufficient working ability to be competent for the project he is leading or involved in

(continued)

Table 3. Continued

Primary Influencing Factors	Secondary Influencing Factors	Content
	Moral credit	A person's level of ethical credibility influences whether or not he will behave in a way that abandons his trust for profit when faced with profit
	Financial condition	Poor financial status may be a breach of trust such as jerry-building, accepting bribes, and other breaches of contract
Environment	Social environment	Social environment influences include the construction of laws and regulations and the overall credit construction of society on personal credit
	Economic environment	Practitioners do not exist alone as individuals but are closely linked to the economic environment
	Working environment	The impact of personal work disposition results on personal credit and the contagion of work environment credit
Governance	Advanced supervision	Mainly through the management of practitioner access qualifications, entry restrictions to the industry
	Process supervision	Supervision of individual practitioners in the course of project execution
	Postmortem supervision	Construction of personal credit supervision platform for practitioners, assessment and management of personal credit information

6 ANALYSIS OF INFLUENCING FACTORS OF PERSONAL CREDIT BASED ON THE FUZZY-DEMATEL MODEL

DEMATEL is a systematic analysis method using graph theory and matrix tools. Due to the uncertainty and fuzziness among various influencing factors of personal credit, evaluators often express fuzzy language when making comprehensive judgments [10]. Therefore, this paper adopts the combination of fuzzy theory (Fuzzy) and DEMATEL to quantitatively analyze the relationship between the influencing factors of personal credit of practitioners in the engineering construction field. By introducing the triangular fuzzy number method, the single value of the comparison matrix can be extended to a fuzzy interval to give the decision maker a proper judgment space. The Fuzzy-DEMATEL model can perfectly fix the flaws of the traditional model and is more applicable to the study of individual credit in the construction market [11].

$$N = (\alpha, \beta, \lambda). \tag{1}$$

Table 4. Scoring value mapping triangular fuzzy number.

Semantic	Score	Corresponding Triangular Fuzzy Number
No influence	0	(0,0,0.25)
Very weak influence	1	(0,0.25,0.5)
Weak influence	2	(0.25.0.5,0.75)
Strong influence	3	(0.5,0.75,1)
Very strong influence	4	(0.75,1,1)

Based on the constructed influence factor system, 10 experts were invited to score the degree of influence relationship between the 10 secondary influence factors. The scoring standard is: "no influence degree: 0; very weak influence degree: 1; weak influence degree:

2: strong influence degree: 3; very strong influence degree: 4". Thus, the judgment matrix was established. The resulting expert scores were then fuzzed. In the triangular fuzzy number (1), α is the conservative value of the influence degree, β is the possible value of the influence degree, and λ is the positive value of the influence degree. The judgment matrix was processed into an initial direct-relation matrix through matrix standardization using triangular fuzzy numbers (Table 4).

The current matrix was subjected to a series of processing such as reduction of fuzzy number calculation standard value, calculation of 10 experts' overall standardized influence degree value, calculation of standardized influence matrix, and calculation of comprehensive relationship matrix. Personal credit comprehensive influence matrix T (Table 5) was finally obtained, and then calculate the centrality and causality of personal credit influencing factors of practitioners in the field of engineering construction (Table 6) and the relationship between them (Figure 2).

Table 5. Personal credit comprehensive influence matrix T.

Influencing Factors	Influencing Factors									
	a	b	c	d	e	f	g	h	I	j
a	0.64	0.62	0.38	0.48	0.40	0.29	0.61	0.34	0.50	0.49
b	0.99	0.75	0.52	0.71	0.50	0.39	0.85	0.45	0.74	0.73
c	0.92	0.74	0.48	0.60	0.49	0.36	0.83	0.42	0.67	0.66
d	0.91	0.69	0.56	0.54	0.44	0.37	0.73	0.40	0.60	0.59
e	0.92	0.76	0.58	0.59	0.49	0.42	0.85	0.48	0.70	0.69
f	1.06	0.85	0.64	0.71	0.63	0.43	0.97	0.58	0.83	0.82
g	0.97	0.84	0.51	0.68	0.49	0.38	0.78	0.45	0.76	0.70
h	0.92	0.84	0.53	0.63	0.58	0.40	0.92	0.47	0.79	0.78
i	1.07	0.92	0.57	0.70	0.56	0.42	0.96	0.53	0.74	0.85
j	1.09	0.94	0.59	0.71	0.62	0.44	0.99	0.56	0.83	0.75

a. In order to make the table clear, use indexes to replace factors, and the order is the same as the second column of Table 3

Table 6. The centrality and causality of personal credit influencing factors of practitioners.

Secondary Influencing Factors	Direct Impact	Degree of being Affected	Centrality	Causality
Willingness to work	4.75	9.48	14.23	−4.73
Working ability	6.64	7.96	14.6	−1.32
Moral credit	6.17	5.36	11.53	0.81
Financial condition	5.83	6.35	12.18	−0.52
Social environment	6.51	5.20	11.71	1.31
Economic environment	7.52	3.93	11.45	3.59
Working environment	6.57	8.50	15.07	−1.93
Advanced supervision	6.86	4.69	11.55	2.17
Process supervision	7.30	7.16	14.46	0.14
Postmortem supervision	7.52	7.05	14.57	0.47

6.1 *Cause analysis*

According to the calculation results of the cause degree of each influencing factor, the 10 influencing factors can be divided into causal factors (calculated as positive) and outcome

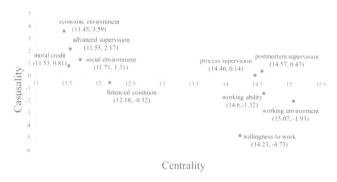

Figure 2. The relationship between influencing factors.

factors (calculated as negative). The economic environment, advanced supervision, social environment, ethics and credit, process supervision, and postmortem supervision belong to the causal factors, which influence other influencing factors more than being influenced; work willingness, work environment, workability, and financial status belong to the result factors.

6.1.1 *Analysis of causal factors*
The economic environment and social environment are the background environment outside the professional work, not easy to be influenced by a person himself, and relatively not easy to be influenced by the regulatory strength of a certain industry, so these two factors are influenced to a lesser extent, while the economic foundation determines the superstructure, the economic environment will have an impact on many factors of the superstructure, and the social environment will also influence other factors implicitly; supervision is the government's regulatory work responsibilities, and regulatory efforts may change from person to person, but the overall approach will not change because of a person, indirectly affected by the economic environment and social environment; moral credit belongs to personal character traits, character traits will occasionally be affected by certain factors and change, but the nature of character is difficult to change, and character will affect other factors of the individual.

6.1.2 *Analysis of outcome factors*
Work willingness is easily affected by the economic environment, social environment, financial situation, workability, supervision, and other factors; the work environment is the environment directly composed of practitioners, and each practitioner's own factors directly affect the work environment in which they work, but also indirectly affected by the economic environment, social environment, and other general backgrounds; financial situation is originally a floating indicator. The overall economic environment of the society is prosperous, the individual's working ability is strong, and the financial situation will be more likely to be good [12].

6.2 *Centrality analysis*
According to the calculation of the centrality of each influencing factor, they can be divided into key influencing factors and non-key influencing factors by combining their degree of influence and degree of being influenced. In the relationship of these influencing factors, work environment, process supervision, postmortem supervision, workability, and work willingness are the key influencing factors with high centrality, both the degree of influence and the degree of being influenced are significant, and they are highly correlated with other

influencing factors; financial status, advanced supervision, social environment, moral and credit, and economic environment are the non-key influencing factors with low centrality, only one of the degree of influence and the degree of being influenced is significant, and they are less correlated with other influencing factors.

6.2.1 *Analysis of key influencing factors*

Work environment is the direct working environment of engineering and construction practitioners, whether it is the disposition of the work of the practitioners or the contagion of the company's credit, it is the direct impact on the personal credit. Supervision of the matter is necessary after practitioners enter the industry and start project work, directly affecting the personal credit of their practice career. Ability and willingness to work are personal attributes of the practitioner that are closely related to the job. The former directly determines whether he or she can do a good job, and the latter directly determines whether he or she wants to do a good job.

6.2.2 *Analysis of non-critical influencing factors*

Although the economic environment and social environment are both causal factors, they are the general background of individual survival and industry development rather than the direct environment and are generally indirectly influenced. Advanced supervision is mainly achieved through registered examinations and the regulation is low, thus the degree of influence is low. Financial status and moral credit are traits in individuals that may influence but not necessarily affect personal credit, so it is reasonable that they belong to non-critical influencing factors.

7 CONCLUSIONS

This paper adopts the grounded theory to establish the concept of personal credit of practitioners in the field of engineering construction and considers that the concept of its institutionalization is that the credit rating of individuals is evaluated, recorded, and filed at any time according to the basic personal information, work status, property status, and credit status of practitioners so that the suppliers of credit can decide their disposal.

Through the identification of personal credit influencing factors of practitioners in the engineering construction field, this paper identifies three influencing dimensions, namely, individual, environment, and supervision, and from these three primary influencing factors, 10 secondary influencing factors, such as willingness to work, are separately identified and the influencing factor system is constructed. Using the Fuzzy-DEMATEL method on the calculation software, we can conclude that the economic environment, advanced supervision, social environment, ethics and credit, process supervision, and postmortem supervision belong to the causal factors. Moreover, work environment, process supervision, postmortem supervision, workability, and work willingness are the key influencing factors. The research in this paper lays the foundation for establishing a credit assessment model for practitioners in the construction market. It also can be used as a reference for enterprises, industry associations, and government regulators to make decisions to regulate the order of the construction market, such as the differentiated supervision of practitioners.

ACKNOWLEDGMENT

The financial support is provided by the Foundation of National College Students Innovation and Entrepreneurship Training Program (No.202110294103) and the National Science Foundation of China (No. 2019B19614).

REFERENCES

[1] Abramowitz, A.J. (1998) Speak Out: A Lawyer Finds that Architects, in their Intense Pursuit of Ethics, Often Deny Themselves. *Archit. Rec*, 186: 24–26.

[2] Chen, T.Q., Yang, Q.H., Sun, X.Y. (2021) Analysis of Credit Risk Infection Mechanism of Bank Counterparties Based on Credit Correlation Super Network, *Mon. J. Acc*, 4: 118–125.

[3] Chen, Y.Y., Wang, X.Q., Wei, Z. (2014) Credit Evaluation Model of Building Contractors Based on Binary Semantics, *Fuzzy Syst. Math*, vol. 28: 98–105.

[4] Du, Y.L., Kou, X., Ke, D. (2019) Constructing and Dimensional Division of PPP Project Efficiency: An Exploratory Study Based on Grounded Theory, *Sci. Technol. Manage. Res*, 39:229–235.

[5] Fan, Z.Q., Wang, X.Q., Lee, B.L. (2009) Research on Credit Evaluation of Practitioners in Construction Market Based on Object Meta-analysis, *Soft. Sci*, 23: 5–6.

[6] Fang, J., Hu, J., Wei, H., Chen, Z.F., Chen, T. (2020) Identification of Key Factors of Fall from Height Human-caused Safety Risks at Construction Sites, *J. Wuhan. Univ. Technol. Inf. Manage. Eng. Ed*, 42: 409–413.

[7] Fellows, R., Liu, A., Storey, C. (2004) Ethics in Construction Project Briefing. *Sci. Eng. Ethics*, 10: 289–301.

[8] Huang, F.N., Wang, J., Zou, J.C. (2020) Research on Credit Evaluation of Chief Supervision Engineer Under Self-management Mode, *J. Rail Sci. Eng.*, 17: 1318–1326.

[9] Lu, Y.Q., Sun, D.D., Zhao, B. (2020) Identification of Key Influencing Factors of Intergenerational Knowledge Transfer, *Inf. Sci*, 38: 7–8.

[10] Wu, J.M. (2015) Consider the Big Data Credit Investigation from the Connotation and Composition of the Credit, *J. Capi Norm Univ. Social. Sci. Ed*, 6: 66–72.

[11] Wu, Y.P., Mao, Y.H., Chen, L.Y., Shi, D.F. (2009) Reasons and Countermeasures of Credit Deficiency in the Construction Field. *China. Soft. Sci*, S1: 36–39.

[12] Xu, K., Zhou, Z.F., Qian, X., Xie, X. (2020) Study on the Mechanism of Associated Credit Risk Contagion Considering Individual Protection Awareness," *Oper. Res. Manage*, 29: 197–206.

Water Conservancy and Civil Construction – Oke & Ahmad (Eds)
© 2024 The Author(s), ISBN: 978-1-032-58618-2

Comparative analysis of carbon emissions of different asphalt pavement structures

Yin Huai Ma*, Cheng Gang Duan* & Li Jiang*
Hebei Provincial Expressway Yan Chong Management Center Zhangjiakou, China

Yang Li*
Shijiazhuang Tiedao University, Shijiazhuang, China

ABSTRACT: This article investigates and analyzes the differences in material carbon emissions of different asphalt pavement structures from the perspective of construction materials. Taking SMA-13 asphalt pavement and AC asphalt pavement as examples, the main road construction materials of asphalt pavement are analyzed, and the pavement structure inventory of SMA-13 asphalt pavement and AC asphalt pavement is established. A carbon emission calculation model is constructed to calculate the carbon emission of road construction materials of asphalt pavement and to propose relevant energy-saving and emission reduction suggestions. The research results show that SMA-13 asphalt pavement has good performance and produces relatively less carbon emission, which is suitable for wide application.

1 INTRODUCTION

One of the largest sources of energy consumption and carbon emissions in the road construction process is the use of materials, and the types and amounts of materials involved are very complex and important. In the collection and production processing of materials, different types and usage of materials will generate different energy consumption and carbon emissions, which will pollute our environment. According to the "China Construction Materials Industry Carbon Emissions Report (2020 Annual)" released by China Construction Materials Federation, the CO_2 emissions of China's construction materials industry will be about 1.48 billion tons in 2020. This shows that energy saving and emission reduction of materials are the focus of future development.

Asphalt marl gravel mix and asphalt concrete mix are the two most widely used materials during the construction of road pavement surface layers. The former is typically used as the top layer of SMA-13 asphalt pavement, while the second is commonly referred to as AC asphalt pavement. Due to the differences in the structure and mixes of the two, they can also differ in terms of carbon emissions of the materials. Many scholars have studied the carbon emissions of asphalt pavements; Olubukola O. Tokede *et al.* (2020) [1] used LCA to assess the environmental impact of asphalt containing lignin binder and normal asphalt; Gao Fang (2016) [2] analyzed the energy consumption and carbon emissions of two types of asphalt pavements, semi-rigid base, and flexible base, in construction based on LCA; Zhang Haitao *et al.* (2018) [3] assessed the carbon emissions of three different pavement structures of a typical highway based on LCA method; Meng Xiangchen (2020) [4] calculated and analyzed the energy consumption and carbon emissions of the whole life cycle of asphalt pavement surface construction process and proposed a scheme to reduce the carbon emissions of the pavement construction process; Yan Qiang et al [5] (2021) studied the

---

*Corresponding Authors: 360791346@qq.com, 907115826@qq.com, 26347913@qq.com and 813491511@qq.com

DOI: 10.1201/9781003450832-61

asphalt pavement structure type on the construction period carbon emissions of the construction period. Although many scholars have studied the construction materials and construction technology of SMA-13 asphalt pavements, fewer scholars have conducted relevant studies on the carbon emissions of material production of SMA-13 asphalt pavements.

This paper compares and analyzes the types and usage of the main construction materials of different asphalt pavements from the perspective of construction materials. Taking a highway project as an example, based on the analysis of the structural inventory of asphalt pavement, the emission factor method is applied to calculate the carbon emission of construction materials for SMA-13 asphalt pavement and AC asphalt pavement to analyze and study the difference between the two types of asphalt pavement in terms of carbon emission from material production and processing.

2 MAIN CONSTRUCTION MATERIALS FOR ASPHALT PAVEMENTS

In asphalt pavement construction, the main material involved is asphalt, followed by cement, gravel, tap water, mineral powder, etc. SMA-13 asphalt pavement and AC asphalt pavement are very similar in terms of the types of raw materials used, but there are significant differences in the proportion of raw materials used. In general, the ratio of SMA-13 asphalt pavement and AC asphalt pavement materials used [6] is shown in Figure 1.

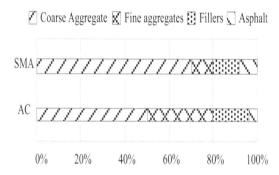

Figure 1. Percentage of SMA and AC asphalt pavement materials.

2.1 Asphalt

Asphalt material is a very important and essential material in asphalt pavement construction. The main types of asphalt often used are petroleum asphalt, emulsified asphalt, modified asphalt, SBS composite modified asphalt, etc. Of course, different types of asphalt produce different carbon emissions. Of course, the carbon emissions from different types of asphalt are different. asphalt marcasite gravel mix is used in SMA-13 asphalt pavement, and the proportion of asphalt in SMA-13 asphalt mix is high, with the amount of 5.7%–6.5% [7].

2.2 Coarse and fine aggregates

Coarse and fine aggregates make up the majority of the material used in asphalt pavements. In SMA-13 asphalt pavement, coarse aggregate is indispensable and it is also the basic skeleton of SMA-13 asphalt pavement. Although the amount of fine aggregate is very small, it plays an important role in SMA-13 asphalt mixes. Coarse and fine aggregates are used in very high amounts in SMA-13 asphalt pavements, up to about eighty percent, which may also lead to a high percentage of coarse and fine aggregates in the total carbon emissions of SMA-13 asphalt mixtures.

It can also be seen from Figure 1 that coarse and fine aggregates account for about 80% of all materials used in AC asphalt pavements, but the percentage of fine aggregates used in AC asphalt pavements is significantly more than the percentage of fine aggregates used in SMA-13 asphalt pavements.

2.3 *Fillers*

Fillers are also essential in asphalt pavement construction. In AC asphalt pavement, mineral powder ground from limestone is generally used, and as shown in Figure 1, the filler also accounts for a large proportion of AC asphalt pavement materials. The material filled in the skeleton of SMA-13 asphalt pavement is the filler, and the filler is also used in a relatively large amount of SMA-13 asphalt mixture, accounting for about 8%–12%, which is about 2–3 times more than the common asphalt mixture [7]. Some projects also mix some slaked lime powder and cement in the mineral powder, which makes the mixture with better adhesion [8].

Of course, in addition to the above materials, it will involve the use of other materials such as water, stone, and coal. Although the use of other materials may account for a relatively small amount, it is not negligible in the calculation of asphalt pavement material use and carbon emissions.

3 STRUCTURAL INVENTORY ANALYSIS OF ASPHALT PAVEMENTS

In the construction of a highway, the SMA-13 asphalt top layer with a thickness of 40mm was applied, and the middle and lower layers were paved with ARHM-20 and AC mix. And the explored AC asphalt pavement surface layer is applied with AC mix, the bedding layer, base layer, and functional layer structure are the same as SMA-13 asphalt pavement. After calculation, the list of pavement structures for constructing each 1 m^2 of SMA-13 asphalt pavement with AC asphalt pavement was obtained, as shown in Table 1 and Table 2.

Table 1.   List of SMA-13 asphalt pavement structures.

Type of structure	Unit	Quantity of work
PC-2 Emulsified Asphalt Permeable Layer	m^2	1.368
PCR-modified emulsified asphalt binder layer	m^2	5.064
Waterproof binder layer	m^2	3.525
Cement Stabilized Crushed Stone Sub-base	m^3	0.294
Cement-stabilized crushed stone sub-base	m^3	0.537
Gravel bedding	m^3	0.040
Binder powder/SBS modified asphalt concrete	m^3	0.143
Fine-Grained Asphalt Concrete	m^3	0.042
Medium Grain Asphalt Concrete	m^3	0.095
Bituminous crushed horseshoe resin	m^3	0.099

Table 2.   List of AC asphalt pavement structures.

Type of structure	Unit	Quantity of work
PC-2 Emulsified Asphalt Permeable Layer	m^2	1.368
PCR-modified emulsified asphalt binder layer	m^2	5.064
Waterproof binder layer	m^2	3.525
Cement Stabilized Crushed Stone Sub-base	m^3	0.294
Cement-stabilized crushed stone sub-base	m^3	0.537
Gravel bedding	m^3	0.040
Binder powder/SBS modified asphalt concrete	m^3	0.143
Fine-grained Asphalt Concrete	m^3	0.042
Medium Grain Asphalt Concrete	m^3	0.095
Fine-grained SBS Modified Asphalt Concrete	m^3	0.048

From the above analysis, it can be found that the materials used in the two types of asphalt pavements and the amount used are different. 0.099 m^3 of SMA-13 asphalt pavement is used for each 1 m^2 of asphalt pavement, while 0.048 m^3 of fine-grained SBS-modified asphalt concrete is used for each 1 m^2 of asphalt pavement. Although there is little difference between the use of asphalt marble mix and fine-grained SBS-modified asphalt concrete per 1 m^2 of asphalt pavement construction, the difference between the two types of constituent materials and the proportion of material usage can lead to a significant difference in the carbon emissions of the two types of asphalt pavements.

4 CALCULATION AND ANALYSIS OF CARBON EMISSIONS FROM ASPHALT PAVEMENTS

Based on the analysis of the structural inventory of two different asphalt pavements, it is clear that the types of materials used and the proportion of material used vary for different types of asphalt pavements. Based on the analysis of the asphalt pavement structure inventory, the carbon emissions of the asphalt pavement materials were calculated.

4.1 *Calculation of carbon emissions of asphalt pavement materials*

The material usage of SMA-13 pavement and AC asphalt pavement is calculated according to the project volume, and then the carbon emission of each material is calculated according to the emission factor method, and the carbon emission calculation formula for material production and processing is shown in (1).

$$C_{co_2} = \sum_i Q_i \times EF_i \qquad (1)$$

where Q_i is the consumption of the ith material, and EF_i is the carbon emission factor of the ith material.

The calculated material carbon emissions released per 1 m^2 of SMA-13 pavement and AC asphalt pavement constructed are shown in Table 3.

Table 3. Carbon emissions of major construction materials for SMA-13 asphalt pavements and AC asphalt pavements.

Name of material	Carbon emissions from SMA-13 asphalt pavement material(kg)	AC asphalt pavement material carbon emissions(kg)
32.5 grade cement	68.563	68.563
Sand	0.542	0.582
Stone	5.469	5.223
Mineral powder	3.928	4.595
Asphalt concrete mixes	629.583	730.515
Emulsified asphalt	18.138	18.138
Petroleum asphalt	106.041	106.041
Modified bitumen	43.881	18.006
Coal	0.002	0.002
Total	876.148	951.666

From Table 3, it can be seen that the types of materials used to construct 1 m^2 of SMA-13 asphalt pavement are very similar to those used to construct 1 m^2 of AC asphalt pavement. However, in terms of total material carbon emissions, SMA-13 asphalt pavement emits less CO_2

than AC asphalt pavement, and the materials used to construct 1 m² of SMA-13 asphalt pavement emit about 75.518 kg less CO$_2$. However, the modified asphalt used in SMA-13 asphalt pavement emits more carbon than the modified asphalt used in AC asphalt pavement.

4.2 *Analysis of carbon emissions from asphalt pavement materials*

The calculation of carbon emissions of the two asphalt pavement materials shows that SMA-13 asphalt pavement has fewer carbon emissions than AC asphalt pavement materials. After analyzing the carbon emissions of each material, a comparative analysis of the carbon emissions of the two different asphalt pavement structures is shown in Figures 2, 3, and 4.

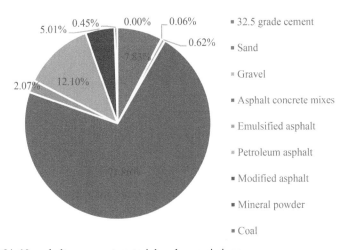

Figure 2. SMA-13 asphalt pavement material carbon emissions.

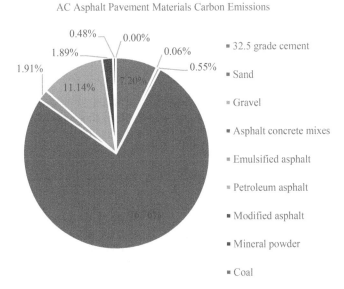

Figure 3. AC asphalt pavement material carbon emissions.

From Figures 2 and 3, it can be seen that SMA-13 asphalt pavement and AC asphalt pavement have a common point on both sides of material carbon emissions, that is, asphalt concrete mixture carbon emissions are the main source of material carbon emissions for both asphalt pavement structures, which is about more than 70% of all material carbon emissions. The carbon emissions of asphalt concrete mix in AC asphalt pavement are more, about 4.9% more carbon dioxide is released than that of asphalt concrete mix in SMA-13 asphalt pavement.

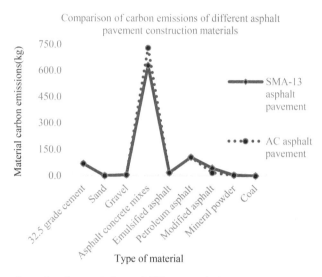

Figure 4. Comparison of carbon emissions of different asphalt pavement construction materials.

It is clear from the analysis of Figure 4 that the main reason for the lower carbon emissions of SMA-13 asphalt pavement than AC asphalt pavement materials is the difference in carbon dioxide emissions from the asphalt concrete mixes. It can be seen that asphalt mastic mix is more energy efficient than SBS-modified asphalt concrete. From a comprehensive point of view, SMA-13 asphalt pavement produces relatively less carbon emissions and is suitable for widespread application.

5 CONCLUSION

This paper focuses on the material carbon emissions of SMA-13 asphalt pavement and AC asphalt pavement, two different pavement structures, from the perspective of carbon emissions from construction material production. The main conclusions obtained are as follows.

In general, the material used in SMA-13 asphalt pavement emits less CO_2 than the material used in AC asphalt pavement carbon emissions. However, from the comparative analysis of carbon emissions of individual materials, the stone and modified asphalt in SMA-13 asphalt pavement emit more CO_2 than the stone and modified asphalt in AC asphalt pavement. And the SMA-13 asphalt pavement produces less carbon emission than the materials used in AC asphalt pavement mainly in the material of asphalt concrete, sand, and mineral powder. In addition, SMA-13 asphalt pavement has very good excellent properties such as wear resistance, high temperature rutting resistance, and effective in reducing crack generation at low temperatures.

To reduce carbon emissions of highway pavement materials, we should increase the investment in energy-saving and emission-reduction technologies in the construction of highway pavement; for example, the application of high-performance concrete materials, cold mix cold paving asphalt mixes, and the adoption of vibration mixing technology measures, etc.

REFERENCES

[1] Tokede O O., Whittaker A., Mankaa R, Traverso M. (2020) *Life Cycle Assessment of Asphalt Variants in Infrastructures: The Case of Lignin in Australian Road Pavements. Structures*, 25:190–199.

[2] Gao F. (2016) *Research on Energy Consumption and Emission Analysis of Asphalt Pavement Construction Period Based on LCA.* Chongqing Jiaotong University.

[3] Zhang H T, Meng L, Lv L H. (2018) Impact of LCA-based Asphalt Pavement Design Parameters on Carbon Emissions. *Highway and Transportation Technology*, 35(02):1–7.

[4] Meng X C. (2020) *Research on Energy Consumption and Carbon Emission During the Construction Period of Asphalt Pavement.* Beijing Jiaotong University.

[5] Yan Q, Yi K L, Li P. (2021) Analysis of the Impact of Asphalt Pavement Structure Type on Carbon Emission During the Construction Period. *Chinese and Foreign Highways*, 41(02):41–48.

[6] Li L. (2015) Comparison of the Application Performance of SMA and AC in the Rehabilitation of Old Roads in Fujian. *Fujian Transportation Science and Technology*, (04):4–6 + 10.

[7] Lai F J. (2020) Discussion on the Quality Control of SMA-13 Asphalt Pavement Construction. *Green Environmental Protection Building Materials*, (10): 91–92.

[8] Xu J Y. (2018) Quality Control of Raw Materials and Construction of SMA Pavement for Highway. *Shanxi Construction*, 44(22):152–154.

Water Conservancy and Civil Construction – Oke & Ahmad (Eds)
© 2024 The Author(s), ISBN: 978-1-032-58618-2

Low carbon strategies based on a steel prefabricated hotel in Shenzhen

Siyu Jiao, Hao Lu* & Dongshan Ding
China Construction Science and Industry Corporation Ltd, Shenzhen, China

ABSTRACT: Taking a prefabricated steel structure hotel in Shenzhen as the research object, this paper collected the actual bill of quantities and energy consumption data, calculated the total carbon emission and carbon emission intensity of the building in the whole life cycle, and discussed the low carbon emission reduction strategies. The purpose was to provide research data on the energy-saving and emission-reduction benefits of prefabricated buildings and to provide technical support for the subsequent popularization and application of prefabricated buildings.

1 INTRODUCTION

In the last twenty years, a series of phenomena, such as global warming, glacier melting, and rising sea levels, have indicated that climate change caused by greenhouse gases is seriously affecting human survival (Laine 2020). As countries become more concerned about global climate change, a series of carbon emission reduction plans are gradually implemented. Most developed countries achieve carbon emission reduction mainly through carbon taxes, carbon emissions trading, and financial subsidies (Wu 2022). China takes more effective measures, such as promoting prefabricated buildings, to peak carbon dioxide emissions before 2030 and achieve carbon neutrality by 2060 (Lu 2022).

The 14th Five-Year Plan for the Development of the Construction Industry, issued by the Ministry of Housing and Urban-Rural Construction, points out that assembly buildings should be vigorously developed, and the proportion of assembly buildings in new buildings should reach more than 30% by the end of the 14th Five-Year Plan. Under the guidelines of energy savings and emission reduction, reducing the cost and energy consumption is very important to promote the sustainable development of building industrialization in the future. In the process of national policy and industrialization, prefabricated buildings will be popularized and applied on a large scale (Zhang 2020). Therefore, it is particularly important to calculate carbon emissions in the whole life cycle of prefabricated buildings. Taking a steel prefabricated emergency quarantine hotel in Shenzhen as the research object, this paper calculated the carbon emissions of the building in the whole life cycle and explored the low-carbon emission reduction strategy, which provided a reference for promoting the sustainable development of prefabricated buildings.

2 PROJECT OVERVIEW

A steel-structure prefabricated hotel in Shenzhen covered an area of 103, 000 square meters, with a total building area of about 305, 900 square meters. There were 24 buildings on the plot. Among them, there were 10 hotels, 2 dormitories, and 12 attached buildings. Taking the B3 building as an

---

*Corresponding Author: luhao094@cscec.com

486

DOI: 10.1201/9781003450818-62

example, the overall structure was a steel structure frame system, and the score of assembly type identification was 100.9. The main prefabricated components included steel columns, steel beams, steel truss floor bearing plates, an integral bathroom, assembly-type decoration, etc.

3 LOW-CARBON STRATEGY PRINCIPLE OF STEEL STRUCTURE PREFABRICATED BUILDING

3.1 *General principles*

To respond to the requirements of Shenzhen for green buildings and in combination with the relevant national carbon emission calculation standards, the project adheres to the principles of low pollution, low energy consumption, and high utilization in the design and construction processes. It adopts prefabricated steel structures, industrialized components, and on-site prefabricated construction to reduce material waste and pollution at the source. Compared with cast-in-place concrete buildings, the total carbon emissions can be reduced by more than 35% by taking corresponding actions in the four stages of the whole life cycle of the building: the production and transportation stage, the construction stage, the operation stage, and the demolition stage (Su 2016).

3.2 *Production and transportation stage*

3.2.1 *Production stage*
During the production stage, two strategies are taken into consideration. On the one hand, most of the buildings in the project adopt the fabricated steel structure frame, which means a large number of prefabricated components can be produced centrally in the factory. As a result, production efficiency improves. Buildings, on the other hand, use the standard modulus of 3.6 m by 9 m with building components whose sizes are commonly used in the market to reduce the types of components in the manufacturing process. As a result, the carbon emissions generated during the production stage are significantly reduced (Ou 2015).

3.2.2 *Transportation stage*
In terms of transportation, the geographical advantages of the project are fully considered. Local materials are used to the greatest extent, and local building material suppliers and equipment factories are selected to reduce carbon emissions from the trans-regional transportation of building materials and equipment. In addition, planning the routes and modes of transport reasonably and using the new energy-powered transport types also reduce carbon emissions.

3.3 *Construction stage*

Compared with the traditional construction method of transporting raw materials to the construction site for concrete pouring, the prefabricated component construction method will advance a large number of on-site tasks under the regulations of the factory process production so that the construction cycle can be greatly shortened. Meanwhile, there is less equipment used on site due to the cancellation of the on-site scaffolding (Alireza 2021). This method has great advantages in artificial carbon savings, equipment power savings, and construction waste reduction.

3.3.1 *Artificial carbon savings*
The construction industry is an industry with a huge number of workers, so the carbon emissions generated by the daily lives of workers in construction cannot be ignored. The fabricated structure has a short construction period and a smaller number of workers because most of the processing work can be completed in the factory, so the carbon emissions generated by manual work are greatly reduced.

3.3.2 *Power saving of equipment*

Due to the reduction of on-site concrete construction operations, the amount or frequency of use of major equipment such as concrete vibrating equipment and mixing equipment on the construction site is reduced, which saves power and reduces carbon emissions.

3.3.3 *Waste and pollution*

In terms of sewage discharge, the construction site is mainly composed of assembly construction, which reduces the amount of wet work in the construction process compared with traditional cast-in-place buildings, greatly reduces water consumption and sewage discharge, and also effectively reduces the noise and dust pollution caused by on-site construction (Kong 2020).

3.4 *Operation stage*

3.4.1 *Envelope system*

The insulation materials used in the project are manufactured in the factory as a whole, which has a longer life than on-site cutting and installation and can reduce the consumption of insulation materials in the whole life cycle of the building (Zhu 2013). The external wall of the project adopts the full unit curtain wall system with a heat transfer coefficient < 3.0 W/$(m^2 \cdot K)$ and a shading coefficient SC $= 0.223$, which is more than 10% higher than the national building energy-saving design regulations. By increasing the area of natural lighting, the carbon emissions generated by the lighting system are considerably reduced.

3.4.2 *Energy saving of equipment*

In terms of equipment selection, the project adopts water, heating, electricity, fire protection, and elevator systems with lower energy consumption and higher transmission efficiency in order to reduce carbon emissions during daily operations.

3.5 *Demolition stage*

3.5.1 *Removal and discharge*

In this project, most of the steel structures are connected by bolts and welding. Compared with traditional concrete structure demolition methods, such as smashing and blasting, the demolition of steel structures is simpler, faster, and more environmental-friendly with less dust and noise pollution (Rubayet 2022).

3.5.2 *Recycling and reusing*

The steel structure and enclosure materials used in the project are green building materials with a high recovery rate (Qiao 2019). The resource recovery rate can be as high as 80% at the end of life. Material consumption and waste can be reduced by using 3R building materials, and thus carbon emissions can be reduced.

4 CARBON EMISSION CALCULATION FOR STEEL STRUCTURE PREFABRICATED BUILDING

4.1 *Calculation of carbon emissions in the production stage of building materials*

Data collection was carried out for the consumption of main building materials according to the actual consumption, procurement, or budgetary estimate of the project. Then the carbon emissions were calculated according to the formula, and the results are shown in Table 1.

$$CSC = \sum_{i=1}^{n} M_i \times F_i \tag{1}$$

where CSC is the carbon emission in the production stage of building materials (kg CO_2e); M_i is the consumption of the ith main building material; F_i is the carbon emission factor of the ith main building material (kg CO_2e/unit building material quantity);

Table 1. The consumptions and carbon emissions of the main building materials.

Building Materials	Consumption of Building Materials (kg/m^2)	Carbon Emission Factor of Building Materials (kg CO_2e/t)	Carbon Emission (kg CO_2/m^2)
Steel	144.61	2380	344.17
Curtain Wall Steel	6.18	3680	22.75
Curtain Wall Aluminum Plate	2.75	20300	55.74
Curtain Wall Aluminum Profile	7.77	20300	157.64
Decorative Aluminum Profiles	3.27	20300	66.32
Glass	9.46	1130	10.69
Rebar	15.06	2340	35.23
Concrete	482.84	735	354.89
Total			1047.43

Compared with the traditional cast-in-place concrete structure, the steel recovery rate of the main building materials of the steel structure can reach 80%, while the concrete materials are not recyclable. Therefore, when calculating the carbon emissions of building material production, it is necessary to consider the recycling and reuse of building materials. The recovery rate of building materials is shown in Table 2, and the reduction of carbon emissions in the production stage is shown in Table 3.

Table 2. Recovery rate of building material.

Material	Steel	Aluminum	Rebar
Recovery Rate	0.8	0.85	0.4

Table 3. The reduction of carbon emissions in the production stage.

Building Materials	Building Material Recovery (kg/m^2)	Carbon Emission Factor of Building Materials (kg CO_2e/t)	Carbon Reduction (kg CO_2/m^2)
Steel	115.688	2380	275.34
Curtain Wall Steel	4.944	3680	18.19
Curtain Wall Aluminum Plate	2.3375	20300	47.45
Curtain Wall Aluminum Profile	6.545	20300	132.86
Decorative Aluminum Profiles	2.7795	20300	56.42
Rebar	6.024	2340	14.10
Total			544.36

According to the table above, the carbon emission per square meter is 503.07 kg/m^2, reduced by 51.97% when considering the recovery.

4.2 Calculation of carbon emissions in the transportation stage of building materials

Data collections were conducted for the types of vehicles, the tonnage of load, the number of vehicles used, and the transportation distance, etc. in the transportation stage. These can be referred to as the construction schedule or the actual vehicle dispatch table of the hired transportation company. Then the carbon emission during the transportation of building materials was calculated according to the following formula:

$$C_{ys} = \sum_{i=1}^{n} M_i \times D_i \times T_i \tag{2}$$

where C_{ys} is the carbon emission during transportation of building materials ($kgCO_2e$); M_i is the consumption of the ith main building material (t); D_i is the average transportation distance of the ith building material (km); and T_i is the carbon emission factor of unit weight transportation distance under the transportation mode of the ith building material ($kg\ CO_2e/t\cdot km$);

The carbon emissions of transportation tools from the origin of building materials and mechanical equipment factories to the construction site must be taken into account. The project's construction site was in Shenzhen, and the factories that selected the building materials and equipment were in nearby or nearby cities. Thus, the mode of transport was mainly road transportation, with an average transportation distance of about 50 km. The types and quantities of vehicles were obtained from the vehicle registration center. The carbon emissions are shown in the following table:

Table 4. The carbon emissions in the transprotation stage.

Mode of Transport	Transportation Tonnage (t)	Transportation Distance (km)	Quantity	Carbon Emission Factor (kg CO_2e/ (t·km)	Carbon Emission (kg)
Light Petrol Truck	2	50	2087	0.334	69705.8
Medium Diesel Goods Vehicle	8	50	308	0.179	22052.8
Heavy Diesel Goods Vehicle	10	50	46	0.162	3726
Heavy Diesel Goods Vehicle	18	50	827	0.129	96014.7
Heavy Diesel Goods Vehicle	30	50	4238	0.078	495846
Total					687345.3
Carbon Emission Per Unit Building Area					2.25 (kg CO_2/m^2)

4.3 Calculation of carbon emission in the operation stage

The carbon emission in the operation stage must be calculated based on the energy consumption of various systems and the carbon emission factors of various types of energy. The total carbon emission per square meter was calculated according to the following formula:

$$C_M = \frac{\left(\sum_{i=1}^{n} (E_i \times EF_i) - C_p \right) \times y}{A} \tag{3}$$

$$E_i = \sum_{i=1}^{n} \left(E_{ij} - ER_{ij} \right) \tag{4}$$

where C_M is the carbon emission per unit building area in the building use stage, $kgCO_2/m^2$; Ei is the annual consumption of building type I energy, unit/a; EFi is the Class I energy carbon emission factor (see Appendix A); Eij is the Type I energy consumption of the Type J

490

system, unit/a; ERij is the Type J system consumes the amount of Type I energy provided by the renewable energy system, unit/a; CP is the annual carbon reduction from building a green space carbon sink system, $kgCO_2/a$; Y = Design life of the building, years; A is the Building area, m^2.

According to the green building energy consumption indicators, calculation methods, data results, and corresponding requirements in national and local standards, the building model was designed and simulated by using digital building energy consumption simulation software and parametric design optimization tools. The effect of emission reduction measures was analyzed, and the energy consumption and carbon emissions in the operation stage were predicted. The results are shown in the following table:

Table 5. The energy consumption and carbon emissions in the operation stage.

Project	Annual Energy Consumption Per Square Meter (kWh/a·m^2)	Carbon Emission Per Square Meter (kgCO$_2$/a·m^2)
Refrigeration	14.42	3.570
Domestic Hot Water	16.25	4.024
Lighting	7.62	1.887
Elevator	0.97	0.240
Electrical Equipment	12.59	3.117
Exhaust	3.11	0.770
Totally	54.96	13.608

According to the Statistical Report on Energy Consumption of Civil Buildings in Shenzhen in 2010 and the survey of the actual hotel project, the annual power consumption of a three-star hotel was 220 kWh/a·m^2. During the serving period, the carbon emissions of the buildings in this project were reduced by 75.02%.

The greening area of this project was 33000 m^2 in total and adopted the planting method of dense planting of large and small trees, which had a good ability to absorb carbon dioxide. The carbon sequestration was calculated and shown below.

Table 6. The carbon sequestration of the project.

Greening Area (m^2)	Carbon Sequestration (kg/m^2·a)	Total Carbon Fixation (kg)	Carbon Sequestration Per Cubic Meter (kg CO$_2$/a·m^2)
33000	22.5	742500	2.3952

In conclusion, the annual carbon emission per square meter in the operation stage of the building is about 11.21 kg CO_2/a·m^2. Assuming the service life of the building in this project is 50 years, the total carbon emission per square meter is 560.64 kg CO_2/m^2.

4.4 Carbon emission summary calculation

According to the Civil Building Energy Consumption Standard and the Implementation Plan for Pilot Construction of a Near Zero Carbon Emission Zone in Shenzhen, any building belonging to the hotel building category B below three stars and located in the area of hot summers and warm winters has a constraint value of energy consumption of 150 kWh/(m$_2$·a) and a guide value of 110 kWh/(m$_2$·a). Since the actual energy consumption of the building in this project was 54.96 kWh/(m$_2$·a) which met the requirement that the actual data was 20% lower than the guide value, the building was a near-zero carbon building. The data for various stages of a life cycle are shown below.

Table 7. Carbon emission in different stages and energy consumption in operation stage.

Project	Carbon Emission (kg CO_2/m^2)	Carbon Reduction (kg CO_2/m^2)	Carbon Emission Per Square Meter (kg CO_2/m^2)	Annual Energy Consumption Per Unit Area (kWh/a·m²)
Production Stage	1047.43	544.36	503.07	/
Transportation Stage	2.25	0	2.25	/
Operation Stage	680.40	119.76	560.64	54.96
Total Carbon Emissions	1730.08	664.12	1065.96	/

5 CONCLUSION

Taking a steel-structure prefabricated hotel in Shenzhen as the research object, the low-carbon emission reduction strategies of a steel-structure prefabricated building were explored, and the carbon emission was calculated. By selecting green materials with a high recycling rate and improving the standardization and production process of prefabricated parts, carbon emissions in the production stage were reduced significantly. Building components shall be collected locally, and transportation efficiency shall be improved to reduce energy consumption in the transportation stage. During construction, the procedures shall be optimized, management shall be strengthened, construction efficiency shall be improved, and waste generation shall be reduced. With the effort expended in all of those stages, the orderly development of prefabricated buildings will be promoted.

REFERENCES

Aisan Kong, Haibo Kang, Siyuan He, Na Li, Wei Wang (2020). Study on the Carbon Emissions in the Whole Construction Process of Prefabricated Floor Slab [J]. *Applied Sciences*, 10 (7).
Chunzhen Qiao, Peihao Hu, Qi Pan, Jialin Geng (2019). Research on CO2 Emission Reduction of a Steel Structure Prefabricated Building Considering Resource Recovery [J]. *IOP Conference Series: Earth and Environmental Science*, 237 (2).
Jani Laine, Jukka Heinonen, Seppo Junnila (2020). Pathways to Carbon-Neutral Cities Prior to a National Policy [J]. *Sustainability*, 12 (6).
Jianjun Zhu, David A.S. Chew, Sainan Lv, Weiwei Wu (2013). Optimization method for Building Envelope Design to Minimize Carbon Emissions of Building Operational Energy Consumption Using Orthogonal Experimental Design (OED) [J]. *Habitat International*, 37.
Karim Rubayet, Nakade Koichi (2022). A Literature Review on the Sustainable EPQ Model, Focusing on Carbon Emissions and Product Recycling [J]. Logistics, 6 (3).
Lu Chenrui, Wang Bing, Chen Tinggui, Yang Jianjun (2022). A Document Analysis of Peak Carbon Emissions and Carbon Neutrality Policies Based on a PMC Index Model in China [J]. *International Journal of Environmental Research and Public Health*, 19 (15).
Tabrizikahou Alireza, Nowotarski Piotr (2021). Mitigating the Energy Consumption and the Carbon Emission in the Building Structures by Optimization of the Construction Processes [J]. *Energies*, 14 (11).
Wei Zhang, Jing Li, Guoxiang Li, Shucen Guo (2020). Emission Reduction Effect and Carbon Market Efficiency of Carbon Emissions Trading Policy in China [J]. *Energy*, 196 (C).
Wu Zhen, Huang Xianjin, Chen Ruishan, Mao Xiyan, Qi Xinxian (2022). The United States and China on the Paths and Policies to Carbon Neutrality [J]. *Journal of Environmental Management*, 320.
Xiaoxing Ou, Dezhi Li (2015). Research on Techniques of Reducing the Life Cycle Carbon Emission at Building Design Stage [J]. *Applied Mechanics and Materials*, 3843 (744–746).
Xing Su, Xu Zhang (2016). A Detailed Analysis of the Embodied Energy and Carbon Emissions of Steel-construction Residential Buildings in China [J]. *Energy & Buildings*, 119.

Research on the measurement of headlight downtilt

Zheng Tao*

China Automotive Engineering Research Institute Co., Ltd, Chongqing, China

ABSTRACT: The vehicle low beam lamp is an active safety part, which provides front lighting for the vehicle. Its beam illumination position directly affects the road lighting. The measurement of headlamp inclination is the most important way to measure the beam irradiation position of headlamps. This paper describes how to calculate the headlamp inclination.

1 INTRODUCTION

The low beam lamp is an important active safety part of a car, which provides the driver with front lighting during road driving (Zhang et al. 2019). The position of the dipped beam has a great influence on the driving comfort of drivers and the safety of road users such as cars and drivers. The downward inclination of the headlamp is an important indicator of the beam irradiation position of the low beam, that is, the height of the light-dark boundary when the beam is projected on the curtain wall. The height of the light-dark boundary is shown in Figure 1. The height of the cut-off line of the passing beam can indicate the distance of the passing beam on the road. The light-dark boundary is too high, which is easy to make the opposite car driver or pedestrian dazzling. If the light-dark boundary line is too high or too low, it will shine too close to ensure the safe distance ahead is visible to the driver. Therefore, the height of the near light-dark boundary is crucial for safe driving and driving experience (Li 2017).

Figure 1. The height of the light-dark boundary.

Figure 2. Schematic diagram of tilt measurement under the headlight.

At present, many domestic scholars have studied headlights. Zheng Zhijun et al. explored the offset of the light-dark cutoff line of automobile headlights (Zheng & Hu 2012). Liu Fatai et al. proposed an evaluation index for the flatness of the light-dark cutoff line of headlights near light (Liu et al. 1997), and He Dong proposed points for attention in the

---

*Corresponding Author: 2819120567@qq.com

detection based on the analysis of the pass rate of light detection (He 2015). Ding Wenrong studied the influence of focal length on the gradient value of light and dark cutoff lines (Ding 2021). Sun Rui *et al.* solved the problem of low illumination of a vehicle model at night through bench tests (Sun *et al.* 2021). These studies have played an important role in driving headlight lighting and the driving experience, but little research has been done to measure and calculate downhill. Therefore, how to measure and calculate headlight downtilt is a problem worth paying attention to.

2 DEFINITIONS

2.1 *Initial inclination*

The initial inclination value of the passing beam specified by the vehicle manufacturer shall be used as the reference value for calculating the allowable change. According to GB4785-2019, the manufacturer shall display the initial downward inclination of the light-dark boundary of the dipped beam lamp with the specified symbol on the manufacturer's name-plate or near the headlamp of each vehicle with an accuracy of 0.1%, which is clear and durable (GB4785-2019). This value is measured when there is an unloaded vehicle with one driver in the driver's seat. This regulation specifies the production requirements of the manufacturer for the initial downward inclination of the dipped beam lamp and urges it to meet this condition when the vehicle is sold. At the same time, the initial downward inclination also provides the basis for the measurement and judgment of the detection mechanism. Especially if the market supervision bureau inspects the sample vehicle, it will compare the report of the inspection agency with the initial downward inclination of the light-dark boundary of the finished vehicle's low beam lamp. Therefore, it aims to further promote the inspection agency to use the standard method to test the initial downward inclination of the low beam lamp.

2.2 *Dipped beam inclination*

This angle is the deviation angle between the light direction at the intersection of the light-dark boundary of the passing beam and the reference horizontal plane of the passing beam. The tangent of the commonly used above angle is expressed in percent inclination.

The schematic diagram for measuring this angle is shown in Figure 2. The calculation method is shown in the following equation:

$$\text{The inclination} = \frac{h1 - h2}{L}$$

Where:

h1: on the vertical curtain wall L away from the front of the vehicle (this curtain wall is perpendicular to the longitudinal symmetry plane of the vehicle and horizontal-vertical), the height of the low beam dark boundary from the ground is measured.

h2: it represents the height from the reference center to the ground (the reference center is located on the reference plane of the passing beam and is the reference axis of the headlamp intersection of the mirror surface). The difference between the height "h2" of the lower edge of the lamp in GB 4785-2019 is noted to avoid measurement errors.

L: it represents the distance between the curtain wall and the reference center of the passing beam.

Note 1: h1, h2, and L are all in mm.

Note 2: In Figure 2, it takes the vehicle M1 as an example, and its principle is applicable to all other vehicles to measure the height of the light-dark boundary of the low beam.

3 MEASURING CONDITIONS

In order to distinguish the dark boundary more clearly, the test shall be conducted in a dark environment and the vehicle shall be driven. The curtain wall shall be in the state shown in the above figure, and the distance from the headlamp reference center shall not be less than 10 m.

When measured, the road surface is horizontal and flat so as to avoid measurement errors caused by uneven roads and ensure the repeatability accuracy of measurement results is $\pm$ 0.05%.

For the selected curtain wall shall also be flat and perpendicular to the ground and perpendicular to the longitudinal symmetry plane of the vehicle, and the repeatability accuracy of the measurement results shall be $\pm$ 0.05%.

During the measurement, the ambient temperature is between 10°C and 30°C.

4 VEHICLE PREPARATION

The driving distance of the vehicle is between 1,000 km and 10,000 km, and 5,000 km is the best choice. It is to ensure the reliability of lamps and consider the impact of tire running. And generally speaking, after the driving mileage reaches 5,000 km, the user will perform a maintenance inspection on the vehicle.

For the tire pressure, it is required to inflate the tire according to factory regulations. Fuel shall be added to at least 90%, and water and lubricating oil as well as all onboard accessories and tools shall be prepared to meet the actual driving conditions.

The car needs to release the parking brake, and the gear transmission will be in neutral to prevent the car from being lifted if the parking brake is tightened.

The vehicle shall be immersed in an environment of 10°C~30°C for at least 8 hours.

To distinguish the dark boundary line more clearly, the headlamp installed on the test vehicle shall have clear light emission and a strong sense of the dark boundary line of the low beam, or the method of removing the lens of the headlamp shall be adopted to make the dark boundary line of the low beam clearer.

5 TEST METHOD

5.1 *General rules*

Since the car has two low-beam lamps, both lamps shall be measured when we measure the inclination of the low beam. Under all loading conditions specified in GB4785-2019, the measured results of left and right headlamps shall be within the limits specified in 4.5. Under different loading conditions, the vehicle needs to be loaded. Gradual loading is used to ensure that the vehicle will not be impacted too much.

5.2 *Determination of initial inclination*

After the vehicle is prepared according to Item 4, it shall be loaded according to GB4785-2019. Before the measurement after each loading, the vehicle shall shake in the manner specified in 5.4. Three measurements are required for each loading condition.

The maximum deviation between each measurement result and the arithmetic mean value is 0.2% inclination, then this mean value can be used as the final result.

If the deviation between any measurement result and the arithmetic mean exceeds 0.2% inclination, it is necessary to continue to test and collect 10 measurement results, and the recorded output of the arithmetic mean of the 10 measurement results is the final result.

5.3 *The method of vehicle handling under each loading condition*

As the suspension or any other part of the vehicle will affect the inclination of the passing beam, the following methods shall be used to drive to avoid inaccurate measurement.

5.3.1 *Vehicles of category M1 fitted with conventional suspension systems*

The vehicle shall experience at least three continuous shakings after loading. The shaking method is to push down the rear end of the vehicle each time and then push down the front end in the same way. This method is mainly used to restore the mobility of the loaded rear suspension. In addition to pushing, the vehicle can also be driven backward for at least one wheel circumference distance and then forward at the same distance to achieve the effect of the same shaking. Because some vehicles cannot recover suspension mobility through simple pushing and short-distance driving, it is necessary to place the vehicle on a movable platform to increase the shaking range. After shaking, the vehicle is short-standing.

5.3.2 *Vehicles of categories M2, M3 and N fitted with conventional suspension systems*

1) In 5.3 a), the above processing method of the M1 vehicle can be used to adjust the suspension mobility. Because such vehicles are generally large in size and mass, the method of using M1 vehicles is not easy to adjust, and the following 5.3 b) 2) or 5.3 b) 3) can be used.
2) After the wheels are on level ground, the purpose of shaking the vehicle is achieved by increasing the load.
3) The vibration device can be used to drive the vehicle suspension system. If the vehicle can be placed on the vibration platform, suspension mobility can be restored. After the vibration, the vehicle is short-standing.
4) For vehicles with unconventional suspension systems, the vehicle will reach a stable state after starting the engine, and the dipped beam inclination will be measured directly.

5.4 *Measurement*

For different loading conditions, the change of dipped beam inclination relative to the measured initial inclination shall be evaluated (as specified in 5.2) as follows:

a) At the beginning, measurement corresponding to each loading condition shall be taken. If for all loading conditions, the variation of inclination is within the calculated limit of 0.4% inclination, the requirement will be met.
b) If any measurement result is not within the safety limit of 0.4% inclination or exceeds the limit value, three measurements shall be made under the corresponding loading conditions, and the results shall comply with 5.4 c).
c) Determination of measurement results
   1) If the deviation between the three measurement results and their arithmetic mean does not exceed 0.2% inclination, the arithmetic mean shall be regarded as the final result.
   2) If the deviation between any measurement result and its arithmetic mean exceeds 0.2% inclination, the test shall be continued, a total of 10 measurements shall be completed, and the arithmetic mean of 10 measurements shall be taken as the final result.

6 TEST AND DATA ANALYSIS

6.1 *Test*

The laboratory shall confirm non-standard methods, the methods developed by the laboratory, and use beyond the intended scope or other modified standard methods (Song 2022). In order to confirm whether the method of headlight downtilt measurement in this paper is accurate and reliable, which can be used reasonably and legally, and the same method is used for repeated detection for verification. Given the different loading of passenger cars and freight cars in the headlight downhill test, this paper only conducted the test and data analysis on the no-load state of multi-purpose passenger vehicles with one passenger in the cab.

The sample vehicle of the M1 class of a certain company is selected for the test. The reassembly mass of the sample vehicle is 1,550 kg, the fixed tire pressure of the tire factory is 230 kPa, and the headlight meets the relevant standard requirements (GB 25991-2010). The sample vehicle was run for 5,000 km before the test, the fuel was filled up, and the tire pressure and all accessory tools of the vehicle were checked, which all met the requirements. The laboratory of the National Motor Vehicle Quality Inspection and Testing Center (Chongqing), which has the national test qualification certification, can select a 20 m wide and 30 m long longitudinal 0 slope test road against the wall, fully meeting the test requirements. The test vehicle was placed in the laboratory for 8 h, and the temperature in the laboratory fluctuated between 18°C and 22°C within 8 h, meeting the test conditions. The photos of the test sample vehicle and test site are shown in Figure 3, and the equipment required for the test is shown in Table 1.

Figure 3. Laboratory and test sample vehicle.

Table 1. Test equipment.

No.	Name	Specification and model	Accuracy rating	Equipment No.	Unit of verification	Verification validity time
1	Digital hygrometer	Testo622	0.1	ZZC164-5	Shenzhen Tiansu Measurement and Testing Co., Ltd	2022-12-13
2	Tire pressure gauge	(35~1,400) kpa	0.05	ZZC161-4	Shenzhen Tiansu Measurement and Testing Co., Ltd	2023-8-15
3	Steel tape	7.5 M	±0.1 mm	ZZC163-1	Shenzhen Tiansu Measurement and Testing Co., Ltd	2023-7-25
4	Portable axle load meter	STW-18	±1%	ZZC176-3	Shenzhen Tiansu Measurement and Testing Co., Ltd	2023-8-3

Before the test, a horizontal mark (already marked in the laboratory) was made on the position 10 m away from the curtain wall. The test vehicle was driven slowly into the road, and the headlight was placed on the horizontal mark. Then, the gear position of the vehicle was placed in neutral, and when the near-light was turned on, a driver of 75 kg was kept in the cab. The test was divided into two groups of people and each group of people was carried out 10 times. Before each measurement, the same driver drove the vehicle out and in. The test results are recorded in Tables 2 and 3.

Table 2. The first group of tests (base center height of near light h2 = 760 mm, initial factory-determined inclination-1.00%).

Number of measurements "i"	Height of the terminatorh1 (mm)	Degree of inclination	The difference from the factory set initial inclination
1	656	−1.04%	−0.04%
2	654	−1.06%	−0.06%
3	655	−1.05%	−0.05%
4	656	−1.04%	−0.04%
5	658	−1.02%	−0.02%
6	658	−1.02%	−0.02%
7	660	−1.00%	0%
8	662	−0.98%	0.02%
9	661	−0.99%	0.01%
10	660	−1.00%	0%
Arithmetic mean value	658	−1.02%	−0.02%

Table 3. The second group of tests (the base center height of near light h2 = 760 mm, initial factory-determined inclination −1.00%).

Number of measurements "i"	Height of the terminator h1 (mm)	Degree of inclination	The difference from the factory set initial inclination
1	655	−1.05%	−0.05%
2	654	−1.06%	−0.06%
3	655	−1.05%	−0.05%
4	656	−1.04%	−0.04%
5	654	−1.06%	−0.06%
6	657	−1.03%	−0.03%
7	658	−1.02%	−0.02%
8	656	−1.04%	−0.04%
9	657	−1.03%	−0.03%
10	658	−1.02%	−0.02%
Arithmetic mean value	655	−1.05%	−0.05%

6.2 Data analysis

6.2.1 Standard uncertainty class A assessment

The standard deviation of the experiment was calculated by using Besser's formula (the first group):

$$s_1(h1_i) = \sqrt{\frac{1}{n-1} \sum_{i=1}^{n} (h1_i - h1)^2} = 2.57 \ mm \tag{1}$$

The standard deviation of the experiment was calculated by using Besser's formula (the second group):

$$s_2(h1_i) = \sqrt{\frac{1}{n-1} \sum_{i=1}^{n} (h1_i - h1)^2} = 1.41 \ mm \tag{2}$$

The two groups repeatedly measured the height above the ground of the light and dark boundary of the near light lamp 10 times respectively, and the arithmetic means the value of the 10 measurements was used as the measurement result. Therefore, the standard uncertainty A of repeatability for measuring the height above the ground of the terminator of the near-light lamp is (the first group):

$$U_{Ah1_1} = \frac{S_1(h1_i)}{\sqrt{10}} = \frac{2.57}{\sqrt{10}} = 0.812 \ mm. \tag{3}$$

the standard uncertainty A of repeatability for measuring the height above the ground of the terminator of the near-light lamp is (the second group):

$$U_{Ah1_2} = \frac{S_2(h1_i)}{\sqrt{10}} = \frac{1.41}{\sqrt{10}} = 0.447 \ mm. \tag{4}$$

6.2.2 *The standard uncertainty UB of the steel tape measure*

Standard uncertainties due to allowable errors and resolution of steel tape measures. By referring to the verification certificate of the test instrument, it can be seen that the maximum allowance of the steel tape measure ZZC163-1 is 0.1mm, then the Class B standard uncertainty component of the measurement result brought by the allowable error of the steel tape measure is:

$$U_{B1} = \frac{0.1mm}{\sqrt{3}} = 0.058 \ mm. \tag{5}$$

The uncertainty caused by the resolution of the steel tape measure is uniformly distributed. The resolution of the steel tape is 1 mm, and the interval half-width $a2 = 0.5$ mm, and $k2 = \sqrt{3}$. Its standard uncertainty:

$$U_{B2} = \frac{0.5mm}{\sqrt{3}} = 0.289 \ mm. \tag{6}$$

6.2.3 *Uncertainty of synthetic standard*

According to 6.2 a) and 6.2 b), considering that each uncertainty component is independent of the other, the uncertainty of the synthesis standard can be obtained as follows:

$$\text{(the first group):} \quad U_{C_1} = \sqrt{U_{Ah11}^2 + U_{B1}^2 + U_{B2}^2} = \sqrt{0.812^2 + 0.058^2 + 0.289^2} = 0.864 \text{ mm}$$

$$\text{(the second group):} \quad U_{C_2} = \sqrt{U_{Ah12}^2 + U_{B1}^2 + U_{B2}^2} = \sqrt{0.447^2 + 0.058^2 + 0.289^2} = 0.535 \text{ mm}$$

6.2.4 *Extended uncertainty*

The inclusion factor $k = 2$ is taken, and the height of the near-light light terminator above the ground (including two groups) is respectively

$$\begin{aligned} U_1 &= kU_{C_1} = 2 \times 0.864 = 1.728 \ mm \\ U_2 &= kU_{C_2} = 2 \times 0.535 = 1.070 \text{ mm} \end{aligned} \tag{7}$$

According to the personnel comparison method, the En value was used to evaluate the result: personnel comparison

$$E_n = \frac{|\mathbf{y1} - \mathbf{y2}|}{\sqrt{U_1^2 + U_2^2}} \tag{8}$$

In Equation (8), y1 and y2 stand for the measurement results of the first and second groups.

When | En | ≤ 1, it shows that "satisfied", and doesn't need to take further measures.

When | En | > 1, it suggests that "not satisfied", and measures the signal.

When getting | En | = 0.98, it indicates that the result is satisfied.

It can be seen from the comparison of the two groups of test results that the uncertainty ratio is satisfactory, the test method is reliable, and the reproducibility is strong. In addition, as can be seen from Tables 2 and 3, the difference between the measured and calculated incline of the two groups and the factory-determined incline is 0.02% and 0.05% respectively, which can well reflect the real incline of the car in the factory within the acceptable range.

7 CONCLUSION

Headlights are important for driving safety and driving experience. By studying the headlight downhill measurement test, this paper analyzed the test conditions, vehicle preparation, and test equipment needed for the test, which sorted out the headlight downhill test process, and verified the test method through personnel comparison. This method can meet the requirements of the headlight downtilt measurement, which not only ensures authenticity but also ensures recurrence. The measurement and calculation method of near-light downdip provides an important reference for the design stage of new vehicle models and the verification and judgment stage of detection institutions. In the later production consistency process, it can also be promptly tracked and confirmed, which is convenient for manufacturers and drivers to adjust in time, so as to provide better road vision for drivers and enhance the safety of road driving.

REFERENCES

Ding Wenrong. *Research on LED Headlight Optics and Gradient Design of Light and Dark Cutoff Line [D]*, Jiangsu University, 2021.10

GB 25991-2010, *LED Headlights for Automobiles* [S]

GB4785-2019. *Provisions on the Installation of Exterior Lighting and Light Signaling Devices for Automobiles and Trailers* [S]

He Dong. A Small Discussion on the Vertical Deviation of Light in Vehicle Detection [J]. *Proceedings of the 12th Automotive Academic Conference of Sichuan Province*, December 2015.

Li Xiangbing. Research on the Method of Determining the Low Beam Dark Boundary of Automobile Headlamps [J] *Automotive Electrical Appliances*, 2017.1

Liu Fatai, Wu Xiqi, and Rao Wenbi. Quality Evaluation of Headlamp Near Light Cutoff Line [J]. *China Lighting Electrical Appliances*, 1997 (5): 25–29

Sun Rui, Li Zhigang, and Ma Wenfeng, et al. Rectification Method of Low Headlight High Beam Illumination of a Vehicle Model [J]. *Automotive Electrical Equipment*, 2021.8

Song Huanting. Discussion on Selection, Validation, and Verification of laboratory Verification, Calibration, and Inspection Methods [J]. *China Inspection and Testing*, 2022.5

Zhang Yaohu, Wang Xin, and Zheng Ying. *Automobile Construction [M]*, Tsinghua University Press, 2019

Zheng Zhijun and Hu Yongliang. Research on the Offset of Light and Dark Cutoff Lines of Automobile Headlights [J]. *Journal of Lighting Engineering*, 2012, 23 (2):108–111

Water Conservancy and Civil Construction – Oke & Ahmad (Eds)
© 2024 The Author(s), ISBN: 978-1-032-58618-2

Experimental study on group root cooperative slope protection at normal temperature in Xining area

Fanxing Meng*, Hui Li* & Ningshan Jiang
College of Civil Engineering, Qinghai University, Qinghai, China

Chengkui Liu
Qinghai Building and Materials Research Co., Ltd, Qinghai, China
The Key Lab of Plateau Building and Eco-community in Qinghai, China

Gencheng Liu
Zhongyu Hengxin Engineering Consulting Co., Ltd, Henan, China

ABSTRACT: Three groups of root-soil complex samples, each containing two different roots, are selected for direct shear tests at room temperature at Qinghai University in Xining, Qinghai Province The three groups of root-soil complex samples are Alkaline grass and North China rice wormwood, Alkaline grass and Caragana, and North China rice wormwood and Caragana. By analyzing the three sets of data of the direct shear test, shear strength, internal friction angle, and cohesion of the three groups of samples at different root contents, and the differences between the three sets of data and the corresponding indexes of loess, the promotion or inhibition effect of the three indexes of each group of root-soil composite samples on loess is analyzed, and after plotting the relationship curves, the slope protection effect of the three groups of samples at room temperature is studied.

1 INTRODUCTION

In recent years, the planning and construction of slope engineering pay more and more attention to ecological environment protection, and vegetation slope protection technology has begun to be widely used in slope protection and reinforcement, using the principle of vegetation culvert water solid soil prevention of slope protection and landscaping, thus achieving effective protection management slope disaster. With the growth of plants and plant reproduction, the effect of reducing slope instability and soil erosion becomes greater. In 2006, Cheng, H (Cheng & Yan 2006) found that the root-soil composites differed by analyzing the root fixation effect of eight different herbaceous vegetation, and the differences were mainly due to the different friction between the root system and the soil, resulting in different soil fixation capacity of different herbal vegetation. In 2007, Deng, W. D. (Deng & Zhou 2007) found the role of vegetative root shearing mixed root systems in soil-soil composites. He used an indoor direct shear test and found its soil stability, while the different distribution and growth positions of soil roots also affected the soil consolidation effect, vegetation planted on the upper slope can have a better consolidation effect than planted on the middle and lower slopes. In 2008, Chen, L. H. (Chen 2008) found that the anchoring effect of vegetation roots in the soil can increase the slip resistance of the soil and thus reduce

---

*Corresponding Authors: 1075377822@qq.com and 365329508@qq.com

DOI: 10.1201/9781003450832-64

the deformation. In 2002, Xie, C. H. (Xie & Jie 2020) studied wax gourd poplar, Emei fir, and alpine willow in dark coniferous forests in the upper Yangtze River and found that the root consolidation ability of the three plants was mainly related to the root structure. Depending on the environment, the effects of root system diameter and trunk base diameter on root stretching capacity were more complex. In 2005, Xu, Z. M. (Xu & Huang 2005) studied the limitations of deep landslides in terms of plant protection and cultivation and pointed out that although the root fixation method was effective, it can only contain surface erosion and shallow landslides and have little effect on deep landslides. From 1999 to 2001, Zhou, Y. (Zhou & Chen 1999; Zhou & Luo 2000; Zhou & Zhang 2001) conducted a systematic study on the oblique support and horizontal traction of trees in Yunnan Province, that is, the vertical root system added shear strength to the rhizosphere to reinforce the soil. And it was concluded that plant slope consolidation was mainly achieved by the mechanical action of anchoring and supporting between roots and soil. In 1981, Ziemer (1981) made an in situ direct shear test of pine root soil, and the test results showed that an increase of 1 kg of root system per unit volume increased the shear strength of the root-soil complex by about 3.5 kPa. Therefore, the relationship between the root content in the soil and the increase in soil shear strength can be evaluated. In 2008, Mickovski (Mickovski & Hallett 2008) used willow to reinforce the soil and found a strong correlation between soil cover and shear strength.

2 DIRECT SHEAR TEST OF ROOT-CONTAINING LOESS AT ROOM TEMPERATURE

2.1 *Physical property index of soil body*

The soil sample selected in this experiment is on a hillside outside the south gate of Qinghai University in Chengbei District, Xining City, Qinghai Province. The soil type is silt and the soil property indexes are shown in Table 1.

Table 1. Soil property indicators.

							Soil particle composition(%)				
Depth/ cm	Average dry density/ g·cm³	Average moisture content l%	Liquid limit w_l%	Plastic limit w_p%	Liquid limit index I_l	Plastic limit index I_p	0.25-0.075 mm	0.075-0.005 mm	<0.005 mm	Nonuniformity coefficient C_u	Soil type
0-100	1.25±0.04	9.11	24.8	16.8	−0.96	8.0	29.8	62.8	7.4	3.92	silt

2.2 *Selecting materials*

According to the characteristics of the environment, climate, and soil of the study area, three types of loess with plant roots (Alkaline grass, caragana, and North China Artemisia Sativa) and original loess without plant roots are selected with good cold tolerance, well-developed root systems, and strong water absorption.

2.3 *Experimental methods*

(1) Soils with an average water content of 9.11% within the surface of 0-100 cm depth below the slope in the test area are selected to prepare the samples of plain soil and root content of different vegetation. The root content of the root-soil complex is set to be 1%, 2%, and 3%, respectively, with three different root content gradients and a 1% difference between adjacent root contents.

(2) Three kinds of root-soil complex with different root-bearing gradients, including plain soil and radix Chinensis, radix Chinensis and radix Chinensis, and Radix Chinensis, respectively. The prepared samples should be consistent with the soil dry density and particle grading. The samples are prepared as follows. First, the rootless slope soil sealed back in the test area is dried at 106°C and passes through a soil sieve with a pore size of 0.25 mm, distilled water is prepared to mix fully with three groups of plant roots with different root-containing gradients in a certain ratio, and finally, three groups of composite samples with different gradients and rootless soil samples are prepared.

(3) The test sample preparation method without rootless soil: we select the compaction cylinder with an inner diameter of 61.8 mm and a height of 125 mm, call a certain amount of rootless soil, divide it into four layers for compaction, and the number of compaction of each layer should be consistent. After compaction, we take out the soil after the compaction, place the ring knife with an inner diameter of 61.8 mm and a height of 20 mm blade down on the soil sample, press down vertically on the soil ring knife, trim the ring knife, clean the ring knife ring wall residual soil particles, keep the sample intact, weigh the total mass of the ring knife and the soil sample, repeat the above steps, and finally prepare a set of soil samples without root element straight shear test.

(4) The root-soil complex sample preparation method: First, we take out and rinse the roots of three groups of plants, then weigh a certain amount of rootless soil, and then cut the roots of three groups of plants evenly into 20 mm root sections of each group (1%, 2%, and 3%), and mix them fully with plain soil to make root-soil complex samples.

(5) The indoor direct shear test adopts the ZJ type strain control direct shear instrument of Qinghai University, and multiple sets of tests are carried out under 4 grades of positive pressure of 50 kPa, 100 kPa, 150 kPa, and 200 kPa, and the shear strain rate is 2.4 mm/min. The coefficient of the dynamometer rate is 1.706, and a set of dynamometer readings are obtained every 10 s when the specimen is sheared until the sample is completely destroyed by shear. The shear strength of each specimen is calculated, the vertical pressure is used as the abscissa, the shear strength is used as the ordinate, and the relationship curve is drawn to obtain the value of cohesion and internal friction angle.

3 RESULTS AND ANALYSIS OF INDOOR TESTS UNDER NORMAL TEMPERATURE CONDITIONS

From the indoor direct shear test, four sets of data, that is, cohesion and internal friction angle at shear strength P = 100 kPa, P = 200 kPa, are selected for the study. Three sets of data of the root-soil complex samples with 1%, 2%, and 3% are different from the root-soil without the root system. It is discussed whether the four groups are enhanced or inhibited with increasing root content. When the result of making a difference is equal to zero, it means that there is no effect on the soil under this root content. When the poor result is less than zero, it indicates that there will be an inhibitory effect on the soil under this root content. Conversely, when the result of making a difference is greater than zero, it indicates that there will be an enhancement in the soil under this root content. The results obtained are to obtain the effect of three sets of root-soil composite samples on the soil at room temperature.

3.1 *Changes of three root-soil complex specimens with the increase of root content at room temperature*

The blue line is a mixed sample of Alkaline grass and North China rice wormwood, the orange line is a mixed sample of Alkaline grass and Caragana, and the pink line is a mixed sample of North China rice wormwood and Caragana.

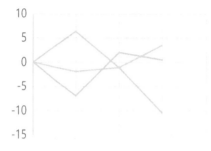

Figure 1. Friction angle change diagram.

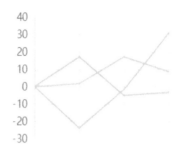

Figure 2. Change of cohesion.

Figure 3. Change of shear strength p = 100 kPa.

Figure 4. Change of shear strength p = 200 kPa.

3.2 *Analysis*

The changes in shear strength and internal friction angle of the mixed specimens of alkali grass and Artemisia annua North China increased with the increase in root content between 1% and 3%. There is a minimum value and its value is less than zero when the root content is 1%, at which time the mixed sample has the greatest inhibitory effect on the shear strength and internal friction angle of the soil. Among them, the change of shear strength between 1% and 2% of root content and the change of internal friction angle between 2% and 3% of root content have a gradually weakening inhibitory effect on the soil, reaching zero at a certain point, followed by an increasing enhancement effect on the soil. The cohesion variation of mixed specimens shows a trend of increasing and then decreasing with the root content between 1%-3%. The cohesion of mixed specimens has an enhancing effect on the soil body all the time, and there is a maximum value when the root content is 2%, that is, there is an optimum root content with the most obvious enhancing effect on the soil body.

At P = 100 kPa, the changes in shear strength and cohesion of mixed specimens of alkali grass and Caragana increase with the increase in root content. There is a minimum value of 1% of root content and its value was less than zero. At this time, the shear strength and cohesion of the soil at P = 100 kPa have the greatest inhibitory effect. The inhibition effect between 2% and 3% of the root content gradually weakens and reaches zero at a certain point, and the enhancement effect on the soil becomes more obvious. At P = 200 kPa, the change of shear strength and internal friction angle decreases with the increase in the root content from 1% to 3%. When the root content is 1%, there is an optimal root content whose value is greater than zero. At this time, the enhancement effect on the soil is the greatest. After that, it gradually decreases and reaches zero at a certain point between 1% and 2% of the root content, and the inhibitory effect on the soil gradually increases.

At P = 100 kPa, the variation of shear strength and cohesion of mixed samples of Artemisia annua North China and Caragna decreases with the increase in root content from

1% to 3%. The variation of shear strength is always less than zero at P = 100 kPa, and the inhibitory effect on soil remains constant and gradually increases. However, the maximum value of the cohesion change is 1%, that is, there is an optimal root content whose value is greater than zero. At this time, the enhancement effect on the soil is the largest, the enhancement effect gradually decreases and reaches zero at a point between 1% and 2%, and the inhibition effect becomes more obvious. At P = 200 kPa, the change of shear strength and internal friction Angle increases and then decreases with the increase in root content. When the root content is 2%, the maximum value is the optimal root content and the value is slightly greater than zero. At this time, the enhancement effect on soil is the largest and the enhancement effect gradually weakens and gradually tends to zero.

4 CONCLUSION

(1) The optimum root content was found in the variation of internal friction angle and cohesion of the three mixed specimens at room temperature, that is, this root content had the greatest enhancement effect on the soil.
(2) The optimal root contents of three groups of mixed samples are 2%, 1%, and 1%, respectively, in the mixed samples of Alkaline grass and North China rice wormwood and in the mixed samples of Alkaline grass and Caragana and in the mixed samples of North China rice wormwood and Caragana.
(3) The internal friction angle of the mixed specimens of Alkaline grass and North China rice wormwood and the mixed specimens of North China rice wormwood and Caragana does not change significantly with the increase in root content, while the internal friction angle of the mixed specimen of Alkaline grass and Caragana has an increasingly obvious inhibitory effect on the soil.
(4) With the increase in positive pressure, there is a significant increase in shear strength at room temperature.
(5) Subsequently, the ABAQUS numerical simulation of the internal friction angle and cohesion force values under the optimal root content of the three groups of mixed samples is carried out. The equivalent plastic strain diagram of the simulated three groups of mixed samples is obtained. The value of the equivalent plastic strain is obtained that the deeper along the slope surface, the greater the value. Both the cohesion force and the internal friction angle influence the size of the equivalent plastic strain, and the smaller the cohesion force, the larger the internal friction angle, the larger the equivalent plastic strain, and the smaller the resulting deformation modulus.

ACKNOWLEDGMENTS

This work was financially supported by the Fund of Qinghai Science and Technology Department 'Study on disaster mechanism of meltwater precipitation on alpine loess slope and synergistic slope protection with mixed vegetation', the fund number is 2020-ZJ-718; 'Frost resistance of recycled concrete foam concrete with waste concrete', the fund number is KLKF-2020-007. Thanks for the guidance of Dr. Hui Li, Professor Ningshan Jiang, and Chengkui Liu.

REFERENCES

Cheng, H. & Yan , C. S. (eds.). *Experimental Study on Soil Mechanism Model and Mechanics of Root Network of Herbaceous Plants* [J]., 2006(01):62–65.
Chen, L.H. *Mechanical Mechanism of Solid Soil in Forest Roots*. Beijing: Science Press, 2008:7.

Deng, W. D. & Zhou, Q. H. (eds.) Experiment and Calculation of Slope Fixation Effect of Plant Roots [J]. *China Journal of Highway and Transport*, 2007(05): 7–12.

Mickovski, S. B. & Hallett, P. D (eds.) Mechanical Reinforcement of Soil By Willow Roots: Impacts of Root Properties and Root Failure Mechanism [J]. *Ecological Engineering*, 2008, 14(5): 304–310.

Xie, C. H. & Jie, W. (eds.) Study on the Multidimensional Structure of the Dominant Tree Species Emei Fir in the Dark Coniferous Forest in the Upper Reaches of the Yangtze River [J]. *Chinese Journal of Applied Ecology*, 2002(07): 769–772.

Xu, Z. M. & Huang, R. Q. (eds.). Limitations of Vegetation Slope Protection and its Contribution to the Breeding of Deep Landslides [J]. *Chinese Journal of Rock Mechanics and Engineering*, 2005(03): 438–450.

Zhou, Y. & Chen, X. P. (eds.). A Preliminary Study on the Horizontal Traction Effect of Pine Lateral Roots on Shallow Soils in Yunnan [J]. *Chinese Journal of Plant Ecology*, 1999, 05:458–465.

Zhou, Y. & Luo, S. H. (eds.). Oblique Support Efficiency and Slope Stability Significance of Trees [J]. *Journal of Mountain Science*, 2000, 04:306–312.

Zhou, Y. & Zhang, J. (eds.). Significance of Lateral Root Strength of Pine and Qinggang Trees in the Role of Soil Consolidation and Slope Protection in Shelter Forest [J]. *Chinese Journal of Plant Ecology*, 2001, 01:105–109.

Ziemer, R. R. Roots and the Stability of Forested Slopes [J]. *Inc Association of Hydrologic Science*, 1981, 18 (3): 343–362.

Water Conservancy and Civil Construction – Oke & Ahmad (Eds)
© 2024 The Author(s), ISBN: 978-1-032-58618-2

Modeling and analysis of vehicle infrastructure cooperative industry based on tripartite evolutionary game theory

Yuan Yuan, Linheng Li, Xu Qu* & Bin Ran
School of Transportation, Southeast University, Nanjing, Jiangsu, China

ABSTRACT: The Vehicle Infrastructure Cooperative System is an inevitable development trend in the future. However, its business model and development path are still unclear and the participants of the industry lack decision-making references, which can greatly affect its further deployment and application. To address these issues, we construct a three-party evolutionary game model involving the government, vehicle companies, and consumers to study the development model and evolutionary process of the collaborative vehicle infrastructure industry. Firstly, based on building the utility function of each participant, the evolutionary stability strategy of the Vehicle Infrastructure Cooperative industrial system is solved by solving the replication dynamic equation set and the Jacobian matrix. Secondly, under the condition of ideal evolution results (1, 1, 1), numerical simulations are carried out by the MATLAB software and the initial participation ratio of each participant and the size of parameter values in the function are changed to study its influence on the system's evolution results. The results show that the evolution process of the government, vehicle enterprise, and consumers is affected by the initial participation ratio of the other two parties. And the government can accelerate the development of the industry by changing its strategy to stimulate and guide the evolution of vehicle enterprises and consumers. This paper provides economic policy references for government departments to actively promote the implementation of the Vehicle Infrastructure Cooperative industry, which can effectively help to accelerate industrial development.

1 INTRODUCTION

Vehicle Infrastructure Cooperative System (VICS) is a frontier technology and an inevitable development trend in the field of transportation (Zhang *et al.* 2021). It realizes two-way and real-time interaction of road information through intelligent roadside systems and intelligent in-vehicle systems, thereby improving traffic efficiency, ensuring traffic safety, and reducing energy consumption. Compared with single-vehicle intelligence, VICS cannot only effectively save production and research and development costs but also is expected to achieve global optimization of the transportation system through the information interaction between vehicles and roads (Zhang *et al.* 2021). Therefore, VICS has become the latest development direction of intelligent transportation systems. In recent years, countries have successively proposed research on key technologies and standards for VICS. At present, the vehicle infrastructure cooperative technology has been perfected and many demonstration areas have been built in China. However, in actual implementation, it is still difficult to popularize and commercialize. Under the condition that the government undertakes the

---

*Corresponding Author: quxu@seu.edu.cn

DOI: 10.1201/9781003450832-65

construction, operation, and maintenance of intelligent roadside equipment, how to encourage more vehicle enterprises to produce connected vehicles and more consumers to purchase connected vehicles is the key issue that needs to be solved to promote the rapid development of the VICS.

The promotion process of the VICS involves the strategic choices of multiple parties. Thus, game theory is used in research. Evolutionary game theory assumes that all parties are limited rational and they adjust their strategies through trial and error until reaching an equilibrium state during the game process, which is widely used in biology, economics, finance, and other fields. To solve the problem of insufficient safety supervision in the road transportation industry, Jiao *et al.* constructed an evolutionary game model between the government and road transportation enterprises (Jiao *et al.* 2021). Jia *et al.* established a shared parking evolutionary game model for travelers and parking space owners (Jia *et al.* 2022). Jiang *et al.* constructed an evolutionary game model between vehicle manufacturers to research the influence of government subsidies on the new energy vehicle manufacturer's choice of independent research and development or technology introduction (Jiang *et al.* 2020). Nie *et al.* applied carbon trading subsidies to the public transport industry and explored the stability strategy of public transport operators and the government (Nie *et al.* 2022). Based on evolutionary game theory, Zhang and Guan analyzed the decision-making of travelers between public transportation and private transportation (Zhang & Guan 2019), while Zhang and Li used the same method to explain the dynamic selection process of travelers' choice between private cars and carsharing (Zhang & Li 2022). When there are more than two participants, scholars will choose the three-party evolutionary game model or even the four-party evolutionary game model for research. For example, for the hydrogen fuel cell vehicle industry, Huang *et al.* explored the decision-making mechanism of the central government, local governments, and enterprises by using evolutionary game theory (Huang *et al.* 2022). Based on the knowledge of game theory, Zhang and Lu constructed an evolutionary game model of the government, enterprises, and consumers in the post-subsidy era of new energy vehicles (Zhang & Lu 2020). Cao *et al.* established an evolutionary game model for the new energy vehicle industry, including the government, vehicle manufacturers, and consumers (Cao *et al.* 2018). For the VICS industry, Vignon *et al.* established the profit function of vehicle manufacturers, infrastructure support service providers, and users, which was the first research on economic analysis in the field of vehicle-infrastructure cooperatives (Vignon *et al.* 2022).

The evolutionary game model can well simulate the evolution process of the participants in the system and well predict the final direction of the industry in different scenarios. Therefore, this paper introduces the three-party evolutionary game theory into the promotion of the VICS industry, aiming to explore the stabilization strategies of all parties and the influence mechanism of different initial ratios and policy choices on the promotion process of the industry.

2 MODEL CONSTRUCTION

2.1 *Basic assumptions*

The VICS is mainly composed of two core parts, intelligent roadside systems and intelligent in-vehicle systems, which realize the interaction of information between vehicles and roads through Internet technology. Intelligent roadside equipment, such as sensing devices and decision-making devices, are constructed by the government. Vehicles that are equipped with intelligent in-vehicle equipment are defined as connected vehicles, which is produced by vehicle enterprise. And the consumer is the purchaser of the VICS service. Thus, the game model of the CVIS industry includes three parties, the government, vehicle enterprise, and consumer. Each party has two strategies. The government can adopt two strategies, The first is to actively promote the CVIS industry, or it can passively promote it. Second, vehicle companies can choose to produce connected vehicles or regular vehicles, while consumers can choose to buy connected vehicles or regular vehicles. The following games exist between the three parties.

(1) *The game between the government and vehicle enterprise*

The government plays the role of advocate and supervisor. When the government chooses the strategy to actively promote the connected vehicle industry, it can give certain financial support to the vehicle enterprises that produce connected vehicles for research and development of intelligent in-vehicle equipment and punish the vehicle enterprises that have fraudulent behavior. If the vehicle enterprise chooses to produce ordinary fuel vehicles, the government can impose a certain environmental pollution tax on it.

(2) *The game between vehicle enterprises and consumers*

Vehicle enterprises and consumers are the suppliers and demanders of connected vehicles, respectively. Compared with ordinary vehicles, enterprises can get additional benefits by seizing the market when producing and selling connected vehicles. However, high research and development costs also need to be invested. Consumers need to pay a higher price when purchasing connected vehicles. However, correspondingly he/she can obtain higher commodity value in terms of travel efficiency, travel safety, and energy consumption.

(3) *The game between the government and consumers*

The benefit of consumers purchasing connected vehicles is directly proportional to the construction degree of intelligent roadside equipment in the city. The government's policy of improving the construction degree of roadside equipment and providing tax incentives for connected vehicles increase consumers' benefit from purchasing connected vehicles, thereby stimulating and guiding consumers to purchase and use connected vehicles.

The symbol settings and descriptions are shown in Table 1.

Table 1. Description of the symbols.

Symbol	Description
x	The probability of the government actively promoting the VICS industry
y	The probability of the vehicle enterprise producing connected vehicles
z	The probability of consumers purchasing connected vehicles
I	The government's financial support for vehicle enterprise
F	Construction costs for roadside equipment of the government
M	Fines for vehicle enterprise's fraudulent behavior
P	Environmental pollution tax of ordinary vehicles
R_1	The benefits of vehicle enterprises producing connected vehicles
R_2	The benefits of vehicle enterprises producing ordinary vehicles
B	The market opportunities obtained by vehicle enterprises for producing connected vehicles
C	The research and development cost of producing connected vehicles
R_3	The benefits of consumers purchasing connected vehicles
R_4	The benefits of consumers purchasing ordinary vehicles
ΔR	Additional benefits for consumers purchasing connected vehicles (such as efficiency and safety)
T	The government's tax incentives for connected vehicles
∂	The construction degree of roadside equipment, which is defined as the ratio of the length of roads with intelligent roadside equipment to the total length of roads in the city)
β	The government's supervision of fraudulent behavior

2.2 *Payment matrix*

According to the above assumptions, the three-party game payment matrix of government, vehicle enterprise, and consumers is constructed, which is shown in Table 2. The three formulas in each cell in the table represent the utility function of the government, vehicle enterprise, and consumers, respectively.

Table 2. Payment matrix of tripartite evolutionary game.

		Consumer	
		Purchase connected vehicles (z)	Purchase ordinary vehicles (1-z)
Government	Vehicle enterprises		
Actively promote (x)	Producing connected vehicles (y)	$M - I - F - T$ $R_1 + I + B - M - C$ $R_3 + \Delta R + T$	$M - I - F$ $R_1 + I - M - C$ R_4
	Producing ordinary vehicles (1-y)	$P - F$ $R_2 - P$ 0	$P - F$ $R_2 - P$ R_4
Passively promote (1-x)	Producing connected vehicles (y)	$-I - F - T$ $R_1 + I + B - C$ $R_3 + \Delta R + T$	$-I - F$ $R_1 + I - C$ R_4
	Producing ordinary vehicles (1-y)	$-F$ R_2 0	$-F$ R_2 R_4

We assume that G_1 and G_2 are the expected benefits of the government's choices of active promotion strategy and passive promotion strategy, respectively. Then the replication dynamic equation for government $G(x)$ can be derived as follows.

$$\begin{cases} G_1 = yz(M - I - F - T) + y(1 - z)(M - I - F) + (1 - y)z(P - F) + (1 - y)(1 - z)(P - F) \\ G_2 = yz(-I - F - T) + y(1 - z)(-I - F) + (1 - y)z(-F) + (1 - y)(1 - z)(-F) \end{cases}$$

$$\tag{1}$$

$$G(x) = x(1 - x)(G_1 - G_2) = x(1 - x)[P + y(M - P)] \tag{2}$$

We assume that E_1 and E_2 are the expected benefits of vehicle enterprises to produce connected vehicles and ordinary vehicles, respectively. Then the replication dynamic equation for vehicle enterprises $E(y)$ can be derived as follows.

$$\begin{cases} E_1 = xz(R_1 + I + B - M - C) + x(1 - z)(R_1 + I - M - C) + \\ (1 - x)z(R_1 + I + B - C) + (1 - x)(1 - z)(R_1 + I - C) \\ E_2 = xz(R_2 - P) + x(1 - z)(R_2 - P) + (1 - x)zR_2 + (1 - x)(1 - z)R_2 \end{cases} \tag{3}$$

$$E(y) = y(1 - y)(E_1 - E_2) = y(1 - y)[R_1 + I - C - R_2 + x(P - M) + zB] \tag{4}$$

We assume that C_1 and C_2 are the expected benefits of consumers to purchase connected vehicles and ordinary vehicles, respectively. Then the replication dynamic equation for consumers $C(z)$ can be derived as follows.

$$\begin{cases} C_1 = xy(R_3 + \Delta R + T) + (1 - x)y(R_3 + \Delta R + T) \\ C_2 = xyR_4 + x(1 - y)R_4 + (1 - x)yR_4 + (1 - x)(1 - y)R_4 \end{cases} \tag{5}$$

$$C(z) = z(1 - z)(C_1 - C_2) = z(1 - z)[-R_4 + y(R_3 + \Delta R + T)] \tag{6}$$

3 STABILITY ANALYSIS

As shown in Equation (7), the equilibrium point of the evolutionary game can be solved by setting the replication dynamic equation of each party equal to 0. It is obtained that there are 8 special equilibrium points in the tripartite evolution game process by solving the equation set, namely $E_1(0,0,0)$, $E_2(1,0,0)$, $E_3(0,1,0)$, $E_4(0,0,1)$, $E_5(1,1,0)$, $E_6(1,0,1)$, $E_7(0,1,1)$, and $E_8(1,1,1)$.

$$\begin{cases} G(x) = x(1-x)[P + y(M-P)] = 0 \\ E(y) = y(1-y)[R_1 + I - C - R_2 + x(P-M) + zB] = 0 \\ C(z) = z(1-z)[-R_4 + y(R_3 + \Delta R + T)] = 0 \end{cases} \quad (7)$$

The stability points obtained by solving the replication dynamic equation system are not necessarily the evolutionary stability strategy of the system. The progressive stabilization strategy can be further determined according to the eigenvalues of the Jacobian matrix of the system. If all eigenvalues are negative, the point is a progressive stability point; If the eigenvalues have both positive and negative values, the point is a saddle point; If all eigenvalues are positive, the point is unstable. The Jacobian matrix J of the system is shown in Equation (8).

$$J = \begin{bmatrix} \dfrac{\partial G(x)}{\partial x} & \dfrac{\partial G(x)}{\partial y} & \dfrac{\partial G(x)}{\partial z} \\[2mm] \dfrac{\partial E(y)}{\partial x} & \dfrac{\partial E(y)}{\partial y} & \dfrac{\partial E(y)}{\partial z} \\[2mm] \dfrac{\partial C(z)}{\partial x} & \dfrac{\partial C(z)}{\partial y} & \dfrac{\partial C(z)}{\partial z} \end{bmatrix}$$

$$= \begin{bmatrix} (1-2x)[P + y(M-P)] & x(1-x)(M-P) & 0 \\ y(1-y)(P-M) & (1-2y)[R_1 + I - C - R_2 + x(P-M) + zB] & y(1-y)B \\ 0 & z(1-z)(R_3 + \Delta R + T) & (1-2z)[-R_4 + y(R_3 + \Delta R + T)] \end{bmatrix}$$

$$(8)$$

The above eight equilibrium points are brought into the Jacobian matrix to solve the corresponding eigenvalues λ_1, λ_2, and λ_3. The results are shown in Table 3. It can be inferred that the system may have three evolutionary stabilization strategies according to the above judgment method, namely (1, 0, 0), (1, 1, 0), and (1, 1, 1). The corresponding conditions of the three evolutionary stabilization strategies are shown in Table 4.

Case 1: When the comprehensive benefit of vehicle enterprises producing connected vehicles is less than that of ordinary vehicles, its evolutionary stability strategy is (1, 0, 0). At this time, the government chooses to actively promote the CVIS industry and vehicle enterprises and consumers still choose the strategy of producing ordinary vehicles and purchasing ordinary vehicles, respectively.

Case 2: When the comprehensive benefit of vehicle enterprises producing connected vehicles is greater than that of ordinary vehicles and the comprehensive benefit of consumers purchasing connected vehicles is less than that of ordinary vehicles, its evolutionary stability strategy is (1, 1, 0). At this time, the government chooses to actively promote the CVIS industry, vehicle enterprises choose to produce connected vehicles, and consumers choose to purchase ordinary vehicles.

Case 3: When the comprehensive benefit of vehicle enterprises producing connected vehicles is greater than that of ordinary vehicles and the comprehensive benefit of consumers purchasing connected vehicles is greater than that of ordinary vehicles, its evolutionary stability strategy is (1, 1, 1). At this time, the government chooses to actively promote the CVIS industry, vehicle enterprises choose to produce connected vehicles and consumers choose to purchase connected vehicles.

Table 3. Eigenvalues for 8 equilibrium points.

Equilibrium point	λ_1	λ_2	λ_3
$E_1(0,0,0)$	P	$R_1 + I - C - R_2$	$-R_4$
$E_2(1,0,0)$	$-P$	$R_1 + I - C - R_2 + P - M$	$-R_4$
$E_3(0,1,0)$	M	$-R_1 - I + C + R_2$	$-R_4 + R_3 + \Delta R + T$
$E_4(0,0,1)$	P	$R_1 + I - C - R_2 + P - M$	R_4
$E_5(1,1,0)$	$-M$	$-R_1 - I + C + R_2 - P + M$	$-R_4 + R_3 + \Delta R + T$
$E_6(1,0,1)$	$-P$	$R_1 + I - C - R_2 + P - M + B$	R_4
$E_7(0,1,1)$	M	$-R_1 - I + C + R_2 - B$	$R_4 - R_3 - \Delta R - T$
$E_8(1,1,1)$	$-M$	$-R_1 - I + C + R_2 - P + M - B$	$R_4 - R_3 - \Delta R - T$

Table 4. Conditions for each evolutionary stabilization strategy.

Stabilization Strategy	Conditions
$(1,0,0)$	$R_1 + I - C - M < R_2 - P$
$(1,1,0)$	$R_1 + I - C - M > R_2 - P, R_4 > R_3 + \Delta R + T$
$(1,1,1)$	$R_1 + I + B - C - M > R_2 - P, R_4 < R_3 + \Delta R + T$

4 SIMULATION ANALYSIS

MATLAB software is used to numerically simulate the process of the evolution game. We set the initial values $I = 7, F = 200, M = 10, P = 1, R_1 = 5, R_2 = 3, B = 1, C = 5, R_3 = 1.2, R_4 = 1, \Delta R = 1.5, T = 1, \partial = 0.5, \beta = 0.5$ to evolve the model to the ideal state (1, 1, 1). Under the above conditions, the results of evolution are shown in Figure 1 as each party takes different initial participation ratios. It is obvious that the system eventually reaches the evolutionary steady state at (1, 1, 1).

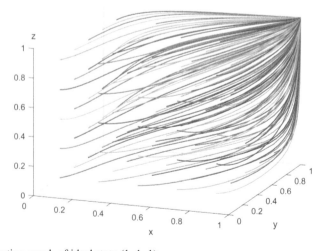

Figure 1. Evolution graph of ideal state (1, 1, 1).

4.1 Influence of initial ratios

To analyze the influence of different initial participation ratios of various parties on the evolutionary game, the participation ratios of the other two parties are changed under the conditions of x, y, and z fixed, respectively.

Given that the initial value x is 0.5, the change of x overtime is investigated by changing the initial values of y and z. The results in Figure 2(a) indicate that increasing the initial values of y and z can accelerate the rate of convergence of x to a certain extent but in at the end, they converge to 1.

Given that the initial value y is 0.5, different initial values x and z are set to research their effect on the evolution of y over time. According to the results of Figure 2(b), y eventually converges to 1 and the initial value of x and z has little effect on the rate of convergence of y.

Similarly, given the initial value z of 0.5, the effect of the initial value of x and y is investigated on the evolution of z and the corresponding results are as shown in Figure 2(c). When the initial value of x and y is low, the change of z with time is not monotonous. There is a decline in the early stage of evolution but it still convergences to 1 in the end. And when the initial values x and y increase, the evolution speed z can also be significantly accelerated.

From the above analysis, it can be concluded that the evolution process of the strategy selection of all parties is affected by the initial value of the other two parties.

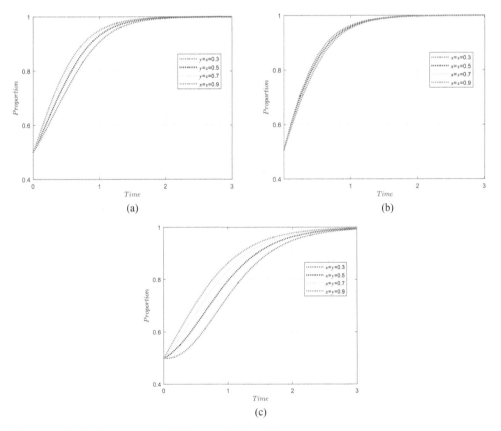

Figure 2. Evolutionary graph of different initial conditions.

4.2 Influence of parameter values

Given that the initial participation ratios of x, y, and z are (0.5, 0.5, 0.5), the parameter values are changed to analyze their influence on the strategy selection of the three parties.

(1) *The evolution trajectory of vehicle enterprise when I changes*

To investigate the impact of financial support on the process and result of the enterprise's evolution, the values of I are assigned values of 7, 9, and 11, respectively. As shown in Figure 3(a), the government's increased financial support for vehicle

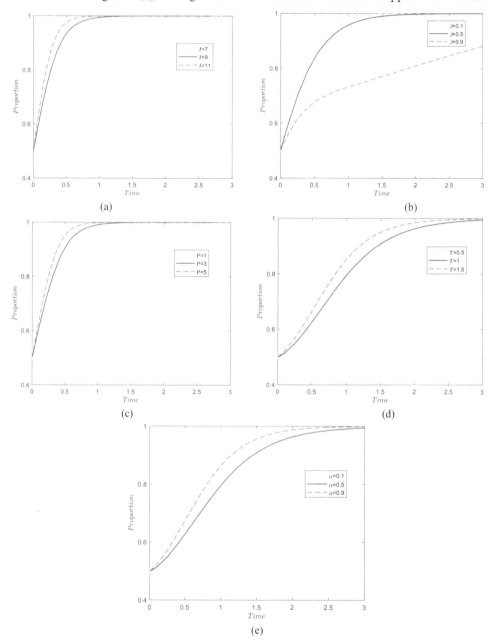

Figure 3. The evolution trajectory when the parameter changes.

enterprises has effectively accelerated the evolution of the enterprise's strategy to "producing connected vehicles". If the government gives more financial support, the research and development costs of vehicle enterprises can be reduced, thereby encouraging the production of connected vehicles.

(2) *The Evolution trajectory of vehicle enterprise when β changes*

To investigate the influence of penalties for fraudulent behavior on the process and result of the enterprise's evolution, the values β are assigned values of 0.1, 0.5, and 0.9, respectively, and the final results are shown in Figure 3(b). According to the results, the greater the government's penalties for fraudulent behavior, the more difficult it is for vehicle enterprises to evolve into a stability strategy. Excessive supervision can make enterprises in a wait-and-see attitude, which greatly slows down the evolution speed of vehicle enterprise.

(3) *The evolution trajectory of vehicle enterprise when P changes*

To investigate the influence of the environmental pollution tax on the evolution of enterprises, the value P is assigned to 0.5, 1, and 1.5, respectively, whose results are shown in Figure 3(c). The government's increase in taxes can slow down the speed of vehicle enterprises to reach a stable state. Appropriately increasing environmental pollution tax can promote the transformation of vehicle enterprises to the connected vehicle industry while alleviating environmental pollution problems.

(4) The evolution trajectory of vehicle enterprise when T changes

The values T are set to 0.5, 1, and 1.5, respectively, to analyze the influence of the tax incentive for purchasing connected vehicles on the evolution game process of the consumer. As shown in Figure 3(d), the greater the government's tax incentives, the faster the evolution speed of consumers choosing "purchasing connected vehicles". In the early stage of evolution, the government can provide consumers with tax incentives for purchasing connected vehicles to stimulate consumers' desire to purchase. After the production cost of connected vehicles gradually decreases, the government can gradually cancel the policy of tax incentives.

(5) *The evolution trajectory of consumers when α changes*

To investigate the influence of the construction degree of roadside equipment on the evolution process and results of consumers, the values α are assigned values of 0.1, 0.5, and 0.9, respectively, and the evolution results are displayed in Figure 3(e). The construction degree of roadside equipment has an important role in promoting the evolution process of consumers. The higher the construction degree of the roadside equipment, the higher the benefit obtained by the consumer, so it is more inclined to purchase the connected vehicles for consumers.

5 CONCLUSIONS

To accelerate the implementation of the VICS industry and promote the large-scale application of connected vehicles, this paper constructs an evolution game model for the promotion of the industry in which the government, vehicle enterprise, and consumers participate together. Through the above research, this paper draws the following conclusions: First, the strategic choice and evolution path of any party in the evolutionary game process is related to the initial participation ratio of the other two parties. Second, the government can accelerate the development of the industry by changing its strategy to stimulate and guide the evolution of vehicle enterprises and consumers. This paper provides policy references for relevant participants by modeling the CVIS industry, which is helpful to promote the development of the industry.

However, there are still some shortcomings in this paper, such as the failure to fully use real data for the model in the numerical simulation process due to the limitations of some data's availability. In addition, the payment matrix established in this paper fails to take all the influencing factors into account, which further affects the accuracy of the model. In the next step of research, the above two points can be further improved.

REFERENCES

Cao, X., Xing, Z. Y. and Zhang, L. P. "An Evolutionary Game Analysis of New Energy Vehicle Industry Development under Government Regulations," *Management Review*, 30(09), 82–96 (2018).

Huang, T. F., Hu, C. H., He, Q. Y., Yang, D. X., He, T. and Fu, Y. "A Coordination Analysis of Stakeholder Interests on the New Subsidy Policy of Hydrogen Fuel cell Vehicles in China: From the Perspective of the Evolutionary Game Theory," *International Journal of Hydrogen Energy*, 47(58) (2022).

Jia, F. Q., Li, Y. Z., Yang, X. F., Ma, C. X. and Dai, C. J. "Shared Parking Behavior Analysis Under Government Encouragement Based on Evolutionary Game Method," *Journal of Transportation Systems Engineering and Information Technology*, 22(01), 163–170 (2022).

Jiang, C. L., Zhang, Y., Li, W. W. and Wu, C. "Evolutionary Game Research between Government Subsidy and R&D Activities of New Energy Vehicle Enterprise," *Operation Research and Management Science*, 29 (11), 22–28 (2020).

Jiao, P., Zhang, S. and Zhao, X. M. "Evolutionary Game of Safety Supervision Between Government and Road Transport Enterprises Based on Third-party," *Journal of Chang'an University (Natural Science Edition)*, 41(03), 106–115 (2021).

Nie, Q. Y., Zhang, L. H., Tong, Z. H. and Hubacek, K. "Strategies for Applying Carbon Trading to the New Energy Vehicle Market in China: An Improved Evolutionary Game Analysis for the Bus Industry," *Energy*, 259 (2022).

Vignon, D. A., Yin, Y. F., Bahrami, S. and Laberteaux, K. "Economic Analysis of Vehicle Infrastructure Cooperation for Driving Automation," *Transportation Research Part C*, 142 (2022).

Zhang, Y., Yao, D. Y., Li, L., Pei, H. X., Yan, S. and Ge J. W. "Technology and Applications for Intelligent Vehicle-infrastructure Cooperation Systems," *Journal of Transportation Systems Engineering and Information Technology*, 21(05), 40–51 (2021).

Zhang, X. Q., Mao, W., Luo, Y. L., Du, Z. H. and Wang, G. "Modeling and Gender Difference Analysis of Acceptance of Cooperative Vehicle Infrastructure System," *China Journal of Highway and Transport*, 34 (07), 177–187 (2021).

Zhang, X. J. and Guan, H. Z. "Research on Travel Mode Choice Behaviors Based on Evolutionary Game Model Considering the Indifference Threshold," *IEEE Access*, 7, 174083–174091 (2019).

Zhang, Y. and Li, L. M. "Research on Travelers' Transportation Mode Choice Between Carsharing and Private Cars Based on the Logit Dynamic Evolutionary Game Model," *Economics of Transportation*, 29 (2022).

Zhang, Y. and Lu, C. X. "Evolutionary Game Analysis of Government, Enterprises and Consumers in Post Subsidy Era of New Energy Vehicles," *Journal of Chongqing Jiaotong University (Natural Science)*, 39(05), 38–48 (2020)

Author index

Bao-yin, M. 299
Boukari, M. 456

Cao, R. 280
Cao, S. 21
Che, M. 3
Chen, D. 92
Chen, J. 50
Chen, L. 223
Chen, X. 27, 192
Chen, Y. 124
Chunjuan, G. 199
Cui, S. 139

Deng, S. 378, 395
Ding, D. 486
Dong, R. 439
Duan, C.G. 479

Fan, L. 3
Feng, J. 231
Feng, L. 57, 156

Gao, J. 171
Gao, Z. 21
Gong, S. 431
Guan, S. 280
Guo, H. 431
Guo, J. 362
Guo-sheng, D. 299

Hao, A. 280
Hao, D. 325

Hao, Y. 3, 156
Hu, C. 57, 65, 378, 395
Hu, D. 325, 338
Hu, T. 74
Huan, H. 338
Huang, H. 124
Huang, J. 82
Huang, X. 74
Huguette Maeva, M.N. 456
Hui, L. 332

Ji, C. 124, 345
Ji, J. 439
Jia, B. 245
Jia, N. 245
Jiang, K. 231
Jiang, L. 479
Jiang, N. 501
Jiang, S. 124
Jiang, X. 217
Jiao, S. 486
Jin, J. 185

Kalissa, F.K. 456
Keliang, W. 199

Lei, L. 299
Li, C. 34, 110
Li, D. 325
Li, F. 362
Li, H. 501
Li, J. 98

Li, L. 57, 110, 371, 507
Li, P. 171
Li, W. 274
Li, X. 132, 403
Li, Y. 185, 252, 439, 479
Li, Z. 325, 338
Liang, C. 294
Liang, X. 371
Liang, Y. 313
Lin, B. 445
Lin, C. 92
Lin, X. 362
Liu, C. 501
Liu, G. 156, 501
Liu, H. 238
Liu, J. 11
Liu, L. 319
Liu, M. 238
Liu, Y. 3, 50, 132, 150
Liu, Z. 177, 313
Liyun, L. 332
Long, Y. 289
Lu, H. 486

Ma, Q. 139
Ma, Y.H. 479
Meng, F. 501
Ming, Y. 378, 395

Pan, Y. 44
Pang, J. 456

Pei, X. 231
Peng, X. 92
Pu, C. 252

Qi, S. 223
Qi, X. 82
Qiao, W. 362
Qu, X. 110, 507

Ran, B. 507
Ren, Z. 305
Rong, W. 208, 451

Shang, M. 274
Shen, R. 124
Sheng, X. 274
Shi, X. 3
Shi, Y. 44
Shi, Z. 98
Shoujia, Z. 208, 451
Song, M. 27
Song, S. 289
Song, X. 3
Su, J. 3
Su, T. 467
Sun, A. 325, 338
Sun, F. 139
Sun, J. 74

Tan, M. 185
Tang, Z. 305
Tao, B. 98
Tao, H. 289

Tao, X. 199
Tao, Z. 493
Tian, Y. 362
Tu, Q. 378, 395

Wang, C. 124, 345
Wang, D. 124
Wang, G. 139
Wang, H. 21, 82, 110, 274, 345, 362, 431
Wang, J. 3
Wang, L. 74
Wang, R. 185
Wang, S. 274
Wang, X. 231, 420
Wang, Y. 305
Wei, H. 245
Wei, L. 11
Wei, P. 238
Wen, J. 252
Wu, D. 156
Wu, W. 44
Wu, Y. 156
Wu, Z. 413

Xiangyu, H. 451
Xie, Z. 192
Xu, K. 345
Xu, X. 110
Xue, D. 388
Xue, L. 345
Xue, S. 467

Yan, S. 431
Yan, Y. 192
Yang, J. 110
Yang, Q. 420
Yang, S. 362
Yefu, L. 208, 451
Yu, S. 294
Yu, Y. 27, 338
Yuan, W. 420
Yuan, Y. 507

Zhang, H. 65, 294
Zhang, J. 150
Zhang, L. 98, 362
Zhang, R. 420
Zhang, T. 231
Zhang, W. 82, 313
Zhang, X. 388
Zhang, Y. 74, 139, 420
Zhang, Z. 1
Zhao, W. 21
Zhao, X. 82
Zheng, C. 403
Zheng, S. 3
Zhong, X. 204
Zhou, Y. 467
Zhu, C. 371
Zhu, D. 204
Zhu, J. 150
Zhu, W. 34
Zhu, X. 388